HEAT AND MASS TRANSFER SOURCEBOOK:

Fifth All-Union Conference, Minsk, 1976

HEAT AND MASS TRANSFER SOURCEBOOK:

Fifth All-Union Conference, Minsk, 1976

Edited by

Mikhail A. Styrikovich, *Moscow*
Algirdas A. Žukauskas, *Vilnius*
James P. Hartnett, *Chicago, Ill.*
Thomas F. Irvine, Jr., *Stony Brook, N.Y.*

s/p

1977

Scripta Publishing Co
Washington, D.C.

A HALSTED PRESS BOOK

JOHN WILEY & SONS
New York   Toronto   London   Sydney

Publishers, Scripta Publishing Co
a division of Scripta Technica, Inc.
1511 K Street, N.W., Washington, D.C. 20005

Distributed solely by Halsted Press Division,
John Wiley & Sons, Inc., New York

Library of Congress Cataloging in Publication Data:

Main entry under title:

Heat and mass transfer sourcebook.

Includes abstracts and selected papers of the 5th National Conference on Heat and Mass Transfer, Minsk, 1976.
   1. Heat—Transmission—Congresses. 2. Mass transfer—Congresses. I. Styrikovich, Mikhail Adol'fovich. II. Vsesoiuznoe
soveshchanie po teplo 1 massoobmenu, 5th, Minsk, 1976.
QC319.8.H415   536'.2   77-22337
ISBN 0-470-99234-4

Composition by **Isabelle Sneeringer**, Scripta Technica, Inc.

CONTENTS

EDITORS' PREFACE

In May 1976, the Fifth National Conference on Heat and Mass Transfer was held in Minsk, the capital of the Belorussian Republic of the Soviet Union. Beginning in 1960, on the initiative of the late Professor A. V. Luikov, these meetings have been held every four years.[1] Although the primary emphasis of the conferences has been on presentation of the most recent work of Soviet engineers and scientists, invitations have always been tendered to a number of scientific visitors from outside the Soviet Union. Thus, these meetings have had an international character and have been extremely useful in catalyzing personal contacts among research workers from a variety of countries.

The format of the conferences has always been the "Reporter Technique," whereby a variety of papers on similar topics is summarized and compared by a reporter, after which the remainder of the session is devoted to questions and discussion. A complete set of the conference papers (in Russian) was available before the meeting, so that the discussion could focus on both the details and general conclusions of the papers under consideration.

The number of participants and the size of the proceedings are generally formidable, to say the least. The Russian language proceedings of the most recent conference, for example, consist of nine volumes, which include 580 technical papers, both Soviet and non-Soviet. The translation and publication of such a large amount of information does not appear to be practical under today's conditions. However, since the All-Union Conferences have established an excellent reputation for the quality of the technical papers and the timeliness of the information, the present volume was conceived as a means of

[1] They have been supported and organized by the USSR Academy of Sciences, the scientific acadamies of the various Soviet Union Republics and the USSR National Committee on Heat and Mass Transfer.

making the conference results available to the international heat-and-mass transfer community.

The conference was organized into technical sessions by topical sub-disciplines of heat and mass transfer. These include:

1. Convection
2. Heat and Mass Transfer with Chemical Reactions
3. Change of Phase
4. Two-Phase Flows
5. Capillary-Porous Bodies
6. Heat and Mass Transfer in Dispersed Systems
7. Rheological Systems
8. Radiative Heat Transfer and Associated Problems
9. Heat Conduction
10. Experimental Techniques

Included in the present "Sourcebook" is an English translation of all the abstracts of the conference papers, arranged in the order of the original conference sessions. The name and affiliation of each author is listed with the abstract. Following the abstract is an indication of where further translated information can be found for each paper. These further sources consist of two categories.

1. A complete English version of selected non-Soviet papers in the present volume.
2. An indication if the English version of a Soviet paper will be published in either of the two journals: *Heat Transfer—Soviet Research* or *Fluid Mechanics—Soviet Research.*

In the cases where papers do not fall into either of the above categories, the authors may be contacted directly by interested readers. A complete authors index at the back of the volume should also prove helpful in locating particular papers.

The editors sincerely hope that the present volume will be of service in calling attention to information on current research and engineering developments in heat and mass transfer both within the Soviet Union and in other parts of the world.

T. F. Irvine, Jr., for the Editors

EVALUATION OF MEASUREMENTS OF THE LOCAL AND TOTAL HEAT TRANSFER FROM SMOOTH AND ROUGH SURFACE CYLINDERS IN CROSS FLOW*

E. Achenbach
Institut für Reaktorbauelmente
Jülich, West Germany

The flow and heat transfer phenomena occurring in the boundary layer and in the separated region of smooth and rough surface cylinders are studied. The results can be used to improve the heat transfer of heat exchangers in nuclear reactors.

The experiments were conducted in the Reynolds number range $2 \times 10^4 <$ Re $< 4 \times 10^6$. The roughness parameter k_s/d, where k_s means the equivalent sand grain roughness and d the diameter of the cylinder, was varied from 0 to 900×10^{-5}. The roughness was realized by a knurling process. Thus a pyramidal roughness pattern was produced. The test cylinders had a diameter of $d = 0.150$ m and a length of $L = 0.5$ m. The span ratio was $L/d = 3.3$. As the test section was 0.5×0.9 m^2, the blockage ratio was $d/B = 1 : 6$. The turbulence level of the incident flow was $Tu = 0.45\%$.

As a first step, the flow was examined for a better understanding of the heat transfer results. The local static pressure distribution was determined as a function of the Reynolds number and of the four roughness parameters. An example of the pressure distribution is given in Fig. 1 for the roughness parameter $k_s/d = 75 \times 10^5$ at various Re-numbers. The Re-numbers were selected so that each of the four flow ranges is represented by one experimental pressure distribution. The four flow ranges are denoted: subcritical, critical, supercritical, and transcritical. A detailed description of the boundary layer phenomena is found in [2] and [3].

Figure 2 represents the distribution of the local heat transfer coefficient for the same roughness parameter $k_s/d = 75 \times 10^{-5}$. The cross symbols denote the subcritical flow range. Near the separation point of the boundary layer ($\varphi = 80°$) the heat transfer coefficient exhibits a minimum value. The critical flow range is characterized by a minimum value near the laminar intermediate

---

*The abstract of this paper appears on page 323.

1

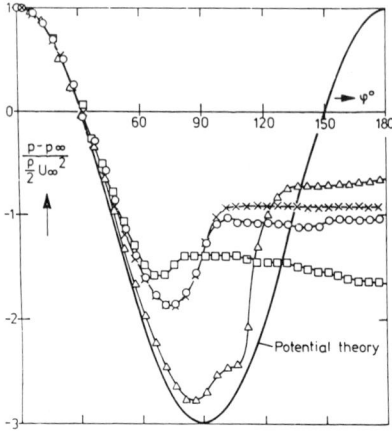

Fig. 1

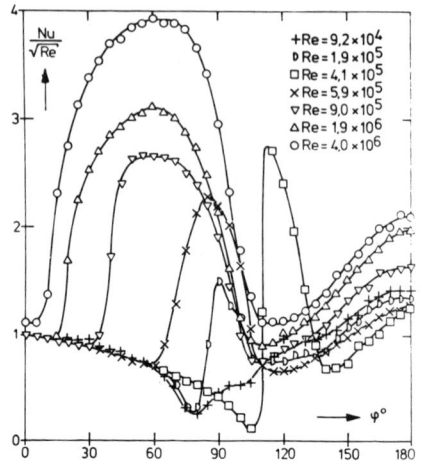

Fig. 2

FIG. 1. Static pressure: $k_s/d = 75 \times 10^{-1}$; □, Re = 8.3 × 10⁴; △, Re = 2.2 × 10⁵; x, Re = 5.8 × 10⁵; ○, Re = 4.1 × 10⁶.

FIG. 2. Local heat transfer.

separation occurring at $\varphi = 100°$. Downstream the heat transfer coefficient increases considerably as an indication of the transition to a turbulent boundary layer. The second relative minimum occurs at $\varphi = 140°$ and denotes the final separation of the turbulent boundary layer. It is obvious that the transition point from a laminar to a turbulent boundary layer shifts upstream with increasing Re-number (cross and triangle symbols). The transition point is indicated by a rapid increase of the heat transfer coefficient. The curve drawn through the circle symbols represents the transcritical flow range. It is seen, that the heat transfer already increases by a factor of 4 in the immediate vicinity of the front stagnation point. This factor is about 8 for the largest roughnesses tested. As can be noted, the results exhibit an improvement of the heat transfer with increasing roughness parameter.

The integration of the local pressure distribution yields the drag coefficient c_d of the cylinder under the condition that the friction forces are of the order of 1 or 2 percent [3]. Figure 3 shows the results for the four roughness parameters. At subcritical flow conditions the drag coefficient is independent of the roughness parameter. With increasing roughness, the value of the critical Re-number, $\mathrm{Re_{crit}}$, defined as the Re-number of the minimum drag coefficient c_d, decreases. In the transcritical flow regime the drag coefficient is independent of the Re-number and increases with increasing roughness parameter. It is remarkable that the transcritical drag coefficient is identical for the two largest roughness parameters $k_s/d = 300 \times 10^{-5}$ and $k_s/d = 900 \times 10^{-5}$. It appears that there exists a maximum value in the transcritical flow regime, which is not exceeded though the roughness parameter is increased.

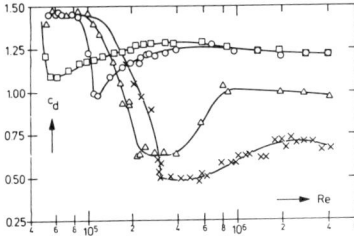

Fig. 3

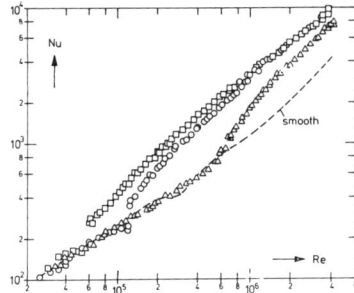

Fig. 4

FIG. 3. Drag coefficient: x, smooth; $\triangle$, $k_s/d = 75 \times 10^{-5}$; $\circ$, $k_s/d = 300 \times 10^{-5}$; $\square$, $k_s/d = 900 \times 10^{-5}$.

FIG. 4. Total heat transfer: Pr = 0.72; symbols as in Fig. 3.

In Fig. 4, the total heat transfer from a circular cylinder is plotted as a function of the Re-number and the roughness parameter for the case of air cooling (Pr = 0.72). It is obvious that, for subcritical conditions, the four curves become the same. The heat transfer suddenly increases while exceeding the critical Re-number. At Re = 10^6 the heat transfer of the rough cylinder is improved by a factor of 2.7 compared with the smooth cylinder. Concerning the heat transfer, it can be noticed that also the heat transfer curves collapse for transcritical flow conditions. It seems that the heat transfer can no longer be improved by further augmentation of the roughness height. The upper limit of the effective roughness height is 3%d.

In Fig. 5, a comparison is made between the results of Daujotas et al. [4] and the present data for the smooth cylinder. The departure is up to about 8 percent in the subcritical and supercritical flow range. The discrepancy is

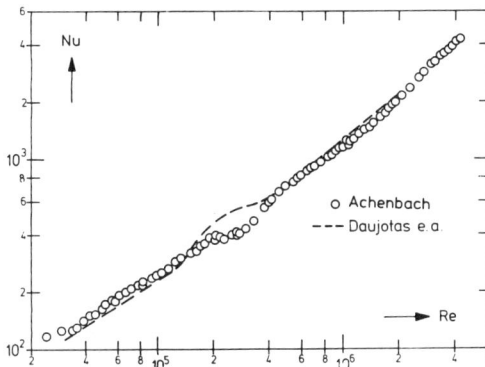

FIG. 5. Total heat transfer; smooth cylinder.

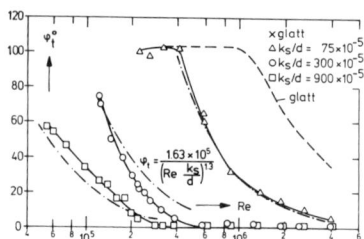

Fig. 6

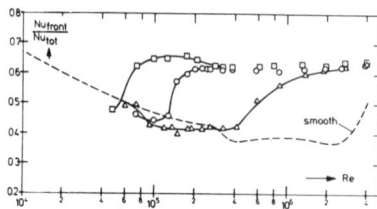

Fig. 7

FIG. 6. Transition point φ_t, laminar–turbulent; symbols as in Fig. 3.

FIG. 7. Total heat transfer; symbols as in Fig. 3.

somewhat larger in the critical flow regime. However, in general, the agreement is satisfactory, considering that the experiments were carried out under different boundary conditions (constant flux vs. constant temperature).

The local heat transfer distribution enables one to determine the position of the transition from a laminar to a turbulent boundary layer as a function of the Re-number and the roughness parameter. The result is given in Fig. 6. It is seen that the shifting of the transition point to the front stagnation point takes place in a rather narrow range of Re-number. The position of the transition point can be given as $\varphi_t = 1.63 \times 10^5/[\text{Re}(k_s/d)]^{1.3}$ where φ_t is measured in angular degrees.

By means of the local heat transfer distribution, it is possible to determine the ratio of heat transferred from the front part of the cylinder (Nu_{front}) to the total heat (Nu_{tot}) as a function of the Re-number and roughness parameter. For subcritical conditions, the ratio $\text{Nu}_{front}/\text{Nu}_{tot}$ decreases with increasing Re-number. In the critical flow regime, this ratio is independent of Re (45%). For transcritical conditions, 65% of the heat is transferred from the front part independent of Re. This can be explained by the improvement of frontal heat transfer caused by the turbulent boundary layer.

REFERENCES

[1] Achenbach, E.: Fifth International Heat Transfer Conference, Tokyo (1974), **II**, paper FC 6.1, 229–233.
[2] Achenbach, E.: J. Fluid Mech., **34**, (1968), 626–639.
[3] Achenbach, E.: *Ibid.*, **46**, (1971), 321–335.
[4] Daujotas, P., J. Žiugžda, and A. Žukauskas: Acad. Sci. Lithuanian SSR, Ser. B 3 (76) t. (1973), 99–109.

TRANSIENT NATURAL CONVECTION OF A HEAT GENERATING FLUID WITHIN AN ENCLOSURE*

R. Viskanta and R. O. Johnson
School of Mechanical Engineering
Purdue University

INTRODUCTION

Natural convection driven by energy released from distributed volumetric energy sources plays a fundamental role in a variety of engineering processes. Examples of such processes include exothermic chemical reactions [1], nuclear reactors [2-4], liquid radioactive-waste heat removal [5], and nuclear reactor design and safety [6]. In liquid radioactive-heat disposal systems and nuclear reactors, correct process design often requires accurate prediction of temperature distribution and heat transfer. For example, the assessment and design for postulated post-accident heat removal in LMFBR (liquid metal fast breeder reactor) nuclear reactor core meltdowns has led to consideration of natural convection heat transfer from molten fuel layers with internal heat generation.

In spite of the technological importance of free convection in cavities containing fluids with internal heat sources there has been a limited treatment of the problem. Steady-state free convection in a layer of heat generating fluid has been studied experimentally [7-11] and analytically [6, 9]. To the authors' knowledge, no transient results have been reported in the literature. A number of theoretical investigations of free convection in fluids without internal heat sources have been published [12-14], but in these studies only steady-state results were reported.

The objective of this work is to gain understanding of transient free convection flow and heat transfer which occurs in a cavity containing a heat generating fluid. To this end, numerical solutions of the governing conservation equations have been obtained for a cavity of rectangular cross section using finite difference techniques. The enclosed fluid is modeled as having a time-dependent or constant, uniformly distributed heat source. For

---

*The abstract of this paper appears on page 328.

certain processes such as fermentation, liquid radioactive–waste heat removal, radioactive heat decay, or certain chemical reactions, this is a reasonable approximation of reality.

ANALYSIS

Physical Model

A fluid initially at rest confined in a rectangular cavity, $0 \leqslant x \leqslant L$ and $0 \leqslant y \leqslant H$, with rigid walls is illustrated in Fig. 1. There is internal heat generation in the fluid at the rate $\dot{q}$ per unit volume. The fluid initially has a temperature distribution given by the steady one-dimensional solution in a heat conducting and heat generating fluid with a temperature T_O at the upper surface and T_H at the lower surface. The lateral boundaries at $x = 0$ and $x = L$ are rigid insulators. A small flow disturbance is introduced in the central region and allowed to develop until a steady-state free convection circulation is established or until all motion begins to die out.

The flow is considered laminar and two-dimensional with density variations only in the buoyancy term. The fluid is assumed to be incompressible with constant physical properties, and the Boussinesq approximation is valid. Two-dimensional circulation has been observed experimentally, and therefore the predictions should have physical significance. We confine our study to two-dimensional flows primarily because of time and storage restrictions of the computer. In addition, the two-dimensional approach is invaluable as a necessary step to understanding the transport processes. Three-dimensional calculations are feasible [13, 15] by an extension of the methods presented herein, but a significant increase (several orders of magnitude) in computer time would be required.

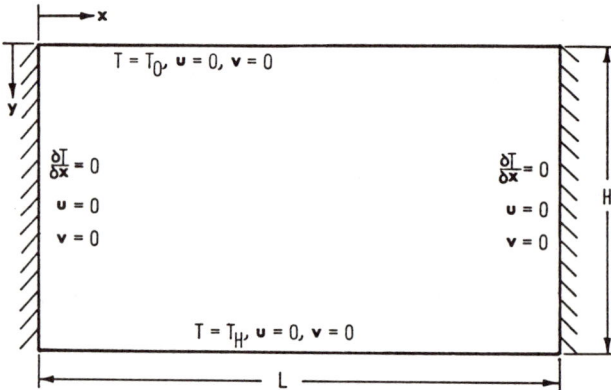

FIG. 1. Physical model and coordinate system.

Governing Equations

If the pressure terms in the x- and y-momentum equations are eliminated by cross differentiation and introduction of vorticity, the dimensionless equations governing the problem become

$$\frac{\partial \Omega}{\partial \tau} + \left(\frac{H}{T}\right)U\frac{\partial \Omega}{\partial X} + V\frac{\partial \Omega}{\partial Y} = \left(\frac{H}{T}\right)^2\frac{\partial^2 \Omega}{\partial Y^2} + \frac{\partial^2 \Omega}{\partial Y^2} - \text{Gr}\frac{\partial \Theta}{\partial X} \tag{1}$$

$$\frac{\partial \Theta}{\partial \tau} + \left(\frac{H}{L}\right)U\frac{\partial \Theta}{\partial X} + V\frac{\partial \Theta}{\partial Y} = \frac{1}{\text{Pr}}\left[\left(\frac{H}{L}\right)^2\frac{\partial^2 \Theta}{\partial X^2} + \frac{\partial^2 \Theta}{\partial Y^2} + \gamma\right] \tag{2}$$

The vorticity Ω has been expressed in terms of the velocity components U and V as

$$\Omega = \left[\left(\frac{H}{L}\right)\frac{\partial V}{\partial X} - \frac{\partial U}{\partial Y}\right] = -\left[\left(\frac{H}{L}\right)^2\frac{\partial^2 \Psi}{\partial X^2} + \frac{\partial^2 \Psi}{\partial Y^2}\right] \tag{3}$$

where the stream function Ψ is defined by

$$U = \frac{\partial \Psi}{\partial Y} \qquad V = -\frac{\partial \Psi}{\partial X} \tag{4}$$

In the foregoing formulation the equations have been made dimensionless through the use of the following variables:

$$\tau = \frac{\nu t}{H^2} \quad X = \frac{x}{L} \quad Y = \frac{y}{H} \quad U = \frac{u}{\nu/H} \quad V = \frac{\upsilon}{\nu/H}$$

$$\Omega = \frac{\omega}{\nu/H^2} \quad \Psi = \frac{\psi}{\nu} \quad \Theta = \frac{T - T_0}{T_H - T_0} \quad \text{Pr} = \frac{\nu}{\alpha} \tag{5}$$

$$\text{Gr} = \frac{g\beta(T_H - T_0)H^3}{\nu^2} \quad \gamma = \frac{\dot{q}H}{k(T_H - T_0)/H}$$

where g is the gravitational body force, β is thermal expansion coefficient, ν is the kinematic viscosity and α is the thermal diffusivity.

The initial and boundary conditions may now be written in the form

$$\Psi(X, Y, 0) = f(X, Y) \qquad \Theta(X, Y, 0) = Y + \tfrac{1}{2}\gamma Y(1 - Y)$$

$$\Theta(x, 0, \tau) = 0 \qquad \Theta(X, 1, \tau) = 1 \qquad \frac{\partial \Theta}{\partial X}\bigg|_{X=0} = \frac{\partial \Theta}{\partial X}\bigg|_{X=1} = 0 \tag{6}$$

$\Psi = \dfrac{\partial \Psi}{\partial n} = 0$ on all surfaces where n is the direction normal to the surface.

Equations (1) through (4) together with the boundary conditions given by Eq. (6) are sufficient to solve the problem.

After the temperature distribution has been determined, the local and average Nusselt numbers at the upper surface ($Y = 0$) are evaluated from

$$\text{Nu} = \frac{hH}{k} = -\left.\frac{\partial \Theta}{\partial Y}\right|_{Y=0} \quad \text{and} \quad \overline{\text{Nu}} = \frac{\bar{h}H}{k} = -\int_0^1 \left.\frac{\partial \Theta}{\partial Y}\right|_{Y=0} dX \quad (7)$$

Method of Solution

The solution of the conservation equations was carried out numerically. The elliptic equation, Eq. (3), for the stream function was rewritten as an unsteady equation of parabolic type which was easier to solve [15]. Numerical integration in time was carried out until the change in the Ψ field became negligible. An alternating direction implicit (ADI) method was employed to solve the energy, vorticity, and stream function equations in sequence. The ADI algorithm is well known [16] and details need not be repeated for the sake of brevity. Basically, the ADI algorithm is a two-step method employing two finite difference equations which are used at successive time steps of increment $\Delta\tau/2$. The first equation is written explicitly in the X-direction while the second is written explicitly in the Y-direction so that the results of the first time step are utilized in the second time step.

The effects of the grid size and time step on the transient solutions were investigated. Since the computing time is very sensitive to the number of nodal points [16], the grid spacing of $\Delta X = \Delta Y = 0.05$ was chosen as a reasonable compromise between accuracy and computer time. All figures and results presented herein are for these grid sizes. A time step of $\Delta\tau = 0.001$ was used for most calculations. However, for high values of Gr, Pr, and γ the time steps had to be reduced to eliminate numerical instabilities in the vorticity equation. The computational time required to reach steady state was typically from 1 to 5 minutes on the CDC 6500 digital computer.

RESULTS AND DISCUSSION

There are four dimensionless parameters governing the problem, Gr, Pr, γ and L/H, and a large number of calculations for a range of these parameters has been obtained. Unfortunately, the transient nature of the problem makes the presentation of the results rather difficult. In addition, because of the space limitations it is possible to include in the paper only some of the results which illustrate the salient features of the free convection heat transfer.

Each calculation was started with an arbitrary initial flow disturbance (stream function) and computations performed until the local temperature and the average Nusselt number $\overline{Nu}$ no longer varied significantly with time. For the limiting case of no heat generation, $\gamma = 0$, the average Nusselt number, $\overline{Nu}$, was found to be in good agreement with the values reported in the literature [14]. An energy balance made on the entire system showed that the heat transfer rate at the cold wall ($Y = 0$) was equal to within a fraction of one percent the heat transfer rate at the hot wall ($Y = 1$) plus the rate of energy generated in the cavity. This finding was encouraging in that it indicated validity and accuracy of the numerical algorithm.

The transient development of the temperature and horizontal velocity is illustrated in Fig. 2. The results indicate that the temperature and velocity fields develop slowly for early times ($\tau < 0.1$) and for late times ($\tau > 0.2$). For $\tau = 0.1$ the departure of the temperature from the one-dimensional heat conduction solution is relatively small. The results also show that the temperature and flow fields change very slowly as the steady-state solution is approached asymptotically at time $\tau = \tau_\infty$.

All the results obtained indicate some "overshooting" and "undershooting" of the temperature in the vicinity of the two isothermal walls. This is illustrated in Fig. 3 where the local Nusselt number (Nu) at several horizontal locations and the average Nusselt number ($\overline{Nu}$) are presented as a function of time. In the vicinity of the cool upper wall ($Y = 0$), the temperature increases above or decreases below the steady-state values and this results in local maximum or minimum heat transfer coefficients. The "overshoot" in the heat flux before equilibrium is obtained has also been noted by others [12]. The transient response curves from about $\tau = 0$ to $\tau = 0.08$ are due primarily to purely conductive heat transfer. The combined effects of conductive and convective heat transfer produce the rest of the response curve.

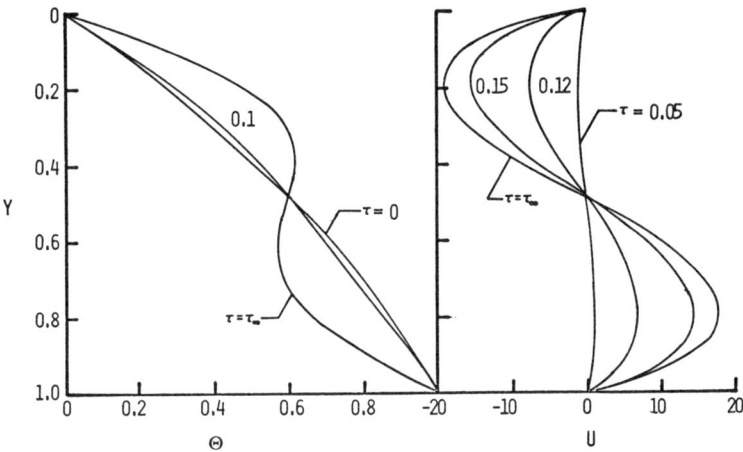

FIG. 2. Transient variation of temperature (a) and horizontal velocity component (b) at the midplane ($X = \frac{1}{2}$): Pr = 1.0, Gr = 10^4, $\gamma = 1.0$, and $L/H = 1.0$.

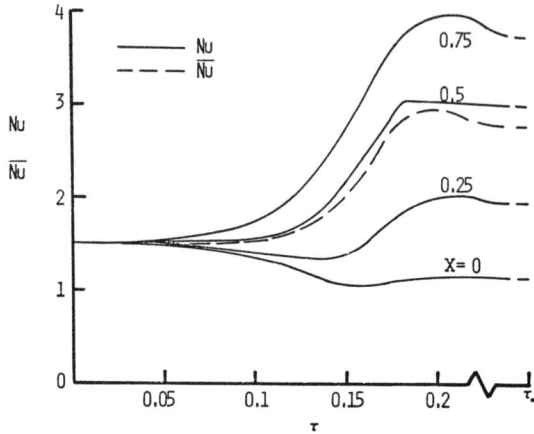

FIG. 3. Transient variation of local and average Nusselt numbers at the upper wall (Y = 0): Pr = 1.0, Gr = 10^4, γ = 1.0, and L/H = 1.0.

At low Rayleigh numbers (Ra = GrPr) the isotherms were found to be nearly straight, horizontal lines while the streamlines were nearly circular. However, as Ra was increased the isotherms became increasingly distorted, and a nearly isothermal core developed. In the core, conduction is negligible in comparison to convection heat transfer. This is illustrated in Figs. 4 and 5 for a square enclosure (L/H = 1) where the streamlines and isotherms are plotted for a typical case. The natural circulation flow predicted is similar to the roll

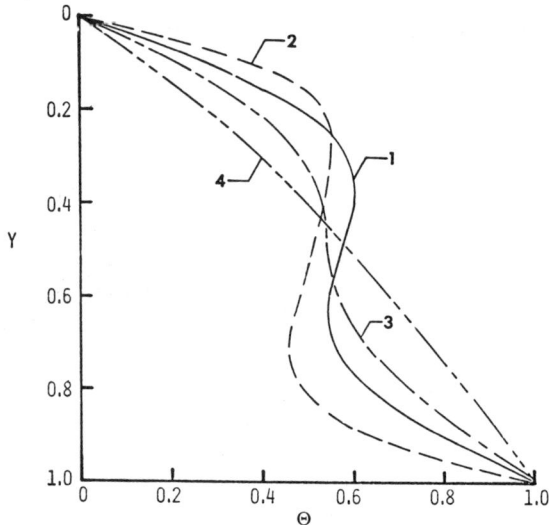

FIG. 4. Effect of governing parameters on the steady-state temperature profiles at the midplane (X = ½) for γ = 1.0, and L/H = 1.0: (1)–Pr = 1.0 and Gr = 10^4, (2)–Pr = 1.0 and Gr = 10^5, (3)–Pr = 0.1 and Gr = 10^5, (4)–Pr = 0.01 and Gr = 10^6.

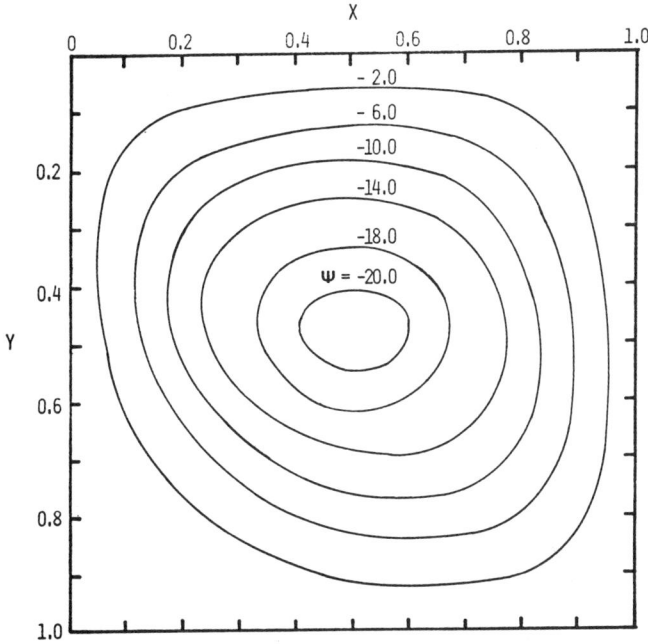

FIG. 5. Steady-state streamlines for Pr = 1.0, Gr = 10^5, γ = 1.0, and L/H = 1.0.

pattern obtained in the absence of internal heat generation [12, 14]. The streamlines are distorted only slightly while the temperatures are distorted significantly. Reversal of temperature gradients develops in the central part of the cavity. For a fixed Pr an increase in the Ra increases the size of the core region. For large Ra the temperature distribution and heat transfer are primarily controlled by the boundary layers formed on the two horizontal walls. Fixing the Ra and decreasing the Pr results in decreased distortion of the temperature field. This is clearly demonstrated in Fig. 6. The results presented in the figure show that unlike fluids having a large Prandtl number [14], the temperature distribution and heat transfer depend strongly not only on the Rayleigh but also on the Prandtl number. The temperatures predicted are in agreement with experimental observations on a plane layer of heat generating fluid maintained between two rigid walls at the same temperature [9].

The effect of the heat generation parameter γ on the steady-state temperature distribution and the local Nusselt number are illustrated in Figs. 7 and 8, respectively. The results indicated in Fig. 7 show that for sufficiently large values of γ there is reversal of heat transfer at the hot wall ($Y = 1$). For the particular values of parameters considered in the figure this occurs at about $\gamma = 5.0$. The results presented in Fig. 8 show that the local Nusselt number at the cool wall ($Y = 0$) is quite sensitive to the heat generation parameter γ and varies significantly with the distance along the walls. The

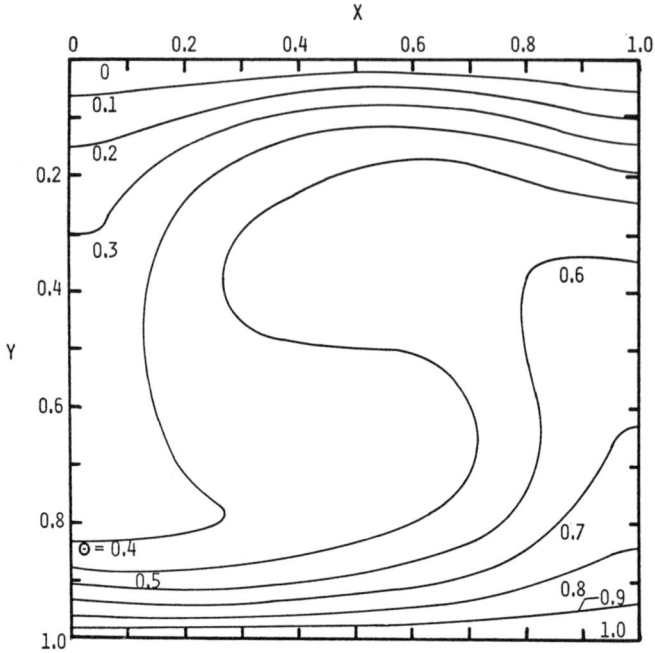

FIG. 6. Steady-state isotherms for Pr = 1.0, Gr = 10^5, γ = 1.0, and L/H = 1.0.

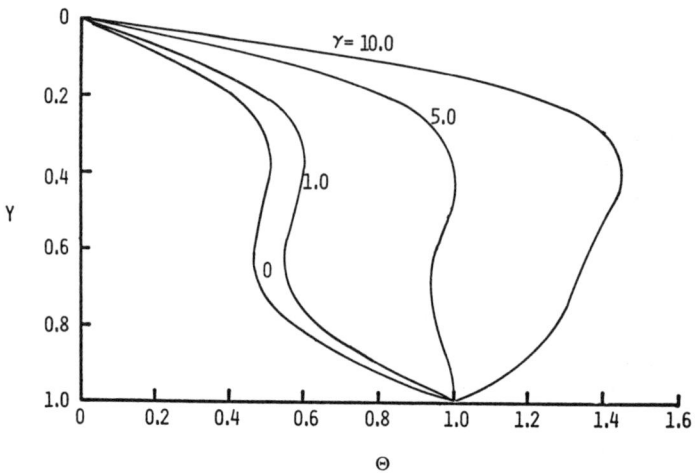

FIG. 7. Effect of heat generation parameter γ on steady-state temperature profiles at the midplane (X = ½) for Pr = 1.0, Gr = 10^4, and L/H = 1.0.

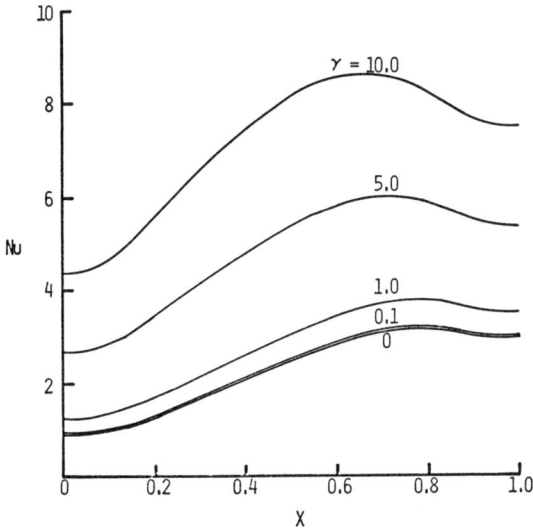

FIG. 8. Effect of heat generation parameter γ on the steady-state local Nusselt number at the upper wall ($Y = 0$) for Pr = 1, Gr = 10^4 and $L/H = 1$.

maximum heat transfer coefficient occurs in the vicinity of the stagnation point. For a given Gr and Pr with the increase in γ, the stagnation point is shifted toward the center of the wall. The results show that the flow field is rather insensitive to the heat generation parameter γ. For example, with Pr = 1.0 and Gr = 10^4 at steady state the maximum value of the U component velocity is increased only about 5 percent between $\gamma = 0$ and $\gamma = 10$.

Numerical solutions have also been obtained for rectangular cavities having aspect ratio L/H of 2 and greater. However, because of space considerations, detailed discussion of the results cannot be included. It was found that steady-state flow pattern (number of cells) was sensitive to the type and magnitude of the initial flow disturbance. For example, when an initially symmetric flow disturbance was used in the central region of a cavity with $L/H = 2$ a single steady cell resulted, but if the disturbance was nonsymmetrical, a steady-state solution with two cells was obtained. The results are in agreement with the findings of others [12, 14] who have also noted metastable behavior. Because of the limited number of solutions obtained the results are inconclusive, and the determination of the preferred mode of natural convection circulation is a fruitful area of future research. This is important because the cell pattern has an important bearing on the local and average heat transfer coefficients. For example, for the case Pr = 1, Gr = 10^4, $\gamma = 1$, and $L/H = 2$ a two-cell pattern at the cold wall yielded a steady-state value of $\overline{Nu} = 3.54$ while a three-cell pattern resulted in $\overline{Nu} = 3.44$.

A difficulty limiting the present numerical calculations occurs in relation to flow separation which could not be predicted. Until a numerical procedure can be developed to account for the effects of three-dimensionality some of the observed details of the flow characteristics must, of course, remain unexplained.

SUMMARY

Convective flow fields, temperature distribution, and heat transfer resulting from heat generation have been computed for a range of dimensionless parameters. In particular, natural convection in an enclosed horizontal region of rectangular cross section which is heated from below has been studied and the applicability of the numerical finite-difference techniques has been demonstrated.

Future work in this area should consist further of both analytical effort and experimental verification. Additional analytical results could include more general boundary conditions at the horizontal walls, cooled or heated vertical walls, as well as nonuniform heat generation; the effect of a free boundary at the upper free surface needs to be considered.

ACKNOWLEDGMENTS

The authors wish to acknowledge the assistance of Mr. R. E. Daniel in the numerical computations and Ms. P. J. Bishop in the preparation of the paper.

REFERENCES

1. Jones, D. R.: Int. J. Heat Mass Transfer, 1974, **17**, No. 1, pp. 11–21.
2. Martin, B. W.: Proc. Roy. Soc., Series A, 1967, **309**, No. 1466, pp. 327–341.
3. Murgatroyd, W., and A. Watson: J. Mech. Eng. Sci., 1970, **12**, No. 5, pp. 354–363.
4. Watson, A.: *Ibid.*, 1971, **17**, No. 3, pp. 151–156.
5. Van Sant, J. H. W.: Nuc. Eng. Des., 1969, **10**, No. 3, pp. 349–355.
6. Catton, I., and A. J. Suo-Anttila: in *Heat Transfer 1974*, JSME/MSChE, Tokyo, 1974, **3**, pp. 69–73.
7. Tritton, D. J., and M. N. Zarraga: J. Fluid Mech., 1967 **30**, Part 2, pp. 21–31.
8. Kulacki, F. A., and R. J. Goldstein: *Ibid*, 1972, **53**, Part 2, pp. 271–287.
9. Jahn, M., and H. H. Reinecke: in *Heat Transfer 1974*, JSME/JSChE, Tokyo, 1974, **3**, pp. 74–78.
10. Kulacki, F. A., and R. J. Goldstein: *Ibid.*, JSME/JSChE, Tokyo, 1974, **3** pp. 64–67.
11. Kulacki, F. A. and M. E. Nagle: *J. Heat Transfer*, 1975, **97**, No. 2, pp. 204–211.
12. Fromm, J. E.: *Phys. Fluids,* 1965, **3**, No. 10, pp. 1757–1769.
13. Aziz, K., and J. D. Hellums: *Ibid.*, 1967, **10**, No. 2, pp. 314–324.
14. Samuels, M. R., and S. W. Churchill: AIChE Journal, 1967, **13**, No. 1, pp. 77–85.
15. Chen, T. -S., and R. E. Goodson: Glass Tech., 1972, **13**, No. 6, pp. 161–167.
16. Roache, P. J.: *Computational Fluid Dynamics*, Hermosa Publishers, Albuquerque, New Mexico, 1972.

TRANSIENT HEATING OF A PLATE-SYSTEM IN A GAS FLOW*

V. Hlavačka
National Research Institute for Machine Design
Prague, Czechoslavakia

The development of unsteady methods of finding the surface heat transfer coefficients in various heat transfer matrices is connected with solutions of interesting problems relating to heat conduction in solids. One of them is the transient heating of plates in a gas flow.

The differential equations and boundary conditions describing the temperature distribution in the gas and the solid

$$\frac{\partial t}{\partial \xi} + \frac{1}{v}\frac{\partial t}{\partial \tau} = \frac{\alpha F}{WL}(t_{MO} - t) \qquad 0 \leqslant \xi \leqslant L \tag{1}$$

$$\frac{\partial t_M}{\partial \tau} = a\frac{\partial^2 t_M}{\partial r^2} \qquad 0 \leqslant r \leqslant R \tag{2}$$

$$t(0,\tau) = A_1 \cos \omega\tau$$

$$\left.\frac{\partial t_M}{\partial r}\right|_{r=0} = 0 \qquad \lambda_M \left.\frac{\partial t_M}{\partial r}\right|_{r=R} = \alpha(t_{MO} - t)$$

are based on the following assumptions:

—the heat transfer coefficient and physical properties of the fluid and plates are constant throughout the system;
—the fluid velocity profile is uniform in an arbitrary cross section;
—the effect of longitudinal conduction of heat is negligibly small.

In order to obtain the asymptotic solution the fluid temperature may be considered to be

$$t(\xi,\tau) = A_1 \exp(p\xi + i\omega\tau) \tag{3}$$

---

*The abstract of this paper appears on page 330.

Then the surface temperature of plates is given by

$$t_{MO} = t_M(R, \tau) = A_1 N \exp{(p\xi + i\omega\tau)} \tag{4}$$

where [1]

$$N = \frac{\cosh\sqrt{(i\omega/a)}\,R}{\cosh\sqrt{(i\omega/a)}R + (\lambda_M/\alpha)\sqrt{(i\omega/a)}\,\sinh\sqrt{(i\omega/a)}R}$$

Substitution of (3) and (4) in Eq. (1) results in

$$p = -\frac{\alpha F}{WL}(1 - N) - \frac{i\omega}{v}$$

Consequently, the outlet fluid temperature

$$t(L, \tau) = A_2 \cos{(\omega\tau + \varphi_c)}$$

and the temperature response will be determined by the expressions

$$\ln\frac{A_2}{A_1} = -\frac{\alpha F}{W}\,\mathrm{Re}(1 - N) \qquad \varphi_c = -\frac{\alpha F}{W}\,\mathrm{Im}(1 - N) - \frac{\omega L}{v}$$

After finding the real Re and imaginary Im parts of $(1 - N)$ the results may be written in the form

$$\ln\frac{A_2}{A_1} = -\frac{\alpha F}{W}\,\frac{(x/\sqrt{2})\,\mathrm{Bi}\,S^-(x) + x^2 C^-(x)}{\mathrm{Bi}^2 C^+(x) + x\sqrt{2}\,\mathrm{Bi}\,S^+(x) + x^2 C^-(x)} \tag{5}$$

$$\varphi_c = -\frac{\alpha F}{W}\,\frac{(x/\sqrt{2})\,\mathrm{Bi}\,S^+(x)}{\mathrm{Bi}^2 C^+(x) + x\sqrt{2}\,\mathrm{Bi}\,S^-(x) + x^2 C^-(x)} \tag{6}$$

where the functions C^+, C^-, S^+, and S^- are defined as follows ($x = \sqrt{\omega/aR}$)

$$C^+(x) = \cosh x + \cos x \qquad C^-(x) = \cosh x - \cos x$$
$$S^+(x) = \sinh x + \sin x \qquad S^-(x) = \sinh x - \sin x$$

Denoting

$$Z = \frac{W}{W_M\omega}\,\ln\frac{A_1}{A_2} \qquad H = \frac{\alpha F}{W_M\omega}$$

then $x = \sqrt{Bi/H}$ and the dimensionless parameter Z will only depend on Bi and H. Figure 1 shows the relationship $Z = f(Bi, H)$ at low values of H; this region is important for the measurement of heat transfer coefficients.

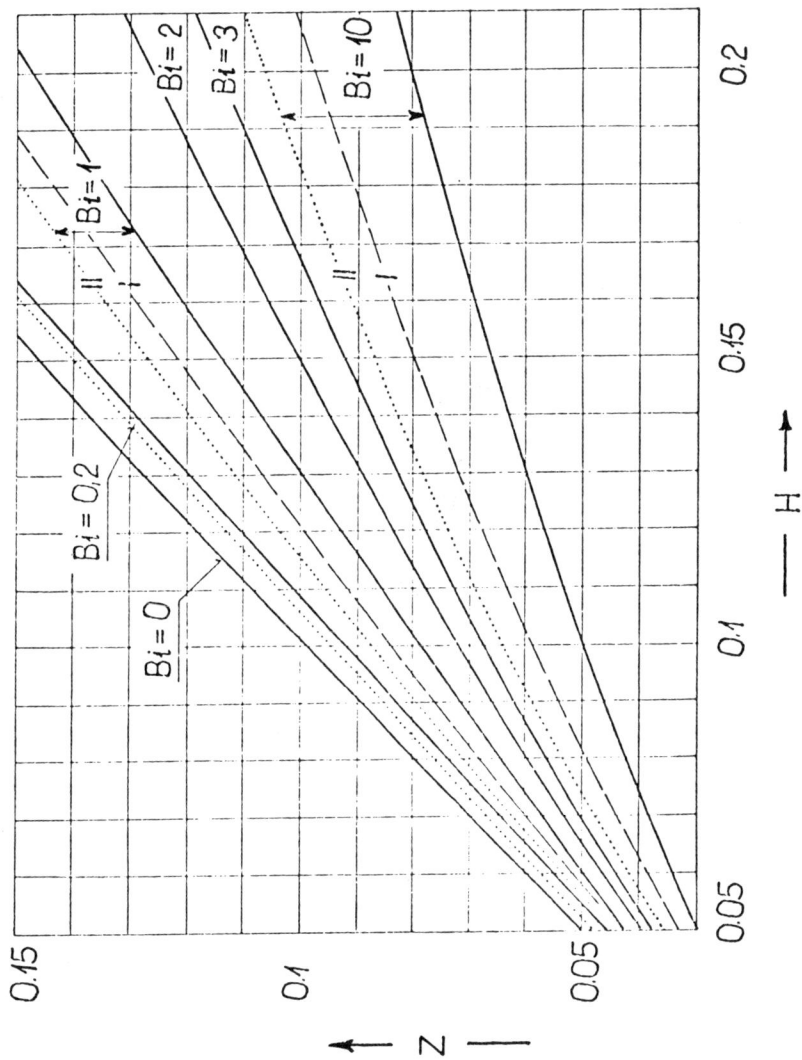

FIG. 1. Relationship of $Z = f(Bi, H)$ for a plate system and its comparison with cylindrical rods (I) and fixed beds of spheres (II).

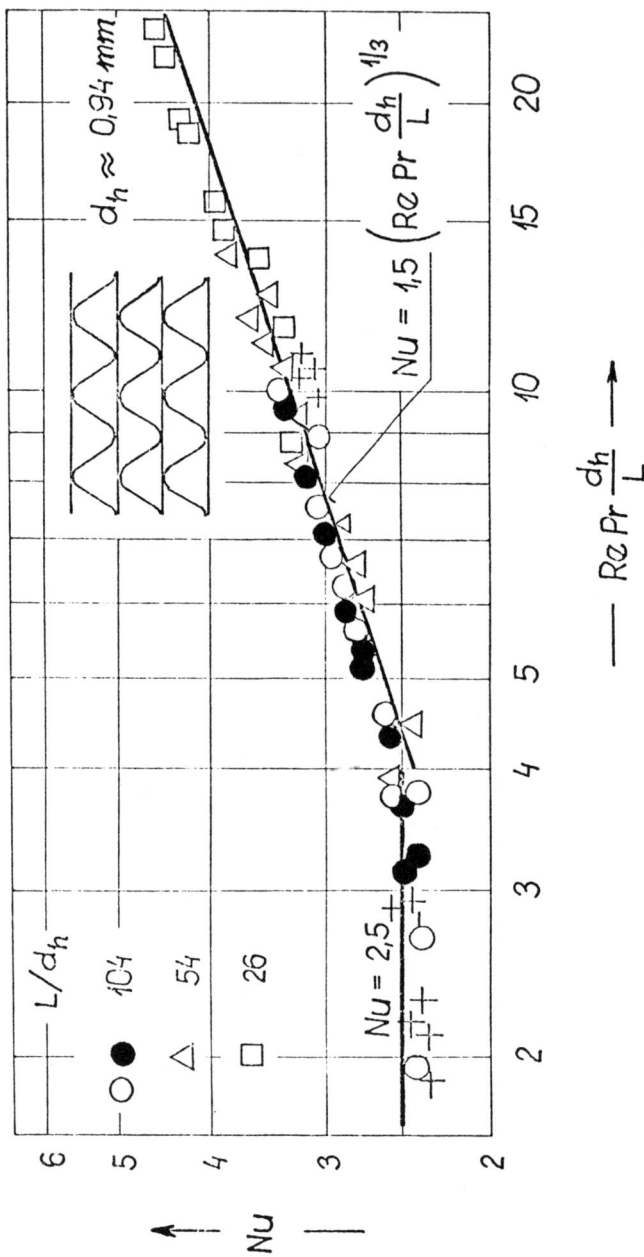

FIG. 2. The Nusselt number obtained by cyclic method vs. the Reynolds number for a rotary regenerator matrix ($\circ$, $\triangle$, $\square$); comparison with the data obtained by the singel "blowdown" technique ($\bullet$) and those from regenerator model tests (+). The geometrical form of the passages with $d_h = 0.92$ mm is shown in the top part of the figure.

Compared with the results obtained recently in [2] and [3] for the matrix of cylindrical rods and a fixed bed of spheres the increase of the Biot number shows a large deviation from the equation

$$Z = \frac{H}{1 + H^2} \tag{7}$$

which characterizes the limiting case of Bi = 0. Expressions (5) and (6), derived for a system of plates augment the background of the cyclic temperature variation method for ascertaining the surface heat transfer coefficient mentioned in [2] and [3] and discussed in references of these papers. It has been shown that the accuracy of this method will be satisfactory at values of H from 0 to 0.3.

To illustrate the quality of the results obtained in terms of the cyclic method, Fig. 2 presents the Nu data for a thin walled rotary regenerator matrix with approximately triangular passages. The experimental apparatus and technique has been described in [3]. The values of Nu calculated from Eq. (5) or (7) and based on the runs carried out at air temperature ranging from 40 to 60°C and at the amplitudes of the harmonic temperature fluctuations from 10 to 15°C shows reasonable agreement with the well-known equation

$$Nu = 1.5 \left(RePr\frac{d_h}{L} \right)^{1/3}$$

and also with the minimal value Nu = 2.5 for laminar flow.

Nomenclature

A_1, A_2	amplitude of gas temperature wave (inlet, outlet)
a	thermal diffusivity
$Bi = \dfrac{\alpha R}{\lambda_M}$	Biot number
d_h	hydraulic diameter of matrix passages
F	heat transfer surface area
$H = \dfrac{\alpha F}{W_M \omega}$	dimensionless parameter
L	length of plates, matrix
$Nu = \dfrac{\alpha d_h}{\lambda}$	Nusselt number
$Pr = \dfrac{\nu}{a}$	Prandtl number
R	one half of plate thickness
$Re = \dfrac{v d_h}{\nu}$	Reynolds number
t, t_m	gas, plate temperature
t_{mo}	temperature of plate surface

v	gas velocity
W	heat capacity of fluid
W_m	heat capacity of plates
$Z = \dfrac{W}{W_M \omega} \ln \dfrac{A_1}{A_2}$	dimensionless parameter
α	surface heat transfer coefficient
λ, λ_m	heat conductivity of gas, plate
ν	kinematic viscosity
ξ, r	space coordinate
τ	time
φ_c	total phase lag
ω	angular frequency of temperature oscillations

REFERENCES

1. Luikov, A. V.: Theory of Heat Conduction. GITTL, Moscow 1952 (in Russian).
2. Hlavačka, V.: Journal of Engineering Physics, 1973, **24**, 1, p. 70–74 (in Russian).
3. Hlavačka, V.: *Ibid.*, 1975, **28**, 4, p. 604–608 (in Russian).

HEAT TRANSFER AND PRESSURE DROP OF IN-LINE BANKS OF TUBES WITH ARTIFICIAL ROUGHNESS*

H. G. Groehn and F. Scholz
Institut für Reaktorbauelmente
Jülich, West Germany

The variation of the flow through a heat exchanger due to the surface roughness of the tubes is generally associated with high Reynolds numbers. The change of characteristics of heat transfer and pressure drop for tube bundles of customary steel tubes is expected at $6 \cdot 10^4 \leqslant \mathrm{Re} \leqslant 2.5 \cdot 10^5$. As conventional heat exchangers operate at lower Re numbers, the surface roughness effect is of no importance. This paper is concerned with the elaboration of heat transfer and pressure drop characteristics for recuperators of gas-cooled nuclear reactors which operate at very high Re-numbers depending on the kind of cooling gas chosen.

The possibility of affecting heat transfer and pressure drop of tube bundles in cross flow by surface roughening is described in [1]. In this investigation, the effect of surface roughness was studied by means of a rough separately heated cylinder in a bank of smooth tubes. The most important result of that investigation was the evidence that for an in-line arrangement of tubes, the heat transfer can be considerably improved by an artificial enhancement of the surface roughness height without increasing the pressure drop. As it is difficult to transfer the results from a single cylinder to a complete heat exchanger, some models were tested in a Re-number range of $2.5 \cdot 10^3 < \mathrm{Re} < 10^6$. The experiments were conducted in the high pressure wind tunnel of the Institut für Reaktorbauelemente of the Kernforschungsanlage Jülich. Two types of artificial surface roughnesses with a similar ratio of roughness height k to tube diameter D were investigated, Fig. 1. The tubes of 4 in-line banks with straight tubes were knurled (pyramidal roughness) as shown in Fig. 1 (right). The knurling technique was developed by the "Wieland Werke," Germany. The tube represented at the left hand side of Fig. 1 was equipped with 20 small longitudinal ribs uniformly distributed around the circumference. These tubes were arranged in a heat exchanger having a multistart helices design. The 33

---

*The abstract of this paper appears on page 330.

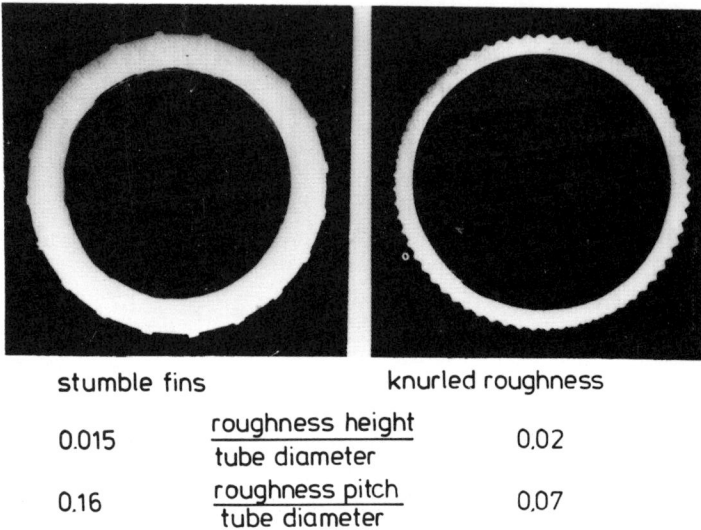

stumble fins   knurled roughness

| 0.015 | roughness height / tube diameter | 0.02 |
| 0.16 | roughness pitch / tube diameter | 0.07 |

FIG. 1. Roughness types investigated.

tubes of this heat exchanger were arranged with 7 tube layers each consisting of from 3 to 5 coils. Heat exchangers of this design are generally used for the steam generators of gas-cooled nuclear reactors.

Both roughness types have been investigated previously [1]. Though the flow mechanism is different in both cases, nearly the same heat transfer was observed. Concerning the pyramidal roughness, the idea is that premature transition from a laminar to a turbulent boundary layer is caused by disturbances generated by the roughness. Thus the flow around the tubes is transcritical. In case of the finned tube, a free shear layer is formed downstream of the particular fins. As it turns turbulent a considerable exchange of mass perpendicular to the main flow direction occurs. Thus the heat is not transfered entirely through a boundary layer.

Figure 2 shows the experimental results for heat transfer and pressure drop as a function of the Re number for an in-line tube bank with straight knurled tubes. The parameter of the pyramidal roughness was $k/D = 0.017$. The additional curves refer to the pyramidal roughness $k/D = 0.03$ and to technically smooth steel tubes with $k/D = 6 \cdot 10^{-4}$.

As a consequence of the increased surface roughness, the critical Re number decreases from Re $\approx 2 \cdot 10^5$ for the technically smooth steel tubes to Re $\approx 2.5 \cdot 10^4$ for the knurled tubes with $k/D = 0.017$. The heat transfer coefficient of the rough tube bank is improved up to 50 percent for Re $< 2.5 \cdot 10^4$ due to an early boundary layer transition. Simultaneously, the pressure drop decreases slightly compared with those of technically smooth steel tubes. The reason for this is a minor contraction and expansion in the gaps between the tubes as demonstrated by Achenbach [5]. The boundary layer separates at

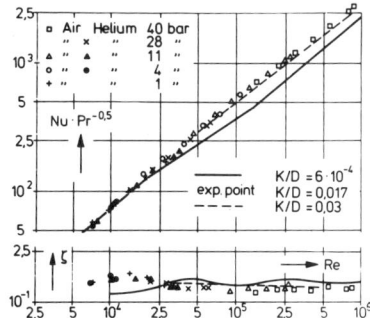

FIG. 2. Heat transfer and pressure drop of in-line tube banks with straight knurled tubes: D = 25 and 51 mm; $S_q \approx 2.1\ D$; $S_L \approx 1.4\ D$; Z = 10.

a lower angle of circumference while the point of impact shifts to a more downstream position. The efficiency coefficient for heat exchangers (defined as the ratio of transferred heat to the expended pumping power at constant Re number and the same temperature conditions) is greater by a factor of 1.5 ... 1.6 for the knurled tubes compared with the same bank with smooth steel tubes. At subcritical conditions, the boundary layer disturbances due to the surface roughness are damped. It is expected that there will be no influence on heat transfer and pressure drop. Regarding the heat transfer results, this assumption is confirmed by experiment. However, it does not hold for the flow resistance. The evidence that the experimental curves do not collapse may be explained by the sensitivity of the flow through the smooth tube bundle. With the view to Zukauskas' [2] work, a considerable scatter of results is observed. His fitted curve comes close to our values for knurled tubes.

As concluded from Fig. 2, the optimum effect of surface roughening on the improvement of heat transfer is already exceeded for k/D = 0.03. The critical Re number decreases insignificantly. Heat transfer is even less than that of k/D = 0.017 at Re $> 7 \cdot 10^4$. However, it cannot be excluded that the increase of roughness pitch as a consequence of the increased roughness height has an influence on the decrease of the heat transfer coefficient.

A corresponding effect as discussed by means of Fig. 2 for tube pitches of $S_q/D \approx 2.1$ and $S_L/D \approx 1.4$ was observed for tube banks with the pitches $S_q/D \approx 1.7$ and $S/D \approx 1.3$ at the same roughness conditions. Figure 3 shows a comparison of the results for k/D = 0.03 and technically smooth copper tubes $(k/D < 2 \cdot 10^{-5})$.

Earlier investigations [4], [3] carried out on two heat exchangers of the multistart helices design have shown that this type of heat exchanger can be treated as an in-line tube bundle. Therefore, this seems to suggest that the advantages obtained for low finned tube banks with straight tubes [1] also exist for helical heat exchangers. Figure 4 points out the difference actually obtained for heat transfer and pressure drop between smooth and finned

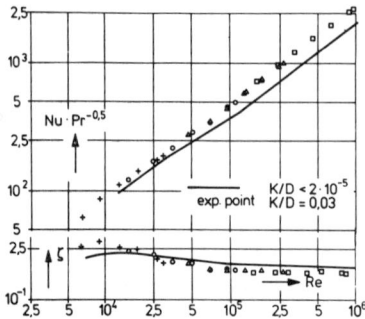

Fig. 3

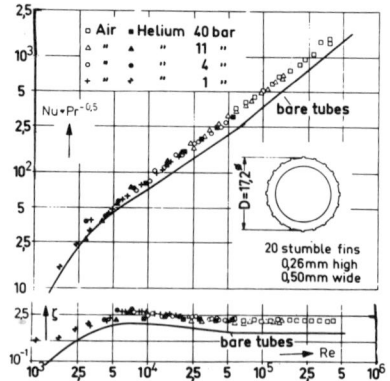

Fig. 4

FIG. 3. Heat transfer and pressure drop of in-line tube banks with straight knurled tubes; $D = 25$ mm; $S_q \approx 1.7\,D$; $S_L \approx 1.3\,D$; $Z = 10$.

FIG. 4. Heat transfer of a coiled heat exchanger with finned tubes: $D = 17.2$ mm; $S_q = 1.68\,D$; $S_L = 1.28\,D$; $Z = 28$.

helical heat exchangers. The fins cause the critical Re number to decrease from Re $\approx 6 \cdot 10^4$ to Re $\approx 5 \cdot 10^3$. Over a large flow range (Re $> 2.5 \cdot 10^4$) the heat transfer coefficient is about 30 percent greater than for a tube bundle of the same geometry but consisting of technically smooth steel tubes. In contradiction to the expectation the flow resistance also increases by about 30 percent. From this evidence it is concluded that the flow through an in-line tube bank of straight tubes and a helical heat exchanger is different than initially expected. The advantage of finned tubes compared with knurled tubes becomes relevant at the lower Re-number range where an improvement of heat transfer is observed for equal values of k/D. The efficiency coefficient is equal to that found for smooth steel tubes.

REFERENCES

1. Groehn, H. G., and F. Scholz: Heat Transfer Conference 1970, **III,** paper FC 7.10.
2. Zukauskas, A., V. Makarevicius, and A. Slanciauskas: Heat Transfer in Banks of Tubes in Crossflow of Fluid, Mintis, Vilnius, 1968.
3. Scholz, F., Th. Meis, and H. G. Groehn: Jül-649-RB (1972).
4. Groehn, H. G., and F. Scholz: Component Design in High Temperature Reactors, London, 1972, paper 6.
5. Achenbach, E.: Wärme- und Stoffübertragung, **4,** (1971), pp. 152–155.

HEAT TRANSFER AND FLUID FRICTION FOR WATER FLOW IN TUBES WITH SUPERCRITICAL PRESSURES*

S. ISHIGAI, M. KAJI AND M. NAKAMOTO
Faculty of Engineering, Osaka University,
Osaka, Japan

INTRODUCTION

In order to analyze convective heat transfer problems for supercritical fluids, it is important to know the correlation between heat transfer and fluid friction. Over the last 20 years, a considerable amount of data for convective heat transfer in the pseudocritical region have been reported by many investigators. For the friction factor, however, only few data were reported [1, 2]. This paper describes the pressure drop and the heat transfer data for supercritical water flow in tubes, simultaneously measured for the same test loop, and discusses the correlation between them.

EXPERIMENTAL ARRANGEMENT

The experimental apparatus was composed of a oncethrough monotube loop which is commonly used. The test section consisted of electrically heated stainless steel tubes, 3.92 or 4.44 mm in ID and 625 or 868 mm in heated length for a vertical upward flow or a horizontal flow, respectively. The inner surface was carefully polished to make it hydraulically smooth. Wall temperatures were measured at 10 cross sections along the test tube by chromel-alumel thermocouples and fluid temperatures at the inlet and outlet of the test section were measured by sheath thermocouples. The total pressure drop in the heated section was measured by a pressure-difference transducer.

The test conditions were as follows: pressure—250, 300, 400 ata, mass flow rate—500, 1000, 1500 kg/m^{2}s, heat flux—up to 1.4×10^6 kcal/m^{2}h. The experiments were carried out with an emphasis on the near-critical region (250 ata) for vertical upward flow.

---

*The abstract of this paper appears on page 336.

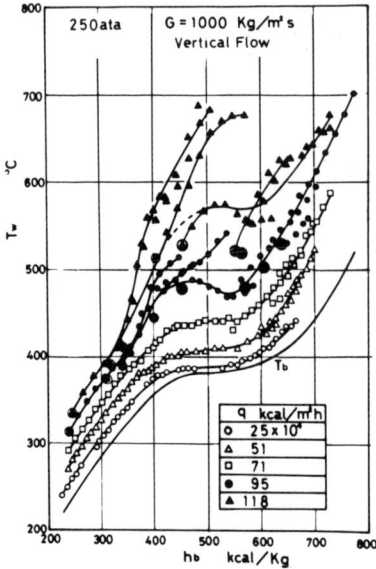

FIG. 1. Wall temperatures.
(Vertical Upward Flow)

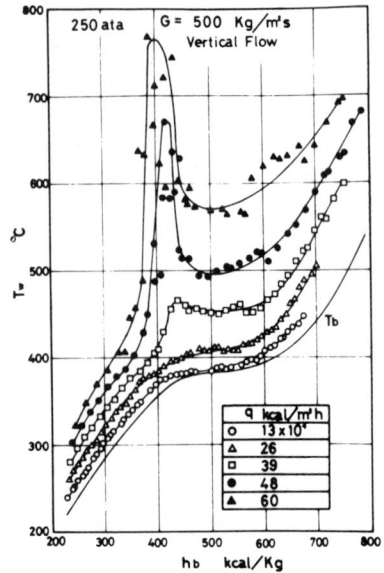

FIG. 2. Wall temperatures.
(Vertical Upward Flow)

RESULTS AND DISCUSSIONS

Typical wall temperature distributions are shown in Figs. 1–4. When the pressure was close to the critical pressure, a local anomaly was detected even with a relatively high mass flow rate for high heat flux conditions (Fig. 1). In this case, the wall temperature distributions versus the average bulk fluid temperature at a given heat flux varied with the inlet fluid temperature. For

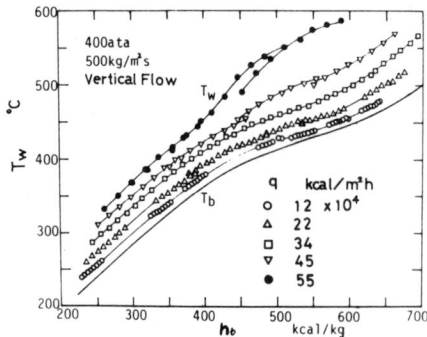

FIG. 3. Wall temperatures.
(Vertical Upward Flow)

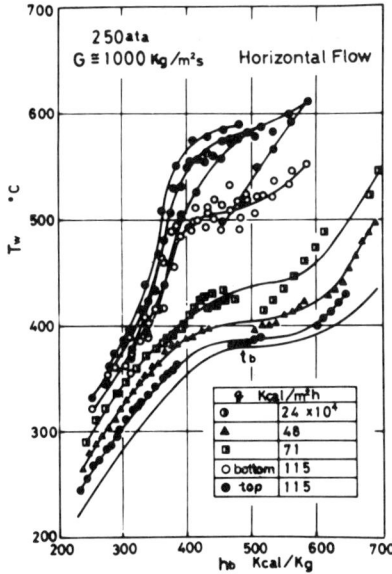

FIG. 4. Wall temperatures.
(Horizontal Flow)

vertical upward flow, such a deterioration is attributed to the combined effects of temperature-dependent parameter distributions in the cross section, the laminarization of the boundary layer due to the acceleration of the flow, and the buoyancy. The deterioration is remedied with an increase in pressure mainly because of the less significant effect of temperature on water properties (Fig. 3).

For horizontal flow, some difference of the wall temperatures between the top and bottom of the tube wall was observed at the same cross section, although the inner diameter of the test tube was very small. This difference increased with an increase in heat flux due to buoyancy effects. The average heat transfer coefficient over the cross section was higher than that for vertical upward flow under the same conditions.

Friction pressure losses were calculated from the measured pressure-drop data by subtracting acceleration and elevation losses (the latter was excluded for the horizontal flow), evaluated on the assumption of homogeneous flow. Heating decreases the friction loss. The mechanism of the decrease was investigated by numerical computations based on the isothermal friction factor. In Fig. 5 experimental and calculated results are shown versus mean bulk enthalpy, i.e., mean value of the inlet and outlet bulk fluid enthalpy. The decrease cannot be accounted for by the longitudinal fluid temperature distribution. The decrease is insignificant in the single phase region, but has a maximum at the pseudocritical temperature.

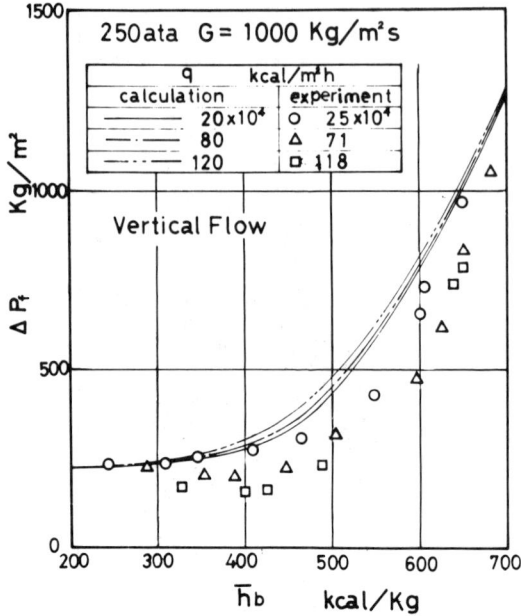

FIG. 5. Friction pressure drop.

In Figs. 6-9 the friction factor ratio, i.e., the ratio of the friction factor for the heated flow to that for isothermal flow, is presented for various flow conditions including the mass flow rate, the heat flux, and the pressure. It was found that the ratio is less than unity near the pseudocritical temperature and decreases as the heat flux increases. In the case of vertical upward flow, the decrease is remedied by increasing the flow rate and the pressure. The friction factor ratio for vertical flow is less than that for horizontal flow, for which

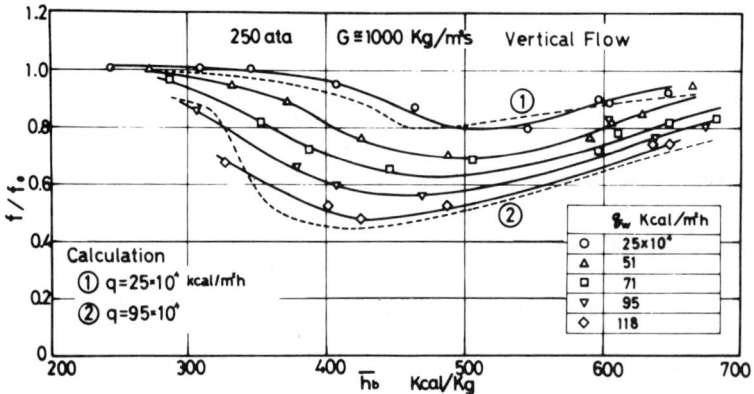

FIG. 6. Friction factor ratio.
(Vertical Upward Flow)

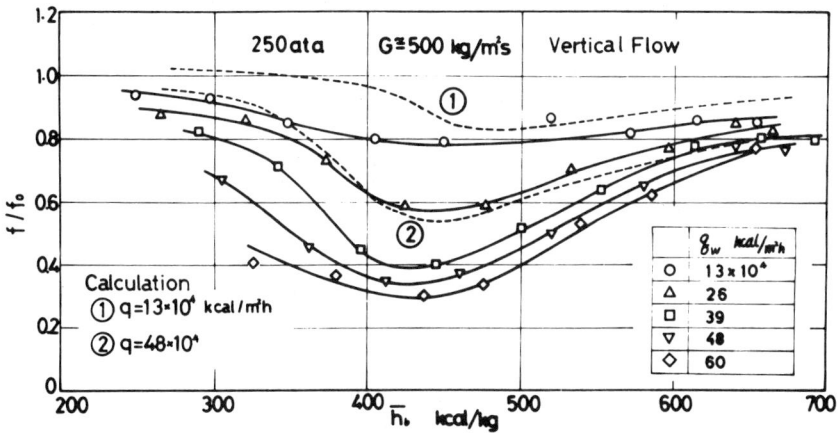

FIG. 7. Friction factor ratio. (Vertical Upward Flow)

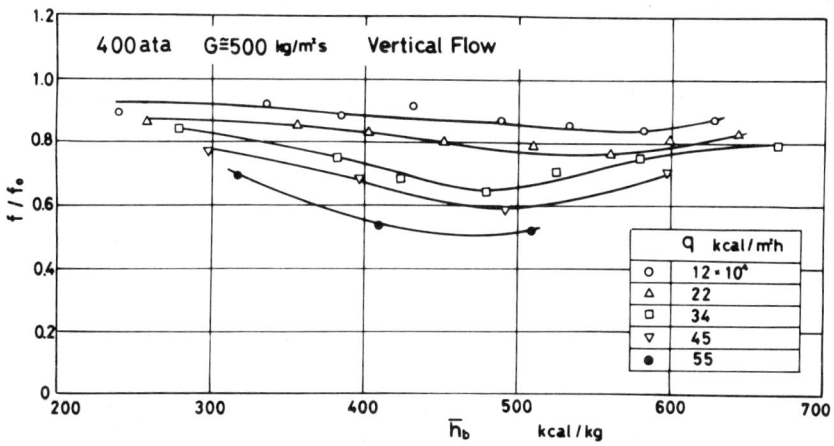

FIG. 8. Friction factor ratio. (Vertical Upward Flow)

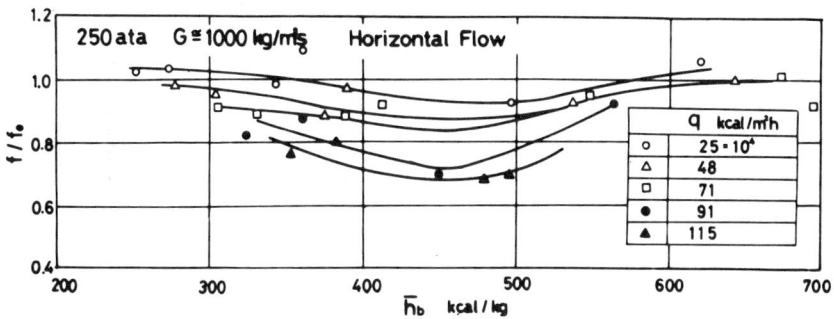

FIG. 9. Friction factor ratio. (Horizontal Flow)

the friction factor ratio is less affected by the heat flux and the flow rate (Fig. 9).

The results of numerical computations based on the Prandtl's mixing length theory are shown by the dotted lines in Figs. 6 and 7. With low mass flow rates, the theoretical result is higher than the experimental, and with high mass flow rates the opposite occurs. In the computation, Reichardt's equation (4) of eddy diffusivity for momentum was used, and the ratio of eddy diffusivities for heat and momentum was assumed to be unity ($\epsilon_H/\epsilon_M = 1$).

The friction factor is affected by the viscosity of the fluid in the vicinity of the wall. The friction factor ratios are plotted against the viscosity ratio (μ_b/μ_w) in Figs. 10 and 11. The correlation by Tarasova et al. [1]

$$\frac{f}{f_0} = \left(\frac{\mu_b}{\mu_w}\right)^{-0.22}$$

is shown in these figures. For the horizontal flow our experimental results agree well with their correlation, but for the vertical upward flow our results show lower values especially in low mass flow rate.

In correlating the forced convection heat transfer and fluid friction data, Colburn's j factor [3] ($j = St \cdot Pr^{2/3}$) is often used successfully in the range

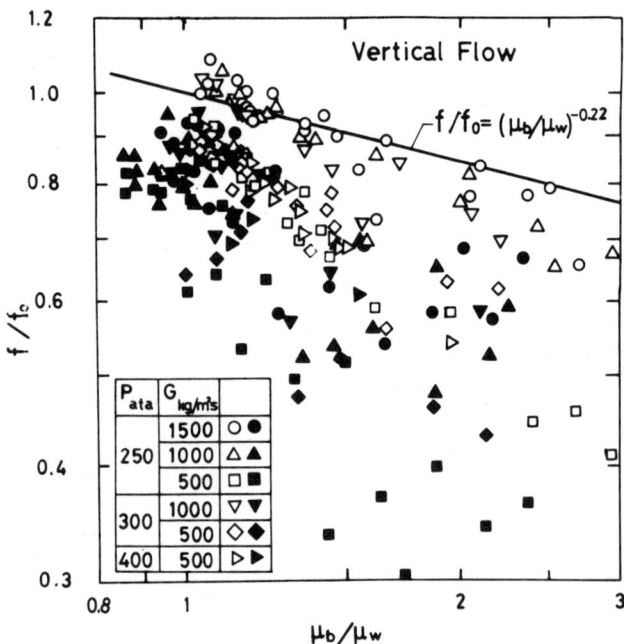

FIG. 10. Friction factor ratio versus viscosity ratio. (Vertical Upward Flow)

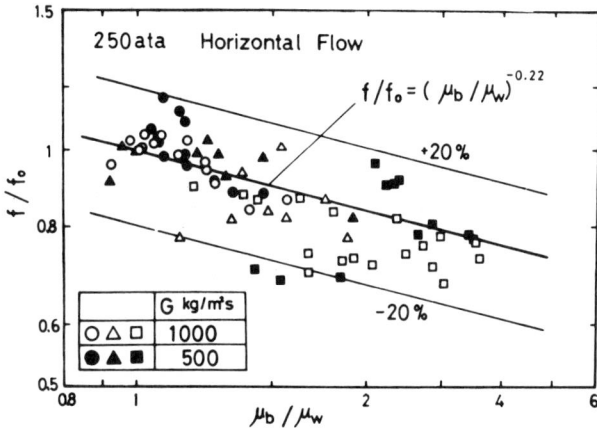

FIG. 11. Friction factor ratio versus viscosity ratio. (Horizontal Flow)

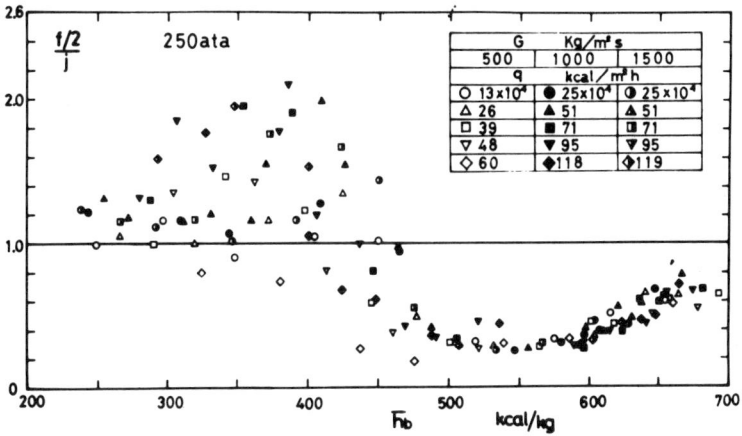

FIG. 12. Correlation of heat transfer and fluid friction. (by Colburn's j Factor)

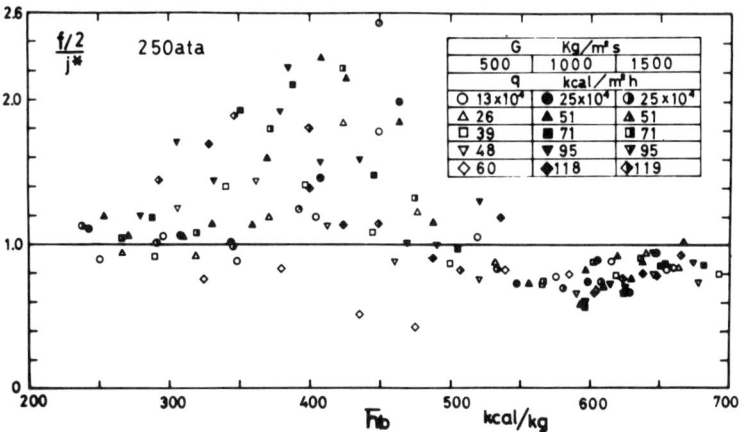

FIG. 13. Correlation of heat transfer and fluid friction. (by modified j^* Factor)

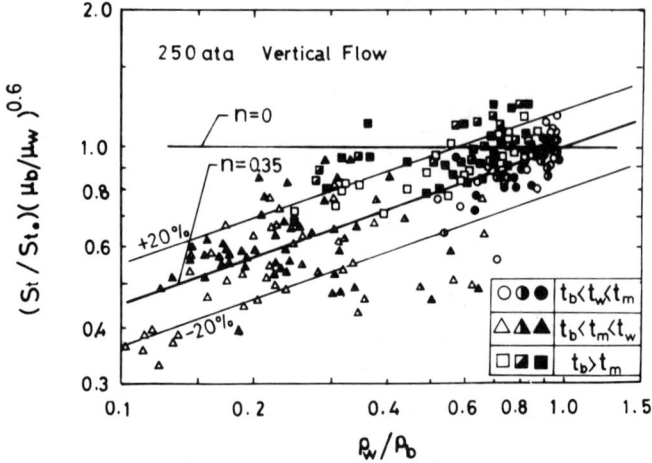

FIG. 14. Correlation of heat transfer data. (Vertical Upward Flow)

$$St/St_0 = (\mu w/\mu b)^{0.6} (\rho w/\rho b)^n$$
$$St_0 = 0.023 \, Re_b^{-0.2} \, Pr_b^{-0.2}$$

of Re = 5,000–200,000. The Stanton number, St, is calculated using the average heat transfer coefficient in the heated length of the tube. In Fig. 12, the correlation of the j factor and friction factor $(f/2)$ for vertical upward flow is shown in the form of a ratio. It is found that the experimental results are not highly scattered but deviate from unity. Therefore, a modified factor $j^* = St \cdot Pr^{0.2}$ is proposed, which yields good results except for the near-critical region (Fig. 13).

Following the above discussion, a dimensionless equation was derived for vertical upward flow heat transfer not using the Nusselt number but the Stanton number (Fig. 14).

$$St = 0.023 \, Re_b^{-0.2} Pr_b^{-0.2} \left(\frac{\mu_w}{\mu_b}\right)^{0.6} \left(\frac{\rho_w}{\rho_b}\right)^n$$

$$n = 0.35 \quad \text{for} \quad t_m \geqslant t_b$$
$$n = 0 \qquad \text{for} \quad t_m < t_b$$

The effects of temperature-dependent parameters are considered in terms of viscosity and density.

NOMENCLATURE

f	friction factor	t_b	bulk fluid temperature
h_b	bulk fluid enthalpy (average)	μ	viscosity
j	Colburn's j factor	ρ	density
t_m	pseudocritical temperature	St	Stanton number ($=Nu/Re \cdot Pr$)

REFERENCES

1. Tarasova, N. V. and A. P. Leont'yev: Teplofizika vysokikh temperatur, **6, 4**, 775, 1968.
2. Krasyakova, L. Yu., I. I. Belyakov, and N. D. Fefelova: *Teploenergetika,* **20**, 31, 1973.
3. Colburn, A. P.: *Int. J. Heat Mass Transfer,* **7**, 1359, 1964.
4. Reichardt, H.: Recent Advances in Heat and Mass Transfer, 1961, p. 223, McGraw-Hill Book Company, New York.

THE EFFECT OF LARGE TEMPERATURE DIFFERENCE ON THE TURBULENT HEAT AND MOMENTUM TRANSFER IN AN AIR FLOW INSIDE A CIRCULAR TUBE*

T. Mizushina, T. Matsumoto and S. Yoneda
Kyoto University, Japan

I. INTRODUCTION

Many investigators have studied the problems of turbulent heat and momentum transfer in a gas flow with a steep temperature gradient, as seen in the papers of Petukhov [7], McEligot et al. [6], and [5] [4]. Both the results of experiments and analyses, in the case of heating, show that the effect of high temperature differences on the local heat-transfer coefficient is correlated by an exponent function of T_b/T_w with an exponent ranging from 0.5 to 0.7.

The aforementioned analytical results predict a temperature dependence also in the case of cooling, while the date of semilocal heat-transfer coefficients by Wolf [13] and Rozhdestvenski [10] indicate no dependence on temperature, when the results are correlated with the properties evaluated at the bulk temperature. This discrepancy between the analytical and experimental results for gas cooling could be resolved by observing that the velocity profile interacts with the heat transfer, and by improving the model of turbulent flow. However, there have been a few experiments on turbulent gas flow cooled in a tube with a steep temperature gradient, except as repeated in the works of Back and Massier [1] and Cebeci [2] in turbulent boundary layer flows.

This paper reports the results of analytical and experimental investigations of the mean structure of turbulent heat and momentum transfer, which interact with each other owing to the variations of fluid properties resulting from the steep temperature gradient in a quasi-developed region of a hot air flow through a cooled circular tube.

---

*The abstract of this paper appears on page 343.

II. EXPERIMENTAL APPARATUS AND PROCEDURE

Indirectly heated air was forced to flow turbulently in the range of Re_{be} = 6 X 10^3 ∿ 2 X 10^4 in a cooling tube (ID = 32 mm, OD = 38 mm, L = 1800 mm) made of stainless steel, and cooled under conditions of T_{be}/T_w = 1.01 ∿ 3.6 at the entrance of the cooling tube. The flow at the inlet of the cooling passage was hydraulically well developed and was maintained at a uniform temperature by means of a calming section. The cooling section was composed of a combination of six pieces of short tube, 300 mm in length, to provide a location for a traversing device for temperature and velocity probes, and was submerged in well agitated cooling water at constant temperature.

Velocity and temperature profiles in the cooling passage were determined by using a Pitot tube (ID = 0.35 mm, OD = 0.7 mm) made of a quartz tube and by using a thermocouple of platinum–platinum-13-percent-rhodium having a sphere-like junction of 0.8 mm in diameter. The correction on the measurements of temperature for radiation effects was determined by calculation assuming that the Nusselt number and the blackness of the thermocouple junction were Nu = 2.0 + $Re^{0.6}Pr^{1/3}$ and 0.15, respectively. The maximum value of the correction is −23°C for the reading of 973°C.

The flow rate was measured by means of an orifice meter after being cooled down to a calibrated condition, and was also confirmed by integration of the velocity profiles for each run. Static pressure distributions along the cooling section were detected with taps of 0.8 mm drilled on the tube wall at several locations.

Axial variations of profiles were determined by remounting the traversing device into next location, being 30 cm apart from previous one, and by repeating the experiment under the same conditions as before.

Inside wall temperature was determined by extrapolating the readings of two thermocouples embedded in the wall 2 mm apart radially from each other. The local heat flux across the wall was also determined by these two thermocouple readings.

III. FIELD EQUATIONS AND ANALYSIS

The basic equations may be reduced to:

$$\tau = -\frac{1}{2}r\frac{dP}{dx} - \frac{1}{r}\frac{\partial}{\partial x}\int_0^r r\rho u^2\,dr + \frac{u}{r}\frac{\partial}{\partial x}\int_0^r r\rho u\,dr \tag{1}$$

$$q = -\frac{1}{r}\frac{\partial}{\partial x}\int_0^r r\rho c_p uT\,dr + \frac{c_pT}{r}\frac{\partial}{\partial x}\int_0^r r\rho u\,dr \tag{2}$$

and

$$\tau = -\left(\mu \frac{\partial u}{\partial r} - \rho \overline{u'v'}\right) \tag{3}$$

$$q = -\left(\lambda \frac{\partial T}{\partial r} - \rho c_p \overline{v'T'}\right) \tag{4}$$

For the expression of turbulent diffusion, introducting the Mixing-length concepts

$$\rho \overline{u'v'} = -\rho l_M{}^2 \left|\frac{\partial u}{\partial y}\right|^2 \tag{5}$$

$$\rho c_p \overline{v'T'} = -\rho c_p l_H l_M \left|\frac{\partial u}{\partial y}\right|\left|\frac{\partial T}{\partial y}\right| \tag{6}$$

into Eqs. (3) and (4), we get

$$\frac{\partial u_w^+}{\partial y_w^+} = \frac{-(1 - \beta t_w^+)^{0.68} + [(1 - \beta t_w^+)^{1.36} + 4(\tau/\tau_w)l_{Mw}^{+}/(1 - \beta t_w^+)]^{1/2}}{2(l_{Mw}^+)^2/(1 - \beta t_w^+)} \tag{7}$$

$$\frac{\partial t_w^+}{\partial y_w^+} = \frac{q/q_w}{(1 - \beta t_w^+)^{0.68}/\mathrm{Pr}_w + l_{Mw}^+ l_{Hw}^+ (\partial u_w^+/\partial y_w^+)/(1 - \beta t_w^+)} \tag{8}$$

where the temperature dependence of physical properties are approximated as

$$\frac{\rho}{\rho_w} = \frac{T_w}{T} = (1 - \beta t_w^+)^{-1} \tag{9}$$

$$\frac{\mu}{\mu_w} = \frac{\lambda}{\lambda_w} = (1 - \beta t_w^+)^{0.68} \tag{10}$$

The boundary conditions are

$$u_w^+ = t_w^+ = 0 \quad \text{at} \quad y_w^+ = 0 \tag{11}$$

Since it may be acceptable in the quasi-developed region to neglect radial velocity components, τ/τ_w and q/q_w in Eqs. (7) and (8) are reduced to

$$\frac{\tau}{\tau_w} = \frac{r_w^+}{r_{ow}^+}\left\{1 + \frac{1}{2r_{ow}^+}\frac{\partial}{\partial(x/D)}\left[\int_0^{r_{ow}^+} \frac{r_w^+ u_w^{+2}}{1 - \beta t_w^+} dr_w^+ \right.\right.$$

$$\left.\left. - \left(\frac{r_{ow}^+}{r_w^+}\right)^2 \int_0^{r_w^+} \frac{r_w^+ u_w^{+2}}{1 - \beta t_w^+} dr_w^+\right]\right\} \tag{12}$$

$$\frac{q}{q_w} = \frac{1}{2r_{ow}^+ \beta} \frac{\partial}{\partial(x/D)} \left(\frac{1}{r_{ow}^+} \int_0^{r_w^+} r_w^+ u_w^+ \, dr_w^+ \right)$$ (13)

Thus, by using the numerical solutions of Eqs. (7) and (8) f_b, Nu_b, and Re_b may be calculated as

$$f_b = \frac{2\tau_w}{\rho_b u_b{}^2} = (1 - \beta t_{bw}^+) \frac{2}{(u_{bw}^+)^2}$$ (14)

$$\mathrm{Nu}_b = \frac{q_w D}{(t_b - t_w)\lambda_b} = (1 - \beta t_{bw}^+)^{-0.68} \frac{2\mathrm{Pr}_w r_{ow}^+}{t_{bw}^+}$$ (15)

$$\mathrm{Re}_b = \frac{\rho_b u_b D}{\mu_b} = (1 - \beta t_{bw}^+)^{-1.68} 2r_{ow}^+ u_{bw}^+$$ (16)

where the reference quantities are

$$u_b = \frac{\displaystyle\int_0^{r_0} 2\pi r u \, dr}{\displaystyle\int_0^{r_0} 2\pi r \, dr}$$ (17)

$$t_b = \frac{\displaystyle\int_0^{r_0} 2\pi r \rho u \, c_p t \, dr}{c_p \displaystyle\int_0^{r_0} 2\pi r \rho u \, dr}$$ (18)

In modeling the representation for the profiles of mixing length, we assume that

(1) In the turbulent core, the mixing lengths are not influenced by temperature, being equal to isothermal ones as

$$\frac{l_M}{r_0} = \frac{k}{2(1+n)} \frac{[1 + n(1 - y/r_0)^2][1 - (1 - y/r_0)^2]}{\sqrt{1 - y/r_0}}$$ (19)

$$\frac{l_H}{r_0} = \frac{k'}{2(1+n')} \frac{[1 + n'(1 - y/r_0)^2][1 - (1 - y/r_0)^2]}{\sqrt{1 - y/r_0}}$$ (20)

where n, k, n', and k' are empirical constants, respectively.

(2) In the wall region, the following extenstion of Van Driest's [12] expression for isothermal flow is presumed: The basic equation and boundary condition for the second problem of Stokes are taken as

$$\rho \frac{\partial u}{\partial t} = \frac{\partial}{\partial y}\left(\mu \frac{\partial u}{\partial y}\right) \tag{21}$$

$$u(0,t) = u_0 \cos \omega t \tag{22}$$

where we assume that $\rho = \rho(y)$ and $\mu = \mu(y)$. Introducing the transformation

$$y_* = \int_0^y \frac{\rho}{\rho_w} \, dy$$

into Eq. (21), we get

$$\frac{\partial u}{\partial t} = \nu_w \frac{\partial}{\partial y_*}\left[\left(\frac{\rho}{\rho_w}\right)^2 \frac{\nu}{\nu_w} \frac{\partial u}{\partial y_*}\right] \tag{23}$$

where suffix w denotes the condition of wall. Though $(\rho^2 \nu)/(\rho_w{}^2 \nu_w) = (T_w/T)^{0.32}$, we will assume that this term in Eq. (23) is constant near the wall. Then the solution satisfying Eq. (22) is

$$u = u_0 \exp\left(-\sqrt{\frac{\omega}{2\nu}} \frac{\rho_w}{\rho} \int_0^y \frac{\rho}{\rho_w} \, dy\right) \cos\left(\omega t - \sqrt{\frac{\omega}{2\nu}} \frac{\rho_w}{\rho} \int_0^y \frac{\rho}{\rho_w} \, dy\right) \tag{24}$$

Hence, from a similar argument as Van Driest the damping factor is taken as

$$D_F = 1 - \left(\exp - \frac{\rho_w}{\rho} \int_0^{y_w^+} \frac{\rho_w}{\rho} \, dy_w^+ \frac{\sqrt{\omega\nu_w/2}/\sqrt{\tau_w/\rho_w}}{\sqrt{\nu/\nu_w}}\right) \tag{25}$$

By assuming an isothermal temperature profile, the following term in Eq. (25) can be fairly approximated as

$$\frac{\rho_w}{\rho} \int_0^{y_w^+} \frac{\rho}{\rho_w} \, dy_w^+ = (1 - \beta t_w^+) \int_0^{y_w^+} \frac{dy_w^+}{1 - \beta t_w^+} \cong C y_w^+ \tag{26}$$

where $C = 1.2$, and the magnitude of $-\beta$ and y_w^+ are less then 0.1 and 300, respectively. Finally, by taking $C\sqrt{\omega\nu_w/2}\,\sqrt{\tau_w/\rho_w} = 1/26$, the expression for the modified damping factor becomes

$$D_F = 1 - \exp\left(-\frac{y_w^+}{26\sqrt{\nu/\nu_w}}\right) \tag{27}$$

Thus, by multiplying Eq. (27) by Eqs. (19) and (20) the mixing lengths near the wall are obtained.

IV. RESULTS AND DISCUSSION

Radial Distributions of Shear Stress and Heat Flux

In Figs. 1 and 2, the experimental results, which were determined by using the measured values in Eqs. (1) and (2), are shown. Since the shear stress balances the axial pressure gradient and the inertial force caused by the axial deceleration of velocity, the profiles of shear stress are a little different from the isothermal case and slightly curved under conditions of large thrmal loading, as seen in Fig. 1. The heat fluxes, however, are not different from those of small thermal loading as seen in Fig. 2.

Mixing Length of Heat and Momentum

Experimental results of the mixing lengths defined by Eqs. (5) and (6) are shown in Fig. 3. This figure shows that the both profiles of mixing length for heat and momentum in the turbulent core region are not influenced by the effect of cooling. The empirical constants in Eq. (19) agree with Reichardt's (8); $k = 0.4$, $n = 2$. For the heat transfer, $k' = 0.44$ and $n' = 1.0$ will be employed as the constants in Eq. (20).

Reliable values of mixing length near the wall were not obtained in this experiment, so that Eq. (27) will be examined in its integrated results affecting the predictions of velocity and temperature profiles by Eqs. (7) and (8).

Universal Correlation of Temperature and Velocity

The experimental and analytical correlations of velocity are shown in Fig. 4. As seen in Fig. 4, the expression of Eq. (27) is comparatively good. This graph also shows that the data in the boundary layer obtained by Back and Massier agree well with the authors' results at a similar value of β. For the correlation of temperature, the abbreviated analogous results to Fig. 4 are shown in Fig. 5.

Some of previous results to correlate velocity and temperature profiles for high thermal loading are also compared in Fig. 5. The correlation of Coles [3] by assuming that $\overline{y_t^+} = r_{ow}^+$, and those of Van Driest [11] and Deissler [4] do not fit the experiments at large value of $-\beta$.

Local Heat Transfer Coefficients and Friction Factors

In Fig. 6, the experimental and analytical results are compared. The correlating curves were calculated by eliminating the parameter r_{ow}^+, u_{bw}^+, t_{bw}^+ in Eqs. (14)–(16). This figure shows that the calculated curves as well as the experimental values do not depend so strongly on the parameter β as

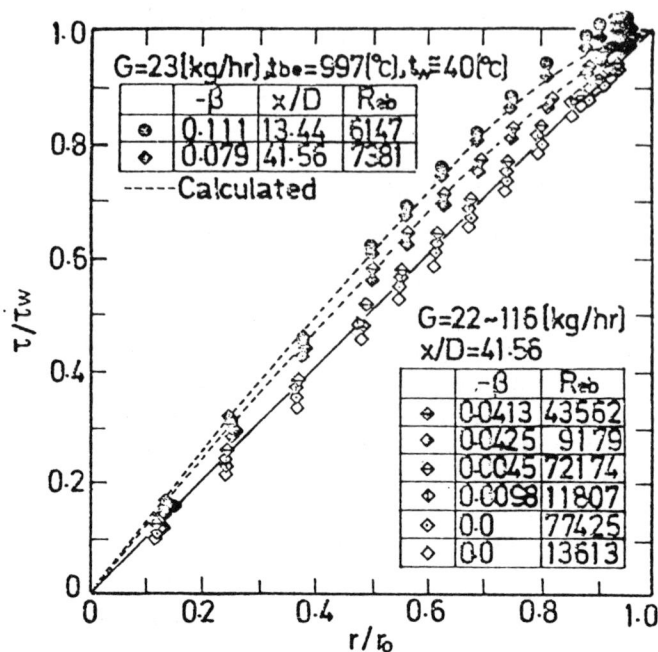

FIG. 1. Shear stress distribution profile.

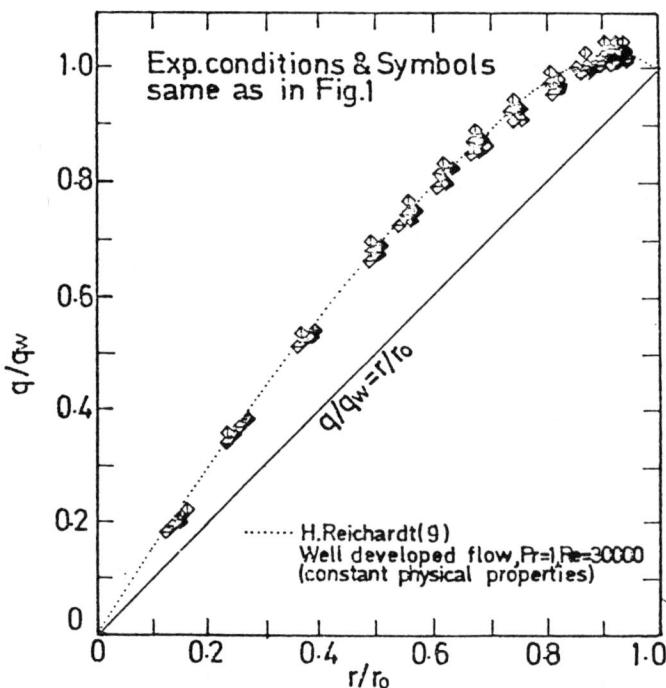

FIG. 2. Heat flux distribution profile.

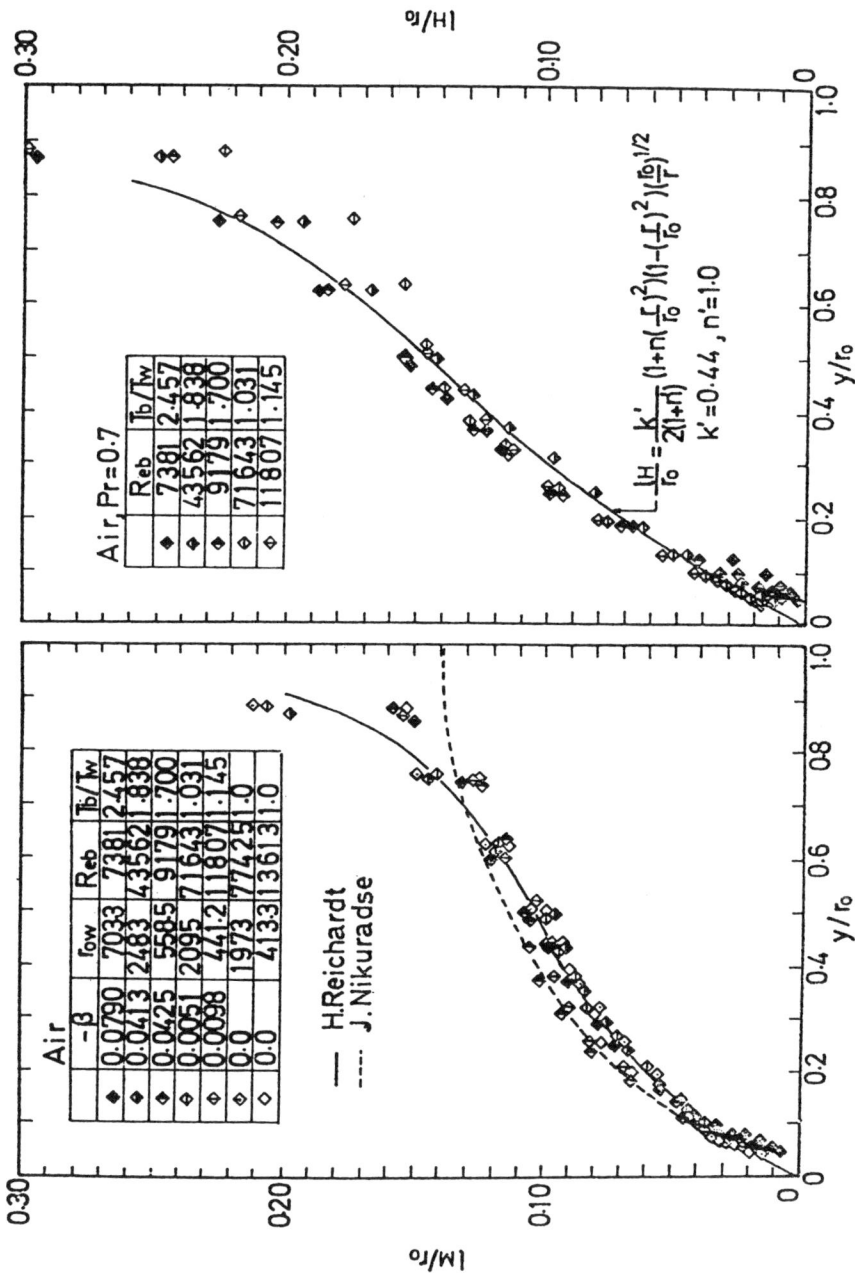

FIG. 3. Mixing length distribution.

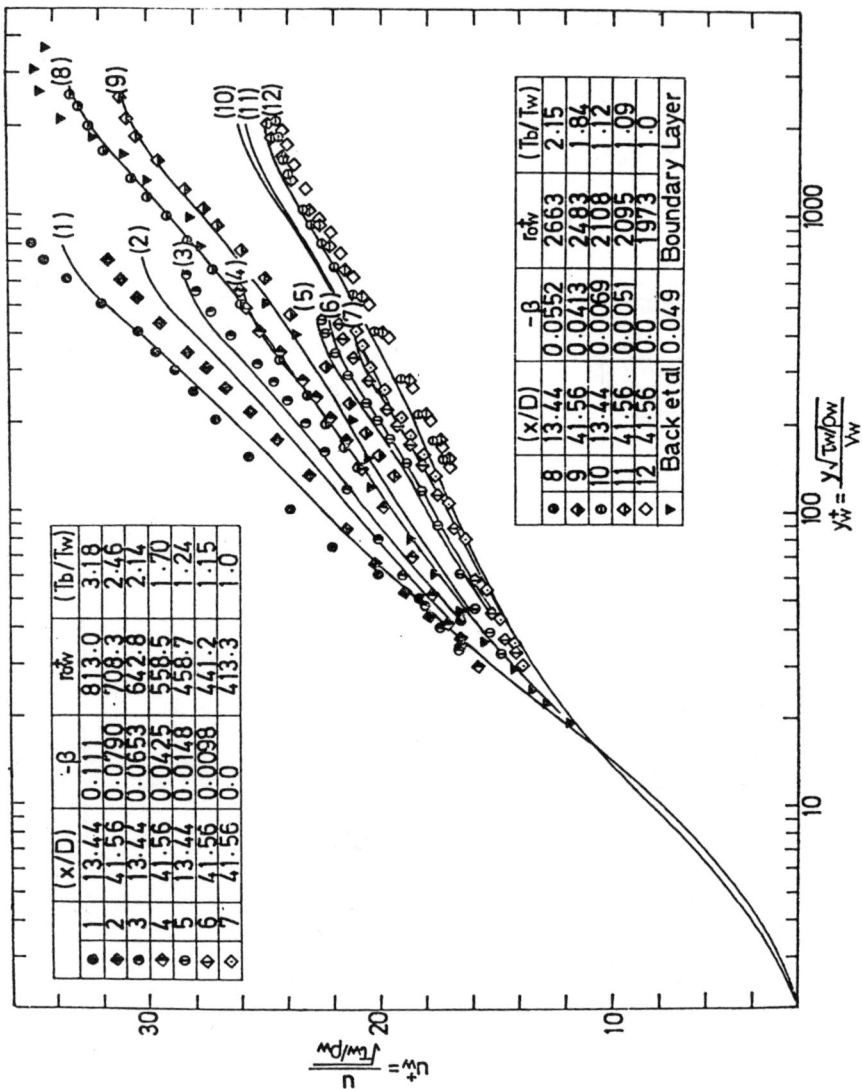

FIG. 4. Universal velocity profile.

	(x/D)	-β	r₀⁺w	(Tb/Tw)
● 1	13·44	0·111	813·0	3·18
◆ 2	41·56	0·0790	708·3	2·46
● 3	13·44	0·0653	642·8	2·14
◑ 4	41·56	0·0425	558·5	1·70
◒ 5	13·44	0·0148	458·7	1·24
◇ 6	41·56	0·0098	441·2	1·15
◇ 7	41·56	0·0	413·3	1·0

	(x/D)	-β	r₀⁺w	(Tb/Tw)
● 8	13·44	0·552	2663	2·15
◆ 9	41·56	0·413	2483	1·84
◑ 10	13·44	0·0069	2108	1·12
◇ 11	41·56	0·051	2095	1·09
◇ 12	41·56	0·0	1973	1·0
▼	Back et al	0·049	Boundary Layer	

$$u_w^+ = \frac{u}{\sqrt{\tau_w/\rho_w}}$$

$$y_w^+ = \frac{y\sqrt{\tau_w/\rho_w}}{\nu_w}$$

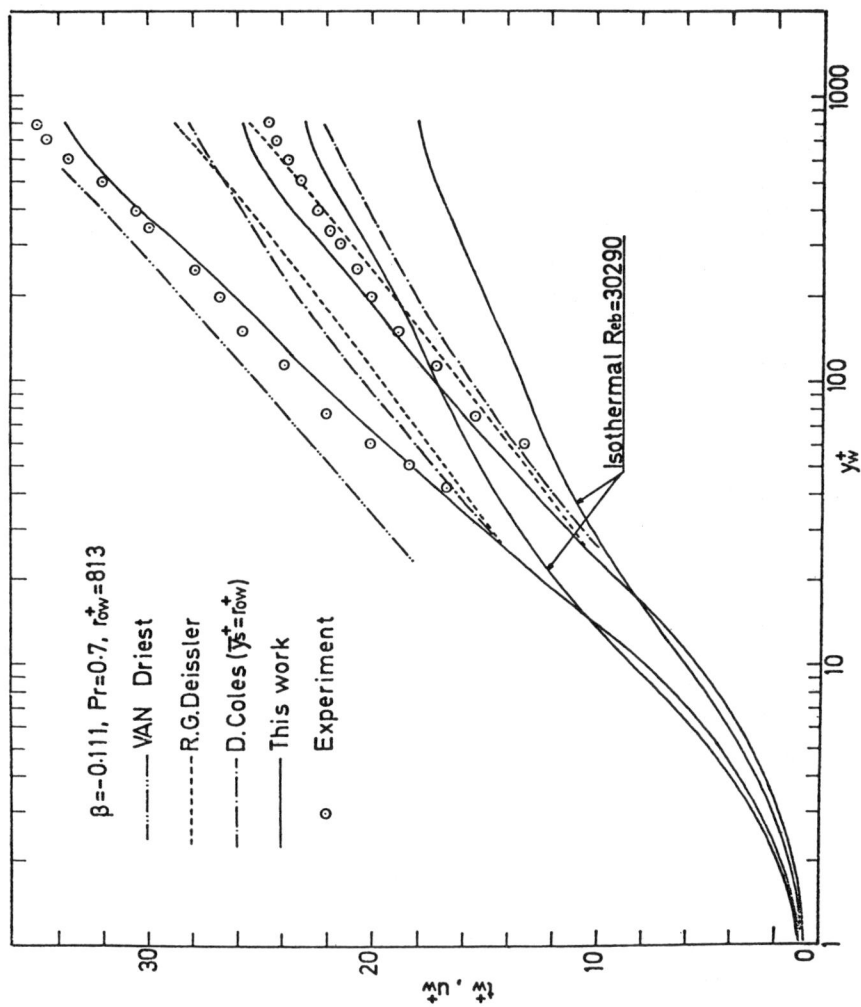

FIG. 5. Universal correlation of velocity and temperature profiles.

43

(for gas)
—·—H.Blasius
- - - -R.G.Deissler
——This work

$\beta=-0.1$

$\beta=-0.05$

$f_b = \dfrac{2\tau_w\rho_w}{\langle\rho u\rangle^2}$

$\beta=0.0$

$\text{Nu}_b = \dfrac{hD}{\lambda_b}$

	$-\beta$	(T_b/T_w)	(x/D)
▽	0.128	3.559	
▼	0.060	2.257	4.06
▽	0.074	2.331	
◉	0.111	3.175	
◑	0.055	2.151	
◓	0.065	2.146	13.44
◒	0.007	1.125	
◐	0.015	1.238	
△	0.097	2.889	
▲	0.050	2.049	22.81
▲	0.016	1.972	
◻	0.088	2.695	
▣	0.046	1.946	32.19
◻	0.049	1.825	
◇	0.079	2.457	
◈	0.041	1.838	
◇	0.043	1.700	41.56
◆	0.005	1.091	
◈	0.010	1.145	

	$-\beta$	(T_b/T_w)	(x/D)
▽	0.068	2.203	
▽	0.037	1.736	50.94
▼	0.036	1.590	
◎	0.0	1.0	
○	0.0	1.0	

$\beta=-0.1$

$\beta=-0.05$ $\beta=0.0$

(Pr=0.7)
—·—A.P.Colburn
- - - -R.G.Deissler
——This work

$\text{Re}_{eb} = \dfrac{D\langle\rho u\rangle}{\mu_b}$

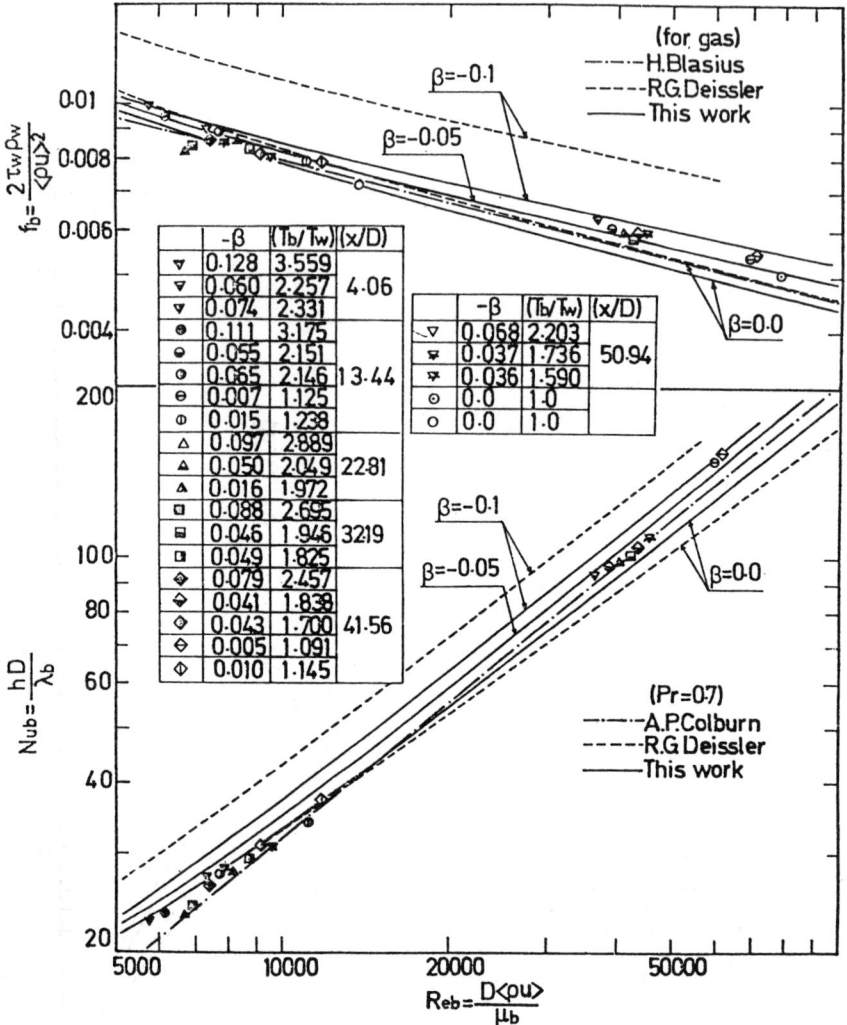

FIG. 6. Friction factor and Nusselt number.

compared to the correlation of Deissler, and that the experimental data are represented fairly well by following the Blasius and Colburn equations, respectively.

$$f_b = 0.079\,\text{Re}_b^{-0.25} \qquad (28)$$
$$\text{Nu}_b = 0.023\,\text{Re}_b^{0.8}\text{Pr}^{1/3} \qquad (29)$$

The analytical results for the deviations of f_b and Nu_b from the isothermal case, at the same value of Re_b, are reduced into the following relations with

T_b/T_w

$$1.0 \leqslant \frac{f_b}{f_{\text{iso}}} \leqslant \left(\frac{T_b}{T_w}\right)^{+0.11} \quad \text{and} \quad 1.0 \leqslant \frac{\text{Nu}_b}{\text{Nu}_{\text{iso}}} \leqslant \left(\frac{T_b}{T_w}\right)^{+0.12}$$

for $\quad 1.0 \leqslant \dfrac{T_b}{T_w} \leqslant 4.0 \quad$ and $\quad 5 \times 10^3 \leqslant 5 \times 10^4$

V. CONCLUDING REMARKS

1. The turbulent diffusion of heat and momentum in the core region were expressed by mixing length models represented by Eqs. (19) and (20) as shown in Fig. 3.

2. The damping of mixing length near the wall, given by Eq. (27), was proposed to fit the analytical results to the universal profiles of measured velocities and temperatures.

3. In low Reynolds number turbulent flow, the dependence of Nu_b and f_b on T_b/T_w is negligibly small when they are correlated with Re_b as shown in Fig. 6.

NOMENCLATURE

c_p specific heat of air at constant pressure
D inner diameter of tube
G mass flow rate of air
P pressure
Pr_w Prandtl number $\left(c_{pw}\mu_w/\lambda_w\right)$

q heat flux toward tube wall
q_w heat flux toward tube wall at the wall surface
r radial distance from tube wall
r_0 inner radius of tube
t Celsius temperature
T absolute temperature
u velocity in axial direction
v velocity in radial direction
x axial distance from entrance
y radial distance from tube wall

β parameter of heat flux $\left(-\dfrac{q_w\sqrt{\tau_w/\rho_w}}{c_{pw}T_w\tau_w}\right)$

λ thermal conductivity
μ viscosity
ν kinematic viscosity
ρ density
τ shear stress
τ_w wall shear stress

Subscripts and Superscripts

e	at entrance condition
b	at bulk temperature condition
s	Coles' sublayer reference temperature
w	at wall condition
$<>$	space averaged in radial direction
$+$	made dimensionless with wall parameter

REFERENCES

1. Back, L. H. and P. F. Massier: *Int. J. Heat Mass Transfer*, **13**, 1029, 1970.
2. Cebeci, T.: *AIAA Journal*, 9(6):1091 (1971).
3. Coles, D.: *Physics of Fluids*, 79:1403 (1964).
4. Deissler, R. G.: *Trans. ASME*, 77:1221 (1955).
5. Goldman, D.: *Chem. Eng. Progr. Symp. Ser., Nucl. Eng. Part 1*, **50**(11):105 (1954).
6. McEligot, D. M., S. B. Smith, and G. A. Bankston: *Trans. ASME, Series C.* 92(1):641 (1970).
7. Petukhov, B. S.: "Advances in Heat Transfer," 7:503 (1971).
8. Reichardt, H.: *Z.angew. Math. Mech.*, **31**:208 (1951).
9. Reichardt, H. and P. A. Schoeck: "Recent Advances in Heat and Mass Transfer," 1961, McGraw-Hill Book Company, New York.
10. Rozhdestvenski, V. I.: *Int. Chem. Eng.*, **10**(2):279 (1970).
11. Van Driest, E. R.: *J. Aeronaut. Sci.*, 18(March):145 (1951).
12. Van Driest, E. R.: *Ibid.*, **23**(Nov.):1007 (1954).
13. Wolf, H.: *Trans. ASME, Series C. 1959*, 81(1):269.

THE EFFECT OF THE PERIODICALLY VARIABLE WALL TEMPERATURE ON HEAT TRANSFER IN LIQUID METALS*

Jitka Mošnerová
National Research Institute for Machine Design
Prague-Béchovice, Czechoslavakia

In fluids with a very low Prandtl number the heat transfer is noticeably affected by the type of boundary conditions. The values of the heat transfer coefficient in liquid metals given in the literature are derived almost exclusively for constant heat flux, or for constant wall temperature, and therefore, their validity is limited. In heat-exchange equipment with liquid metals where temperature instabilities occur, the wall temperature is time and space dependent. To define the effect of the wall temperature distribution on the thermal conditions in liquid metals, we solved the case of heat transfer in turbulent flow through a tube the surface temperature of which is a periodic function of time and changes simultaneously with the axial coordinate.

The solution is based on a simplified energy equation for time dependent heat convection in a turbulent fluid flow with constant thermophysical properties having the form

$$\frac{\partial t}{\partial \tau} + u \frac{\partial t}{\partial x} = a \frac{\partial^2 t}{\partial x^2} + \frac{1}{r} \frac{\partial}{\partial r} \left[r(a + \epsilon_a) \frac{\partial t}{\partial r} \right] \tag{1}$$

In Eq. (1), a zero radial peripheral velocity of the medium has been specified, the viscosity dissipation and the turbulent heat conduction in the axial direction have been neglected, and with respect to cylinder symmetry, there is zero molecular and turbulent heat diffusion in the peripheral direction. By introducing the dimensionless axial coordinate $X = x/2L$, the dimensionless radial coordinate $R = r/r_w$, the dimensionless time $\theta = \tau/T$, the dimensionless velocity $U = u/u_s$, the Reynolds number $\text{Re} = 2r_w u_s/\nu$, the Prandtl number $\text{Pr} = \nu/a$, the Fourier number $\text{Fo} = aT/r_w^2$ and the dimensionless geometrical factor $S = r_w/2L$ it is possible to express the equation in the following form

---

*The abstract of this paper appears on page 345.

$$\frac{\partial t}{\partial \theta} + \tfrac{1}{2}\mathrm{FoSRePr}U\frac{\partial t}{\partial X} = \mathrm{Fo}S^2\frac{\partial^2 t}{\partial X^2} + \frac{\mathrm{Fo}}{R}\frac{\partial}{\partial R}\left(R\frac{a + \epsilon_a}{a}\frac{\partial t}{\partial R}\right) \qquad (2)$$

The following conditions are supplementary to Eq. (2):

a) In the axis of the tube $(R = 0)$ the temperature profile reaches a maximum, i.e.,

$$\frac{\partial t(X, 0, \theta)}{\partial R} = 0 \qquad (3)$$

b) The tube wall temperature $(R = 1)$–Fig. 1– is assumed to have the form of a portion of a sinusoid which moves periodically over the length L, i.e., in the interval $-0.25 < X\epsilon < 0.25$, according to the equation

$$t(X, 1, \theta) = \frac{t_{\mathrm{wh}} + t_{\mathrm{wd}}}{2} - \frac{t_{\mathrm{wh}} - t_{\mathrm{wd}}}{2}\,\sin\frac{\pi}{2K}(X - \tfrac{1}{4}\sin 2\pi\theta) \qquad (4)$$

for $0 < \theta\epsilon < 1$ and $-K \leqslant X - 1/4 \sin 2\pi\theta \leqslant K$. Before as well as behind the region of pulsation, the temperature of the wall is constant, i.e.,

$$\begin{aligned} t(X, 1, \theta) &= t_{\mathrm{wh}} \quad \text{for} \quad X - \tfrac{1}{4}\sin 2\pi\theta \leqslant -K \\ t(X, 1, \theta) &= t_{\mathrm{wd}} \quad \text{for} \quad X - \tfrac{1}{4}\sin 2\pi\theta \geqslant K \end{aligned} \qquad (5)$$

c) At the location, $X = -0.5$, the temperature profile is independent of time and is given by the solution of the energy equation for steady turbulent flow in a tube with a constant temperature of the heat-exchange surface

$$t(-0.5, R, \theta) = t(R) \qquad (6)$$

d) At the location, $X = 0.5$, the axial heat conduction is neglected, i.e.,

$$\frac{\partial^2 t(0.5, R, \theta)}{\partial X^2} = 0 \qquad (7)$$

e) The periodicity of the boundary condition for $R = 1$ results in a periodic change of the medium temperature field, i.e.,

$$t(X, R, 0) = t(X, R, 1) \qquad (8)$$

The variables Fo, Re, Pr, and S are parameters, the values of which depend on the tube radius r_w, a wall temperature pulsation period T, the length L of the pulsation zone, and on the kind of liquid metal. The velocity has been

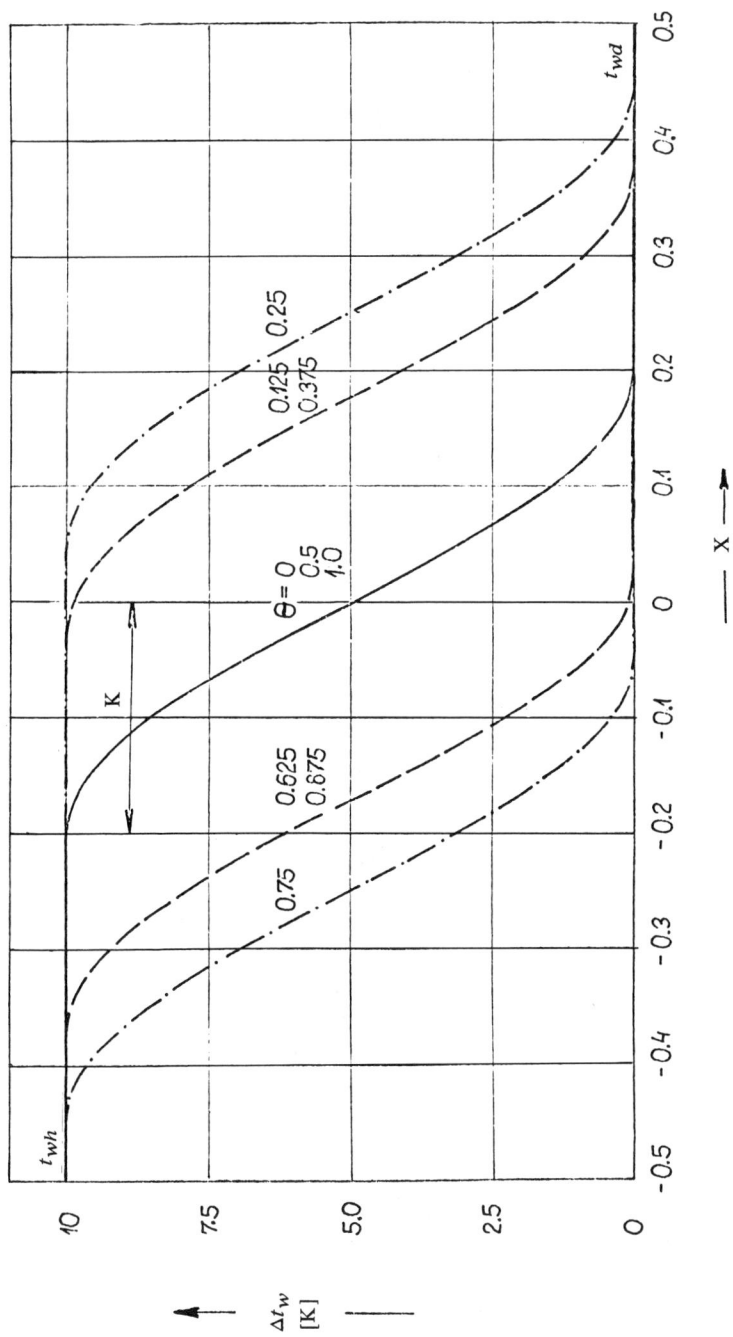

FIG. 1. Distribution of the tube wall temperature.

considered in the form of the universal velocity profile [1]

$$U = \sqrt{\frac{f}{2}} \ 2.5 \ln \frac{\text{Re}}{2} \sqrt{\frac{f}{2}} (1 - R) + 5.5 \tag{9}$$

where the Nikuradse equation

$$\frac{1}{\sqrt{f}} = 4 \log (\text{Re} \sqrt{f}) - 0.4 \tag{10}$$

has been used to calculate the friction factor, f. For the eddy diffusivity for heat transfer, the dependence according to Mizushina [2] has been chosen

$$\frac{\epsilon_a}{a} = 0.219 \left(\frac{\epsilon_v}{\nu} \text{Pr}\right)^2 \left[1 - \exp\left(-\frac{1}{0.416(\epsilon_v/\nu)\text{Pr}}\right)\right] \tag{11}$$

and the eddy diffusivity for momentum has been considered in the following form

$$\frac{\epsilon_v}{\nu} = 0.02556 \text{Re}^{0.898} \left\{(1 - R) \exp\left[-\frac{(1 - R)^2}{0.4517}\right]\right.$$
$$\left. + (1 + R) \exp\left[-\frac{(1 + R)^2}{0.4517}\right]\right\} \tag{12}$$

Equation (2) with boundary conditions (3) to (8) was solved numerically by using the alternating direction method [3], which is one of the modifications of the finite-difference method for multidimensional problems. The systems of linear algebraic equations, with a tridiagonal matrix, having been derived after substitution of the derivations by differences in Eq. (2) were then solved using the factorization method. For computation of the Nusselt number as related to the temperature difference at $X = -0.5$ the following equation holds

$$\text{Nu}(X, \theta) = -\frac{2}{(t_s - t_w)_{X=-0.5}} \left(\frac{\partial t}{\partial R}\right)_{R=1} \tag{13}$$

In the numerical integration, 609 nodal points in the cross section RX were considered in a time step equal to 1/8 of the period and the parametric values were as follows: Fo = 0.1 to 10; Re = 10^5 to 10^6; Pr = 0.001 to 0.05; S = 0.01 to 1; $t_{wd} = 0$; t_{wh} = 5 to 20°C; K = 0.05 to 0.2.

An illustration of the time and local dependent Nusselt number for one combination of parameters is shown in Fig. 2. In comparison with heat transfer with a constant wall temperature, the decrease of the heat transfer

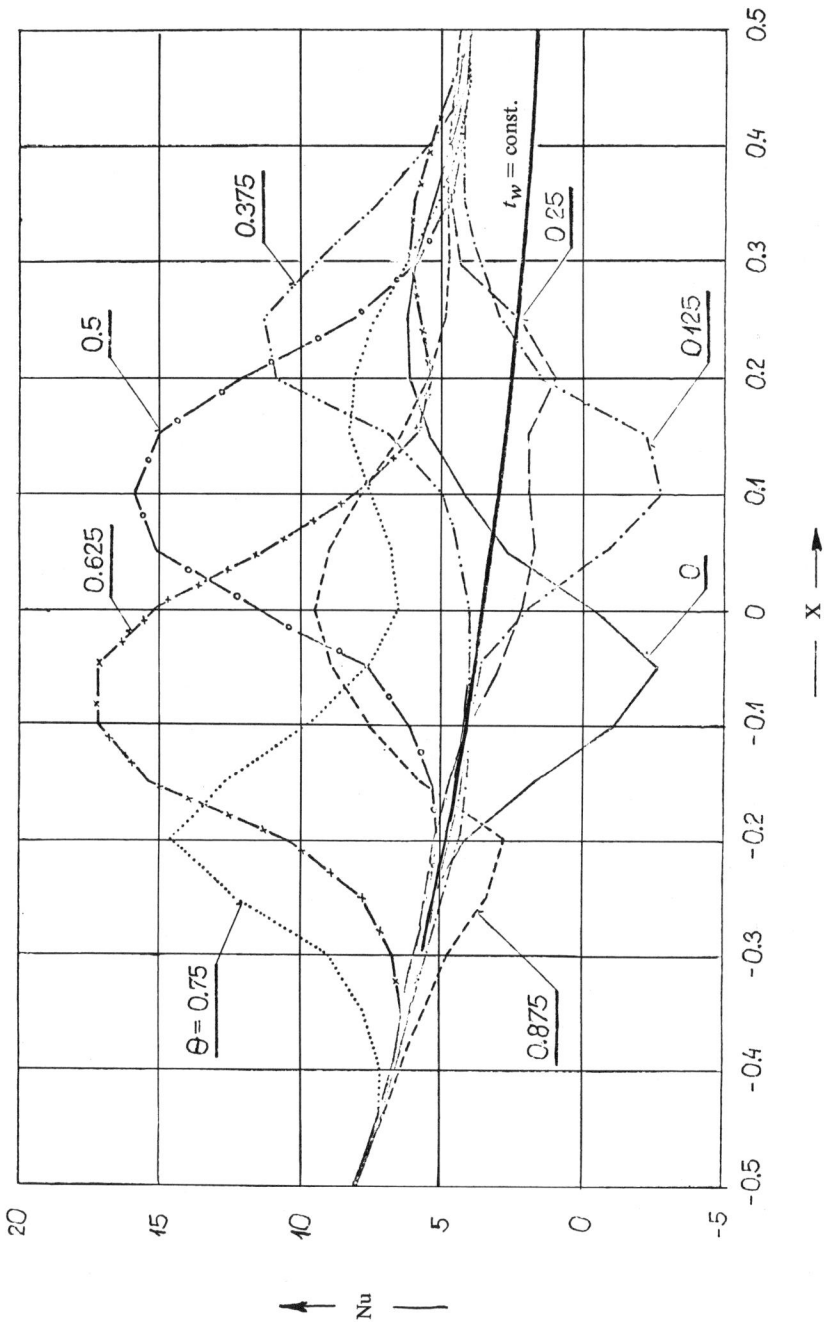

FIG. 2. The spatial and time changes of the Nusselt number at $Re = 10^5$; $Pr = 0.01$; $Fo = 0.1$; $S = 0.01$; $\Delta t_w = 10$ K; $K = 0.2$.

51

occurs both at the beginning and at the end of the period and at the same time the heat flux between the liquid metal and the wall may be reversed (negative values of Nu at low Fo and S, i.e., a high pulsation frequency of the surface temperature and long zones of pulsation). On the other hand, in the middle of the period, an intensive increase in heat transfer occurs.

From the additional calculations the following conclusions may be drawn:

1. The effect of the wall temperature step on the temperature field of the medium decreases with the distance from the wall and the deformation of the temperature field is felt at most as far as the channel axis. Only in the case of small Fo and S, is the temperature field on the axis of the tube not affected.

2. Due to the periodical pulsation of the wall temperature, the heat transfer coefficient is subjected to spatial and time changes approaching the highest value at the higher pulsation frequencies.

3. The maximum deviations in heat transfer with regard to the case of constant wall temperature occur roughly in the right limit position of the temperature change $(X = 0.25)$.

4. With Fo increasing, i.e., with decreasing frequency of the temperature fluctuation, the dependence of heat transfer on time disappears.

5. With the exception of the lowest values of Fo, S, and Re the heat transfer coefficient always grows compared to the case when t_w = const. This increase approaches its highest value when S is high, i.e., when the zones of pulsation are comparatively short relative to the diameter of the tube and when the Nusselt number does not depend on frequency. When Pe = 10^3; Fo = 0.1; $S = 1$, the Nusselt number increases up to 5.8 times compared to its value at the location $X = -0.5$.

6. The higher the Re the lower the effect of frequency on heat transfer.

7. Heat transfer grows simultaneously with the increase of the wall temperature difference $\Delta t_w = t_{wh} - t_{wd}$. For instance, when $\Delta t_w = 20$ K (Pe = 10^3; Fo = 1; $S = 0.01$) the Nusselt number is 18 times higher as compared to the case with t_w = const. $(\Delta t_w = 0)$.

8. A steeper change of the wall temperature, i.e., a smaller K in Eq. (4), results in a larger scatter of the values Nu and more contrasting extremes in the local curve.

9. The effect of Re and Pr on the heat-transfer coefficients may be considered simultaneously an effect of Pe, for the changes of these parameters result in relatively negligible deviations, 10 percent at the maximum. The largest changes of the Nusselt number take place when Pe is small. With increasing Pe, the heat transfer rate grows, whereas the space and time dependences of Nu are more balanced and the discrepancy between them is less apparent.

The results obtained show that local changes of the wall temperature may result in relatively large changes of the heat-transfer coefficient in liquid metals. Similar distributions of the heat-exchange surface temperatures are also likely to be found in heat exchangers where, for instance, the temperature of the heat-transfer surface in contact with liquid metals is affected by the boiling crisis on the water side. It is absolutely necessary to know the thermokinetic conditions in the medium in order to successfully determine thermal stresses in the tube material or in order to predict their service life under conditions of cyclic stress.

NOMENCLATURE

a	thermal conductivity	L	length of the pulsation region
f	friction factor	T	period
r	radial coordinate	ϵ_a	eddy diffusivity of heat
t_s	bulk mean temperature	ϵ_v	eddy diffusivity of momentum
t_w	wall temperature	v	kinematic viscosity
u_s	mean velocity	τ	time
x	axial coordinate		

REFERENCES

1. Kays, W. M.: "Convective Heat and Mass Transfer," McGraw-Hill Book Company, New York, 1966.
2. Mizushina, T.: Proc. 1961 Heat Transfer Conf. ASME, New York, 1963.
3. Samarskij, A. A.: An introduction to the theory of finite-difference schemes, Nauka, Moskva, 1971.

A NONCLASSICAL MODEL OF TURBULENT HEAT TRANSFER*

Gert Naue, Wolfgang Schmidt, and Walter Wolfgang Schmidt
Merseburg Institute of Technology
Merseburg, GDR

I. INTRODUCTION

In our paper we wish to present a model for the calculation of turbulent flows and especially turbulent heat transfer. Turbulent flow in a flat channel is the object of the first application of this model. The velocity distribution being fully developed and the wall temperature being constant, the task consists of determining temperature distributions at different channel locations depending on the specified difference between wall and inlet temperatures.

II. CONCEPTIAL BASIS OF THE MODEL

The starting point of our theory is a generalized continuum characterized by an infinite number of kinematic variables in the form of velocities and spin intensities of different order which satisfy a corresponding set of conservation laws. This is a continuum with microrotations, containing substructures within the fluid, which in turn show further substructures. A kinematic self-induction in the flow field takes place in such a way that spin fields of low order act on spin fields of higher order, i.e., large vortices transfer energy to small vortices which exist in the large ones [1], [2].

Turbulent processes are characterized kinetically by the totality of movements due to velocity, first order spin intensity, and higher order spin intensities. According to this definition, turbulence is also possible without temporal fluctuations of velocity occurring, which will be designated as pulsations. These pulsations stimulate the spin motion, after which they may vanish again. In our theory, we begin from the hypothesis that turbulence consists of elements of different orders of magnitude. The motion of these elements is described by the spin intensities of corresponding order. The spin

---

*The absbstract of this paper appears on page 345.

intensity of the first order describes the movement of the largest turbulence elements designated as the macrostructure of turbulence. For formulation of the energy balance of each spin order s ($s \geqslant 2$)

$$\frac{1}{2}\rho I \frac{d\overset{s}{\beta_i^2}}{dt} = \overline{\rho \overset{s}{\hat{n}}\overset{s-1}{\beta_i}\overset{s}{\beta_i}} + \overline{(\overset{s}{a} + \overset{s}{c})\overset{s}{\beta_i}\partial_{nn}\overset{s}{\beta_i}} + \overline{(\overset{s}{a} + 2\overset{s}{b} - \overset{s}{c})\overset{s}{\beta_i}\partial_{in}\overset{s}{\beta_n}} - \overline{4\overset{s-1}{c}\overset{s}{\beta_i}\overset{s}{\beta_i}}$$

with the hypothesis that the kinetic energy of each spin order be proportional to the kinetic energy of the velocity

$$\overline{\overset{s}{\beta_i^2}} \sim \bar{v}^2$$

and by formal transition from the discrete to the continuous spin order σ with the equation

$$\overline{\overset{\sigma}{\beta_i^2}} = k\bar{v}^2 f^2$$

we obtain a scalar differential equation, the equation of turbulence intensity, for the energy balance of turbulent motion [1].

$$\epsilon \frac{df}{dt} = \partial_{nn}f + \delta_1 f + \delta_2 f^2$$

By the scalar quantity f, turbulence intensity, we shall understand a measure of the kinetic energy of turbulent motion. The turbulence intensity f takes account of the average effect of the spin motions of higher order—microstructure of turbulence—with respect to energy transport and dissipation. The energy transport in turbulent flow is realized in such a way that energy is continually flowing from the basic flow into the largest elements. These in turn pass the energy to smaller elements and so on. This process is continued until the viscous forces become so great that further decomposition is rendered impossible. The energy of these smallest elements then undergoes laminar dissipation [3]. In a figurative sense, a cascade process of further spin motions of higher order begins when the velocity is increased, i.e., in the case of a finite velocity, there is only a finite number of spin motions. The beginning of the spin motion is always connected with the extraction of energy from the spin motion of next lower order. The overall dissipation of a turbulent flow field consists of additive parts of the basic flow, macrostructure, and microstructure, the dissipation of the microstructure being proportional to the third power of the velocity.

$$\phi = 4(\overset{\circ}{a} + \overset{\circ}{c})\omega^2 - 8\overset{\circ}{c}\varphi\omega + 4\overset{\circ}{c}\varphi^2 + (\overset{\circ}{a} + \overset{\circ}{c})[(\partial_1\varphi)^2 + (\partial_2\varphi)^2]$$
$$+ \alpha\{h^2[(\partial_1 f)^2 + (\partial_2 f)^2 + f^2]\}$$

III. MODEL FOR TURBULENT HEAT TRANSPORT

The starting point of our considerations is the first law of thermodynamics, chemical reactions being excluded, i.e., the enthalpy of reaction is not taken into account.

$$c_p \frac{dT}{dt} = -\frac{1}{\rho} \partial_i q_i + \frac{1}{\rho} \phi$$

In the following we examine the heat flux density q_i. The classical theory of heat transfer starts from the phenomenological statement that the heat flux is proportional to the temperature gradient (Fourier's law of thermal conduction).

$$q_i = -\lambda \partial_i T$$

In turbulent processes of heat transfer, the thermal conductivity λ is substituted by an "effective heat conductivity" which can be dependent both on the coordinates and the velocity. This "effective heat conductivity" is expressed by empirical or semiempirical equations which for the most part are applicable only to special problems since a physically based model of turbulent heat transfer is not available.

A contribution to the physical modeling of energy transport in turbulent flow is made by applying the theory of microrotation of turbulence elements, which was presented in Sec. 2, to turbulent heat transfer. Energy transport in the continuum takes place by different transport processes such as convection, conduction, and radiation. In concrete transport processes, special mechanisms are dominant. In turbulent flows, energy is transferred mainly by convection due to the basic fluid motion and by turbulent mixing processes.

For turbulent heat transport, the effects of the microrotation of the continuum are essential. By microrotation in nonisothermal flow, regions of higher temperatures are transported close to regions of lower temperature and vice versa, such that, a temperature equilization is possible. This energy transport is described by an additional heat flux. Therefore, the heat flux of the spin of order s is proportional to the temperature gradient and the local spin intensity. The additional turbulent heat flux is obtained by summation of all orders of spins:

$$\overset{+}{q_i} = \lambda \sum_{s=1}^{n} \lambda^s \left(|\overset{s}{\beta}| \partial_i T - \frac{\overset{s}{\beta_i}}{|\overset{s}{\beta}|} \overset{s}{\beta_m} \partial_m T \right)$$

The second term in parentheses guarantees that a spin in the direction of the temperature gradient does not cause any heat flux. This term vanishes in flat problems since there is only a spin intensity component $\beta_3 \neq 0$, and the temperature gradient in three directions equals zero. In flat flows, the overall heat flux composed additively of the molecular and the additional heat flux is

$$q_i = -\lambda \left(1 - \sum_{s=1}^{n} \lambda^s |\overset{s}{\beta}|\right) \partial_i T$$

In laminar flows, this equation corresponds to Fourier's law, and for isothermal flows gives $q_i = 0$. The influences of the micro- and macrostructure on the heat transport become evident when the equation is written in the following form:

$$q_i = -\lambda \left(1 - \hat{\lambda}|\hat{\beta}| - \sum_{s=1}^{n} \lambda^s |\overset{s}{\beta}|\right) \partial_i T$$

The effect of all spin intensities of higher order

$$\sum_{s=1}^{n} \lambda^s |\overset{s}{\beta}|$$

is taken into account through the turbulence intensity f in the form of

$$\overline{\beta_i^2}^\sigma = k\bar{v}^2 f^2$$

The spin intensity of the first order is proportional to the velocity. The overall heat flux density can be written as follows:

$$q_i = -\lambda(1 + k_1|\varphi| + k_2 f) \partial_i T$$
$$k_1 \sim v$$
$$k_2 \sim v$$

The local energy balance in the form of the first law of thermodynamics then reads

$$\rho c_p \frac{dT}{dt} = \lambda \partial_i [(1 + k_1|\varphi| + k_2 f) \partial_i T] + \phi$$

IV. GOVERNING EQUATIONS

For the calculation of flat steady fields of turbulence the following equations are available:

1. Equation of vorticity

$$\rho v_k \partial_k \omega = (\hat{a} + \hat{c}) \partial_{kk} \omega - \hat{c} \partial_{kk} \varphi$$
$$\omega = -\tfrac{1}{2} (\partial_2 v_1 - \partial_1 v_2)$$
$$\partial_k v_k = 0$$

2. Equation of spin intensity (first stage)

$$\rho \hat{I} v_k \partial_k \varphi = \rho \hat{n} \omega + (\hat{a} + \hat{c}) \partial_{kk} \varphi - 4\hat{c}(\varphi - \omega)$$

3. Turbulence equation

$$\epsilon v_k \partial_k f = \partial_{kk} f + \delta_1 f + \delta_2 f^2$$

4. Energy equation

$$\rho c_p v_k \partial_k T = \lambda \partial_i [(1 + k_1 |\varphi| + k_2 f) \partial_i T] + \phi$$

V. APPLICATION

The model for turbulent heat transfer presented in this paper is first applied to incompressible, steady, flat channel flow. We assume the velocity profile to be fully developed and the temperature of the wall T_w and the fluid temperature at the inlet T_0 to be constant.

The dimensionless temperature profiles are calculated from the energy equation. For this purpose, a numerical integration technique from [4] was used. As a second method of solution, the variational method of Krajewski [5] was used which was developed especially for energy equations with forced convection. In both methods of solution, the dissipation term was neglected with respect to the heat flux term. To determine the influence of the macro- and microstructure on the heat transport, the parameters K_1 and K_2 of the term of heat flux were varied in large regions. In the variational method of Krajewski, the influence of the microstructure of turbulence $K_1 = 0, K_2 > 0$ was taken into account.

It can be shown that the influence of the macrostructure of turbulence upon the heat transport is essential. Figure 1 shows the temperature distribution normal to the flow direction in different axial locations of the channel. Figure 2 shows the development of the center line temperature in the flow direction.

In the center of the flow, the profiles calculated by the variational method are initially below those determined by the numerical method, owing to the error at the point $x = 0$. Due to the chosen boundary conditions in the numerical calculation, $(\partial \tilde{T}/\partial x)|_{x=1} = 0$, a final profile is reached.

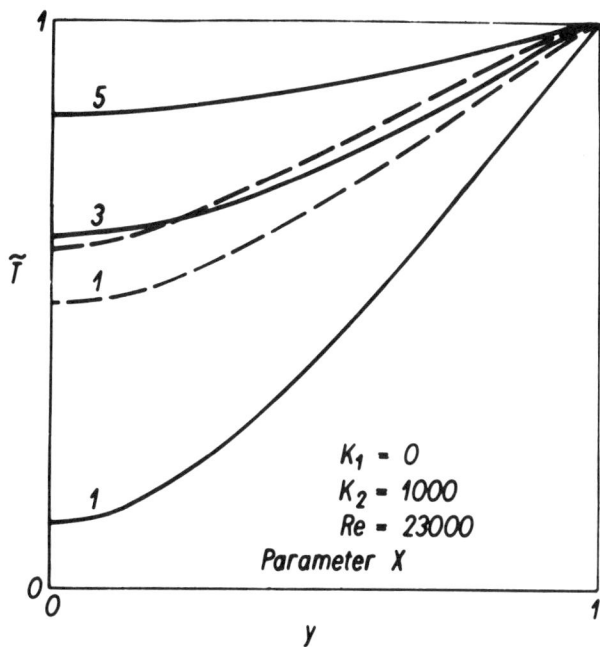

FIG. 1.

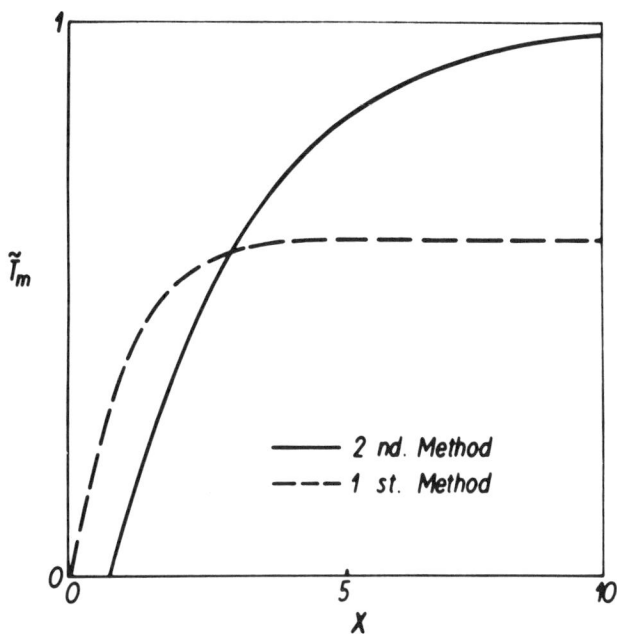

FIG. 2.

After some steps in the x direction, the temperature profiles obtained by the variational method exceed the profiles calculated by the numerical method and tend toward the value 1.

One of the next tasks is to determine the parameters of the heat flux term which depend on the Re number and the average velocity, respectively, connected with the proof that the model parameters to be chosen are optimal for a turbulent heat-transfer process.

SYMBOLS

t	time
x, y	cartesian coordinates (dimensionless with h)
h	half of channel height
v_i	velocity vector
$\overset{s}{\beta_i}$	spin intensity vector of order s
φ	spin intensity vector of order 1
f	intensity of turbulence
T	temperature
$\tilde{T} = \dfrac{T - T_0}{T_w - T_0}$	dimensionless temperature
ϕ	dissipation
ω	vorticity
$\overset{s}{I}, \overset{s}{n}, \overset{s}{a}, \overset{s}{b}, \overset{s}{c}$	coefficients of spin equation
$\epsilon, \delta_1, \delta_2$	coefficients of turbulence equation
$\overset{s}{\lambda}, k_1, k_2, \alpha$	coefficients of energy equation
c_p	specific enthalpy capacity at constant pressure
λ	thermal conductivity
q_i	heat flux density
ρ	mass density
ν	kinematic viscosity
$\mathrm{Re} = \dfrac{u_c \cdot h}{\nu}$	Reynolds number

Subscripts

m center line o at $x = 0$ w wall

ACKNOWLEDGEMENT

The authors would like to express their appreciation to Mrs. Amin for translating and typing the manuscript.

REFERENCES

1. Naue, G.: Strömungsmechanik der Kontinua mit unsymmetrischen Spannungen, unpublished report from the Technical University Leuna-Merseburg (GDR, 1973).
2. Naue, G.: Ergebnisse und Probleme der nichtklassichen Strömungsmechanik, ZAMM 72, 52, T 255–T260.
3. Rotta: Turbulente Strömungen Leitfaden der angewandten Mathematik und Mechanik, Bd. 15, Verlag B. G. Teubner, Stuttgart 1972.
4. Gosman, Pan, Runchat, Spalding, and Wolfstein: Heat and Mass Transfer in Recirculating Flows, Academic Press, London and New York, 1969.
5. Krajewski: Int. J. of Heat and Mass Transfer, 16, 73, 469–483, 19.

HEAT TRANSFER IN A SEPARATED FLOW REGION DOWNSTREAM OF A ROUGHNESS ELEMENT *

Niichi Nishiwaki
Noguchi Institute, Tokyo

Tamotsu Sakuma and Akira Tsuchida
Seikei University, Tokyo

Hiroaki Tanaka
University of Tokyo, Tokyo

There is continuing interest in the mechanism of flows over a rough surface because of the interest in augmentation of heat transfer [1]. A physical flow over roughness elements may be represented by a series of separated and reattached flow regions. There are a number of experimental investigations of separated flows downstream of a rearward-facing step [2, 3]. In those experiments, the boundary layer approaching the step was made turbulent by tripping, resulting in a turbulent separation at the step. The purpose of the present investigation is to determine the heat transfer characteristics of separated flows downstream of a single rib, which may be considered to provide a more appropriate model of flows over a single roughness element than a step. Further, in this investigation, various characteristic Reynolds numbers associated with the approaching and separated flows were made considerably lower than those in previous investigations. Also, the interest was primarily directed to obtaining a criterion to determine whether the separated flow becomes laminar or turbulent.

Figure 1 shows the geometry used in the present experiment. An air jet, of width B = 25 mm, impinged a wedge which was placed symmetrically in the flow at a distance from the nozzle $H = 4B$ = 100 mm. At this distance, the air jet preserved its potential core. In order to secure two-dimensionality, the depth of the system was made 260 mm, being bounded by side plates. The velocity, U_0, of the air jet was made between 17 m/s and 26 m/s. The vertical angle θ of the wedge was varied between 60° and 180°. The free stream velocity U_x varied with the distance x from the leading edge according to the value of θ. A rib made of a plate of 1 mm in thickness was placed at a distance s = 10 to 75 mm downstream of the leading edge. At this distance the boundary layer developing from the leading edge remained laminar. The height δ of the rib was varied between 2.5 and 5 mm, which values were

---

*The abstract of this paper appears on page 346.

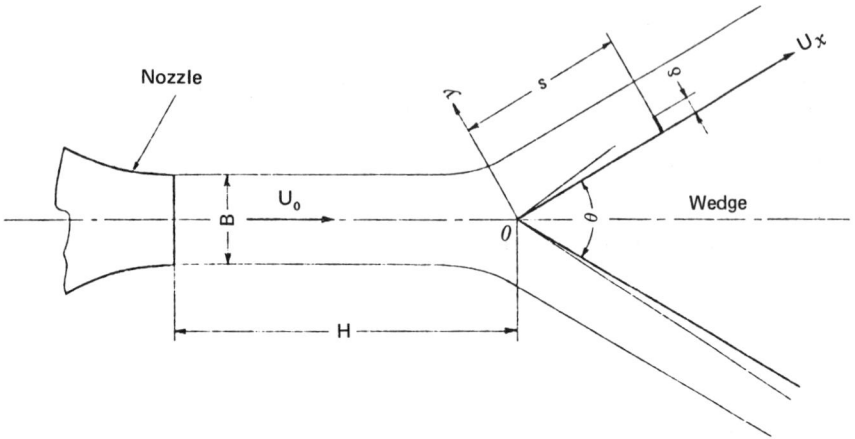

FIG. 1. Geometry of experiment.

larger than the thickness of the boundary layer approaching the rib. A stainless steel plate, 30μ in thickness and 40 mm in width, was fastened along the center line on one side of the wedge surface, and was slightly heated by an alternating current. The heated surface temperatures were measured by thermocouples soldered on the backside of the stainless steel plate. Another rib was placed symmetrically on the other side of the wedge surface, and static pressure distributions were measured with pressure taps placed along the center line on this side.

Preliminary measurements were performed on heat transfer for undisturbed flows without a roughness element for $\theta = 180°$; namely, the air jet impinged normal to the flat plate. The results are shown in Fig. 2 in the form of the local Nusselt number $Nu_x = h_x x/\lambda$ against the local Reynolds number $Re_x = U_x x/\nu$, where h_x signifies the local heat-transfer coefficient and λ and ν, respectively, are the thermal conductivity and the kinematic viscosity of the fluid. In the reduction of the data, the free stream velocity U_x was calculated from the static pressure measurement by using Bernoulli's equation. In the region where $Re_x < 1 \times 10^5$, the boundary layer was considered to be laminar. In this region, the experimental heat-transfer coefficient was a little higher than the theoretical prediction [4]:

$$Nu_x = 0.58 Pr^{1/3} Re_x^{0.5} \tag{1}$$

but almost agreed with McMurray's empirical equation for a water jet [5]:

$$Nu_x = 0.73 Pr^{1/3} Re_x^{0.5} \tag{2}$$

The present experimental results were best correlated by

$$Nu_x = 0.695 Pr^{1/3} Re_x^{0.5} \tag{3}$$

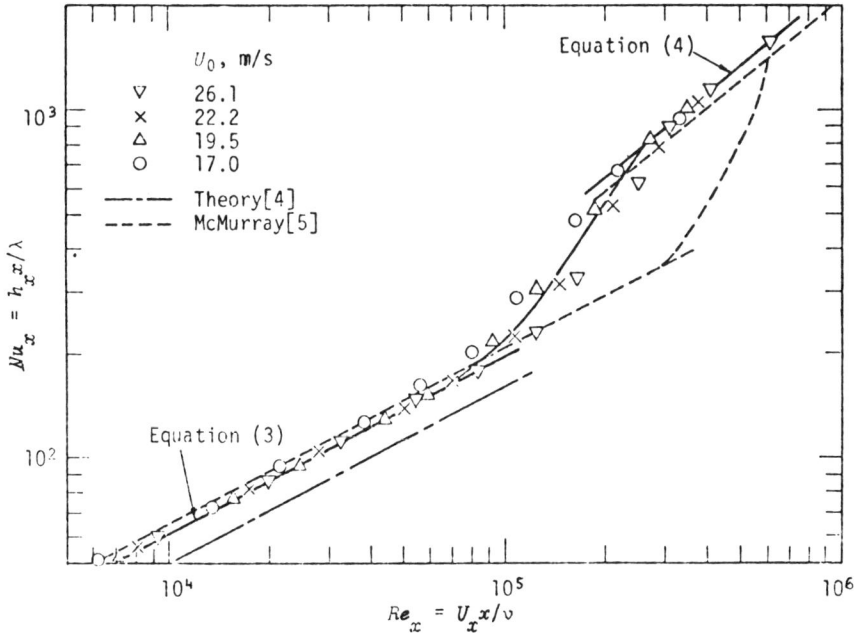

FIG. 2. Heat transfer for undisturbed flows; $\theta = 180°$.

The boundary layer made a transition to turbulence in a range of the Reynolds number 1×10^5 to 2×10^5. Over this transition range, the following experimental relation was obtained:

$$\mathrm{Nu}_x = 0.042\,\mathrm{Pr}^{1/3}\,\mathrm{Re}_x^{0.8} \tag{4}$$

When a roughness element was introduced, the flow separated at the tip of the roughness element, and then reattached at a certain distance downstream. For the case: $U_0 = 17.0$ m/s, $\theta = 180°$, $s = 25$ mm, and $\delta = 5$ mm, the velocity profiles downstream of the roughness element were measured by using impact and static tubes. The results are shown in Fig. 3.

The distributions of the local heat-transfer coefficient downstream of the roughness element are shown in Fig. 4 for various values of s under the same values of U_0, θ, and δ as in Fig. 3. The heat-transfer coefficient shows a sharp maximum at the reattachment point. In the present investigation, special attention was focused on this maximum heat-transfer coefficient h_{max}, which was considered to be representative of the whole physical phenomenon. Whether the separated flow remains laminar or turbulent may be determined by some characteristic Reynolds number Re*. Here it was assumed that

$$\mathrm{Re}^* = \frac{U^* z}{\nu} \tag{5}$$

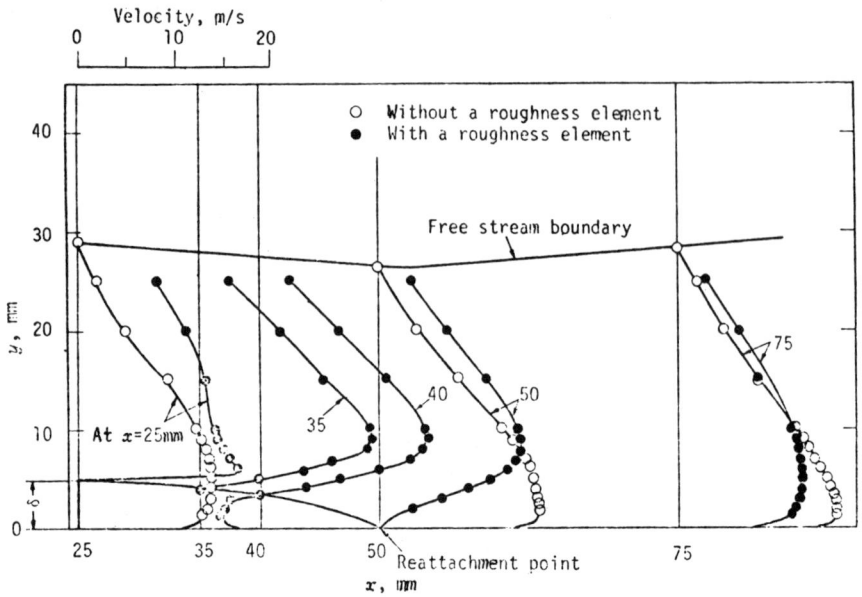

FIG. 3. Velocity profiles downstream of the roughness element; U_0 = 17.0 m/s, θ = 180°, s = 25 mm, δ = 5 mm.

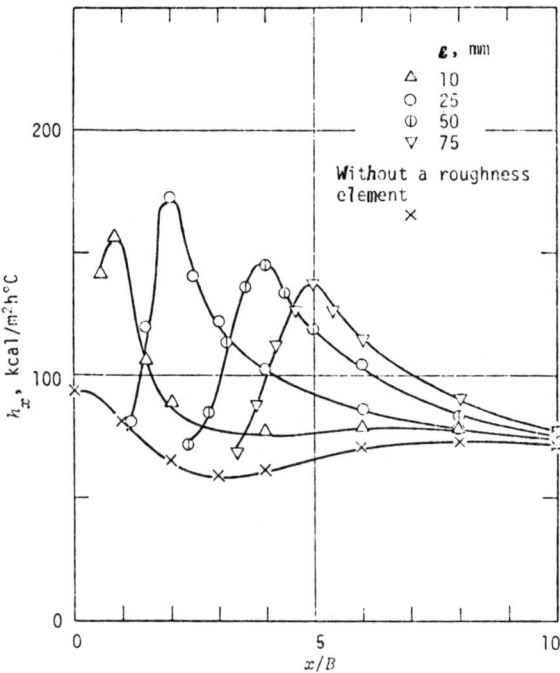

FIG. 4. Distributions of the local heat-transfer coefficient downstream of the roughness element for various s's; U_0 = 17.0 m/s, θ = 180°, δ = 5 mm.

64

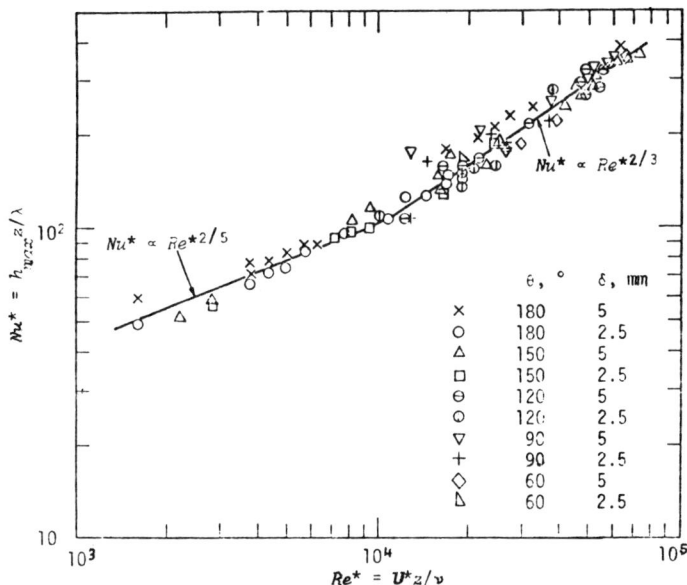

FIG. 5. The heat-transfer coefficient at the reattachment point.

where U^* is the free stream velocity outside the boundary layer at $x = s$ in the case of an undisturbed flow and z stands for the separation length, namely, the distance from the roughness element to the reattachment point. At the same time, we defined the characteristic Nusselt number as

$$Nu^* = \frac{h_{max}z}{\lambda} \tag{6}$$

All the heat-transfer data in the present experiment are plotted in the form of Nu^* vs. Re^* in Fig. 5. From this figure we came to the following conclusions. When $Re^* < 10^4$, the separated flow seems to remain laminar, and the experimental results were well correlated by

$$Nu^* \propto Re^{*2/5} \tag{7}$$

On the other hand, when $Re^* > 10^4$, the separated flow seems to become turbulent, giving an experimental relation:

$$Nu^* \propto Re^{*2/3} \tag{8}$$

It should be noted that, when the height δ of the roughness element is introduced as a characteristic length into Re* and Nu*, such a classification as shown in Fig. 5 was not obtainable, since the separation length z did not enter as a constant times δ but changed according to the magnitude of the values of θ, s, and δ.

REFERENCES

1. Lewis, M. J.: Journal of Heat Transfer, 1975, 97-2, 294.
2. Seban, R. A.: *Ibid.*, 1964, 86-2, 259.
3. Aung, W. and R. J. Goldstein: Israel Journal of Technology, 1972, 10-1, 35.
4. Eckert, E. R. G. and R. M. Drake, Jr.: Analysis of Heat and Mass Transfer, McGraw-Hill Book Company, New York, 1972.
5. McMurray, D. C., P. S. Myers, and O. A. Uyehara: Proc 3d Int. Heat Transfer Conf., 1966, 2, 292.

ON THE EFFECT OF FILM COOLING IN THE SEPARATION ZONE*

Niichi Nishiwaki
Noguchi Institute, Tokyo

Nakaho Numata and Ryo Fujii
Seikei University, Tokyo

Hiroharu Kato
University of Tokyo, Tokyo

1. INTRODUCTION

Recently, many papers have been presented which treated the heat transfer in a separation zone and they showed experimentally that in this region the heat transfer coefficient sometimes becomes higher than that without separation, especially at the reattachment point. This is favorable for the promotion of heat transfer in a heat exchanger. But this could become a problem when thermal insulation is intended in the separation region.

The authors intend to apply film cooling in the separation zone to reduce heat transfer. As far as the authors know, no similar research work has been done in this region.

As a first step of this kind of research, this paper treats film cooling in the separated zone behind a fence on a heated flat plate by injecting hot air normal to the main flow.

2. EXPERIMENTS

Figure 1 shows the arrangement of the experimental apparatus. A flat plate, 120 cm in length, is located in a wind tunnel whose cross section is 70 cm X 22 cm. A thin stainless steel foil is fastened on the plate and is heated by an electric current passing through it. The backside of the flat plate is insulated thermally by an insulation material. Forty thermocouples are soldered on the backside of the stainless steel foil to measure the surface temperatures.

The injection slit is situated at 58 cm from the leading edge of the flat plate. Three kinds of slits whose widths are 1, 2, and 4 mm were used to investigate the effect of injection slit width. The injected air is heated by an electric heater, mixed in a mixing chamber and injected vertically through the slit. At the leading edge of the plate a tripping wire, 1 mm in dia., is located

---

*The abstract of this paper appears on page 346.

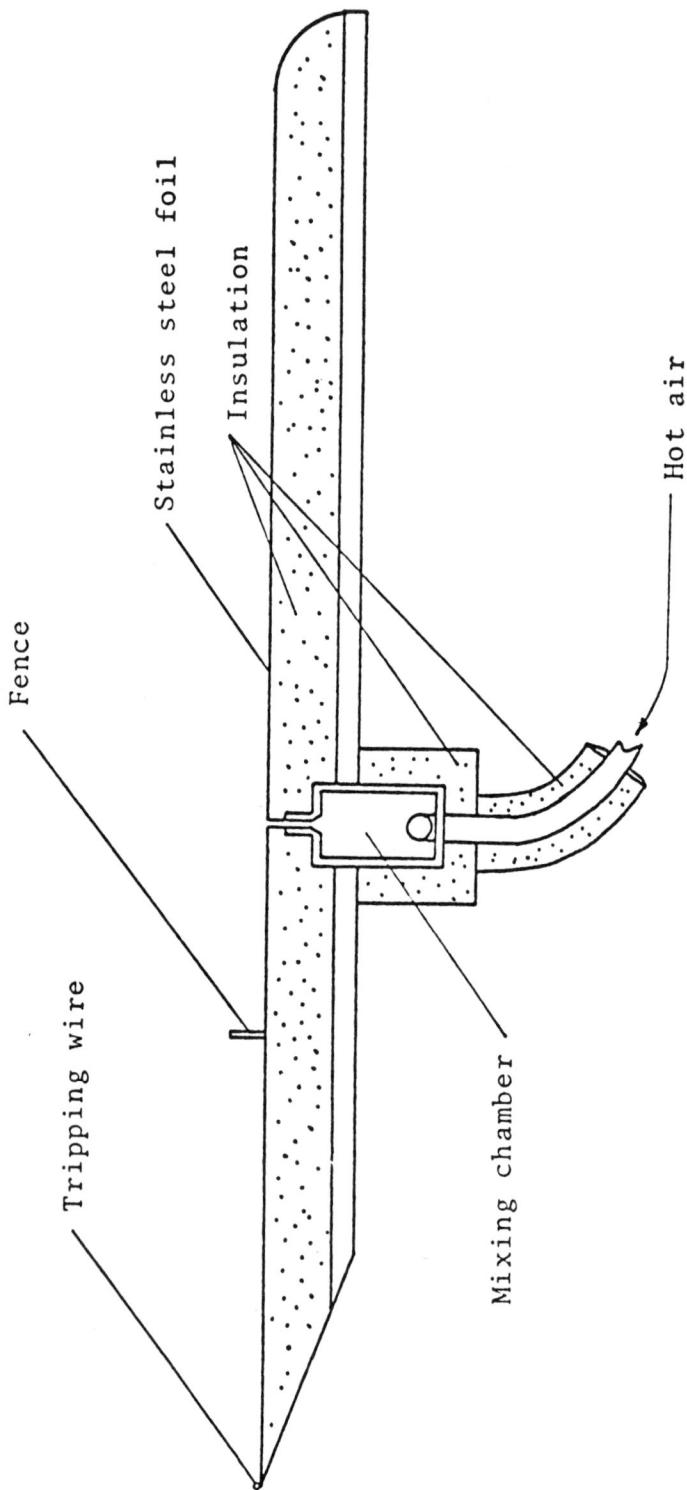

Stainless steel foil

Insulation

Fence

Tripping wire

Hot air

Mixing chamber

FIG. 1. Experimental apparatus.

68

to promote transition. By a preliminary measurement, the existence of a turbulent boundary layer was confirmed at the position of the fence. The uniformity of the flow field in the breadth direction was also confirmed.

A fence whose shape is rectangular, 20 mm in height and 3 mm in thickness was placed on the heated flat plate and the temperature distribution in the separated zone behind the fence was measured. The distance between the fence and the injection slit L, the amount of the injected air Q, and the width of the injection slit S were changed in the experiments.

The experimental conditions were as follows:

$$L/H = 2.5 \sim 11.5 \qquad (H = 20\text{mm})$$

$$Vj/U_\infty = 0.17 \sim 2.25 \qquad (U_\infty = 12\text{m/s})$$

$$S/H = 0.05, 0.1, 0.2$$

$$Tj - T_\infty = 22°\text{C}$$

3. EXPERIMENTAL RESULTS

The coordinate system is shown in Fig. 2. A typical variation of the reduction of heat transfer coefficient is shown in Fig. 3 where $S/H = 0.05$, $L/H = 7.5$, and $Vj/U_\infty = 0.68$. Figure 4 is a sketch of air flow under typical conditions. The injected air flows back to the fence by the counter flow near the surface in the separated zone. The effect of film cooling is predominant near the injection slit and it decreases gradually toward the fence. A part of the injected air, however, flows out of the separated zone, and the film cooling effect can also be seen in the downstream region. It is interesting that at this flow condition the film cooling effect is much larger at the upstream region than at the downstream region.

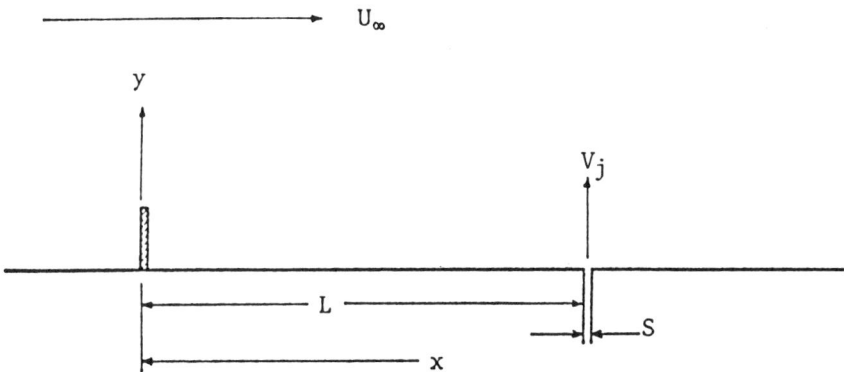

FIG. 2. Coordinate systems.

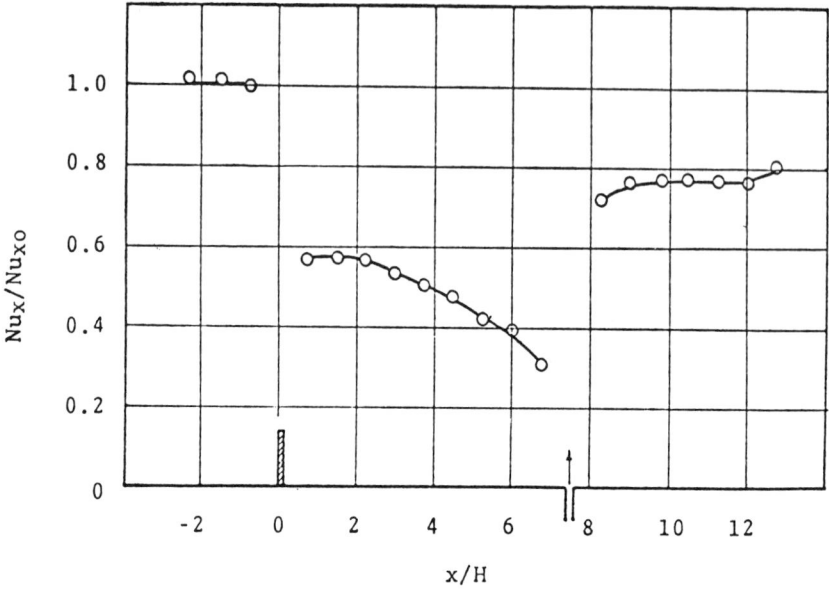

FIG. 3. Effect of film cooling; S/H = 0.05, L/H = 7.5, V_j/U = 0.68.

The effect of the location of the air injection point on film cooling is shown in Fig. 5, where L/H is changed from 4.5 to 11.5, that is, from the inside of the separated zone to the position after the reattachment point. When $L/H \leqslant 75.$, the effect of air injection is more predominant at the upstream zone. The effect reaches a maximum at L/H = 7.5.

At the region $L/H > 7.5$, the effect of air injection decreases in the upstream region. On the contrary, the effect on the downstream region becomes much larger when the injection slit approaches the reattachment point. When L/H becomes large, the air injection effect approaches that of a flat plate without a fence.

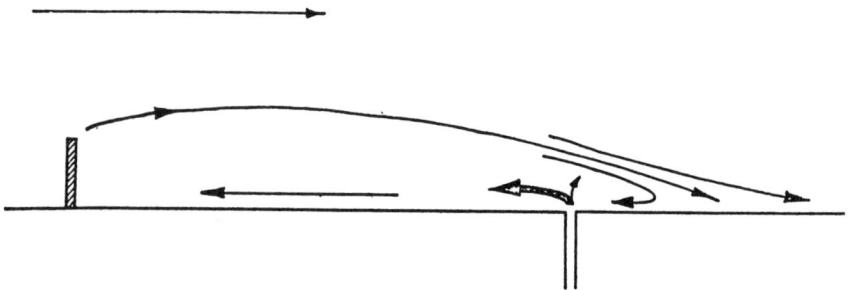

FIG. 4. Flow patterns.

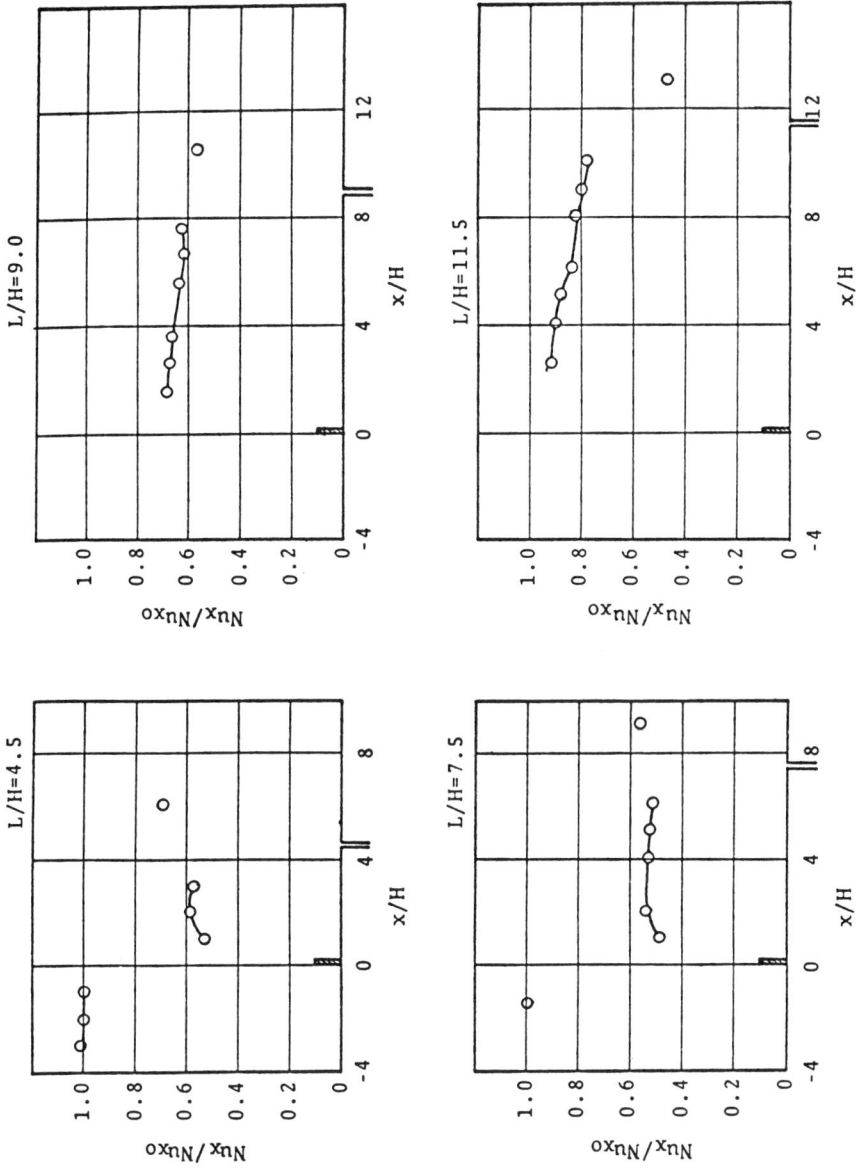

FIG. 5. Effect of film cooling for various *L/H*.

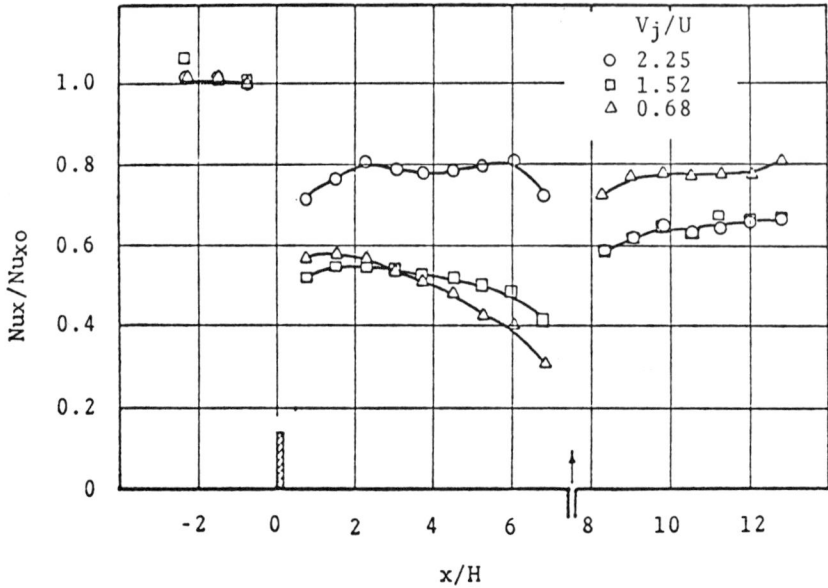

FIG. 6. Effect of film cooling for various injection velocities.

Figure 6 shows the effect of the injection velocity whose tendency at the upstream region is contrary to that of the downstream region. At the upstream region, the lower injection velocity gives a higher reduction of the heat transfer coefficient. Especially at the condition of $Vj/U_\infty = 2.25$, the reduction rate is much less than that of lower injection velocity. This is because the velocity of the injected air is too high and it blows off from the separated zone as mentioned previously. On the contrary, in the downstream region, the higher injection velocity gives a larger reduction of the heat transfer rate.

The effect of the width of the injection slit is complicated as can be seen in Fig. 7. Figure 7 (a) and (b) show the effect of the width of the injection slit for various flow rates of injected air near the injection slit and near the fence, respectively. At lower injection air rates, a narrower slit shows improved characteristics, i.e., the narrow slit of $S = 1$ mm shows very bad characteristics for higher flow rates. The reason for this was mentioned previously. It seems there is an optimum width for a given injection flow rate.

4. CONCLUSIONS

1. The effect of film cooling in the separation zone behind a fence was studied experimentally.
2. By injecting air near the rear end of the separation zone, the film cooling effect was observed over the zone. The reduction of the heat transfer

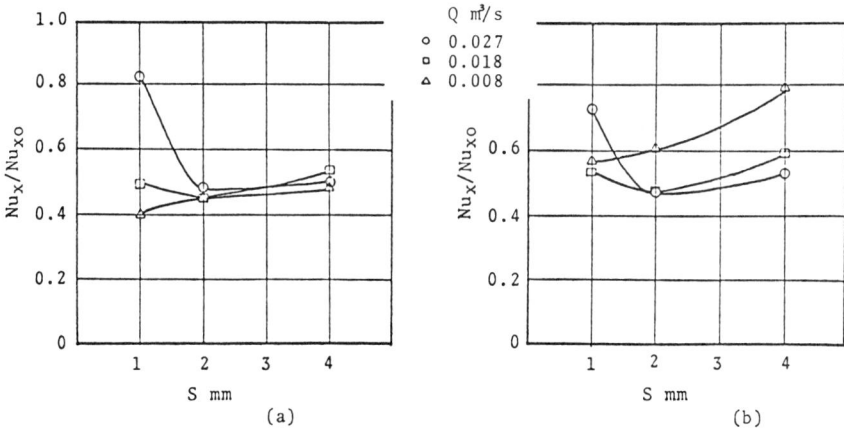

FIG. 7. Effect of the width of injection slit for various flow rate of injected air. (a) x/H = 6; (b) x/H = 0.75.

coefficient reaches approximately a third of that without air injection by this experiment.

3. The best injection condition obtained was at L/H = 7.5 and Vj/U_∞ = 1.
4. The effect of the width of an injection slit is complicated. Roughly speaking, a narrower slit shows better characteristics.

LAMINAR ENTRY LENGTH HEAT TRANSFER IN DUCTS OF RECTANGULAR CROSS SECTION WITH BOUNDARY CONDITIONS OF THE SECOND KIND*

Vera Preiningerová
Research Institute of Electrical Engineering
Praha-Bechovice, Czechoslovakia

In practical calculations of heat transfer with viscous flow of fluids in ducts having noncircular cross sections, frequent use is being made of the relationships derived for a circular tube or a flat duct using the hydraulic diameter of an actual duct.

This work solves the problem of steady heat transfer in the thermal entrance section of ducts of rectangular cross section with laminar flow. It investigates the effect of the cross section side ratio and the geometry of the heated and unheated duct walls with boundary conditions of the second kind. It is assumed that the velocity profile at the entrance is fully developed and that the thermal flux flows from the duct walls into the fluid. The solution is presented both for constant physical properties of the coolant and for a temperature dependent viscosity.

METHOD OF SOLUTION

The velocity and temperature equations have been derived from the general form of the Navier-Stokes equations [1] and the Fourier equation [2] using simplifications which are appropriate for the above assumptions. These simplifications, discussed in more detail in the literature [3], involve first of all the omission of the velocity components perpendicular to the duct axis and the omission of heat conduction and momentum variations in the axial direction.

BASIC SYSTEM OF EQUATIONS

Equation of continuity:

---

*The abstract of this paper appears on page 348.

$$v_0 XY = \int_0^X \int_0^Y v \, dx \, dy \tag{1}$$

Temperature field equation:

$$\frac{\partial^2 \vartheta}{\partial x^2} + \frac{\partial^2 \vartheta}{\partial y^2} = \frac{c\rho}{\lambda} v \frac{\partial \vartheta}{\partial z} \tag{2}$$

Velocity field equation:

$$\frac{\partial^2 v}{\partial x^2} + \frac{\partial^2 v}{\partial y^2} = \frac{1}{\mu(\vartheta)} \frac{\partial p}{\partial z} \tag{3}$$

BOUNDARY CONDITIONS OF THE TEMPERATURE EQUATION

On the heated walls (grad $\vartheta)_w$ = $-w/\lambda$, while on the unheated walls (grad $\vartheta)_w$ = 0. At the entry the profile is isothermal, i.e., ϑ_0 = constant.

BOUNDARY CONDITIONS OF THE VELOCITY EQUATION

On the duct walls the velocity is zero; at the entrance the following equation holds:

$$\frac{\partial^2 v_0}{\partial x^2} + \frac{\partial^2 v_0}{\partial y^2} = \frac{1}{\mu(\vartheta)} \left(\frac{\partial p}{\partial z}\right)_0 \tag{4}$$

In most cases the *physical properties of the fluid* are considered to be constant and are referred to the temperature at the entry ϑ_0. If the temperature dependence of viscosity is taken into account it is expressed by the relation $\mu(\vartheta) = (c_1 \vartheta^2 + c_2 \vartheta + c_3)^{-1}$.

The calculations reported here were performed numerically on a computer using the method of finite differences with a rectangular net made more dense at points with a high temperature gradient, i.e., at heated walls and at the entrance. The parabolic temperature equation was solved using the method of alternating directions; the elliptic velocity equation was solved by block superrelaxation. The equation of continuity serves, when the temperature dependence of viscosity is taken into account, for checking and making corrections to the pressure gradient. For the determination of the entry pressure gradient, corresponding to the mean velocity at the entry $\bar{v}_0$, the linear relation between the different mutually corresponding pairs of values $\bar{v}$ and $\partial p/\partial z$ is used: for an arbitrarily selected value $(\partial p/\partial z)_a$ the values of $v_a(x, y)$

and $$\bar{v}_a = \frac{1}{XY} \int_X \int_Y v_a(x,y)\, dx\, dy$$

are calculated according to Eq. (4). The searched value $(\partial p/\partial z)_0$ is then found using the relation

$$\left(\frac{\partial p}{\partial z}\right)_0 = \frac{(\partial p/\partial z)_a}{\bar{v}_a}\, \bar{v}_0 \tag{5}$$

and the velocity profile at the entry is obtained from the relation

$$v_0(x,y) = \frac{\bar{v}_0}{\bar{v}_a}\, v_a(x,y) \tag{6}$$

A more detailed description of the calculation procedure is given in [3].

RESULTS AND DISCUSSION

The method described above has been used to calculate heat transfer in ducts having a side ratio Y/X ranging from $1/1$ to ∞

a. for equal thermal flux density on all the duct walls, $w_X = w_Y$
b. for heating of only the two longer duct walls, $w_X = 0$
c. for heating of only one of the longer duct walls, $w_X = 0$, $w_{Y1} = 0$

The results of the heat transfer calculations, expressed as functions of the dimensionless quantities $\overline{Nu}_m = f(Z^+)$ and $\overline{Nu}_z = f(Z^+)$ and calculated on the assumption of constant physical properties are shown, within the limits $10^{-3} \leqslant Z^+ \leqslant 10^{-1}$, in Figs. 1 through 3. $\overline{Nu}_m$ is the integrated mean of the Nusselt number over the duct length and periphery, $\overline{Nu}_z$ is the local value of the Nusselt number along the duct length and the mean value around the duct periphery. In the cases a and b of Figs. 2 and 3, both of the Nusselt numbers are referred only to the heated walls and are, therefore, denoted by $\overline{Nu}_m$ and $\overline{Nu}_{zY}$. From the figures it can be seen that in all the cases considered $\overline{Nu}_m$ as well as $\overline{Nu}_z$ increases with increasing ratio Y/X. The curves of the functions, particularly of the minimum value of the Nusselt number corresponding to a thermally developed flow, are different for each individual case.

Within the limits $0 < Z^+ < 10^{-3}$ the values of $\overline{Nu}_m$ and $\overline{Nu}_z$ can be represented by the simple relation

$$Nu = K_n (Z^+)^{-1/3} \tag{7}$$

where K_n are constants depending on the ratio Y/X and on the number and geometry of the heated walls.

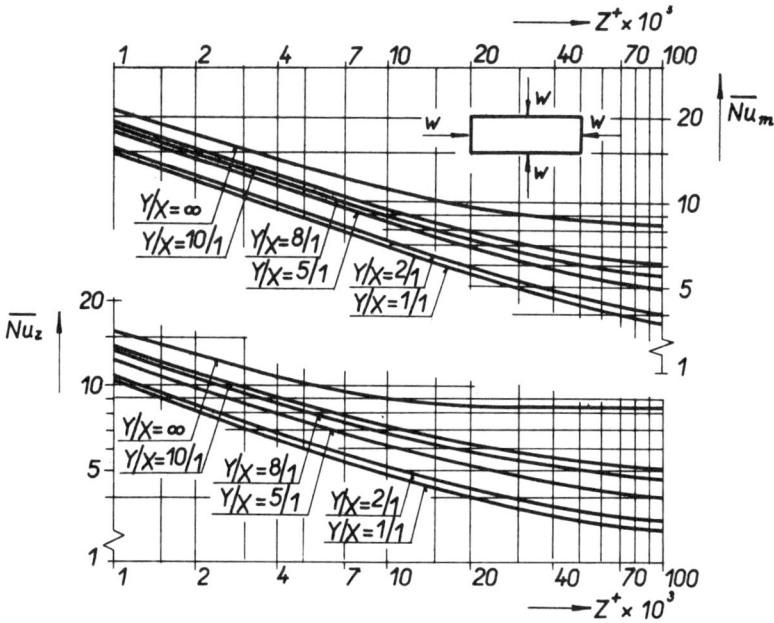

FIG. 1. Relationships $\overline{Nu}_m = f(Z^+)$ and $\overline{Nu}_z = f(Z^+)$ for case a, i.e., when all the walls are heated by an equal thermal density w.

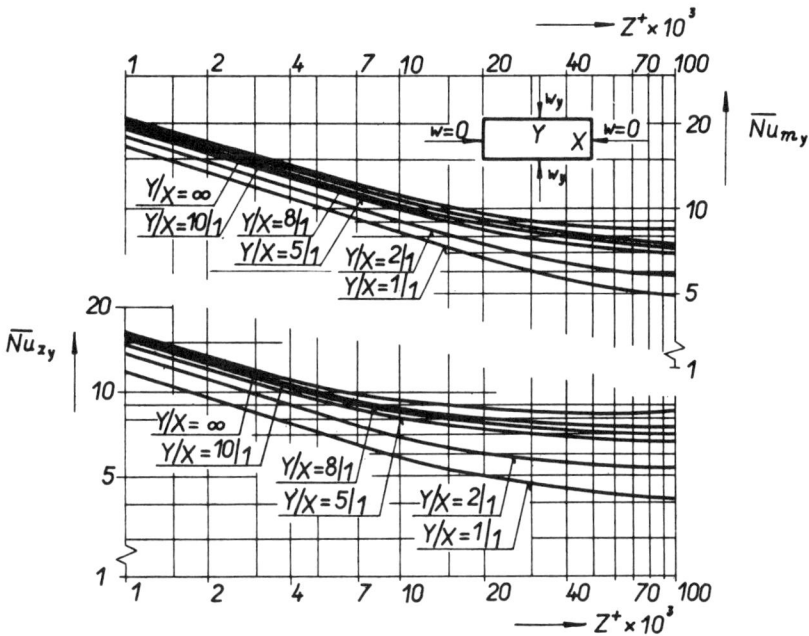

FIG. 2. Relationships $\overline{Nu}_{m_Y} = f(Z^+)$ and $\overline{Nu}_{z_Y} = f(Z^+)$ for case b, i.e., when only the two longer walls are heated.

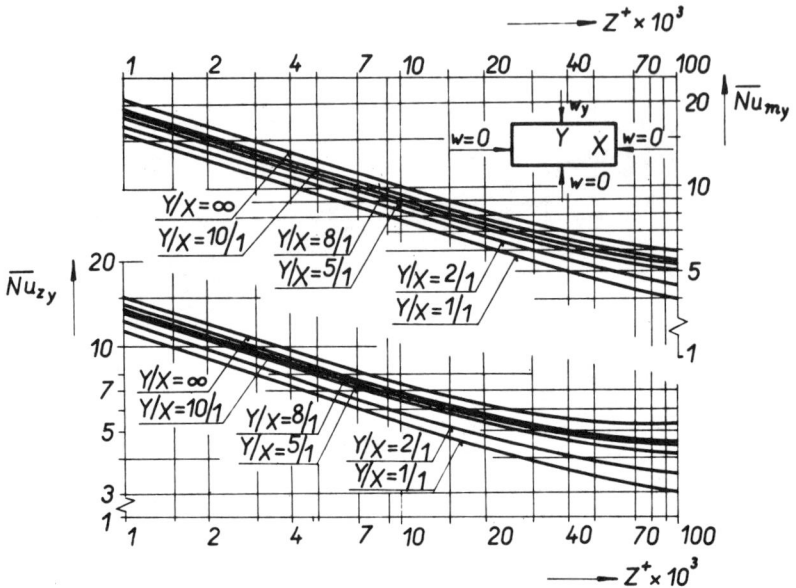

FIG. 3. Relationships $\overline{Nu}_m = f(Z^+)$ and $\overline{Nu}_{zY} = f(Z^+)$ for case c, i.e., when only one of the walls is heated.

Figure 4 shows how the ratio Y/X affects the distribution of both quantities along the duct periphery in the case a at a distance $Z^+ = 10^{-2}$ from the entrance. The wall temperature is expressed by the dimensionless number $(\vartheta_w - \vartheta_b)/\lambda^{-1}wD$, whose reciprocal is equal to the local value of Nu. The development of the peripheral distribution of the wall temperature along a duct having the ratio $Y/X = 1/1$ is shown in Fig. 5.

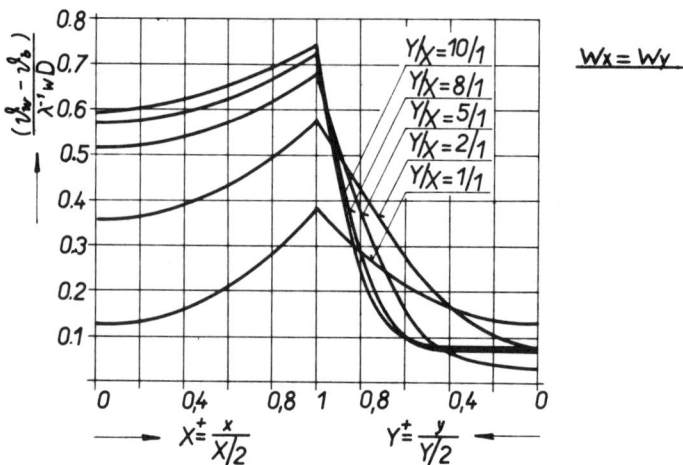

FIG. 4. Distribution of the dimensionless wall temperature $(\vartheta_w - \vartheta_b)/\lambda^{-1}wD$ along the periphery of ducts having various ratios Y/X for case a at distance $Z^+ = 10^{-2}$.

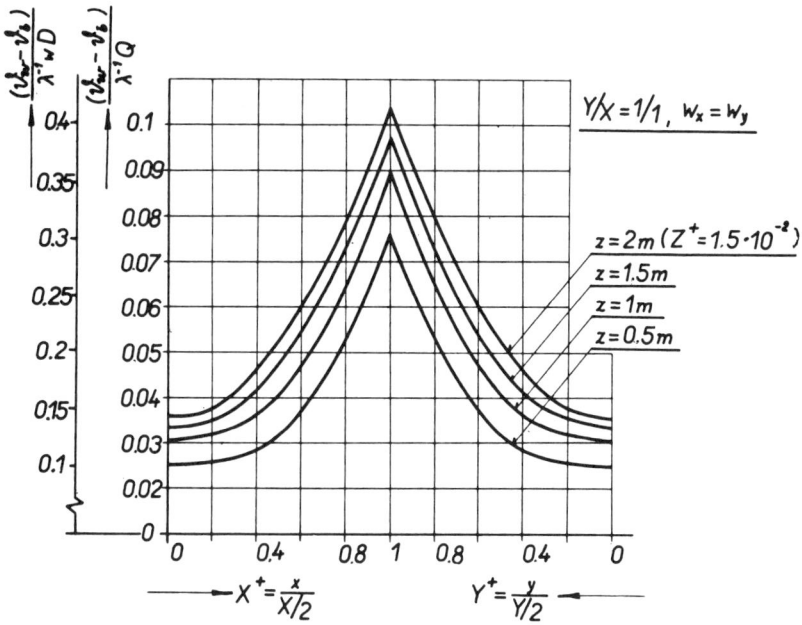

FIG. 5. Development of the peripheral temperature distribution of a wall and the quantity 1/Nu along the length of a duct having the ratio Y/X for case a.

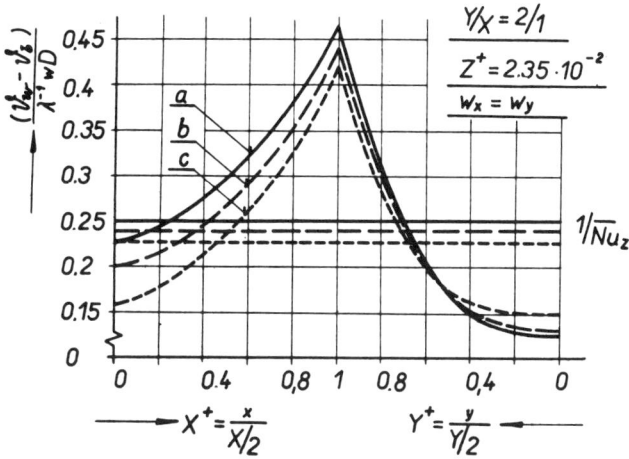

FIG. 6. Effect of variable viscosity on the wall temperature distribution and the quantity 1/Nu along the periphery of a duct having the ratio Y/X and $Z^+ = 2.35 \times 10^{-2}$. Curve a: constant physical properties; curve b: BTS oil, $\mu = f(\vartheta)$, $w = 3000$ W/m²; curve c: BTS oil, $\mu = f(\vartheta)$, $w = 6000$ W/m².

With fluids whose viscosity decreases with increasing temperature as, for instance, in the case of transformer oil, the influence of variable viscosity is shown by smaller peripheral deviations of temperature and by an increase in heat transfer. An example is shown in Fig. 6. However, the condition of constant thermal flux along the duct periphery prevents the effect of variable viscosity from manifesting itself as strongly as in the cases where the temperature along the duct periphery is constant, e.g., in the case of a circular tube or flat duct [4], [5], [6].

In conclusion it may be stated that the ratio Y/X and the geometry of heated and unheated walls of rectangular ducts greatly affects the heat transfer with laminar flow under the conditions considered and that taking account of the cross section shape by a mere introduction of the hydraulic diameter into the relations derived for a flat duct or circular tube is not sufficiently accurate.

NOMENCLATURE

c	specific heat [J/kg · K]	x, y	distances normal to duct axis [m]
D	hydraulic diameter [m]	X	shorter side of cross section [m]
n	distance in the direction of the normal [m]	Y	longer side of cross section [m]
		z	distance along duct axis [m]
p	pressure [N/m²]	λ	thermal conductivity [W/m · K]
Q	thermal flux per unit flux length [W/m]	μ	dynamic viscosity [N · s/m²]
v	velocity along duct axis [m/s]	ρ	density [kg/m³]
w	thermal flux density [W/m²]	ϑ	temperature [°C]

Dimensionless Quantities

Nu	Nusselt number	X^+	dimensionless distance x
Pr	Prandtl number	Y^+	dimensionless distance y
Re	Reynold number	Z^+	dimensionless distance z

Subscripts

a	arbitrarily selected subscript	x	at shorter wall
b	bulk fluid	y	at longer wall
m	integrated mean over duct length	z	at distance z
w	on duct wall	0	at duct entry

The symbol —— denotes integrated mean over duct section $(\bar{v})$ or over duct periphery $(\overline{Nu})$.

REFERENCES

1. Navíer, L. M. H.: Mem. Acad. Sci. (2d series), 1823, **6**, 389.
2. Fourier, J.: Mem. Acad. Roy. Sci. Inst. France, **12**, 1833, 727.
3. Preiningerova, V., P. H. G. Allen: Proc. 5th Internat. Heat Transfer Conference, Tokyo, Japan, 1974, paper NC5.4.
4. Petukhov, B. S.: Teploobmen i soprotivlyenyie pri laminarnom tetshenyi shidkosti v trubakh, 1967, Energoizdat, Moscow.
5. Shukausas, A. and I. Shiugshda: Teplodatsha v laminarnom potoke shidkosti, 1969, Mintis, Vilnius.

A NEW MODEL OF TURBULENT COMBUSTION *

D. Brian Spalding
Imperial College of Science and Technology
London

In practice, flames are nearly always turbulent. This is a consequence of the large sizes and high velocities of industrial and domestic fuel-burning equipment.

The sketch on slide panel 1 illustrates a typical burner for gaseous fuel. The fuel is premixed with air, but not in sufficient quantity for complete combustion; but further air is entrained from the atmosphere. Therefore, *two* flames are present: the premixed flame and the diffusion flame. However, these are not as clearly separated in turbulent flow as they are in laminar flow, because of the fluctuations which are characteristic of turbulence.

There is a great need for methods of computing turbulent flames which are as well-tested, and as successful, as those for laminar ones; but the methods must be simple enough for use.

The present paper describes a simple method, and illustrates its use.

The prediction of fluid-dynamic and heat-transfer phenomena necessitates solving transport equations of the kind shown on slide panel 2. Here ϕ is a general symbol: it can stand for velocity, temperature, enthalpy, concentration, turbulence energy, and many other variables.

Numerical methods of solving such equations are now well known, and widely available (Refs. 1, 2, 3, 4, 5, 6, 7); however, it is necessary to possess means of computing the turbulent exchange coefficients, Γ_ϕ and the source terms, $\overline{S}_\phi$.

The Γ_ϕ's, and some of the $\overline{S}_\phi$'s, are obtainable from turbulence models (e.g., Ref. 8). However, there is great uncertainty about the $\overline{S}_\phi$ for fuel in a turbulent flame.

If the flow were laminar, the vast body of reaction-kinetic information could be employed; this connects the soruce term with the local composition, temperature, and pressure.

---

*The abstract of this paper appears on page 361.

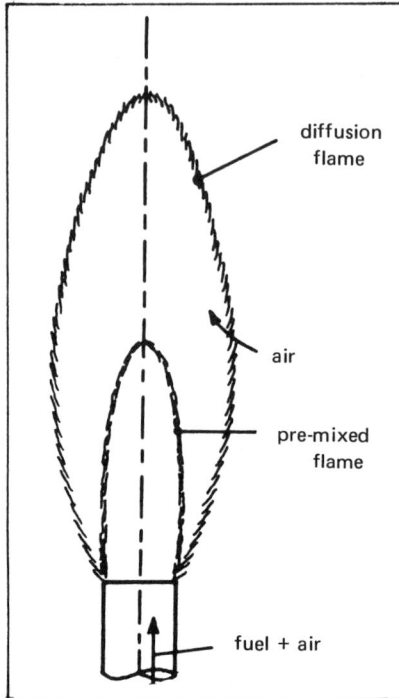

SLIDE 1

Attempts to predict turbulent-flame phenomena by the use of laminar reaction-kinetic expressions and time-average compositions, etc., have failed. Often the reaction-kinetic constants are unimportant; however, the hydrodynamic state of the gas is very important. The problem is: How can these facts be expressed in the $\overline{S}_{fu}$ function?

MATHEMATICAL ASPECTS:

- METHODS EXIST FOR NUMERICAL SOLUTION OF THE TRANSPORT EQUATION: $\rho\frac{D\overline{\phi}}{Dt} = \mathrm{div}(\Gamma_\phi\ \mathrm{grad}\ \overline{\phi}) + \overline{S}_\phi$, WHERE $\overline{\phi} \equiv$ TIME-AVERAGE VELOCITY, ENTHALPY, CONCENTRATION, ETC.
- TURBULENCE MODELS GIVE EXCHANGE COEFFICIENT Γ_ϕ.
- THE PROBLEM: HOW TO CALCULATE S_ϕ.

PHYSICAL ASPECTS:

- FOR LAMINAR FLOW, REACTION KINETICS GIVES:
  $S_\phi = S_\phi\{\text{CONCENTRATIONS, TEMPERATURE, PRESSURE}\}$.
- FOR TURBULENT FLOW, $\overline{S}_\phi \neq S_\phi\{\text{ARGUMENT}\}$; BUT $\overline{S}_\phi$ IS AFFECTED BY TURBULENCE AND FLUID SHEAR.

SLIDE 2. The problem: How to calculate turbulent reaction rates.

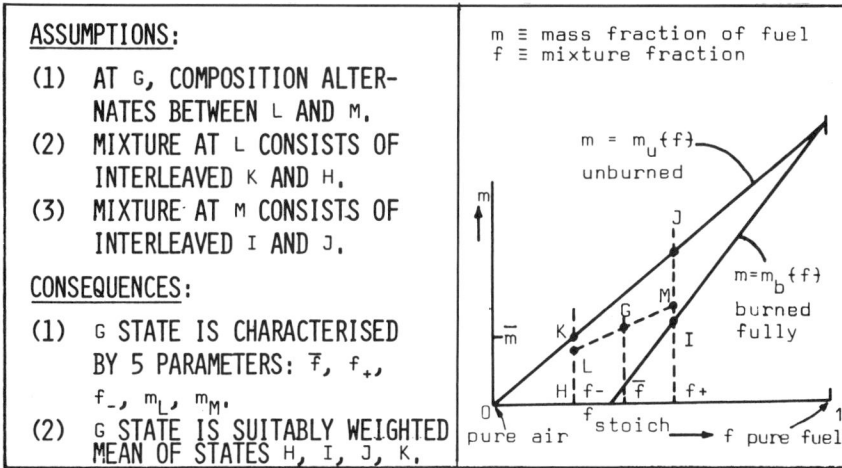

SLIDE 3. The new hypothesis for the structure of the local mixture.

It is possible to make only a brief presentation in the space available. The main ideas are represented on slide panels 3 and 4.

At any position in a turbulent flame, let the *time-mean* gas condition be represented by the point G on the $m \sim f$ diagram in slide 3, wherein m and f are defined. The assumptions of the "Simple Chemically–Reacting System" are implied; a consequence is that G must lie within the triangle, on its upper edge (when no fuel has been burned), or on its lower edge (when as much as possible has been burned).

The slide displays the main assumptions about how the time-mean composition is actually made up. They are crude ones; but they do contain the

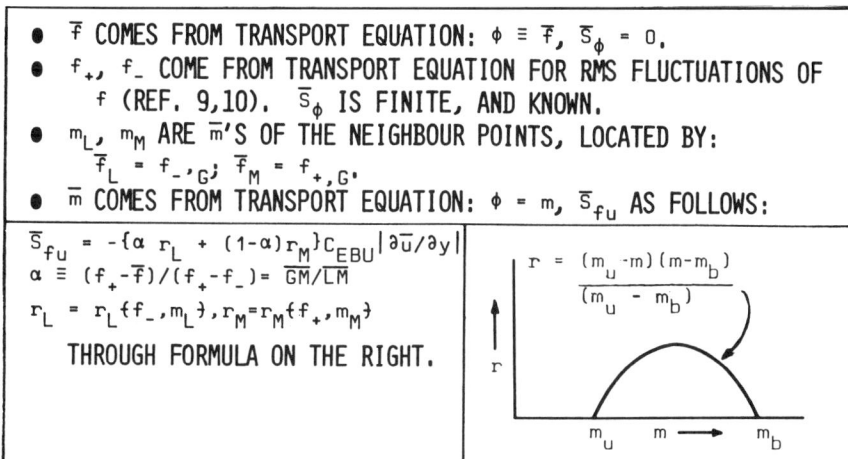

SLIDE 4. The procedure for computing the five parameters.

essential features. The mixture fraction fluctuates between high and low values, f_+ and f_-; so material from state point L alternates, in the G mixture, with material from state point M.

The L and M states are regarded as themselves being mixtures of fully burned and wholly unburned material. Thus, L-state material consists of layers of H-state and K-state material, interleaved.

A consequence is that, to characterize the mixture at G completely, it is necessary to specify five distinct quantities.

How are these quantities to be ascribed values? Slide 4 provides the answer.

The time mean value of the mixture fraction f is obtainable from the solution of the relevant transport equation. This has a zero source; for f measures the composition without regard to the state of chemical aggregation.

To obtain values f_+ and f_-, it is necessary to employ an equation for the root-mean-square fluctuations of f. This has been proved to give rather satisfactory predictions (Refs. 9, 10). No difficulty is presented by this part of the problem.

It is now proposed that the values of m_L and m_M, the fuel mass fraction in the low-f and high-f material present in the mixture represented by G, can be taken as being the time-average values prevailing at the locations where the f_+ and f_- of G are the *local time-mean* values of f. This is how the f_+ and f_- values are actually used, namely, to locate those places.

The time-mean reaction rate of fuel now follows. The formula has two components: a composition expression; and a measure of the rate of distortion of the gas by shearing motion. The slide shows what is recommended.

The r function is worth noting: it shows that, as is necessary, the reaction rate falls to zero when the fuel is fully burned; and it is also zero for completely unburned material.

The thinking which underlies the reocmmended formula is explained at greater length in Refs. 11, 12, 13. The present model is too crude to represent all the features of so complex a phenomenon as turbulent chemical reaction; and even the full version of the theory, when completely worked out, will fall short of reality.

However, the present model is not bad; and, indeed, serviceable results can be obtained from an even simpler one; this is the one to which the present model reduces when the fluctuations of mixture fraction are supposed to be very small. Then it is necessary to compute only two variables, $\overline{m}$ and $\overline{f}$, in order to characterize the state of the mixture.

This extremely simple version is the one which has been built into the published version of the GENMIX computer code (Ref. 7). This code is especially useful for the prediction of two-dimensional steady flames without recirculation. Versions of it exist which contain the present model of turbulent reaction, as well as more complex ones.

A few results of computations will now be presented, to illustrate what the new model implies. The "even simpler version" of Slide 5 is employed.

- **MORE GENERAL VERSION**: THE Γ AND $|\partial\overline{u}/\partial y|$ EXPRESSIONS ARE SIMPLIFICATIONS OF MORE GENERAL ONES (REFS.11,12,13). HOWEVER, THE PRESENT ONES ARE FAIRLY SATISFACTORY.

- **EVEN SIMPLER VERSION**: IF FLUCTUATIONS OF f ARE IGNORED $(f_+ = f_- = \overline{f})$, L AND M COINCIDE WITH G. THEN ONLY TWO PARAMETERS ARE NEEDED: $\overline{f}$ AND $\overline{m}$.

- **COMPUTER CODE**: ● FOR TWO-DIMENSIONAL FLAMES WITHOUT RECIRCULATION, THE GENMIX CODE (REF.7) IS SUITABLE.

- ● THE "EVEN SIMPLER VERSION" APPEARS IN THE PUBLISHED CODE.

SLIDE 5. The new hypothesis; discussion.

Slide panel 6 presents predictions for a turbulent diffusion flame for which extensive experimental data have been reported in Ref. 14. The axial values are plotted, against longitudinal distance x/D_O, of: the dynamic head ρu^2, the fuel mass fraction m_{fu}, the gas temperature T, the radius at which the temperature has its maximum value y_{Tmax}, and the oxygen mass fraction m_{ox}. The mixing-length model of turbulence has been employed; the value of the mixing length has been taken as proportional to the radial thickness of the turbulent region. Details are supplied in the Appendix.

All the results are qualitatively plausible: the fuel concentration falls, and that of oxygen increases, with increasing distance from the nozzle; in the

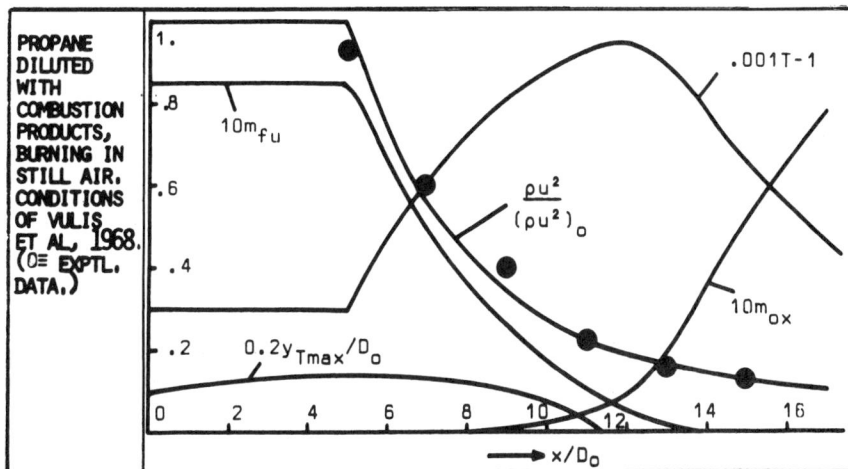

SLIDE 6. The turbulent diffusion flame, 1: Longitudinal profiles of variables on flame axis.

region of maximum temperature, both the fuel and the oxygen have finite concentrations (unlike the situation in a *laminar* diffusion flame).

Quantitative comparisons with the data of Ref. 14 reveal satisfactory agreement, as is indicated by the experimental values (•'s) of ρu^2.

The computations performed by GENMIX produce much detailed information about the spread of the jet. Some is shown in slide panel 7, in the form of profiles, that is to say, of distributions with radius y, at various values of longitudinal distance x.

On the left are plotted values of the dynamic head, nondimensionalized with respect to the axial value. Clearly evident are the spread of the mixing layer, formed at the nozzle lip, inwards to the axis and outwards into the surrounding atmosphere. Taken together with the relevant curve of panel 6, this diagram represents the whole ρu^2 field. Some data points from the Ref. 14 study are included for comparison. The agreement is satisfactory.

On the right are plotted profiles of m_{fu} and m_{ox}, respectively, the mass fractions of fuel and of oxygen. This time, the values are *not* nondimensionalized with respect to axial values. The gradual disappearance of fuel is evident, as x increases; and it is seen how the oxygen concentration rises. Once again, the presence is observed of finite oxygen and fuel concentrations at the same place; this is a result of the fluctuations of concentration.

The new model of turbulent reaction can provide plausible predictions for premixed flames as well as for those in which the fuel and oxygen are supplied separately. Slide panel 8 shows what happens when the fuel-products mixture of the previous example is diluted by being mixed with cold air. The consequences are there displayed in terms of the oxygen mass fraction on the flame axis, and its variation with axial distance.

The parameter is the aeration ratio of the injected gases. The curve marked

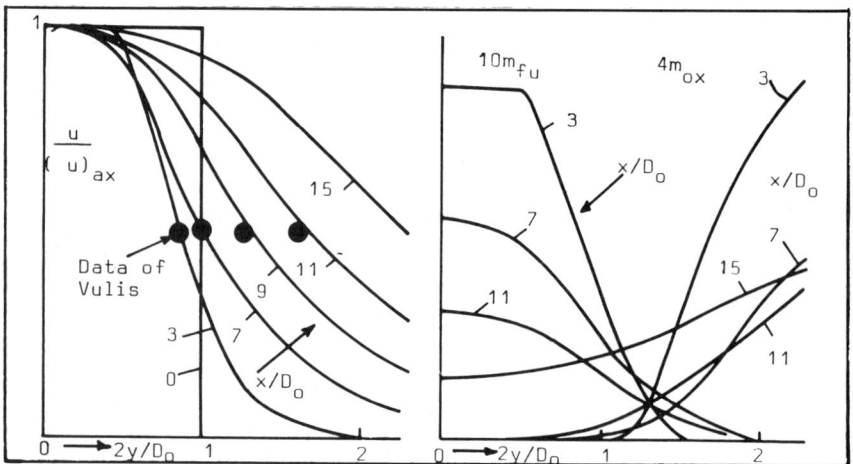

SLIDE 7. The turbulent diffusion flame, 2: Radial profiles at various x/D_0.

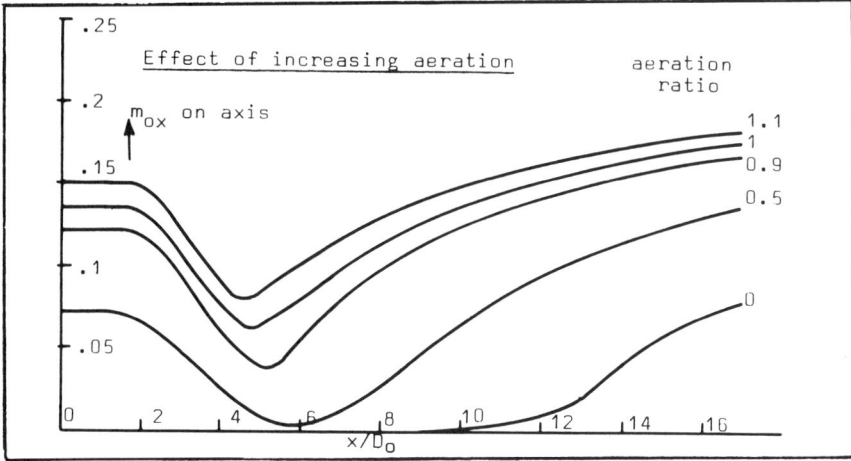

SLIDE 8. The turbulent bunsen burner, 2: Cross-stream profiles of variables at $x/D_0 = 3$.

0 is from panel 6; and others correspond to increasing aeration ratios, up to and beyond stoichiometric (1.0).

The curves demonstrate that, as is easily understood, the supply of oxygen with the fuel significantly shortens the flame (the length of the flame is the x value at the point where m_{ox} starts to increase).

For the aeration ratio of 0.5, the axial oxygen concentration falls almost to zero; this implies that the inner (premixed) flame is distinct from the outer (diffusion) flame. At higher aeration ratios, the premixed flame has about the same length, but the diffusion flame becomes shorter; the result is that the two flames coalesce.

Further predicted results are shown on panel 9, which displays cross-stream profiles at an x/D_0 value of 3. At this location, the mixing region has already spread to the axis.

On the left are shown the mass fractions of oxygen. It is noteworthy that the aerated flames are significantly wider than the pure diffusion flame. The reason is presumably that the fuel burns in a shorter distance, so that the average density of the gases is lower; a larger diameter is needed for the gases to flow in, because the flow velocity is not significantly changed. At the x/D_0 in question, the oxygen concentration nowhere falls to zero for aerated flames. The separation of premixed and diffusion flames is scarcely noticeable, even for the aeration ratio of 0.5.

The right-hand diagram provides further profiles for the latter aeration ratio, namely those of temperature and of fuel concentration. The maximum of temperature occurs, understandably enough, toward the outer edge of the reaction zone, for the mixture ratio is nearest to stoichiometric there, and both fuel and oxygen concentrations are low.

The predictions which have been displayed reveal that the new model of

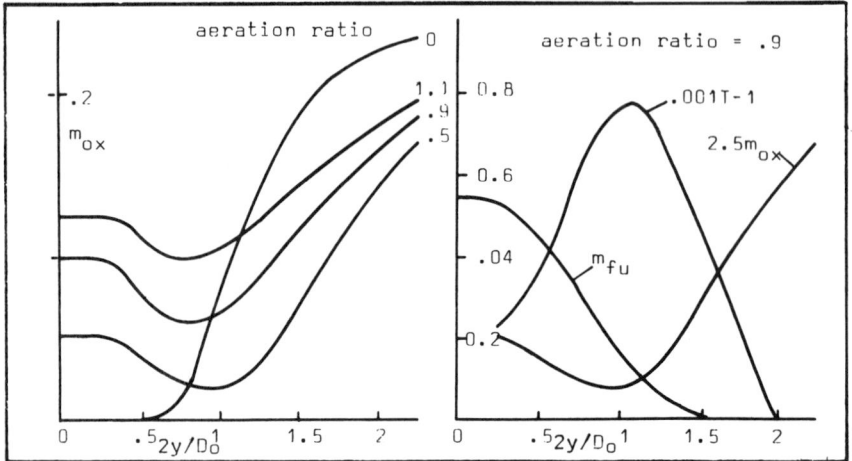

SLIDE 9. The turbulent bunsen burner, 2: Cross-stream profiles at variables at x/D_0 = 3.

turbulent combustion has succeeded in its main task, namely that of representing both premixed and nonpremixed combustion with a single formulation. The results of the computations are plausible in their major features; and agreement with some experimental data has been claimed.

The computations are easy to perform; each flame requires only a few seconds of central-processor time on a CDC 6600 machine.

It must, however, be emphasized that no extensive comparisons with experiment have been made of the predictions based upon the new model. Nor, when they are made, is it to be expected that *very* good agreement will be observed over the whole range of conditions. In particular, as the last

---

ACHIEVEMENTS:

● THE NEW MODEL PROVIDES A UNIFIED APPROACH TO BOTH UNMIXED AND PRE-MIXED FLAMES.
● AGREEMENT WITH EXPERIMENT CAN BE PROCURED WITH APPROPRIATE EMPIRICAL CONSTANTS.

FURTHER DEVELOPMENTS REQUIRED:

● REFINED MODELS TAKE ACCOUNT OF THE MICROSCALE OF THE TURBULENT FLUCTUATIONS, AND ITS INTERACTION WITH THE CHEMICAL KINETICS.
● ALSO CONSIDERED IS THE PROBABILITY-DENSITY FUNCTION OF THE MICROSCALE.
● THESE DEVELOPMENTS WILL PERMIT LOW-RE EFFECTS TO BE INCLUDED.

SLIDE 10. Concluding remarks.

slide implies, the influences of chemical kinetics and of Reynolds number are not at present taken into account.

Developments of a more refined model of turbulence, capable of accounting for Reynolds-number effects and much else besides, are in the course of development, along the lines indicated in Refs. 11, 12, and 13. The present model must be regarded as a simple stopgap, useful for making order-of-magnitude calculations and representing the main effects.

The model can, of course, be empirically adjusted, by choice of the constant C_{EBU} for example; and indeed this "constant" can be made a function of Reynolds number and of other nondimensional parameters. However, there will be little guidance available as to how the adjustments are to be made until the more-refined treatment is better developed.

APPENDIX

The details of the computations presented in slide panels 6, 7, 8, and 9 are as follows. The entries correspond to the major **BLOCK DATA** provisions of GENMIX. Only *deviations* from the standard KASE (KIND = 1) of Ref. 7 are here indicated; and obvious ones (e.g., absence of inner and outer tubes) are omitted.

- The number of cross-stream grid intervals is 20.
- The grid extends at $x = 0$ from $y = .009$ to $y = .01$m. The latter is the nozzle-lip radius.
- The fuel is propane.
- The mixing length is taken as 0.0475 the mixing-layer thickness before the end of the potential core, and 0.065 this thickness after it.
- The injection velocity is 61 m/sec.
- The propane is mixed with combustion products. Its mass fraction at the nozzle is 0.085, and its temperature is $1300°K$.
- The air is about $300°K$. When air is premixed with the fuel and combustion products, the temperature of the mixture is that resulting from adiabatic mixing of streams at $1300°K$ and $300°K$.

NOMENCLATURE

C_{EBU} eddy-breakup coefficient
D_o nozzle diameter
f mixture fraction, i.e., mass fraction of material emanating from fuel stream, regardless of state of chemical aggregation
f_+, f_- maximum and minimum falues of f at a point during fluctuations
$\bar{f}$ time-average value of f
m mass-fraction
m_{fu} mass fraction of fuel
$m_{fu,b}$ m_{fu} for completely-burned mixture

$m_{fu,u}$ m_{fu} for completely-unburned mixture
m_{ox} mass fraction of oxygen
S_ϕ source of ϕ
r part of reaction-rate expression
t time
T temperature
u velocity
u_0 velocity of injected fuel
y radius
y_{Tmax} radius where T has maximum value
x longitudinal distance
α fraction of time in f_- state
Γ exchange coefficient
ρ density
ϕ general fluid variable obeying a transport equation

REFERENCES

1. Patankar, S. V. and D. B. Spalding: "Heat and mass transfer in boundary layers." Intertext Books, London, 2d Edition, 1970.
2. Gosman, A. D., W. M. Pun, A. K. Runchal, D. B. Spalding, and M. Wolfshtein: "Heat and mass transfer in recirculating flows." Academic Press, London, 1969.
3. Patankar, S. V. and D. B. Spalding: "A calculation procedure for heat, mass and momentum transfer in three-dimensional parabolic flows." Int. J. of Heat and Mass Transfer, 15, pp. 1787–1806.
4. Caretto, L. S., R. M. Curr, and D. B. Spalding: "Two numerical methods for three dimensional boundary-layers." Computer Methods in Applied Mechanics & Engineering, 1, pp. 39–57, 1972.
5. Caretto, L. S., A. D. Gosman, S. V. Patankar, and D. B. Spalding: "Two calculation procedures for steady, three-dimensional flows with recirculation." Proceedings of the Third International Conference on Numerical Methods in Fluid Mechanics, II, pp. 60–68. Edited by J. Ehlers, K. Hepp, and H. A. Weidenmüller, Springer-Verlag, Heidelberg, 1973.
6. Harlow, F. H. and A. A. Amsden: "Numerical calculation of almost incompressible flow." J. Computational Physics, 3, 1, 1968.
7. Spalding, D. B.: "GENMIX: A general computer program for two-dimensional parabolic phenomena." Imperial College, Mechanical Engineering Department Report HTS/75/17, 1975.
8. Launder, B. E. and D. B. Spalding: "Mathematical models of turbulence," Academic Press, London and New York.
9. Spalding, D. B.: "Concentration fluctuations in a round turbulent free jet." Chemical Engineering Science, 26, pp. 95–107.
10. Elghobashi, S. E.: "Characteristics of gaseous turbulent diffusion flames in cylindrical chambers—a theoretical and experimental investigation." Ph.D. Thesis, University of London. Imperial College, London, Mechanical Engineering Department.
11. Spalding, D. B.: "Mathematical models of turbulent flames: a review." Imperial College, London, Mechanical Engineering Department Report HTS/75/1, 1975.
12. Spalding, D. B.: "Heat and mass transfer in rivers, bays, lakes and estuaries." Imperial College, London, Mechanical Engineering Department Report HTS/76/7, 1976.
13. Spalding, D. B.: "Further development of the eddy-break-up model of turbulent combustion." Imperial College, London, Mechanical Engineering Department Report HTS/76/9, 1976.
14. Vulis, L. A., Sh. A. Ershin, and L. P. Yarin: "Fundamentals of the theory of the gaseous diffusion flame." Energiya Press, Leningrad, 1968.

TURBULENT FLOW IN COILED TUBES AND THE INFLUENCE OF PRANDTL NUMBER ON ASSOCIATED HEAT CONVECTION*

R. Schiestel and J. Gosse
Conservatoire National des Arts et Métiers, Paris

I. INTRODUCTION

A mathematical model of turbulence [1], [3] involving a set of partial differential transport equations for double velocity correlations and for two effective length scales, constituted the basis for various numerical applications to classical turbulent flows [2], [3], [6]. The level of closure utilized, permits a more general modeling of Reynolds stresses, the model being free from limitations that a turbulent viscosity hypothesis would imply. Thus, more complex flow configurations can be studied. In particular, numerical predictions of turbulent convection in annular ducts [4], [5] have shown that the model allows the treatment of turbulent flows having a strong dissymetry. Using the same mathematical approach, the case of a circular tube with constant curvature is considered, corresponding to the practical case of a helically coiled tube with small pitch compared to its radius of curvature. The influence of curvature and Prandtl number on turbulent convection is examined.

2. PRESENTATION OF THE THREE-DIMENSIONAL MODEL

2.1 The Mean Flow

The detailed turbulent motion of the fluid is described by the Navier-Stokes equations, and the mean flow is governed by the following well-known Reynolds equations:

$$\partial_t \bar{v}_i + \bar{v}^j \bar{v}_{i,j} = -\bar{\pi}_{,i} + \nu g^{jk} \bar{v}_{i,jk} - R^j_{i,j} \tag{1}$$

with
$$\bar{v}^j_{,j} = 0 \tag{2}$$

---

*The abstract of this paper appears on page 352.

2.2 The Reynolds-Stress Equations

The general equations of Reynolds stresses are directly derived from the equations of fluctuating velicities by elementary methods

$$\partial_t R_{ij} + \bar{v}^k R_{ij,k} = -R_i{}^k \bar{v}_{j,k} - R_j{}^k \bar{v}_{i,k} + \overline{\pi'(v'_{i,j} + v'_{j,i})} - T^k_{ij,k}$$
$$- \overline{(v'_i \pi')}_{,j} - \overline{(v'_j \pi')}_{,i} - E_{ij} + vg^{kl} R_{ij,kl} \quad (3)$$

2.3 Closure Assumptions

The unknown correlations appearing on the right hand side of the transport equations are based on two kinds of general assumptions providing approximations of turbulence correlations as empirical functions of mean values. For diffusive type correlations, we assume:

$$\overline{f'v'_i} = -A_f L \sqrt{q} \bar{f}_{,i} \quad (4)$$

and for dissipative correlations, a second form of assumption is formulated:

$$2vg^{ij} \overline{f'_{,i} h'_{,j}} = \frac{v \overline{f'h'}}{s^2} \quad (5)$$

where L and s represent, respectively a turbulence macroscale and a turbulence microscale.

The particular hypothesis referring to the Reynolds stress equations has to take into account the necessary symmetry in indices.

$$T_{ij}{}^k = -A_1 g^{kl} L \sqrt{q} (R_{ij,l} + R_{jl,i} + R_{il,j}) \quad (6)$$

$$\overline{v'_i \pi'} = -B_1 L \sqrt{q} \, q_{,i} \quad (7)$$

$$E_{ij} = \frac{v R_{ij}}{s^2} \quad (8)$$

Modeling the pressure-velocity gradient correlation includes several contributions:

$$\overline{\pi'(v'_{i,j} + v'_{j,i})} = P_{ij}{}^{(1)} + P_{ij}{}^{(2)} + P_{ij}{}^{(3)} \quad (9)$$

$$P_{ij}{}^{(1)} = -\frac{\varsigma}{2} \frac{\sqrt{q}}{L} F (R_{ij} - \tfrac{2}{3} q g_{ij}) \quad (10)$$

with
$$F = \frac{A_0 L \sqrt{q}}{v + A_0 L \sqrt{q}} \quad (11)$$

$$P_{ij}^{(2)} = -\eta \frac{\nu}{q^2} q_{,k}(R_i{}^k q_{,j} + R_j{}^k q_{,i} - \tfrac{2}{3} R^{kl} q_{,l} g_{ij}) \tag{12}$$

$$P_{ij}^{(3)} = -\gamma q(\bar{\nu}_{i,j} + \bar{\nu}_{j,i}) - \alpha F(R_i{}^l \bar{\nu}_{j,l} + R_j{}^l \bar{\nu}_{i,l})$$
$$- \gamma F(R_i{}^l \bar{\nu}_{l,j} + R_j{}^l \bar{\nu}_{l,i}) + \tfrac{2}{3}(\alpha + \gamma) FR^{kl} \bar{\nu}_{k,l} g_{ij} \tag{13}$$

$P_{ij}^{(1)}$ is a term analogous to the one already proposed by Rotta [16] and expresses the tendency to isotropy of turbulence agitation, while $P_{ij}^{(2)}$ and $P_{ij}^{(3)}$ represent, respectively, wall effects (in a low intensity turbulence) and corrective terms suggested by the Poisson equation of the fluctuating pressure. The transport equaiton for R_{ij} can the be written as follows:

$$\partial_t R_{ij} + \bar{\nu}^k R_{ij,k} = -R_j{}^k \bar{\nu}_{i,k} - R_i{}^k \bar{\nu}_{j,k} + P_{ij}^{(1)} + P_{ij}^{(2)} + P_{ij}^{(3)} - \frac{\nu R_{ij}}{s^2}$$
$$+ A_1[g^{kl} L \sqrt{q}(R_{il,j} + R_{jl,i} + R_{ij,l})]_{,k}$$
$$+ B_1[(L \sqrt{q}\, q_{,i})_{,j} + (L \sqrt{q}\, q_{,j})_{,i}] + \nu g^{kl} R_{ij,kl} \tag{14}$$

2.4 Evolution Equation of the Turbulence Microscale

The classical transport equation of turbulence energy decay rate $E = g^{ij} E_{ij}/2$ utilized by Hirt [17], Daly and Harlow [18] and then by Hanjalic and Launder [19] provided a basis for microscale calculations. Similar closure relations, arising from hypothesis (4) and (5) are introduced, and taking into account the contraction of (8) they lead to the following equation.

$$\partial_t s^2 + \bar{\nu}^j(s^2)_{,j} = \omega \frac{s^2}{q} R^{ij} \bar{\nu}_{i,j} + A_s s^4 \left[\frac{L\sqrt{q}\, g^{ij}(s^2)_{,i}}{s^4}\right]_{,j}$$
$$- A_{sq} \frac{s^4}{q} \left(\frac{L\sqrt{q}}{s^2} g^{ij} q_{,i}\right)_{,j} + \nu \Delta + a\nu g^{ij} s^4 \left[\frac{(s^2)_{,i}}{s^4}\right]_{,j} \tag{15}$$

2.5 Evolution Equation of the Turbulence Macroscale

The macroscale was defined by the relationship

$$L(\vec{x}, t) \sim \frac{1}{q} \int_0^\infty \frac{E(k, \vec{x}, t)}{k}\, dk \tag{16}$$

an expression already proposed by Rotta [16], where $E(k, \vec{x}, t)$ is the mean energy spectrum. A phenomenological equation for L can then be constructed from the evolution equation of $E(k, \vec{x}, t)$:

$$\partial_t L + \bar{\nu}^j L_{,j} = \beta \sqrt{q} + A_L(L \sqrt{q}\, g^{ij} L_{,i})_{,j} - \beta \frac{\nu L}{s^2} + b\nu g^{ij} L_{,ij} \tag{17}$$

2.6 Turbulent Heat Transfer

A transport equation for mean enthalpy is obtained by a classical statistical treatment of the instantaneous equation of total energy balance. Noticing that, in the cases under consideration, the thermal problem is largely determined by diffusional fluxes, the heat transfer equation is written in a simplified form.

$$\partial_t \bar{T} + \bar{v}^j \bar{T}_{,j} = \left(\frac{\lambda}{\rho C_p}\right) g^{ij} \bar{T}_{,ij} - g^{ij} \overline{(T'v_i')}_{,j} \tag{18}$$

with the hypothesis

$$\overline{T'v_i'} = -D_t \bar{T}_{,i} \quad \text{and} \quad D_t = \frac{A_0 s^2 q}{\nu} \tag{19}$$

2.7 Determination of Numerical Constants

The values of the essential constants have been previously fixed once and for all by reference to theoretical and experimental data concerning typical flows of different structures (grid turbulence, wall turbulence, uniform shear flow). These values already used in Eq. (5) are utilized here:

$A_1 = 0.10$	$B_1 = 0.01$	$A_s = 0.18$	$A_{sq} = 0.05$	$A_L = 0.08$
$\omega = 0.532$	$\zeta = 5.00$	$\beta = 0.40$	$\Delta = 1$	if $\epsilon_t/\nu > 0.92$
$A_0 = 0.175$	$a = 0.80$	$b = 2.40$	$\Delta = 0.40$	if $\epsilon_t/\nu < 0.92$
$\gamma = 0.30$	$\alpha = 0.30$	$\chi = 0.10$	$\eta = 2.50$	$A_t = 0.175$

3. APPLICATION TO HELICALLY COILED TUBES

3.1 Formulation of the Fundamental Equations

The fundamental transport equations (1), (14), (15), (17), and (18) have been treated in a (r, θ, z) frame, r and θ representing radial and angular position ($\theta = 0$ referring to the centripetal direction) and z is the curvilinear abscissa along the tube axis.

The flow is treated utilizing boundary layer approximation assuming that the diffusion fluxes in the z direction are negligibly small compared with transverse fluxes. The hypothesis of a large radius of curvature $\mathfrak{R}$ relative to the pipe radius h allows us to assume that the effect of curvature is essentially taken into account by adding the centrifugal forces in the mean momentum equations formulated in cylindrical coordinates.

$$\frac{1}{r}\left(\frac{\partial r \bar{v}_r^2}{\partial r} + \frac{\partial \bar{v}_r \bar{v}_\theta}{\partial \theta} + \frac{\partial r \bar{v}_r \bar{v}_z}{\partial z} - \bar{v}_\theta^2\right) + \frac{\cos\theta}{\mathfrak{R}}\bar{v}_z^2 = -\frac{\partial \bar{\pi}}{\partial r}$$

$$-\left(\frac{\partial R_{rr}}{\partial r} + \frac{1}{r}\frac{\partial R_{r\theta}}{\partial \theta} + \frac{R_{rr} - R_{\theta\theta}}{r}\right) + \nu\left[\frac{\partial}{\partial r}\left(\frac{1}{r}\frac{\partial r \bar{v}_r}{\partial r}\right) + \frac{1}{r^2}\frac{\partial^2 \bar{v}_r}{\partial \theta^2} - \frac{2}{r^2}\frac{\partial \bar{v}_\theta}{\partial \theta}\right]$$

$$\frac{1}{r}\left(\frac{\partial r \bar{v}_r \bar{v}_\theta}{\partial r} + \frac{\partial \bar{v}_\theta^2}{\partial \theta} + \frac{\partial r \bar{v}_\theta \bar{v}_z}{\partial z} + \bar{v}_r \bar{v}_\theta\right) - \frac{\sin \theta}{\mathfrak{R}} \bar{v}_z^2 = -\frac{1}{r}\frac{\partial \bar{\pi}}{\partial \theta}$$

$$-\left(\frac{\partial R_{r\theta}}{\partial r} + \frac{1}{r}\frac{\partial R_{\theta\theta}}{\partial \theta} + 2\frac{R_{r\theta}}{r}\right) + \nu\left[\frac{\partial}{\partial r}\left(\frac{1}{r}\frac{\partial r \bar{v}_\theta}{\partial r}\right) + \frac{1}{r^2}\frac{\partial^2 \bar{v}_\theta}{\partial \theta^2} + \frac{2}{r^2}\frac{\partial \bar{v}_r}{\partial \theta}\right]$$

$$\frac{1}{r}\left(\frac{\partial r \bar{v}_r \bar{v}_z}{\partial r} + \frac{\partial \bar{v}_\theta \bar{v}_z}{\partial \theta} + \frac{\partial r \bar{v}_z^2}{\partial z}\right) = -\frac{\partial \bar{\pi}}{\partial z} - \left(\frac{\partial R_{rz}}{\partial r} + \frac{1}{r}\frac{\partial R_{\theta z}}{\partial \theta} + \frac{R_{rz}}{r}\right)$$

$$+ \nu\left[\frac{1}{r}\frac{\partial}{\partial r}\left(r\frac{\partial \bar{v}_z}{\partial r}\right) + \frac{1}{r^2}\frac{\partial^2 \bar{v}_z}{\partial \theta^2}\right] \quad (20)$$

The other equations, including transport of Reynolds stresses, scales and temperature evolution, are treated in classical cylindrical coordinates without additional terms.

3.2 Numerical Procedure

The different equations of the problem are formulated in a nondimensional form by introducing dimensionless groups constructed with h and ν. A previous calculation dealing with a case of curved pipe with R/h ratio of 40 and Re = 25,000 has shown that in the region $10 < (yu_*/\nu) < 30$, near the wall, the logarithmic velocity profiles, for various θ directions, remain sufficiently close one to another. Based on the approximate verification of a universal log-law, we have adopted the method of the wall functions recommended by Patankar and Spalding [20] because of its economy in computer time and storage. Boundary conditions are not taken on the actual wall surface, but near the wall, at a point located at $yu_*/\nu \simeq 20$. These conditions express the joining of the computed profiles to a constant shear stress layer.

The numerical method utilized is derived from the three-dimensional procedure of Patankar and Spalding [7] and the S.M.A.C. method developed by Amsden and Harlow [10]. The finite difference grid presented 15 equal angular divisions of the semicircumference and 15 regular divisions of the radius. The forward step of the marching procedure was chosen equal to 0.5 h and one iteration required about 0.4 s on a CDC 7600 computer.

3.3 Scope

We have considered four cases corresponding to relative curvatures $h/R = 0 -$ $1/18.7 - 1/40 - 1/100$. The Reynolds number Re = 2 hV/ν relative to the mean velocity V has been choosen equal to 25000 and the Prandtl number has varied from a liquid metal one up to middle and high values : $0.015 - 0.7 - 1 - 5.7 - 10 - 100 - 1000$. Initial conditions introduced at $z = 0$ correspond to uniform velocity and temperature profiles. Then, the marching procedure gives the developing-flow and developing-heat transfer solutions in a long coiled pipe. The condition of constant wall temperature only has been considered since it is easier to apply.

3.4 Dynamical Results

The emergence of secondary flow in curved pipes causes frictional resistance to increase. The augmentation of friction factors is found to be in fair agreement with those given by the empirical correlation of H. Ito [11].

$$\lambda_{*c}\left(\frac{R}{h}\right)^{1/2} = 0.029 + 0.304\left[\mathrm{Re}\left(\frac{h}{R}\right)^2\right]^{-0.25} \tag{21}$$

valid for

$$0.034 < \mathrm{Re}\ \frac{h}{R}^2 < 300$$

Calculated $\lambda_{*c}/\lambda_{*s}$ ratios where λ_{*c} stands for the straight tube value are compared in the table below, with corresponding ratios of λ_{*c} (given by H. Ito) to λ_{*s} (Blasius).

R/h	$\mathrm{Re}(h/R)^2$	$(\lambda_{*c}/\lambda_{*s})$calc.	$(\lambda_{*c}/\lambda_{*s})$H. Ito
100	2.5	1.07	1.08
40	15.6	1.13	1.16
18.7	71.7	1.20	1.25

Figure 1 provides a comparison of the predicted longitudinal velocity profiles with the experimental results of Mori and Nakayama [12] for a 1/40 curvature radius ratio. The distribution of turbulent shear stress R_{rz} plotted on the same figure indicates that the point of zero shear stress does not coincide with the maximum of the velocity profile. It has been noticed that a higher degree of curvature not only increases the dissymetry of the flow but causes a steepening of mean profiles near the wall and widens the central concavity of the $(\theta = \Pi/2)$ curve.

3.5 Heat Transfer Results

Various calculated temperature profiles for different Prandtl numbers are given in Fig. 2. Not as good an agreement is obtained with the experimental measurements of Mori and Nakayama taken in an air flow. In fact, the Mori and Nakayama experiments have a constant heat flux boundary condition while the present calculations assume a constant wall temperature. Nevertheless, the influence of wall boundary conditions on mean Nusselt numbers is known to be especially sensitive at low Prandtl numbers [15]. The conjugated influence of Prandtl number and curvature is illustrated in Fig. 3 in terms of

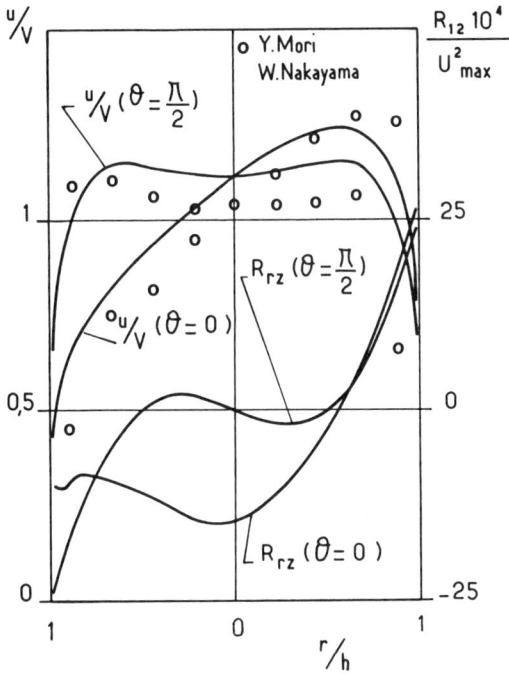

FIG. 1. Velocity profiles.

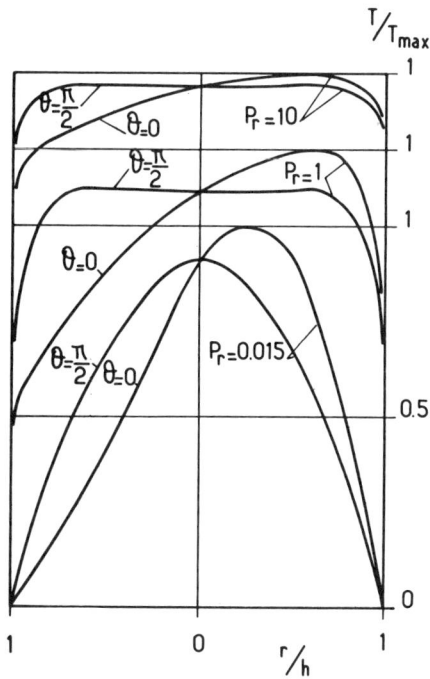

FIG. 2. Temperature profiles.

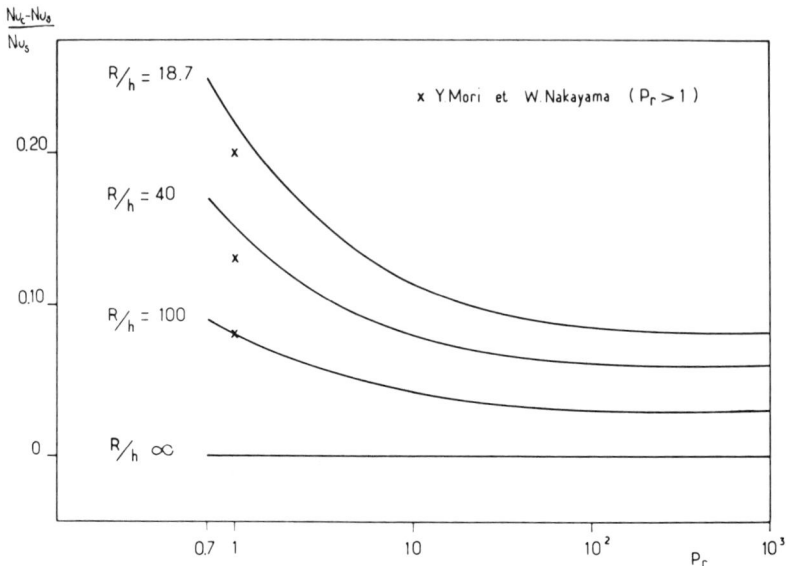

FIG. 3. Relative Nusselt numbers.

$(\text{Nu}_c - \text{Nu}_s)/\text{Nu}_s$ values. It is seen that the ratios $(\text{Nu}_c - \text{Nu}_s)/\text{Nu}_s$ for a Prandtl number near unity reach a value comparable with the friction factor ratio $(\lambda_{*c}/\lambda_{*s})$. Moreover, the heat transfer increase due to curvature is more important for Prandtl numbers near unity. A qualitatively similar behavior was obtained by the analytical study of Mori and Nakayama who predicted a slight decrease of $(\text{Nu}_c/\text{Nu}_s)$, when the Pr increases, according to the relation $\text{Nu}_c/\text{Nu}_s \propto \text{Pr}^{2/3}/(\text{Pr}^{2/3} - 0.075)$. The authors found a rapid convergence of Nusselt ratios to a constant limit for $\text{Pr} > 1$. These values inferred from the analytic results of Mori and Nakayama are plotted on Fig. 3 for comparison at $\text{Pr} = 1$. However, our calculated results show a stronger influence of Prandtl number with asymptotic values not reached before $\text{Pr} \simeq 100$.

For liquid metal heat transfer ($\text{Pr} = 0.015$) we have found a sharp decrease of calculated $(\text{Nu}_c/\text{Nu}_s)$ ratios : 1.020, 1.055, 1.105 for R/h values of 100, 40, 18.7, respectively. In this case, thermal molecular conductivity is so important that additional transfer due to recirculation becomes less important.

On Fig. 4 are plotted local values of friction factors and Nusselt numbers around the circumference. Peripheral variations increase with curvature and are more important at low Prandtl numbers.

4. CONCLUDING REMARKS

The example which was presented of thermal convection in curved ducts allows an illustration of the capabilities of the three-dimensional model for the

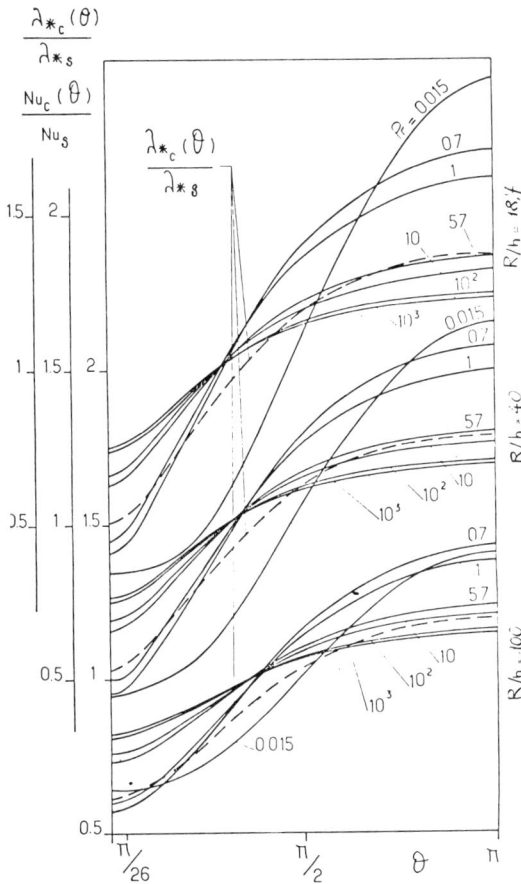

FIG. 4. Peripheral variations of friction factors and Nusselt numbers.

calculation of recirculating flows in complex geometries, a description of the turbulence field, and the evaluation of wall friction and heat transfer. The introduction of wall functions in the numerical procedure, although altering the precision of the results, permits a great economy in computer time. The present example constitutes an illustration of the possible use that the calculation procedure is able to yield in industrial fields.

NOMENCLATURE

$A_f, A_0, A_1, B_1, A_S, A_{sq}, A_L, \omega, a, b, \Delta, \beta, \zeta, \eta, \alpha, \chi, \gamma$	numerical constants
D_t	thermal eddy diffusivity
$E(k, \vec{x}, t)$	mean energy spectrum

E_{ij}	Reynolds stress decay rate $(\nu\, g^{kl}\overline{v'_{i,k}v'_{j,1}})$
g_{ij}	metric tensor
$\bar{f}, f', \bar{h}, h'$	turbulence quantities, mean value and fluctuation
h	pipe radius
k	wave number
L	turbulence macroscale
Nu_c	mean Nusselt number of curved pipe
$\mathrm{Nu}\,(\theta)$	peripheral Nusselt number $(2h/(T_w - T_f)(\partial T/\partial r)_w)$
Nu_s	mean Nusselt number of straight pipe
Pr	Prandtl number
q	mean turbulence energy $(g^{ij}\overline{v'_i v'_j}/2)$
r	radial coordinate
R_{ij}	Reynolds tensor $(\overline{v'_i v'_j})$
R	Radius of curvature
Re	Reynolds number $(2h\,V/\nu)$
s	turbulence microscale
$\overline{T}$	absolute temprature
T_f	bulk temperature
T_w	wall temperature
T_{ijk}	triple velocity correlation $(\overline{v'_i v'_j v'_k})$
$\bar{v}_i, v'_i$	mean velocity and fluctuation
$\bar{v}_r$	radial mean velocity
$\bar{v}_\theta$	circumferential mean velocity
$\bar{v}_z$	longitudinal mean velocity
$v_*(\theta)$	friction velocity $((h\,\partial\bar{v}_z/\partial r)^{\frac{1}{2}}_{\mathrm{wall}})$
V	average flow velocity
z	longitudinal coordinate
λ	thermal conductivity
λ_*	mean friction factor
$\lambda_*(\theta)$	peripheral friction factor $(8(v_*(\theta)/V)^2)$
θ	angular coordinate
ν	kinematic viscosity
Π	pressure density ratio

REFERENCES

1. Gosse, J. and R. Schiestel: C.R. Acad. Sci. Paris, 1972, t. 275, série A, p. 471.
2. Schiestel, R. and J. Gosse: *Ibid.,* p. 1371.
3. Schiestel, R.: Thèse, Nancy, 1974, C.N.R.S. No. A.0.10596.
4. Schiestel, R. and J. Gosse: C.R. Acad. Sci. Paris, 1974, t. 279, serie B, p. 543.
5. Gosse, J. and R. Schiestel: Int. J. Heat Mass Transfer, 1975, **18**, No. 6, p. 743.
6. Gosse, J. and R. Schiestel: 5th International Heat Transfer Conference, Tokyo, Sept. 3-7 (1974).
7. Patankar, S. V. and D. B. Spalding: Int. J. Heat Mass Transfer, 1972, **15**, p. 1787.
8. Patankar, S. V., V. S. Pratap, and D. B. Spalding: J. of Fluid Mechanics, 1975, **67**, pt. 3, p. 583.
9. Patankar, S. V., V. S. Pratap, and D. B. Spalding:*Ibid.,* 1974, **62**, pt. 3, p. 539.
10. Amsden, A. A. and F. H. Harlow: Los Alamos Scientific Laboratory report LA-4370 (1970).
11. Ito, H.: Trans. ASME, J. Basic Engng., 1959, **82**, p. 123.
12. Mori, Y. and W. Nakayama: Int. J. Heat Mass Transfer, 1967, **10**, p. 37; **10**, p. 681.
13. Seban, R. A. and E. F. McLaughlin: Int. J. Heat Mass Transfer, 1963, **6**, p. 387.
14. Rogers, G. F. C. and Y. R. Mayhew: *Ibid.,* 1964, 7, p. 1207.
15. Siegel, R. and E. M. Sparrow: Trans. ASME, J. Basic Engng, 1960, **82**, p. 152.

16. Rotta, J. C.: Zeitschrift für Physik, 1951, **129,**, p. 547; **131,** p. 51.
17. Hirt, C. W.: Int. Seminar of the Int. Centre for Heat and Mass Transfer Herceg Novi, Yugoslavia, September 1969.
18. Daly, B. J. and F. H. Harlow: *Physics Fluids,* 1970, **13,** 11, p. 2364.
19. Hanjalic, K. and B. E. Launder: *J. of Fluid Mech.,* 1972, **52,** pt. 4, p. 609.
20. Patankar, S. V. and D. B. Spalding: Heat and Mass Transfer in Boundary Layers, Intertext Books, London (1970).

INTERACTION OF MOLTEN TIN WITH WATER: EFFECT OF TEMPERATURE STRATIFICATION IN THE COOLANT *

V. H. Arakeri, I. Catton, and W. E. Kastenberg
Energy and Kinetics Department, University of California
Los Angeles

As is well known, laboratory scale experimental studies [1, 2] show that thermal interaction between molten tin and water can result in "vapor explosions" accompanied by fragmentation of the hot material providing certain thermal constraints are met. The objective of the present experimental study is to illustrate that the required thermal constraints to produce vapor explosions can be relaxed through stable temperature stratification (hot to cold in the direction of gravity) of the coolant medium.

Multiflash photographs of the interaction between twenty-five grams of molten tin dropped into a water bath were obtained with two General Radio stroboscopes as light sources. The principal advantage of using this type of photographic technique was in obtaining very sharp and clear pictures in each frame due to the extremely short duration (of approximately one micro-second) flash produced by the stroboscopes. Figure 2 illustrates a typical sequence of photographs obtained by the present technique. In addition, the thermal interaction was sensed with a pressure transducer located about 25 cm away. Details of the experimental techniques may be found in [3].

For tests with water bath at uniform temperatures ranging from 8°C to 52°C the following observations were made. At the two highest initial tin temperatures of 787°C and 676°C consistent vapor explosions accompanied by extensive fragmentation was observed. At an initial tin temperature of 537°C only occasional explosive interaction was detected but consistent fragmentation in porous mossy form was observed. At the two lowest tin temperatures of 426°C and 343°C no explosive interaction was detected and very little fragmentation was observed. On the contrary, with temperature stratification in the coolant bath (45.5°C to 21.5°C in a distance of about 15 cm) explosive interactions were detected at initial tin temperatures of as low as 343°C. The measured peak pressure magnitudes at various initial tin

---

*The abstract of this paper appears on page 363.

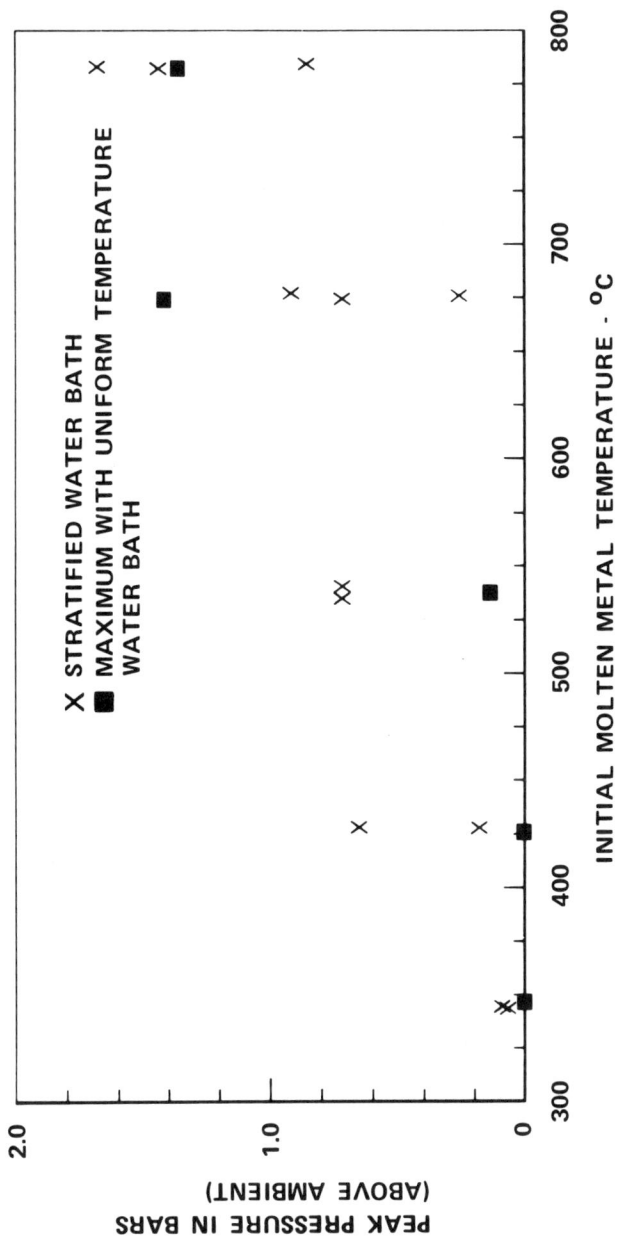

FIG. 1. Peak pressure versus initial tin temperature with and without thermal stratification in the coolant.

103

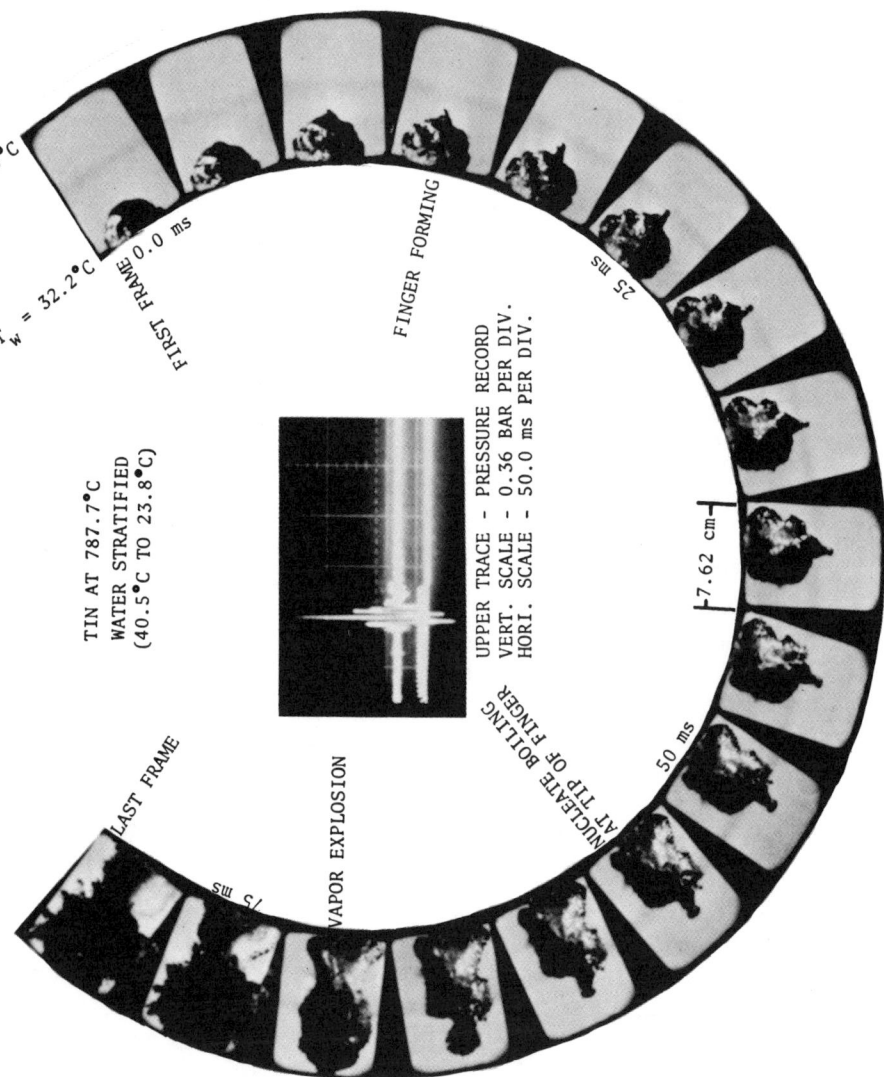

FIG. 2. Sequence of multiflash photographs illustrating the development of transition or

$T_w = 23.8°C$

$T_w = 32.2°C$

FIRST FRAME 0.0 ms

FINGER FORMING

25 ms

TIN AT 787.7°C
WATER STRATIFIED
(40.5°C TO 23.8°C)

UPPER TRACE - PRESSURE RECORD
VERT. SCALE - 0.36 BAR PER DIV.
HORI. SCALE - 50.0 ms PER DIV.

7.62 cm

50 ms

NUCLEATE BOILING
AT TIP OF FINGER

LAST FRAME

75 ms

VAPOR EXPLOSION

temperatures with thermally stratified water bath are shown in Fig. 1. Also shown in the same figure are similar measurements with uniform temperature water bath tests. It is clear from the comparison afforded in Fig. 1, that the effect of temprature stratification in the coolant is to reduce the threshold temperature required for explosive interaction between molten tin and water. In fact, the detected pressure pulse with initial tin temperature of 343°C shows that explosive interactions between two liquids are not only as a consequence of superheating of the coolant to superheat limits associated with spontaneous nucleation [4]. This follows from the observation that, the calculated interface temperature of molten tin at 343°C in contact with a coolant at 45.5°C (maximum temperature within stratification) is about 20° below the homogeneous nucleation temperature of water.

The effect of thermal stratification in the development of vapor explosions may be likened to the collapse of cavitation bubbles as they pass from low pressure regions to high pressure regions. Thus, the thermal stratification may facilitate destabilization of an existing vapor film to promote transition or nucleate boiling. Some evidence for such a happening is presented in the sequence of photographs shown in Fig. 2. The explosive interaction at the finger-like protrusion of molten metal is observed to occur after transition or nucleate boiling on the surface has been established.

REFERENCES

1. Board, S. J., C. L. Farmer, and D. H. Poole: International Journal of Heat and Mass Transfer, 1974, 17, 331.
2. Bjornard, T. A., W. M. Rohsenow, and N. E. Todreas: Transactions American Nuclear Society, 1974, 19, 247.
3. Arakeri, V. H., I. Catton, W. E. Kastenberg, and M. S. Plesset: UCLA Eng. Report No. 7592, 1975, University of California, Los Angeles.
4. Cole, R.: Advances in Heat Transfer, 10, Academic Press, 1974.

EXPERIMENTAL INVESTIGATION OF CONVECTIVE BOILING OF AMMONIA AT HIGH PRESSURES*

G. Barthau
University of Stuttgart, Stuttgart, FRG

Heat transfer in forced convection boiling is more difficult to predict than pool boiling heat transfer. Experimental investigations have been done mostly for water and organic liquids. With comparable reduced pressures, the heat-transfer coefficients of water are about twenty times as high as for organic liquids.

Measurements were made with different pressures and a fixed combination of heating-surface/fluid to investigate the influence of thermodynamic and transport properties on heat transfer. Because of the large variations in thermodynamic and transport properties, the region of high subcritical pressures is of special interest.

Ammonia (p_{k_r} = 113.5 bar, T_{kr} = 405.4 K), for which we have consistent thermodynamic and transport properties [1] was selected as the test fluid. It is comparable to water with respect to its high heat of evaporation and its high polarity.

Short Description of the Experimental Apparatus (Fig. 1)

The experimental apparatus (exactly described in [2], [3]) is a closed loop (V_{total} = 0.320 m^3). The pressure in the loop is produced thermally. The mass content in the loop during experiments was m_{total} = 80 kg.

The test fluid is carried from the pump a through the flow-meter b, and the preheater c to the test section. Between the preheater and the test section, part of the fluid is drawn off and flows through flow-meter e to evaporator f. The ammonia vapor (needed for pressure production in the system) goes from the evaporator to the separator g. The NH$_3$-mean-flow comes from the test section through a viewing station h and to the separator g. From the separator, the NH$_3$ vapor flows to the condenser i and from there (liquified)

---

*The abstract of this paper appears on page 364.

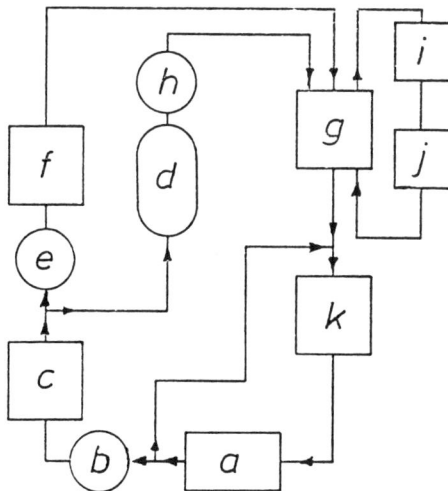

FIG. 1. Experimental loop: a-pump; b, e-flowmeter; c-preheater; d-test section; f-evaporator; g-separator; h-sight-glass; i-condensator; j-collecting vessel; k-cooler.

to the collecting vessel j. From the collecting vessel and separator, the liquid NH_3 flows through cooler k back to the pump.

By means of a bypass, part of the fluid can be drawn off behind the pump and can be carried directly to the cooler. The preheater c and the evaporator f are heated by condensing steam. The cooler k and the condenser i are cooled by water. The plant pressure is regulated by controlling cooling water flow in the condenser.

Description of Test Section

The test section is a vertical pipe with upward flow, heated from outside by pressure-controlled condensing steam. The controlling factor for heat transfer is the overall temperature difference and not the heat-flux. The material of the test section is Nickel (99.6%). The inner and outer walls of the test section are electroplated with gold (0.01 mm thick). The inner diameter of the test section is $d_i = 30.34$ mm, the wall thickness is $s = 4.83 \pm 0.02$ mm, the heated length is $l = 474$ mm, and the roughness of the heating surface is $R_p = 5 \cdot 10^{-6}$ m.

The test section begins after an undisturbed starting section of 1.5 m length. The sight-glass for the observation of flow patterns is 0.5 m behind the end of test section. Fourteen thermocouples (chromel-alumel, diameter 0.5 mm) are used for the measurement of wall temperatures. The distance between the thermocouples is 34 mm. They are arranged spirally along the test section. Additionally, in 3 temperature measurement planes, thermo-couples are arranged to allow local heat-flux measurement (see Fig. 2). The holes for the couples were made by electrical erosion (Funkenerosion).

The ammonia entrance and outlet temperatures were measured with thermocouples installed in perforated disks, which were arranged before the starting section and behind the sight-glass.

The pressure was measured between the starting section and the test section with a high-precision Bourdon pressure gauge (accuracy 0.1).

The heat transferred in the test section was determined by calculating the enthalpy difference of the steam entering the test section chamber and the leaving condensate and by weighing the condensate.

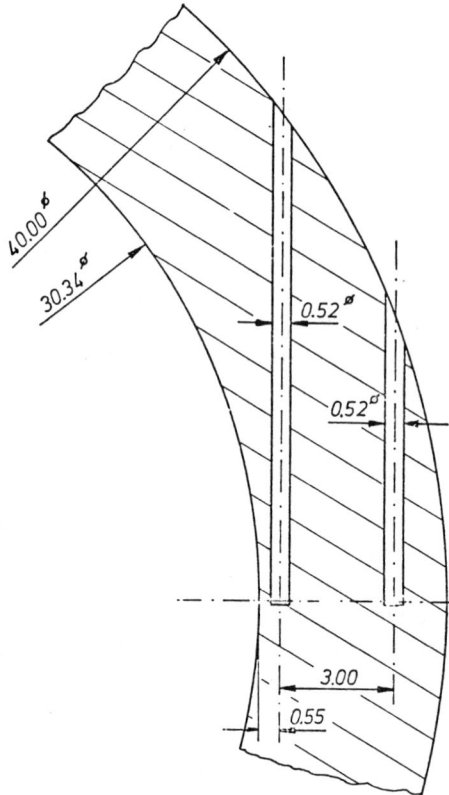

FIG. 2. Arrangement of thermocouples for measuring local heat-flux density.

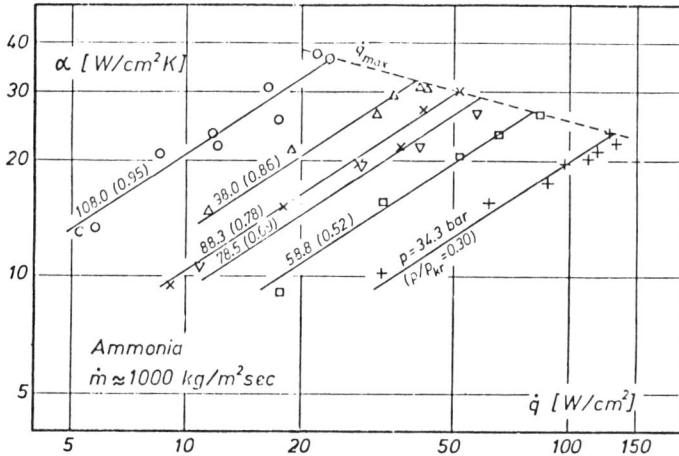

FIG. 3. Heat transfer coefficients α at various pressures p as a function of heat-flux density $\dot{q}$.

Experimental Results

After about 150 h prerunning time at pressures of 60 bar, reproducible conditions of temperature differences and heat-flux were reached in the test section. This was determined by repetitive measurements made at the end of the test program.

All series of measurements were made with decreasing heat-flux, starting from maximum heat-flux at the respective pressure.

Temperature profiles along the test section and the measured local heat-fluxes showed that dropwise condensation occurred on the heating side of the test section during all experiments. For fully developed bubble-boiling at mass velocities of $\dot{m} \approx 1000$ kg/m^2 sec, the heat-transfer coefficients shown in Fig. 3 were obtained. The fluid was always saturated at the entrance of the test section.

The heat-flux density $\dot{q}$ was determined as the arithmetic mean over the whole test section. The temperature difference for the calculation of the heat-transfer coefficient was computed as the arithmetic mean of the 14 local inner wall temperatures and the NH$_3$-outlet temperature.

The maximum heat-flux densities $\dot{q}_{max}$ shown in Fig. 3 (dashed line) are defined as those being just below the occurrence of film boiling at the outlet of the test section.

In all experiments, the flow-patterns observed behind the test section were homogeneous bubbly flow.

SYMBOLS

$\dot{m}$ mass velocity p pressure
$\dot{q}$ heat-flux density p_{kr} critical pressure
α heat-transfer coefficient

REFERENCES

1. Barthau, G. and J. Sohns: Thermische Eigenschaften von Ammoniak. Chemie-Ingenieur-Technik, 1974, **46**, 4, 149.
2. Sohns, J.: Untersuchung des konvektiven Wärmeübergangs an Ammoniak bei Zweiphasenströmung unter hohem Druck. Dissertation Universität Stuttgart, 1972.
3. Sohns, J.: Two-phased forced convective heat transfer at high subcritical pressure. Session B 4.2, 5th Int. Heat Tr. Conf. Tokyo, 1974.

POOL BOILING MAXIMUM HEAT FLUX CORRELATIONS*

J. Berghmans
Katholieke Universiteit Leuven,
Heverlee, Belgium

INTRODUCTION

A very successful approach to the determination of the maximum heat flux occurring during pool boiling has been the "hydrodynamic theory" as developed by Zuber [1] for a flat plate heater and by Sun and Lienhard [2] for a cylindrical heater. In this theory it is assumed that, for a flat plate, the distance between the vapor columns which are formed near q_{max} is equal to the critical wavelength of the plane vapor film formed over the heater during film boiling. However, the critical wavelength is the wavelength of the perturbation with zero growth rate and which thus does not grow in time. Instead the most dangerous wavelength (i.e., the wavelength with the largest growth rate) should be used.

The hydrodynamic theory furthermore assumes that the boiling crisis starts when, due to Helmholtz instability, the vapor columns are destroyed close to the heater such that a vapor blanket forms around the heater. To obtain an estimate of the vapor speed at which this occurs, it is assumed that this is equal to the critical vapor speed U_c for Helmholtz instability of a perturbation of wavelength λ occurring at the interface of a moving vapor and a stagnant liquid as discussed in [3]:

$$U_c = \sqrt{\frac{2\pi\sigma}{\rho_g \lambda}}$$

Both the assumptions of a critical speed (corresponding to a zero growth rate) and a flat interface do not seem to be realistic. Furthermore, the wavelength λ is taken to be equal to the critical wavelength ($2\pi R_c$) for instability of a fluid cylinder under the action of surface tension only (Rayleigh instability). Also this assumption is based upon a critical condition.

---

*The abstract of this paper appears on page 365.

It is the purpose of the present work to derive q_{max} correlations which are not based upon the approximate assumptions mentioned above. To do this an analysis of the Helmholtz instability of a vapor column will be performed. The results of this analysis will be applied to the determination of the maximum heat flux for spherical, cylindrical, and flat plate heaters.

STABILITY ANALYSIS

The stability problem analyzed is that of a fluid cylinder (density ρ_2) of infinite length situated in a fluid of infinite extent and of density ρ_1. A uniform flow of velocity, U, exists in the cylinder while the surrounding fluid is stagnant. Both fluids are assumed to be incompressible, immiscible, and inviscid. The effect of surface tension will be considered while gravitational effects will be neglected.

Assuming a small radial displacement of the column interface of the type

$$\alpha_n = a \, X \, [\alpha_{n-1} \, (1 - (r/R))]^m$$

then the linearized equations of motion can be solved with the boundary conditions

$$\Delta p = \sigma \left(\frac{1}{R_1} + \frac{1}{R_2} \right)$$

at the cylinder to obtain the dispersion relation

$$\alpha \omega'^2 \frac{K_m(kR)}{K_m'(kR)} - (\omega' + kR \, Wb)^2 \frac{I_m(kR)}{I_m'(kR)} - (1 - m^2 - k^2 R^2) \, kR = 0 \quad (1)$$

in which

$$\omega' = \frac{\omega}{\Omega} \qquad (2)$$

$$\Omega = \frac{\sigma}{\rho_2 R^3} \qquad (3)$$

$$\alpha = \frac{\rho_1}{\rho_2} \qquad (4)$$

$$Wb^2 = \frac{\rho_2 U^2 R}{\sigma} \qquad (5)$$

Here K_m' and I_m' denote derivatives of the Bessel functions with respect to their argument kR.

Solving for the imaginary part of ω' yields

$$\omega_i'^2 = \frac{(1 - m^2 - k^2 R^2) kR[\ I_m/I_m' - \alpha(K_m/K_m')] - \alpha k^2 R^2 Wb^2(I_m/I_m')(K_m/K_m')}{[I_m/I_m' - \alpha(K_m/K_m')]^2} \tag{6}$$

This expression shows that real values of ω_i' are possible when the right hand side of Eq. (6) is positive. This corresponds to unstable perturbations. Perturbations for which the right hand side of Eq. (6) is negative are stable. In the present case of a vapor column surrounded by a liquid, α is very large and Eq. (6) can be approximated by

$$\omega_i'^2 = -\frac{1}{\alpha} \frac{K_m'}{K_m} \left[(1 - m^2 - k^2 R^2) kR + Wb^2 k^2 R^2 \frac{I_m}{I_m'} \right] \tag{7}$$

With kR as variable, this last equation can be used to determine the maximum values which ω_i' takes for specified values of Wb and α. The dependence of $\omega_{i,max}'$ upon Wb is shown in Fig. 1 for $m = 0$ and for $\alpha = 1000$. This shows that $\omega_{i,max}'$ increases with Wb.

MAXIMUM HEAT FLUX CORRELATIONS

On approaching q_{max} the average vapor speed U_a in the vapor columns increases and thus also Wb and $\omega_{i,max}'$. The point at which the columns are destroyed, therefore, comes closer to the heater surface. Application of Bernoulli's equation at the bottom of the column, where the speed is U_0, and at a distance L downstream, where the speed is U_L, yields

$$\tfrac{1}{2}\rho_g U_L^2 = (\rho_f - \rho_g)gL + \tfrac{1}{2}\rho_g U_0^2$$

The average speed in a column of length L can then be taken as given by

$$\tfrac{1}{2}\rho_g U_a^2 = (\rho_f - \rho_g)g\frac{L}{2} + \tfrac{1}{2}\rho_g U_0^2$$

This shows the effect of gravitational acceleration upon U_a. By means of the Bond number B_c based upon the column radius R_c, this can be written in dimensionless form as

$$Wb_a^2 = B_c^2 \frac{L}{R_c} + Wb_0^2 \tag{8}$$

with
$$B_c^2 = \frac{(\rho_f - \rho_g)gR_c^2}{\sigma} \tag{9}$$

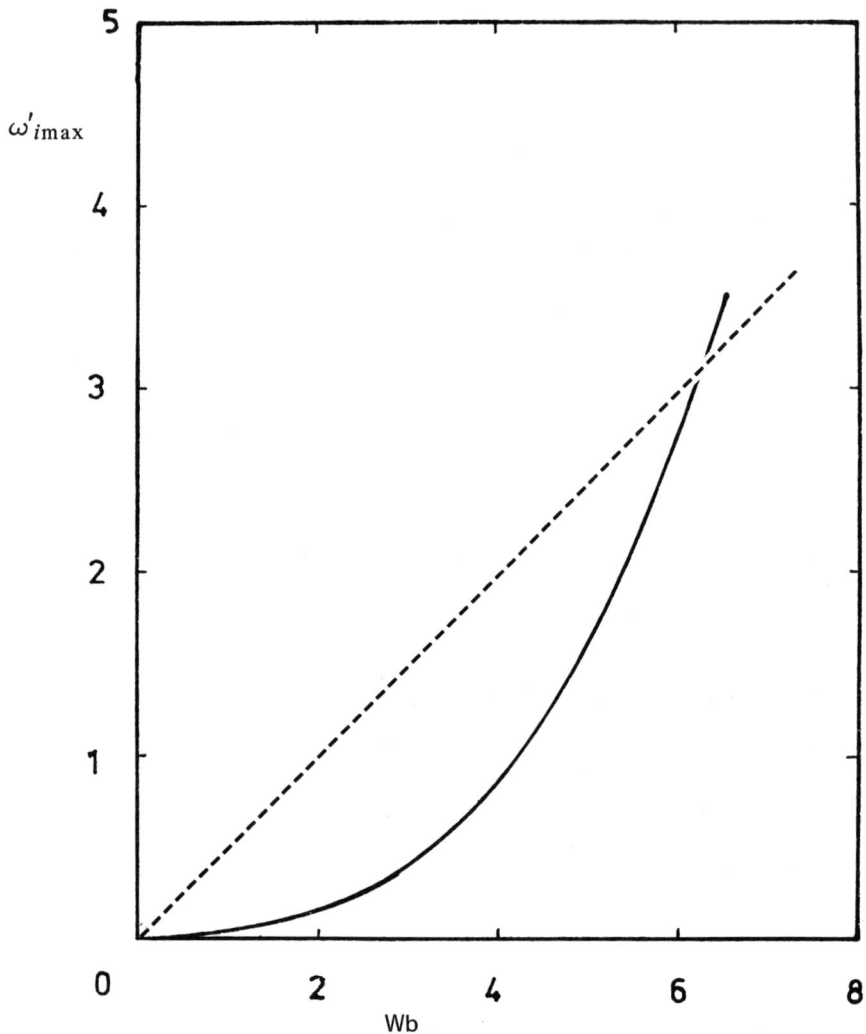

FIG. 1. —— Maximum value of the growth rate from Eq. (7); . . . . . Eq. (10).

A perturbation of amplitude βR_c at the bottom of the column (at the heater) will have reached an amplitude of $\beta R_c e^{\omega_i L/U}$ after traveling a distance L from the heater. The smallest distance L for which this amplitude is equal to the column radius is given by

$$\beta R_c \exp\left(\omega_{i,\,max}\frac{L}{U}\right) = R_c$$

If this is close enough to the heater it may be expected that film formation will occur. Putting $L = \gamma R_c$, the value of the growth rate which follows from above is

$$\omega'_{i,\,max} = \frac{\ln(1/\beta)}{\gamma}\, Wb_a$$

Reasonable values of the parameters are $\beta = 0.2$ and $\gamma = 3$, yielding:

$$\omega'_{i,\,max} = 0.5\, Wb_a \qquad (10)$$

Also this expression has been plotted on Fig. 1. From this figure it can be seen that for Helmholtz instability to be the cause of q_{max} it is necessary that the average Weber number in the vapor columns reaches the value

$$Wb_a = 6.3 \qquad (11)$$

Furthermore, Eq. (8) shows that Wb_0 should also reach this value when the Bond number is small. However, the velocity U_0 at the heater can result only from overpressures acting at the heater. It will be shown that for small Bond numbers, overpressures large enough to give rise to such large Wb_0 values are not present.

a) Small Cylindrical Heater

The maximum vapor velocity at the bottom of the vapor columns for the case of a cylindrical heater is determined by gravitational acceleration and may be estimated by

$$\tfrac{1}{2}\rho_g U_0^2 = (\rho_f - \rho_g)g R_c \qquad (12)$$

The maximum value of Wb_c is therefore

$$Wb_c = \sqrt{2}\,B\sqrt{\frac{R_c}{R}}$$

This shows that for small B the value of 6.3 cannot be obtained and thus the Helmholtz instability cannot explain q_{max} in this case.

The maximum value of U_0 follows from Eq. (12):

$$U_0 = U_c = \sqrt{2}\,B\sqrt{\frac{\sigma}{\rho_g R}}\sqrt{\frac{R_c}{R}} \qquad (13)$$

The maximum heat flux being given by the latent heat transported away in columns a distance λ_d apart, it follows that for a cylindrical heater:

$$q_{max} = \rho_g h_{fg} \frac{\pi R_c^2}{2\pi \lambda_d R} U_c \qquad (14)$$

or since $\lambda_d = 5.54 \, \pi \, R_c$ for small B from [4]:

$$q_{max} = \rho_g h_{fg} \frac{1}{11.08} \frac{R_c}{\pi R} U_c$$

Since $R_c = R + \delta$, this last expression together with Eq. (13) gives

$$q_{max} = 0.89 \left\{ \frac{\pi}{24} \sqrt{\rho_g} \, h_{fg} \, [\sigma(\rho_f - \rho_g)g]^{1/4} \right\} \left[\frac{6}{\sqrt{3}\pi^2} \frac{(B + \delta')^{3/2}}{B} \right] \qquad (15)$$

in which
$$\delta' = \frac{(\rho_f - \rho_g)g\delta^2}{\sigma}$$

This result is the same as the one obtained by Sun and Lienhard [2], except for the coefficient 0.89. With respect to the experimental results quoted by Sun and Lienhard, both correlations agree with the data because of the large scatter in the reported values of q_{max}.

b) Small Spherical Heaters

In case of a small spherical heater the maximum vapor velocity at the top of the sphere can be estimated by

$$\tfrac{1}{2}\rho_g U_0^2 = (\rho_f - \rho_g)gR + \frac{2\sigma}{R} - \frac{\sigma}{R}$$

Here the effect of the spherical shape of the vapor sheet formed at the heater is taken into account. In dimensionless form:

$$Wb_0^2 = 2B^2 + 1$$

which once again shows that for small B the value of 6.3 for Wb_c cannot be reached. For small B a good approximation for the maximum average vapor speed is therefore

$$U_c = \sqrt{\frac{\sigma}{\rho_g R_c}} \qquad (16)$$

In analogy with Eq. (14), the expression for q_{max} now becomes

$$q_{max} = \rho_g h_{fg} \frac{\pi R_c^2}{4\pi R^2} U_c$$

Considering the fact that R_c is usually somewhat smaller than R [5], putting $R_c = 0.9 R$ yields

$$q_{max} = \left\{ \frac{\pi}{24} \sqrt{\rho_g}\, h_{fg}\left[(\rho_f - \rho_g)g\sigma\right]^{1/4} \right\} \left(\frac{1.55}{\sqrt{B}} \right) \tag{17}$$

This is the same correlation as the one obtained by Ded and Lienhard [5] except that they obtained a factor of 1.734 instead of 1.55. Equation (17), however, seems to agree with the data [5] somewhat better. The fact that the Helmholtz instability does not cause q_{max} in this case is also substantiated by Fig. 8 of [5] which shows a wavelength which is much larger than the one which the hydrodynamic theory predicts.

c) Large Heaters

In the limit of large B, all heaters should approach a flat plate heater. However, large B does not mean large B_c since the latter is based upon R_c which is always finite. Assuming with Zuber [1] that the column radius on a flat plate heater is $1/4$ times the column spacing, it follows that

$$R_c = \frac{\lambda d}{4\sqrt{2}} = \frac{\pi\sqrt{3}}{2\sqrt{(\rho_f - \rho_g)g/\sigma}}$$

in which λ_d is the most unstable wavelength on a flat vapor film, as reported by Sernas et al. [6]. From this it follows that $B_c = 2.72$ for a flat plate heater.

Applying Bernoulli's equation at the heater at a distance L away from it yields

$$\tfrac{1}{2}\rho_g U_L{}^2 = (\rho_f - \rho_g)gL$$

This gives as average Weber number over a distance L:

$$Wb_a = B_c\sqrt{\frac{L}{R_c}}$$

The Weber number Wb_0 at the bottom of the vapor column a distance L_0 away from the heater is given by

$$Wb_0 = B_c\sqrt{\frac{L_0}{R_c}} = Wb_a\sqrt{\frac{L_0}{L}}$$

Based upon the phenomena occurring during the impingement of a circular jet upon a flat plate one may safely put $L_0 = 0.15\, R_c$, which together with a previous equation yields $L_0 = 0.03\, L$ so that with a Wb_a value of 6.3:

$$Wb_0 = 6.3\sqrt{0.03} \cong 1.1 \tag{18}$$

For a flat plate heater q_{max} can be written as

$$q_{max} = \rho_g h_{fg} \frac{2\pi R_c^2}{\lambda_d^2} U_c$$

With Eq. (18), this gives

$$q_{max} = \left\{ \frac{\pi}{24} \rho_g h_{fg} [(\rho_f - \rho_g)g\sigma]^{1/4} \right\} \tag{19}$$

which is identical to what Zuber [1] obtained.

CONCLUSIONS

It is found that classical hydrodynamic theory does not give a satisfactory explanation for the q_{max} of small heaters. The occurrence of q_{max} in this case can be explained by the insufficiency of vapor removal from the heater (gravity for cylindrical heaters and surface tension for spherical heaters). For a flat plate heater the hydrodynamic theory can be adapted to explain q_{max}.

NOMENCLATURE

g	gravitational constant
h_{fg}	heat of evaporation
k	wavenumber in the axial direction
m	wavenumber in the azimuthal direction
p	pressure
Δp	pressure difference across the column interface
q_{max}	maximum heat flux
r	radial cylindrical coordinate
t	time
z	axial coordinate
B	Bond number
C	constant
K_m, I_m	modified Bessel functions of order m
L	length
R	heater radius
R_c	vapor column radius
R_1, R_2	principal radii of curvature of the column interface
U	vapor speed
U_c	maximum vapor speed
Wb	Weber number
α	density ratio defined by Eq. (4)
δ	vapor film thickness
η	radial displacement of the column interface
σ	surface tension

λ wavelength
λ_d most dangerous wavelength
ρ density
θ angular coordinate
ω growth rate
ω' dimensionless growth rate
Ω defined by Eq. (3)

Subscripts

0   bottom of the vapor column
1, 2   refer to fluids 1 and 2, respectively
a average
i imaginary part
f liquid
g vapor
max   maximum value

REFERENCES

1. Zuber, N.: Atomic Energy Commission Report No. AECU 4439, Physics and Mathematics, 1959.
2. Sun, K. H. and J. H. Lienhard: *International Journal of Heat and Mass Transfer*, **13**, 1425, 1970.
3. Lamb, H.: Hydrodynamics, Dover, 1945.
4. Berghmans, J.: International Journal of Multiphase Flow, 1975, 2.
5. Ded, J. S. and J. H. Lienhard: AIChE J., 1972, **18**, 337.
6. Sernas, V., J. H. Lienhard, and V. K. Dhir: International Journal of Heat and Mass Transfer, 1973, **16**, 1820.

HEAT TRANSFER AND PRESSURE DROP IN ONCE-THROUGH STEAM GENERATORS*

G. Bianchi and M. Cumo
*Centro di Studi Nucleari
Della Casaccia, Rome*

1. HEAT TRANSFER IN BOILING

1a. Local Correlations

Extensive tests have been performed at full scale (with relation to L.M.F.B.R. once-through steam generators). Both straight vertical test sections (11 m long, 12 mm in inside diameter) and coiled test sections (D_{coil} = 83 cm; D_{tube} = 15 mm; heated length = 62.8 m), uniformly heated by the Joule effect, have been tested. Seventy and 200 wall thermocouples, respectively, were used to measure the wall temperature profiles in the two cases.

Local boiling data, through different two-phase boiling regimes, are vary spread in nature.

Boiling correlations of both types:

$$-h_b = A \cdot \varphi^n \cdot e^{mp}$$

and

$$-h_b = h_1 \cdot B \left[\frac{\varphi}{\lambda G} + A \left(\frac{1}{X_{tt}} \right)^n \right] \tag{1}$$

with

$$h_1 = 0.023 \frac{K_{1s}}{D} (\mathrm{Re}_{1s})^{0.8} \cdot (\mathrm{Pr}_{1s})^{0.3}$$

$$X_{tt} = \left(\frac{X}{1-X} \right)^{0.9} \left(\frac{\rho_1}{\rho_v} \right)^{0.5} \left(\frac{\mu_v}{\mu_1} \right)^{0.1}$$

have been computed and best-fitted through experimental values. No correlation has been found particularly good to correlate the available data. Best

---

*The abstract of this paper appears on page 366.

results have been obtained with Thom's correlation, i.e., Eq. (1) with:

$$A = 0.444$$
$$n = 0.5 \quad \text{(with } \varphi \text{ in W/cm}^2, p \text{ in kg/cm}^2)$$
$$m = 0.01128$$

1b. Overall Correlations

With specific reference to coiled test sections (see temperature profiles of Fig. 1) a mean boiling heat transfer coefficient $\bar{h}_b$ through all the boiling length $(0 < X < X_{DNB})$ may be defined:

$$\bar{h}_b = \frac{1}{X_{DNB}} \int_0^{X_{DNB}} \frac{\varphi}{T_w - T_{sat}} \, dX$$

A best fit through the data (Fig. 2) brings to the correlation:

$$\bar{h}_b = 11.226 \cdot \varphi^{0.6} \cdot e^{0.0132\bar{p}} \quad \text{W/m}^2{}^\circ\text{C}$$

(with p in bar, φ in W/m^2).

Computed and experimental results are compared in Fig. 3, where $a \pm 25\%$ (uncertainty) band practically encloses all the data.

2. BURNOUT POWER

2a. Burnout Power in Straight Vertical Ducts

With reference to the abovementioned straight vertical test sections, the burnout quality in once-through steam generators may differ greatly from the values predicted using well-known burnout correlations experimentally obtained for the reactor core geometry (shorter lengths, higher heat fluxes) (see Fig. 4). Two separate pressure ranges do appear, $60 < p < 90$ kg/cm^2 and $110 < p < 160$ kg/cm^2, separated by a transition region, in which the burnout qualities are substantially different and two different physical burnout mechanisms seem to exist.

From extensive experimentation the following relations have been deduced:

$$X_{B.O.} = 3.758\varphi^{5/13} \cdot G^{-3/5} \cdot p^{-4}/700 \quad \text{for} \quad 50 < p < 90 \text{ kg/cm}^2$$

(with φ in W/cm^2, G in g/cm^{2}s, p in kg/cm^2) and

$$X_{B.O.} = 2.929 \cdot \varphi^{8/9} \cdot G^{-14/17} \cdot p^{-1/7} \quad \text{for} \quad 110 < p < 160 \text{ kg/cm}^2$$

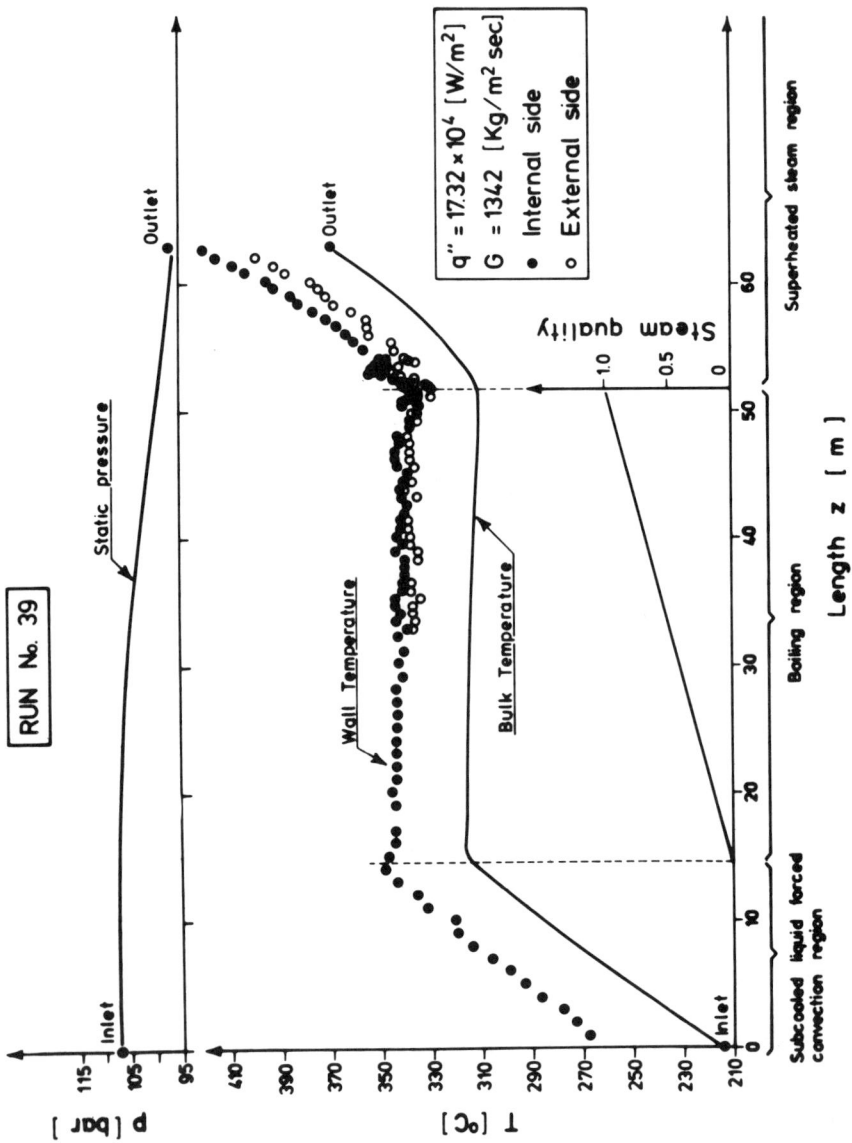

FIG. 1. Example of one of the 120 experimental wall-temperature profiles along the coiled tube. Internal and external side wall-temperatures are clearly distinguished. Small fluctuations are mainly due to the huge experimental dimensions.

122

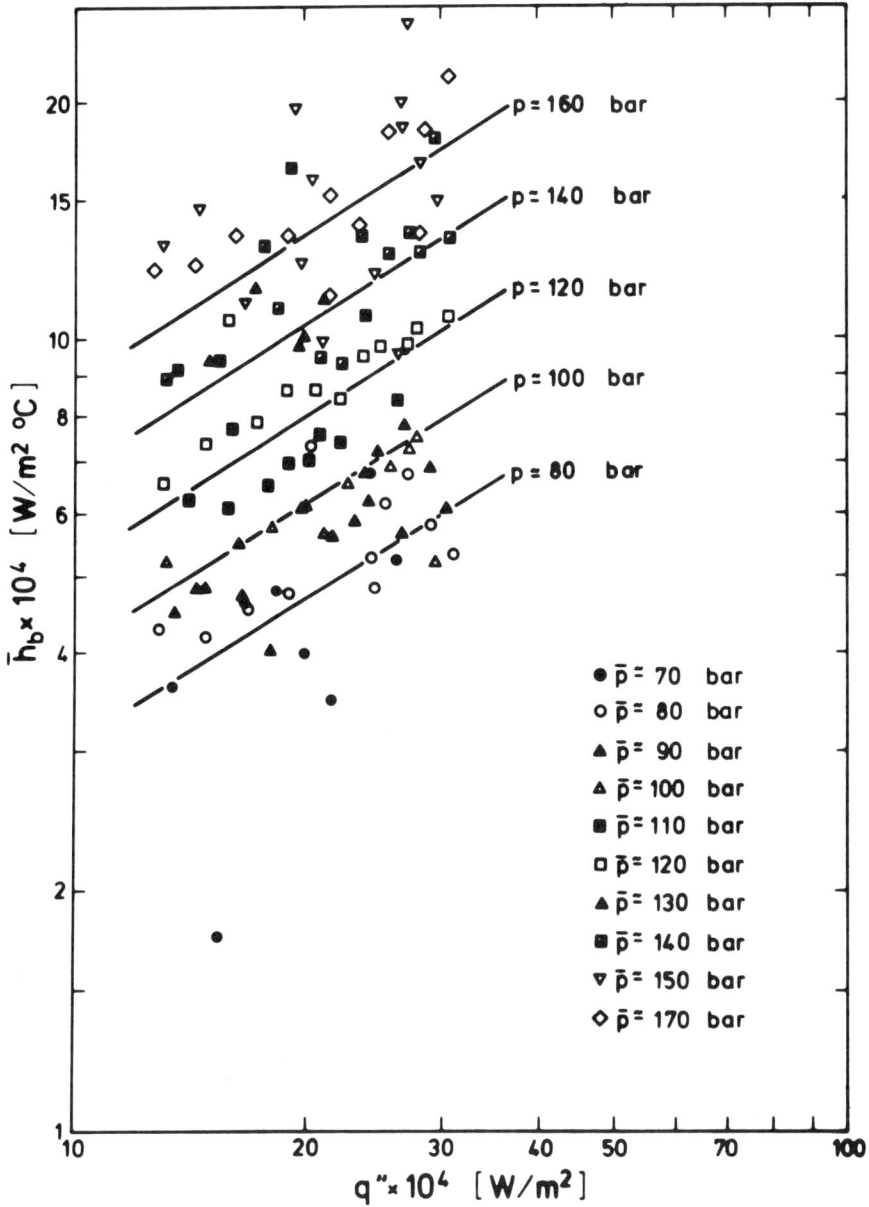

FIG. 2. Mean boiling heat transfer coefficient at different pressures: each dot refers to the whole boiling length for the corresponding run.

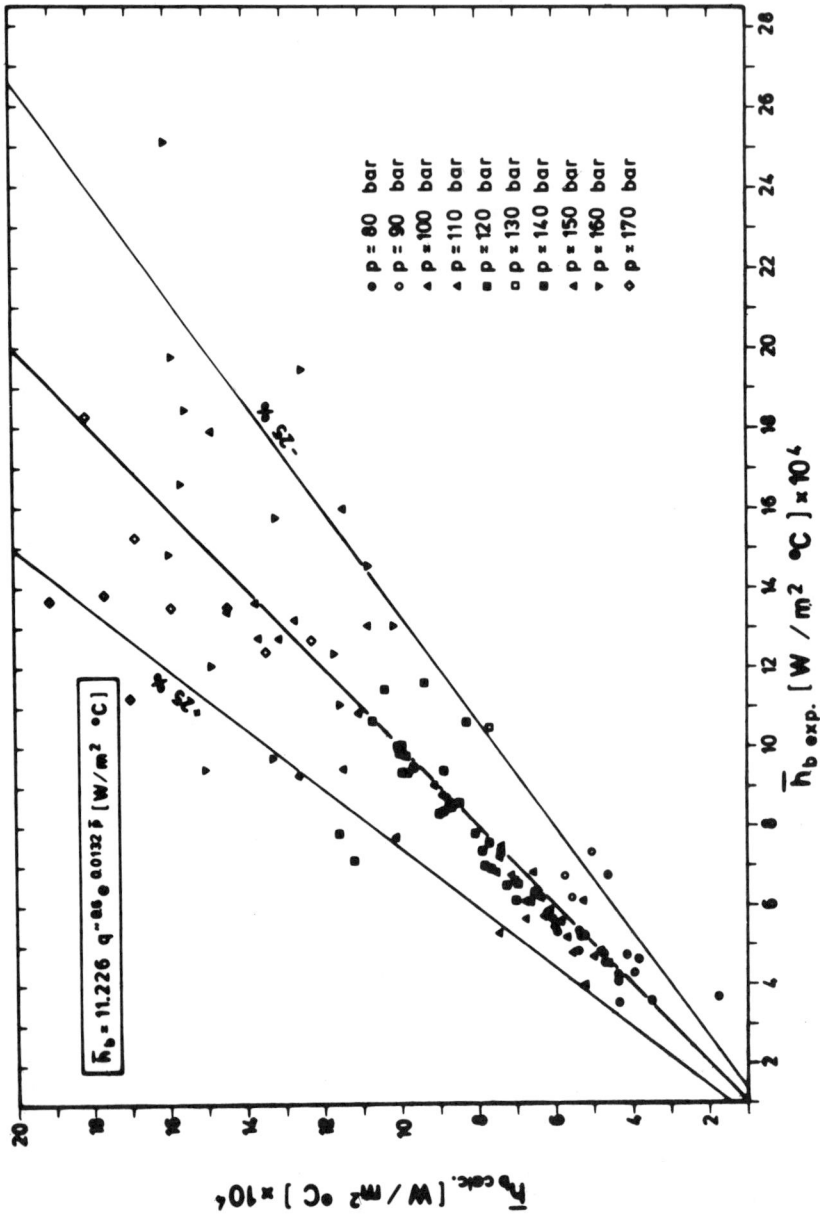

FIG. 3. Comparison of predicted and experimental results for the mean boiling heat transfer coefficient.

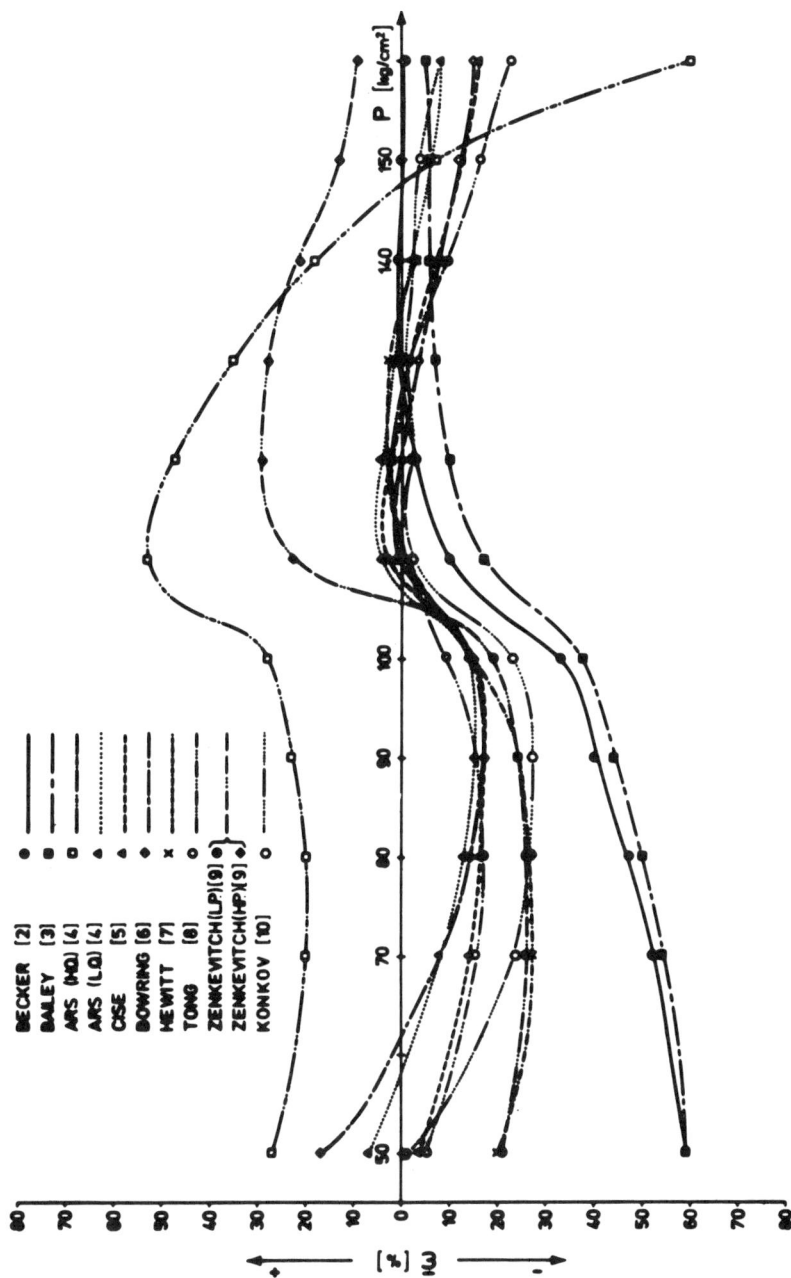

FIG. 4. Comparison of burnout correlations versus CNEN experimental data (percent differences versus p); for $p > 120$ kg/cm²
Becker [2] correlation gives best previsions.

125

A comparison of the computed qualities versus the corresponding experimental values is shown in Figs. 5 and 6.

2b. Burnout Power in Coiled Ducts

Heat transfer tests with the coiled test section, clearly indicate that in a heat flux (q'')–flowrate (G) (Fig. 7) two regions can be distinguished. A region (lower, right) where DNB doesn't occur, and a region (upper, left) where DNB may eventually occur. The separation line (with an uncertainty of $\pm 1 \cdot 10^4$ W/m^2) has the equation:

$$\frac{q''}{G} = 0.13 \text{ kJ/kg} \quad \text{for} \quad 80 < p < 170 \text{ bar}$$

So, the "no DNB" design criterion is:

$$\frac{q''}{G} < 0.13 \text{ kJ/kg}$$

Figure 7 shows, in marked dots, some DNB limiting points which are characterized by very high qualities $(X_{DNB} > 0.7)$ and very limited wall temperature excursions $(T_w - T_{sat} < 20°C)$.

2c. Difference between Indirect and Direct Heating

In actual once-through steam generators, in conventional or nuclear plants, the heat is transferred from hotter fluids (gas, liquid metal, etc.) to the vaporizing water. This way of heating is somewhat different from the heating devices which experimentally are usually employed in the laboratories, i.e., direct Joule heating by electric current.

To look to the differences between the two basic boundary conditions of the heat transfer problems, a number of tests have been undertaken.

A once-through flow of Freon-12 is alternatively heated by D.C. Joule effect and by a counter current of forced convection, pressurized water in an annular gap, in the same thermal situations. In the first case, there is a uniform, axial heat flux; in the second, the axial profile of heat flux follows the heat absorption capacity of the inner fluid, along its different modes of two-phase flow.

The burnout powers (from $X = 0$ to $X = X_{B.O.}$) give different results, thus confirming that laboratory burnout tests require a correction factor, $\psi = \varphi_{B.O.i.h.}/\varphi_{B.O.J.H.}$ (i.h. and J.H., respectively, indicate "indirect heating" and "Joule heating").

Figure 8 shows the ψ trends versus G and reduced pressure π, with a maximum (up to 1.7) close to $\pi = 0.6$.

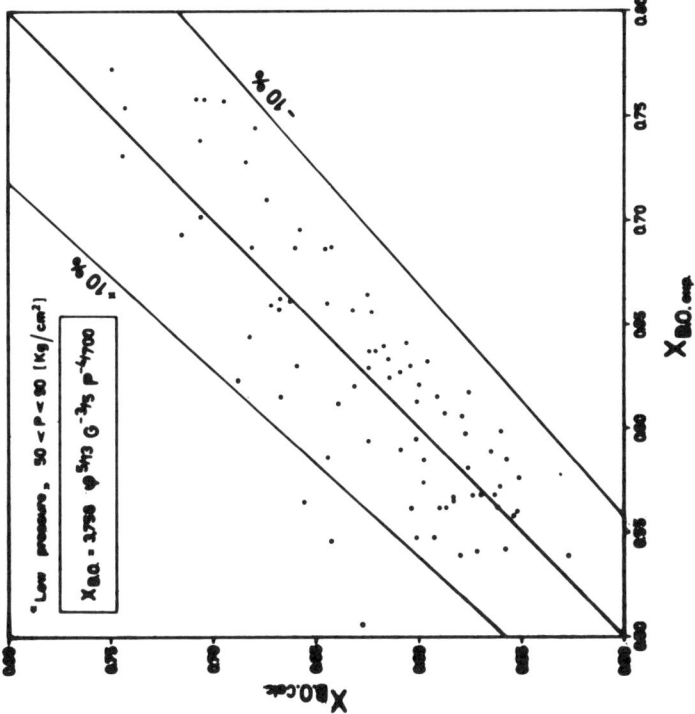

FIG. 6. "High pressure" burnout correlation.

"High pressure, 110 < P < 160 [kg/cm²]

$$X_{B.O.} = 2.929 \, \varphi^{0.45} \, G^{-0.14_{917}} \, P^{-1/7}$$

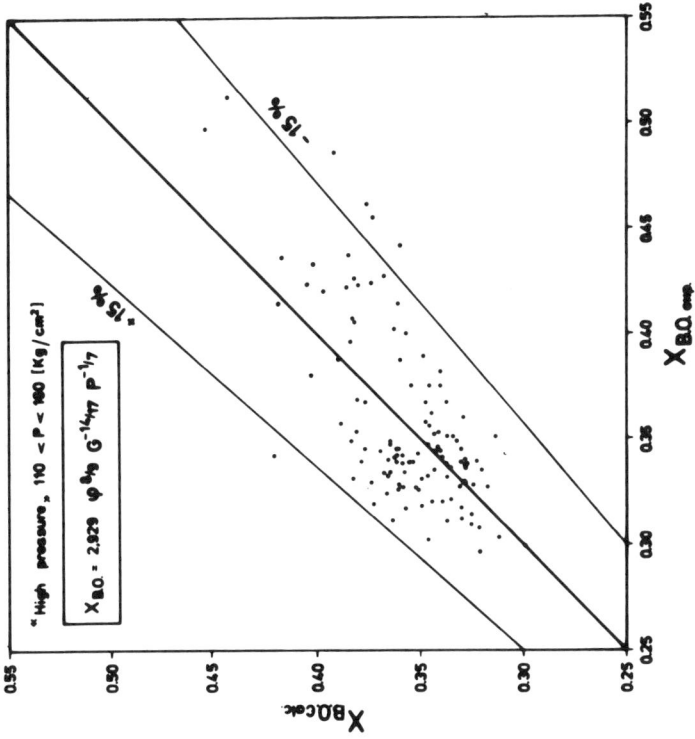

FIG. 5. "Low pressure" burnout correlation.

"Low pressure, 50 < P < 90 [kg/cm²]

$$X_{B.O.} = 3.798 \, \varphi^{5.13} \, G^{-3.25} \, P^{-1/1700}$$

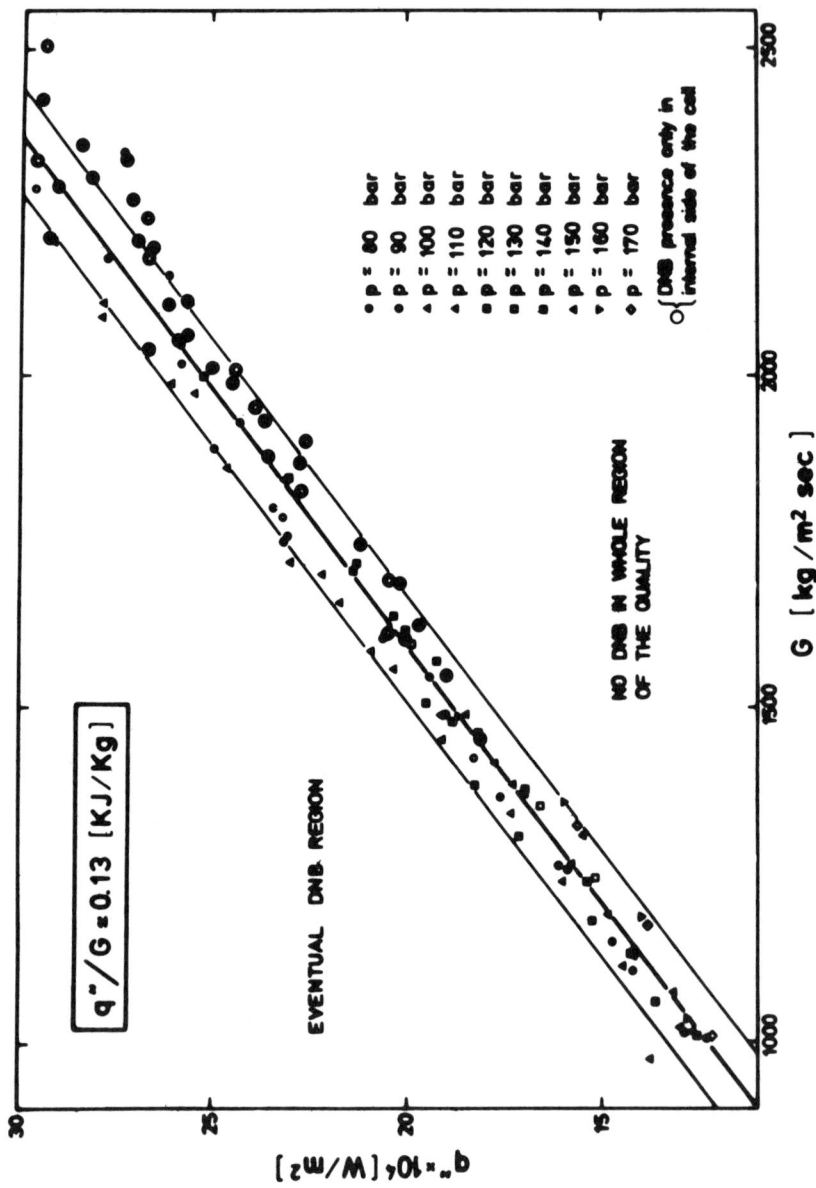

FIG. 7. DNB design criterion in the 80–170 bar pressure range: for $q''/G < 0.13$ kJ/kg the boiling length extends up to 100% quality.

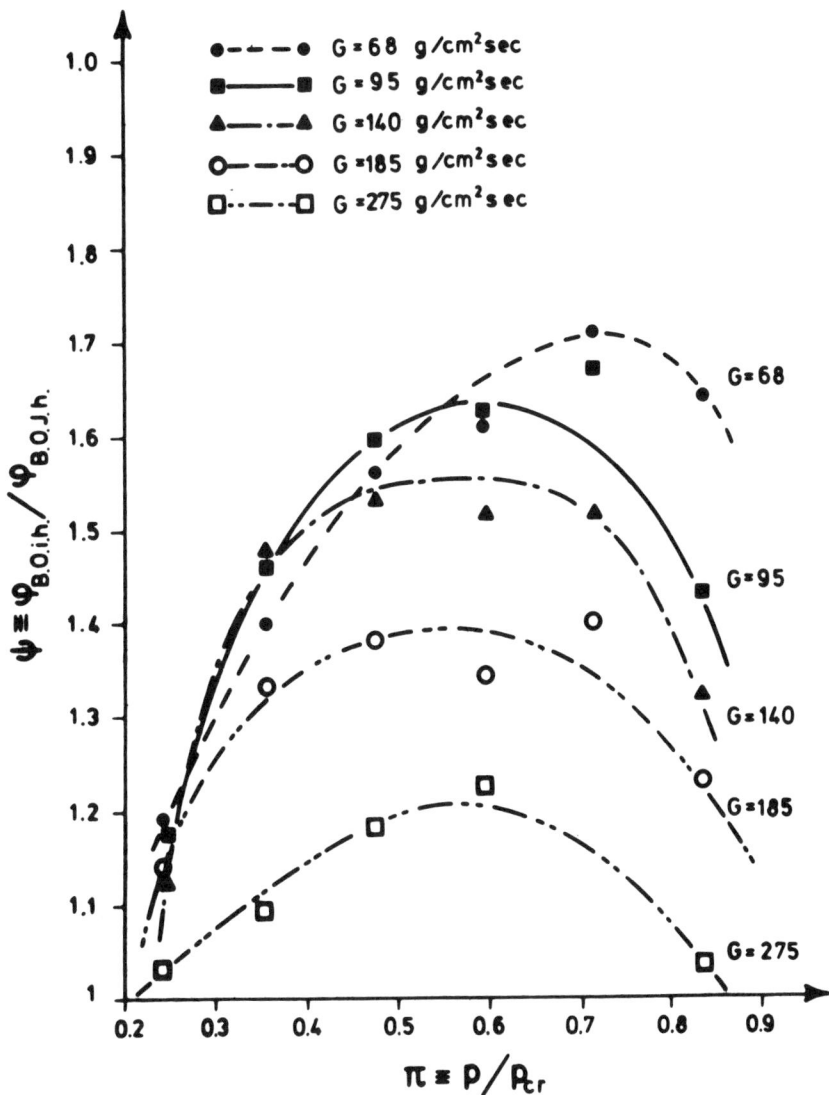

FIG. 8. Ratio of the two burnout heat fluxes versus π: the maxima are approximately located at $\pi = 0.6$. The two burnout heat fluxes clearly cannot be confused.

3. HEAT TRANSFER IN SUPERHEATED STEAM

The following heat transfer correlation is suggested for steam generator design:

$$(Nu)_b = 0.0142 \, (Re_b)^{0.824} \cdot (Pr_b)^{0.98}$$

where the reference temperature for the computation of the physical properties is the bulk temperature (T_b) (see Fig. 9). Design computations are simplified if iterative procedures may be suppressed, i.e., if the reference temperature is T_b rather than, for instance, the wall temperature T_w or the "film" temperature T_f. Figure 10 shows the standard deviations, for all the experimental data, of three correlations whose constants have been best fitted at T_b, T_f, and T_w; slight differences enable the more straightforward use of T_b.

4. PRESSURE DROP DESIGN CORRELATIONS

Experimental pressure drops in coiled, once-through steam generators (2nd test section, $100 < G < 250$ g/cm^{2}s; $60 < p < 180$ kg/cm^2; $10 < \overline{q}'' < 30$ W/cm^2) may be calculated through the well-known homogeneous model ($\Delta p_{\text{h.m.}}$) with a correction factor, $\epsilon(p, G)$, which is a function both of p and G:

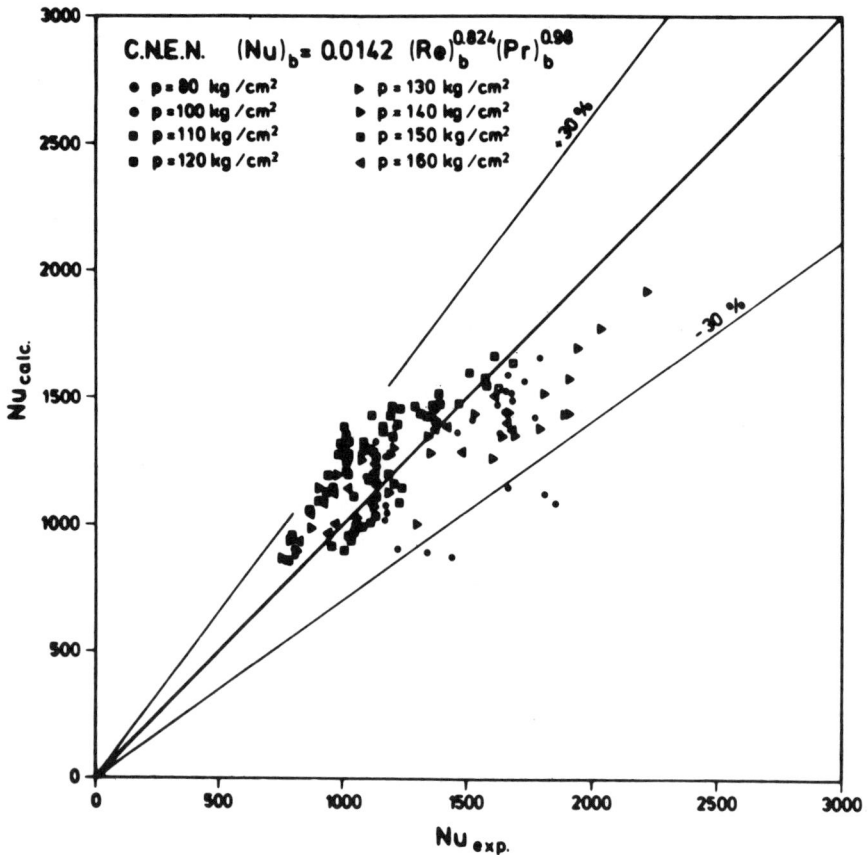

FIG. 9. The suggested CNEN correlation with physical properties computed at bulk temperature versus the experimental data.

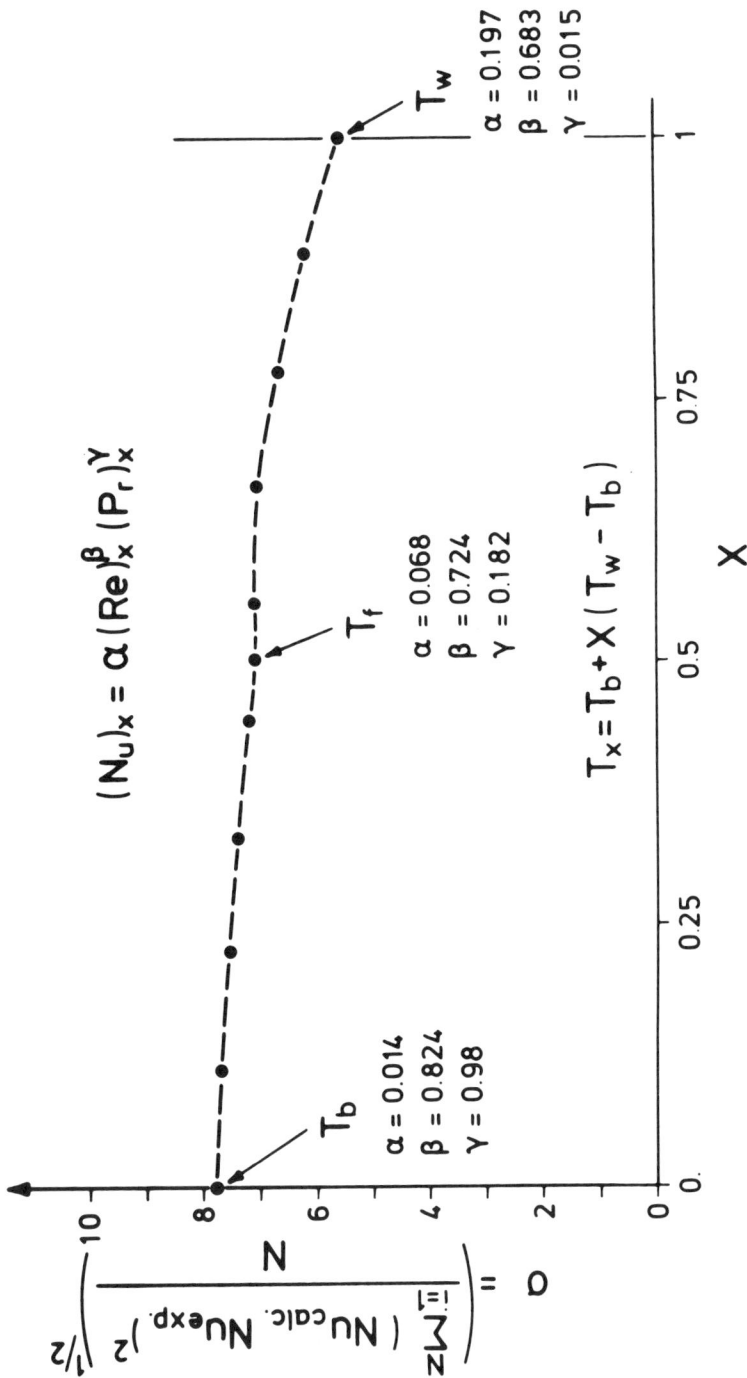

$$(N_u)_x = \alpha (Re)_x^\beta (P_r)_x^\gamma$$

T_w

$\alpha = 0.197$
$\beta = 0.683$
$\gamma = 0.015$

T_f

$\alpha = 0.068$
$\beta = 0.724$
$\gamma = 0.182$

T_b

$\alpha = 0.014$
$\beta = 0.824$
$\gamma = 0.98$

$$T_x = T_b + X(T_w - T_b)$$

$$\sigma = \left(\frac{\sum_{i=1}^{N}(Nu_{calc.} \cdot Nu_{exp.})^2}{N} \right)^{1/2}$$

FIG. 10. Standard deviation of best-fitted correlation from T_b to T_w: the choice of T_b as reference temperature appears justified.

131

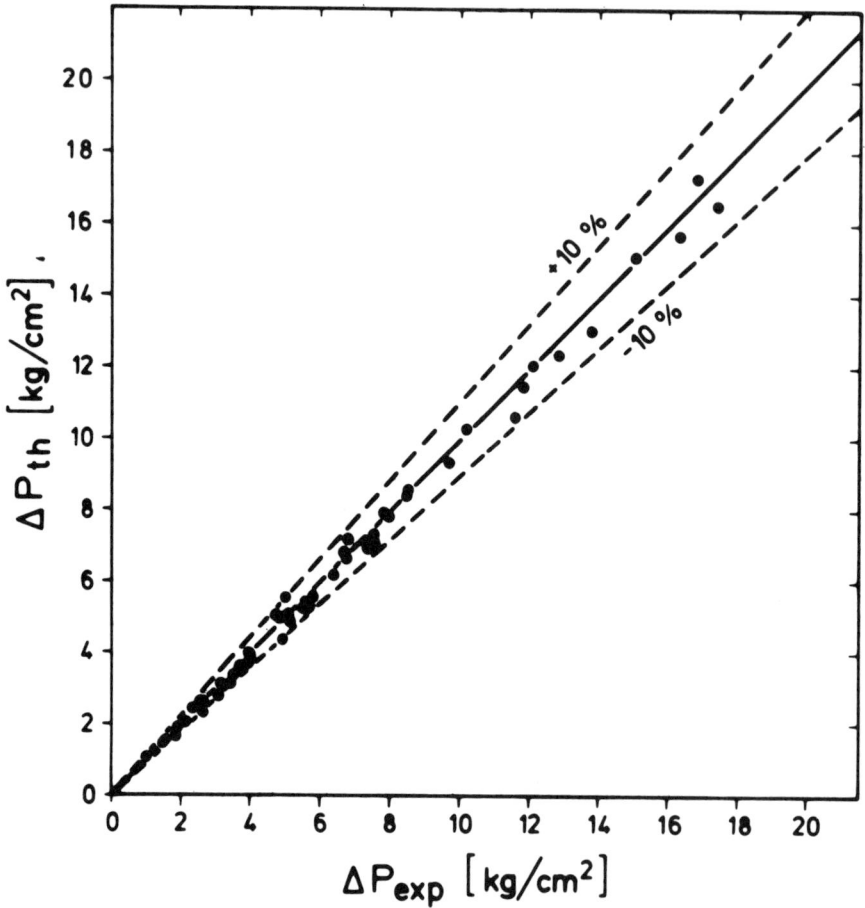

FIG. 11. Comparison between the experimental and the computed data.

$$\Delta p = \frac{\Delta p_{h.m.}}{1 + \epsilon(p, G)}$$

where

$$\epsilon(p, G) = A(p) + \frac{B(p)}{G}$$

with

$$A(p) = -1.0124 \cdot 10^{-4} p^2 + 0.0243 p - 1.757$$

$$B(p) = 0.016 p^2 - 3.238 p + 183.86$$

Figure 11 gives an overall picture of the comparison between computed and experimental data: errors are within ±10%.

LIST OF SYMBOLS

D   test section inner diameter
ϵ correction factor in pressure drop correlation
G   specific mass flowrate
h   heat transfer coefficient
H   enthalpy
H_{fg}, λ latent heat of vaporization
K   thermal conductivity
μ dynamic viscosity
p   pressure
q'', φ heat flux
ρ density
Re   Reynolds number
Pr   Prandtl number
T   temperature
X   steam quality
Z   length

Indexes

b   boiling
calc   calculated, theoretical
DNB
B.O.   departure from nucleate boiling, burnout, dryout, critical
cr
f   film
w   wall
exp   experimental
h.m.   related to the homogeneous model
in   inlet (of the heated test section)
i.h.   indirect heating
J.h.   Joule heating
l   liquid
out   outlet (of the heated test section)
sat   saturation
sub   subcooling
sup   superheating

REFERENCES

1. Bianchi, G., M. Cumo, and A. Merelli: "Le attivita del Centro Studi Nucleari della Casaccia per lo sviluppo dei generatori di vapore per centrali nucleari," 19th Nuclear Congress, Rome, March 21-22, 1974.
2. Cumo, M., G. E. Farello, and G. Ferrari: "Heat transfer in sub-critical once-through steam generators," Polska Akademia Nauk, Nat. Heat and Mass Transfer Symp., Jablonna (Warsaw), November 25-28, 1974.

HEAT TRANSFER INTENSIFICATION DURING DROPWISE CONDENSATION OF ORGANIC LIQUIDS*

V. T. Vylkov
Energoproekt, Sofia, Bulgaria

The phenomenon "dropwise condensation" has been comparatively well studied only in the case of water vapor. Although some experiments of a qualitative nature have been carried out on other liquids, a quantitative study of heat transfer is only available for ethylene glycol [1]. In most organic liquids the vapor-side heat transfer coefficient, with filmwise condensation, is considerably lower than that for water vapor and is comparable in magnitude or even somewhat smaller than the value of the water-side heat transfer coefficient. Under these conditions, dropwise condensation results in the largest increase of the overall heat-transfer coefficient.

The available methods for predicting the type of condensation by measuring the angle of contact of a static drop [2], or by comparing the surface tension σ with the critical surface tension of a lyophobilizer σ_c introduced by Shafrin and Zisman [3], have some disadvantages and do not always describe the actual process. They do not cover the cases of transition from dropwise to filmwise condensation of a number of liquids at higher heat fluxes. Accurate results concerning the possibilities to realize dropwise condensation of liquids can be obtained only by experiments based on the direct observation of the condensation process itself. To this end, a test setup consisting of 8 glass evaporators-condensers was used to study the transition from filmwise to dropwise condensation of a wide range of organic liquids. The experiments were carried out with cyclohexane, methanol, acetone, tetrachloromethane, dichloroethane, benzaldehyde, benzene, chloroform, quinoline, trichloroethylene, ethyl iodide, dimethyl formamide, aniline, pyridine, phenol, furfural, ethylenediamine, dibromomethane, carbon disulphide, bromoform, and glycerine (dynamite type) employing as lyophobilizers, perfluorocapric acid, Teflon, octadecylamine, ceresin, oleic acid, stearic acid, cupric stearate, polished coatings of chrome and nickel, and polished surfaces of stainless

---

*The abstract of this paper appears on page 368.

steel. Copper, stainless steel and brass tubes were used as condensing surfaces. The surface tension σ of the liquids tested varied between 0.0176 and 0.036 N/m, this range covering most of the organic liquids. The lyophobilizers used have also a wide range of σ_c values. The results can be used to predict the type of condensation of other investigated liquids upon the lyophobilizers used here, or upon lyophobilizers with a known σ_c.

Dropwise condensation of pyridine, phenol, carbon disulphide, ethylene-diamine, furfural, dibromo-methane, bromoform and glycerine was obtained by using a Teflon layer as a lyophobilizer. In the case of dynamite-type glycerine ceresin 75 was used as a lyophobilizer. When used with a polished stainless steel surface, this type of glycerine also produced dropwise condensation but of an inferior quality. For a number of liquids, dropwise condensation took place only for certain heat fluxes; beyond a certain maximum heat flux, a transition to filmwise condensation was observed.

The evaluation of the transition-determining parameters and the study of the increase of the overall heat transfer coefficient in dropwise condensation were carried out using an experimental apparatus [4] which allowed a quantitative study of the heat exchange process "steam-cooling water." The transition should be carried out on a lyophobilizer with a certain magnitude of the critical surface tension. To this end, Teflon with $\sigma_c = 0.018$ N/m is

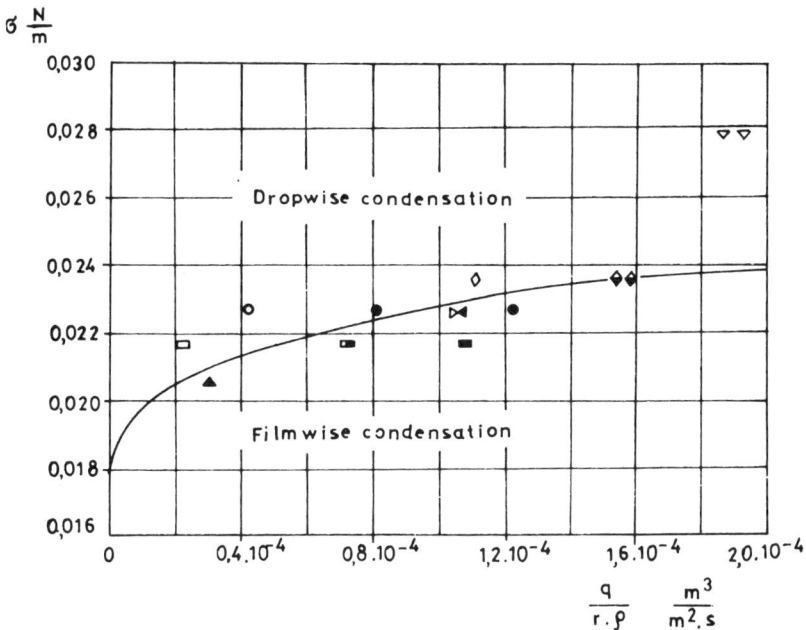

FIG. 1. Regions of dropwise and filmwise condensation at critical surface tension 0.018 N/m—ethyliodide: o - dropwise; ◑ - transition; ● - filmwise—dibromomethane: ▼ - dropwise—quinoline: ▢ - dropwise; ▣ - transition; ■ - filmwise—nitro-benzene: ◄ - transition—dimethyl formamide: ◇ - dropwise; ◈ - transition—dichloroethane [1,1-] : ▲ - filmwise.

used. It has been established that transition is determined by the lyophobilizer's critical surface tension, the liquid's surface tension σ and the
amount of condensate released. Test data are shown in Fig. 1. One of the
coordinates contains the amount of condensate $q/r\rho$ in m^3/m^2s, and the
other—the surface tension σ of the liquids under investigation. The test points
at which there is a transition to filmwise condensation as well as those at
which condensation is only dropwise or only filmwise serve as a dividing line
between the two types of condensation. As seen from the diagram, all liquids
having $\sigma \leqslant 0.018$ N/m will be in the region of filmwise condensation for any
amount of condensate because the value of σ_c for Teflon is 0.018 N/m.

All liquids with σ exceeding $\sigma_c = 0.018$ N/m by very much will always
result in dropwise condensation up to the boundaries for practically attainable
heat fluxes. Such is the case for water where we have experimental data about
dropwise condensation up to $q = 1.8 \cdot 10^6$ W/m^2. All liquids with surface
tension at saturation in the range 0.018–0.024 N/m will result in this case in
dropwise or filmwise condensation depending on the heat flux. In the case
that another lyophobilizer with a different σ_c is used, then we will have a
similar dependence whose quantitative determination at this time is possible
only by experiment. In the case of a liquid whose mode of condensation is
unknown, its own value of σ can be used to determine the amount of
condensate $q/r\rho$ for which only dropwise or only filmwise condensation is
possible; the latter can then be used to calculate the respective heat flux. The
line dividing the two regions in Fig. 1 refers to the case when the
condensation surface is occupied by approximately 50% dropwise and 50%
filmwise condensation.

In the study of the quantitative estimation of the overall heat transfer
coefficient, the case of a solid lyophobilizer was examined; this lyophobilizer
is the most unfavorable one with respect to heat-transfer intensification. Tests
were conducted with short vertical tubes. For the solid lyophobilizer a $4 \cdot
10^{-6}$ m to $7 \cdot 10^{-6}$ m high Teflon layer was used. For ethylene-diamine, the
overall heat-transfer coefficient increased by 26.6 percent at a cooling water
rate of 0.34 m/s and by 34.3 percent at a cooling water rate of 0.98 m/s (the
average temperature of the cooling water was 288°K). The comparable figures
for phenol were 25.5 percent at 0.25 m/s and 31.1 percent at 0.75 m/s
(average temperature of cooling water = 315°K).

The detailed study of the process from the steam side, i.e., the phenomenon "dropwise condensation" itself, was conducted with respect to
carbon disulphide. The basic components of the test apparatus [4] are made
of glass and condensation takes place over a horizontal copper tube. The
average temperature of the overall condensation surface was measured without
disturbing its integrity through the resistance of a wire, $5 \cdot 10^{-5}$ m in
diameter, laid in a screw-like groove with a small pitch and a depth of 0.0003
m; the latter is made on the surface of a thick-walled copper tube with a
0.016 m internal diameter and a 0.032 m external diameter. The tube surface

was then restored by galvanically applying a small-grained copper coating. The noncondensing gases were measured in the various regimes and ranged between 0.0035 and 0.0076 percent (the ratio of air to steam is expressed in weight units). The value 0.015 percent was measured only once and should be accepted as the upper boundary of noncondensing gases content for the test runs. Dropwise condensation was realized with perfluorocapric acid as the lyophobilizer. The relative difference for both design heat fluxes—with respect to condensate and cooling water—for the tests run reached a maximum of ±8.5 percent.

The test results are shown in Fig. 2. Other researchers have established lower test values during filmwise condensation than the analytical predictions

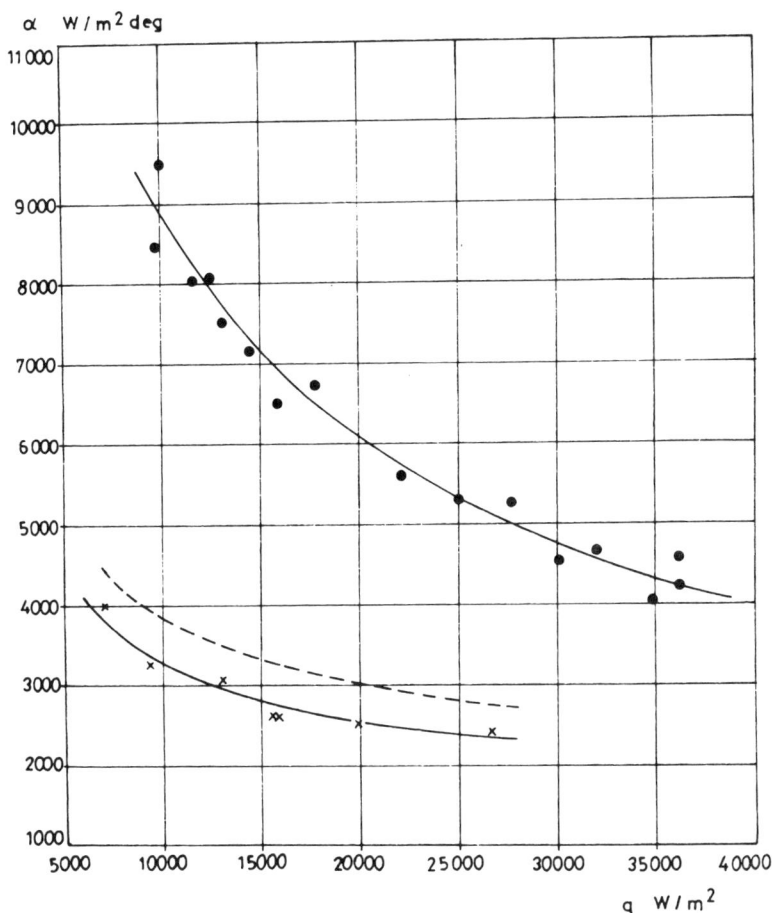

FIG. 2. Coefficients of steam-side heat transfer for condensation of carbon disulphide upon a horizontal tube: ●—dropwise condensation; x—filmwise condensation; ---- -— analytically calculated for filmwise condensation with a correction for the change of physical properties.

for a condensation of different vapors upon a horizontal tube in the absence of a noticeable dynamic effect of the steam upon the condensate layer; however, it is considered that the basic reason is the noncondensing gases. The graphic representation of the data for the steam-side heat-transfer coefficient indicates its large increase during transition to dropwise condensation.

The processing of the test results is determined by the existing analytical relationships for the steam-side heat transfer coefficient. The most appropriate method of obtaining the relationships of α is to investigate the mechanism of the phenomenon and to set up a valid physical and mathematical model of the process. However, the use of the relationships derived by the models of various researchers meets with great difficulties when calculating α for an arbitrary regime and leads to its practical unutilizability. The reasons for this are some values in the relationships which represent the physical system of the process. They can be determined only experimentally with great difficulty and the resulting values differ greatly. At present, results for practical purposes can be obtained only through criterial dependences which describe the phenomenon.

In the analysis, by similarity theory, of Isachenko [5], it is assumed that the heat of the phase transition released on the surface of a truncated sphere is thermally conducted through the drop; this being equivalent to the present understanding of the phenomenon's mechanism. Some additional considerations, based on later investigations of the phenomenon, have been suggested to the critical dependence which describe the phenomenon.

In the analysis, by similarity theory, of Isachenko [5], it is assumed that the heat of the phase transition released on the surface of a truncated sphere is thermally conducted through the drop; this being equivalent to the present understanding of the phenomenon's mechanism. Some additional considerations, based on later investigations of the phenomenon, have been suggested to the criterial dependence derived in [5]; the latter was used subsequently by other authors [1]. Proceeding from the latter results [6, 7], concerning the effect of internal circulation in the drop on heat transfer during dropwise condensation, a suggestion is made to the effect that the criterial dependence should not consider the criterion of thermocapillary motion. Thus, for a given liquid, the criterial dependence will be:

$$\mathrm{Nu}_* = c\mathrm{Re}_*^n \mathrm{Pr}^m$$

where

$$\mathrm{Nu}_* = \frac{\alpha R_c}{\lambda} = \frac{2\alpha\sigma T_s}{\lambda\tau\rho(T_s - T_w)}$$

$$\mathrm{Re}_* = \frac{w_* R_c}{\nu} = \frac{\lambda(T_s - T_w)}{\nu\tau\rho}$$

The processing of the test results for carbon disulphide in this criterial form should be compared with the same processing for another liquid. The comparison is made with water vapor for which there exist many test results from other researchers. Their processing will provide the exact picture of the numerical values of the exponents n and m.

For the case of dropwise condensation of water vapor upon a horizontal tube, a criterial dependence has been derived [4, 8] which permits the practical calculation of the heat-transfer coefficient α. The dependence is derived as follows. A criterial dependence for α for a horizontal tube is derived by processing the available test results. Then the criterial dependence for α for a short vertical plate is derived by processing the available test results for this case. On the basis of this dependence and experimental results [9] for the local heat transfer coefficient for a plate as a function of its slope, one obtains a criterial dependence for α for a horizontal tube. The satisfactory coincidence of both dependences for a horizontal tube proves the correctness of the conclusions made. The processing [4] of the test results by various researchers for pressures between $2.65 \cdot 10^3$ N/m^2 and $1.08 \cdot 10^6$ N/m^2 when deriving the criterial dependence for a short vertical plate resulted in the value $m = -2.655$ for the exponent of Pr. The criterial dependence [4, 8] derived for dropwise condensation of water vapors upon a horizontal tube is

$$Nu_* = 6.023 \times 10^{-7} Re_*^{-1.424} Pr^{-2.655}$$

The experimental results for carbon sulphide were turned into criterial form by the above dependence. The dependence for water vapors under atmospheric pressure is

$$Nu_* = 1.37 \times 10^{-7} Re_*^{-1.424} \qquad Pr = 1.75 = const$$

When processed, the experimental results for carbon disulphide at atmospheric pressure yield

$$Nu_* = 3.6 \times 10^{-7} Re_*^{-1.365} \qquad Pr = 1.89 = const$$

Clearly, there is a satisfactory coincidence of the value n for both liquids.

Processing of data for carbon sulphide according to the dependence $Nu_* = c Re_*^n Pr^{-2.655}$ yields the following criterial dependence for dropwise condensation on a horizontal tube:

$$Nu_* = 1.945 \times 10^{-6} Re_*^{-1.365} Pr^{-2.655}$$

If the criterial dependence describing heat transfer during dropwise condensation of liquids is of the form $Nu_* = f(k_1, k_2, k_3, \ldots, k_n)$, then we can represent it in the form $\mathrm{Nu}_* = \mathrm{Re}_*^n\,\mathrm{Pr}^m f_1(k_i, k_{i+1}, \ldots, k_n)$ where k_i, $k_{i+1}, \ldots, k_n$ are the complex criteria reflecting the peculiar features of heat transfer for a specific liquid. Assuming, as a first approximation, f_1 $(k_i, k_{i+1}, \ldots, k_n)$ being a constant for a specific liquid, then $\mathrm{Nu}_* = c\mathrm{Re}_*^n\,\mathrm{Pr}^m$. On the basis of the above considerations on heat-transfer calculations during dropwise condensation of a liquid upon a horizontal tube, we propose, as a first practical approximation, the dependence $\mathrm{Nu}_* = c\mathrm{Re}_*^{-1.4}\,\mathrm{Pr}^{-2.66}$ where the constant c depends on the specific liquid.

The above dependences do not cover the regimes for which $\mathrm{Re}_* < 1.6\,\cdot\,10^{-3}$; in this region the dependence $\alpha = f(q)$ is increasing. These regimes occur with a small difference in temperatures of the saturated steam and the cooling water. However, in the majority of cases they are of no practical interest.

The above studies on dropwise condensation of organic liquids, in addition to their bearing on the study of the phenomenon, are also of practical interest.

NOMENCLATURE

α	steam-side heat-transfer coefficient in case of dropwise condensation
λ	thermal conductivity of liquid at saturation temperature
ν	kinematic viscosity of liquid at saturation temperature
ρ	liquid density at saturation temperature
σ	surface tension of liquid at saturation temperature
σ_c	critical surface tension of lyophobilizer
q	heat flux
r	evaporation heat at saturation temperature
T_s	temperature of saturated steam
T_w	temperature of surface on which condensation occurs
R_c	critical radius of nucleus
c, n, m	constant
Pr	Prandtl number
k_i	similarity criterion

REFERENCES

1. Peterson, A. C. and J. W. Westwater: Dropwise Condensation of Ethylene glycol, Chemical Engineering Progress, Sympos. series, **62, 64**, 135–142 (1966).
2. Davies, G. A., et al.: Measurement of Contact Angles under Condensation Conditions. The Prediction of Dropwise-Filmwise Transition, Int. J. Heat Mass Transfer, **14**, 709–713 (1971).
3. Shafrin, E. G. and W. A. Zisman: Constitutive Relations in the Wetting of Low-Energy Surfaces and the Theory of the Retraction Method of Preparing Monolayers, Journal of Physical Chemistry, **64**, 519–524 (1960).
4. Volkov, V. T.: Studies of Heat Transfer Accompanying Dropwise Condensation on Liquids in Chemical Manufacturing. Thesis, Sofia, 1974.
5. Isachenko, V. P.: Mechanism and Dimensionless Equations of Heat Transfer in Dropwise Condensation, Teploenergetika, **9**, 81–85, (1962).

6. Lorenz, J. J. and B. B. Mikic: The Effect of Thermocapillary Flow on Heat Transfer in Dropwise Condensation, Journal of Heat Transfer, **92**, 1, 46-52 (1970).
7. Hurst, C. J. and D. R. Olson: Conduction through Droplets during Dropwise Condensation, Journal of Heat Transfer, **95**, 1, 12-22, (1973).
8. Volkov, V. T.: Determination of the Coefficient of Heat Transfer between Vapor and Wall in Dropwise Condensation of Steam on a Horizontal Tube. Godishnik na NIPPIES Energoproyekt, **16**, 127-135, (1972).
9. Tower, R. E. and J. W. Westwater: Effect of Plate Inclination on Heat Transfer during Dropwise Condensation of Steam, Chemical Engineering Progress, Sympos. series, **6**, 102, 21-25 (1970).

EFFECT OF THE GEOMETRICAL FACTOR ON CRITICAL HEAT FLUX IN A ROD BUNDLE*

T. Kobori, A. Kikuchi, T. Obata, and M. Matsuo
Power Reactor and Nuclear Fuel Development Corporation, Japan

1. INTRODUCTION

In the thermal design of liquid-cooled nuclear reactors, an important aspect is to allow a burnout margin based on the limit of critical heat removal by the coolant, with a view to preventing thermal failure of the fuel cladding. These phenomena have been examined by various researchers from different points of view [1]–[3], but so far, most experiments have been done on nominal dimensions of rod bundles. In order to achieve a higher reliability of reactor fuel, several tests were carried out to measure the critical heat flux as one of the limits of reactor safety, simulating actual abnormal conditions in the reactor core [4], [5].

In this report, an experimental study and subchannel analysis is described, regarding the effect of rod bundle eccentricity on critical heat flux in a pressure tube type reactor.

In this type of reactor, it is necessary to provide a gap between the rod bundle and the pressure tube to prevent excessive damage to the surface of the pressure tube, and to facilitate the work of refueling. The maximum gap is generally of the order of 1 mm. In an extreme case, there is some possibility that a rod bundle may move to one side to the extent of the gap. In these experiments, using actual spacers, burnout was found to take place at one side of the rod bundle.

2. EXPERIMENTAL APPARATUS AND METHOD

2.1 Apparatus

The experiment was conducted in a heat transfer loop, with electric power of 14 MW for heating the test section. The loop consisted of a steam drum,

---

*The abstract of this paper appears on page 374.

cooler, test section, circulating pump, high-pressure condenser, heating power supply system, and other components. The maximum temperature, pressure and flow rate were respectively, 310°C, 100 bars, and 22 kg/s. The circulating fluid used was demineralized water with a conductivity of 1 μS-cm.

The system pressure was adjusted by controlling the heat removal rate in the high-pressure condenser. Water separated from the steam-water mixture in the drum, was cooled in the subcooler, circulated by the pump at a measured flow rate and heated by a preheater with a heating capacity of 1.2 MW with adjusted inlet subcooling and then fed into the test section.

The heating power supply system consists of a 15 MVA transformer with an onload tap changer and two units of transformer/SCR rectifier, each with a maximum power of 7 MW.

2.2 Test Section

The full scale bundle of 28 rods used for the experiment was heated by electricity. High electrical current was passed directly through the cladding of the fuel rods to simulate the very high heat generation rates of a power reactor.

In order to control rod bundle eccentricity, the experiment used specially manufactured spacers, which were attached directly to the pressure tube to adjust eccentricity to between 0 and 1 mm with a tolerance of ± 0.05 mm.

Simulated fuel rods with an outside diameter of 16.5 mm, were set in position by the use of the spacers, in three concentric circles 16, 8, and 4 rods from the outside inward. The effective heated length was 3.7 m. Electric bridge type detectors were attached to each rod to detect burnout and to lower and/or break the heating power as quickly as possible to prevent rod failure and to yield data repeatedly and effectively.

2.3 Experimental Method

The critical heat flux was measured in the same way as in the usual steady-state burnout experiment: Heating power was gradually increased, while maintaining the system pressure, flow rate and inlet subcooling at fixed values, until one or more burnout detectors indicated a temperature rise due to burnout. The sensitivity of the detectors was calibrated and the setting point to trip the power was adjusted in advance at an adequate value to prevent rod failure.

Experimental conditions were as follows: pressure 70 bars, mass velocity 1000 ~ 2300 kg/m^{2}s, inlet subcooling 40 ~ 600 J/g. A flow diagram of the loop is shown in Fig. 1.

3. RESULTS AND DISCUSSION

The experimental data are shown in Figs. 2 and 3 for three bundle eccentricities and three mass velocities. These data show that critical heat flux

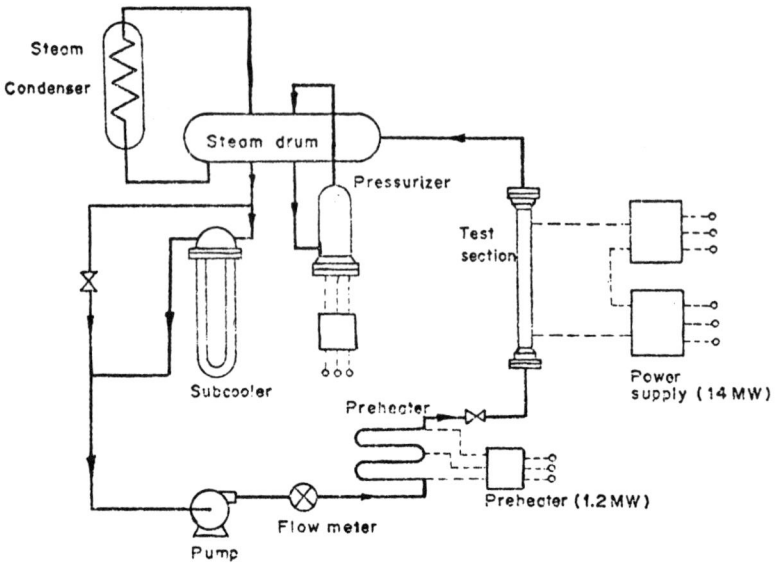

FIG. 1. Flow diagram of 14 MW heat transfer loop.

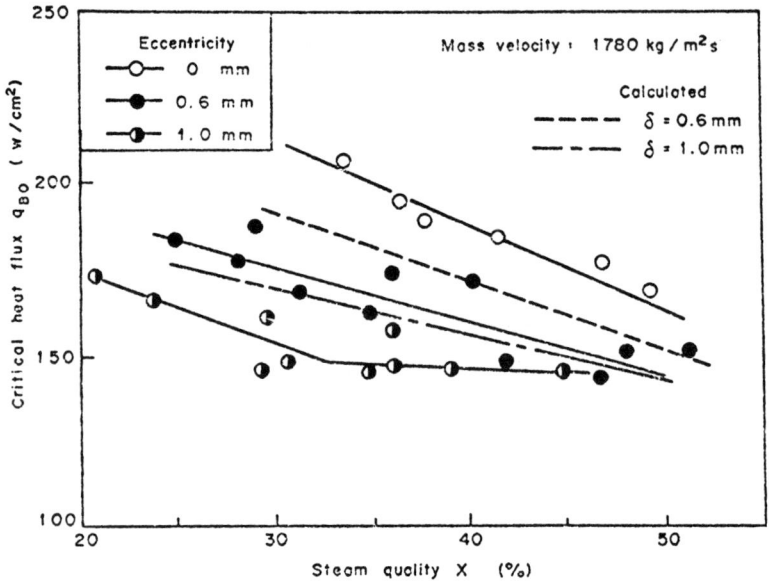

FIG. 2. Effect of rod cluster eccentricity.

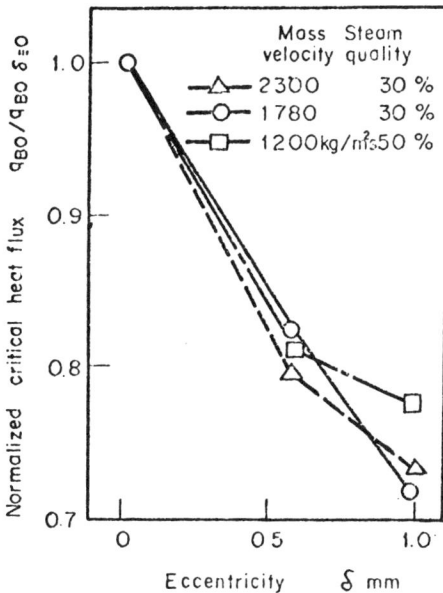

FIG. 3. Effect of rod bundle eccentricity.

is strongly dependent on bundle eccentricity: critical heat flux decreases by 18 ~ 22 percent and 22 ~ 28 percent, respectively, for 0.6 and 1.0 mm of eccentricity, compared with zero eccentricity.

In order to investigate these effects from a subchannel distribution point of view [6], three basic conservation equations for mass, momentum, and energy were solved in the single- and two-phase flow in the rod bundle.

Figure 4 gives a typical example of subchannel distribution of local steam quality deviation from the cross sectional averaged value for 0.6 mm of eccentricity: the analysis shows that the maximum local steam quality in the hottest subchannel reaches 31.3 percent with an average value of 21.9 percent of steam quality. On the contrary, in the concentric case, the maximum local steam quality for the same condition as above is 25.2 percent; therefore 0.6 only of eccentricity is considered enough to produce an increase of 6.1 percent in local steam quality.

Since the critical heat flux decreases, in general, in parallel with steam quality, it becomes clear, from a calculation utilizing the gradient of experimental correlation, that the critical heat flux decreases by approximately 13 percent.

In a similar manner, calculations show that the maximum local steam quality deviation is 12.7 percent for 1.0 mm of eccentricity, which decreases the critical heat flux by 18 percent.

Thus, it was found that the subchannel analysis can describe the effect of rod bundle eccentricity, and the calculated value qualitatively agrees well with the experimental data.

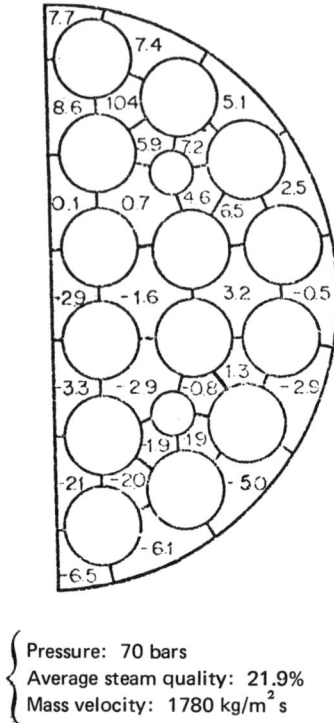

Pressure: 70 bars
Average steam quality: 21.9%
Mass velocity: 1780 kg/m^2 s

FIG. 4. Distribution of local steam quality deviation at burnout (calculated).

After each experiment, the test section was disassembled to investigate the traces of overheating of the rods, and to check the location of burnout. These inspections showed that the burnout locations were also affected by bundle eccentricity.

4. CONCLUSIONS

Steady-state critical heat flux measurements were carried out to study the effect of rod bundle eccentricity in boiling water flow with a full-scale rod bundle for a pressure tube type reactor.

From the experiment, the following conclusions were reached.

1. The critical heat flux is affected considerably by rod bundle eccentricity.
2. The effect is considered to be produced by an increase of deviation in local steam quality from the average value in the rod bundle.
3. The effect of eccentricity should be taken into consideration in the design of water-cooled nuclear power reactors.

ACKNOWLEDGMENT

The authors wish to express gratitude to their colleagues of the Heat Transfer Laboratory, PNC O-arai Engineering Center for operation of the loop, and for the design and manufacture of the test sections.

REFERENCES

1. Tong, L. S.: USAEC Report TID-25887 (1972).
2. Nylund, O., et al: FRIGG-4 (1970).
3. Caspari, G. P., et al: Int. Meeting on Heat Transfer (1973), No. 63.
4. Lahey, Jr., R. T., et al: *Ibid,* No. 5.
5. Israel, S., et al: Trans. ASME, Ser. C (1969) 91-3, 335.
6. Rowe, D. S.: USAEC Report BNWL-1229 (1970).

FILM BOILING OF FREON-114 DROPS AT PRESSURES UP TO CRITICAL*

T. R. Rhodes and K. J. Bell
Oklahoma State University
Stillwater, Oklahoma

I. INTRODUCTION

The first scientific investigation of the film boiling of discrete liquid masses was by J. G. Leidenfrost [7], in 1756. Leidenfrost noted that small spherical droplets formed on a hot smooth iron spoon took a relatively long time to evaporate. In his honor, the stable film boiling of discrete masses is termed the Leidenfrost Phenomenon. The surface temperature at which the minimum heat flux occurs is termed the Leidenfrost Point.

Previous Work

Since Leidenfrost first recorded his observations, measurements of the behavior of discrete masses undergoing film boiling have been made for many fluids and diverse surfaces. Bell [3] has surveyed the literature on the Leidenfrost Phenomenon prior to 1967.

Until recently the bulk of the research has been conducted at atmospheric pressure. Adadevoh [1] has studied the evaporation of benzene, isooctane, and *n*-cetane drops at pressures from atmospheric to 6.9 bar in a nitrogen atmosphere. The results were somewhat anomalous in that there was not a distinct maximum in the lifetime-temperature difference curve as observed by subsequent investigators.

Temple–Pediani [11] has investigated the drop lifetime behavior under pressures from 1–70 bars for *n*-hexane, *a*-methylnaphthalene, and *n*-hexadecane. At subcritical pressures Temple–Pediani's results were similar to previous investigations at atmospheric pressure. The drops show a minimum lifetime at temperature differences corresponding to the maximum heat flux in nucleate boiling, and a maximum lifetime in film boiling corresponding to the Leidenfrost Point. In the supercritical pressure region the drop lifetime

---

*The abstract of this paper appears on page 382.

decreases with increasing temperature until the critical temperature is reached on the hot surface. At a surface temperature of approximately $60°C$ above the critical for the fluids studied the drop lifetime becomes independent of further increases in surface temperature.

The film boiling of n-pentane and n-hexane drops on copper, brass, aluminum, and stainless steel surfaces was studied by Nikolayev, Bychenkov, and Skripov [8]. The investigation covered a pressure range from atmospheric to a reduced pressure of 0.8 in a saturated atmosphere. The evaporation rate for n-hexane on a brass surface was a maximum at a reduced pressure of 0.25. The authors did not present any evaporation rate data for n-pentane nor for any other surface.

Emmerson [4] investigated the film boiling of water drops on various surfaces in a nitrogen atmosphere up to 6.2 bars. For the brass and stainless steel surfaces the Leidenfrost Point increased with increasing pressure, while it was almost constant for the monel surface. The maximum evaporation time was found to be essentially independent of the thermal diffusivity. The author found that the lifetime-plate temperature data agreed with Baumeister's [2] correlation for vaporization times.

II. THEORETICAL DEVELOPMENT

The development of the model closely parallels the formulation presented by Gottfried, Lee, and Bell [6]. The primary differences are in the analysis of the radiant heat transfer and the technique of solving the differential equation for the mass balance. For a detailed discussion the reader is referred to Refs. [6] and [9].

Several physical processes are postulated to occur at the upper and lower surfaces of the drop (Fig. 1). On the lower surface heat is transferred by conduction through the flowing vapor film between the hot surface and the drop. Radiant heat transfer occurs from the hot surface and the surroundings to the entire surface of the drop. Mass is removed from the upper surface by molecular diffusion into the surrounding inert atmosphere. Mass is removed from the lower surface by evaporation into the vapor film. A radial pressure gradient beneath the drop results from the flow of the vapor from the bottom

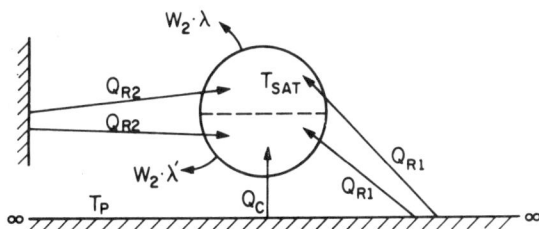

FIG. 1. Heat and mass transfer processes for a spherical drop in film boiling.

of the drop. The integrated pressure gradient, in turn, provides the force necessary to support the drop above the heated surface.

The corresponding conservation equations can be written for the drop. The drop is considered to be spherical and at its saturation temperature throughout its lifetime. The vapor film beneath the drop is assumed to be superheated to a temperature halfway between the saturation and heat surface temperatures.

Q_c is the heat conducted into the bottom of the drop, and Q_{R1} and Q_{R2} are the net heats radiated to the drop from the heated surface and the surroundings, respectively. W_1 is the rate of evaporation from the lower surface and W_2 is the rate from the upper surface. A mass balance on the drop may be written

$$\rho_l \frac{dV}{dt} = -(W_1 + W_2) \tag{1}$$

and a heat balance may be written as

$$Q_c + Q_{R1} + Q_{R2} = W_1\lambda' + W_2\lambda \tag{2}$$

Q_c can be calculated from Fourier's Law, and the height above the plate obtained from the force balance. Q_{R1} and Q_{R2} are calculated from an analysis of the radiant heat transfer to the drop. W_2 can be calculated since the rate of diffusion into a semi-infinite medium is known from Froessling [5]. Thus, in Eq. (2) the only unknown is W_1. With the values of W_1 and W_2, Eq. (1) can be integrated by the method of Runge and Kutta to give the drop volume and diameter as a function of time. Typical results predicted by the model are shown on the graphs of experimental results.

III. EXPERIMENTAL APPARATUS AND PROCEDURE

The experimental cell was made from 316 stainless steel, with a quartz window in the top to permit photographing the drop. On the inside bottom surface of the cell a depression was machined to keep the drop in sight. Five chromel-alumel thermocouples were inserted from the bottom of the cell to within 0.4 mm of the surface. The bottom surface of the cell was polished to 5-10 $\times$ 10^{-6} cm RMS roughness and then oxidized to a dull golden brown. The bottom of the cell was heated by two electric resistance heating units. The cell was pressurized by dry nitrogen. A mercury drive system was used to inject a drop into the surface through a hypodermic needle. Details of the apparatus are given in [9].

The experimental procedure consisted of adjusting the electric heaters to the voltage desired and allowing the cell to heat. Inert gas was then purged through the system to remove any air. After the plate reached the desired

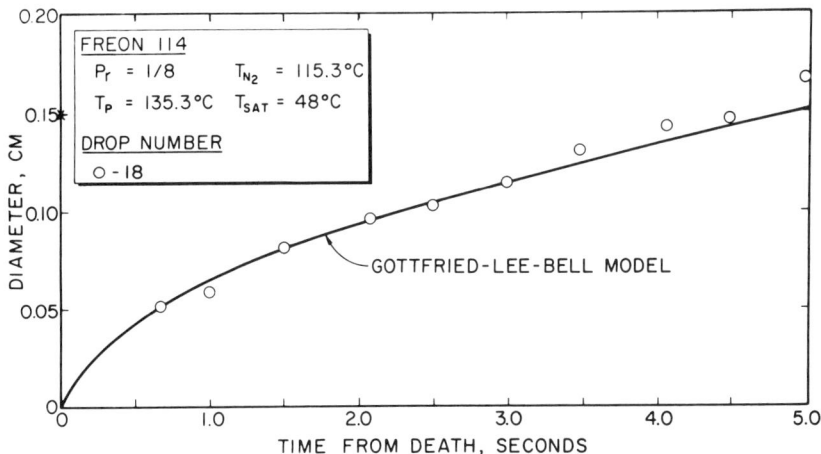

FIG. 2. Freon-114 drop diameter history at $P_r = 1/8$.

temperature, a drop was placed on the surface and allowed to evaporate. During the vaporization process a motion picture camera recorded the drop diameter history.

IV. RESULTS

Complete graphical and tabular representations of results are given in Ref. [9]. A few representative results are given in Figs. 2-5. In Figs. 2-4, the measured diameter of the Freon-114 drop is plotted vs. time from death (i.e., from the complete evaporation of the drop). The asterisk on the ordinate indicates the diameter above which the drop may no longer be considered

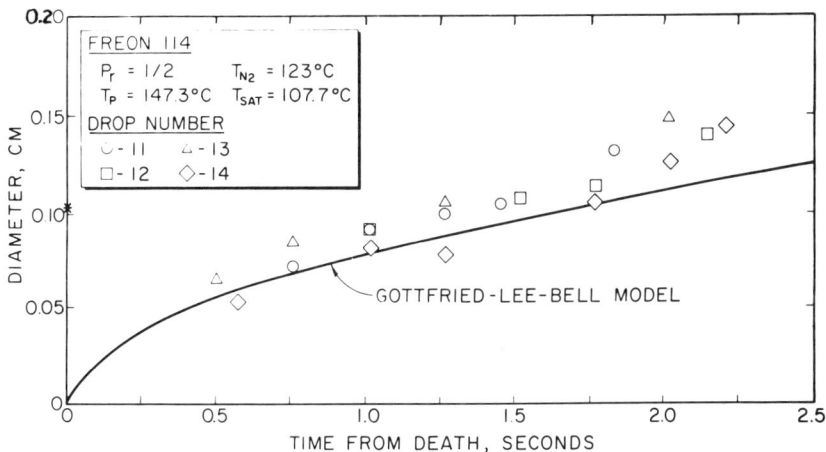

FIG. 3. Freon-114 drop diameter history at $P_r = 1/2$.

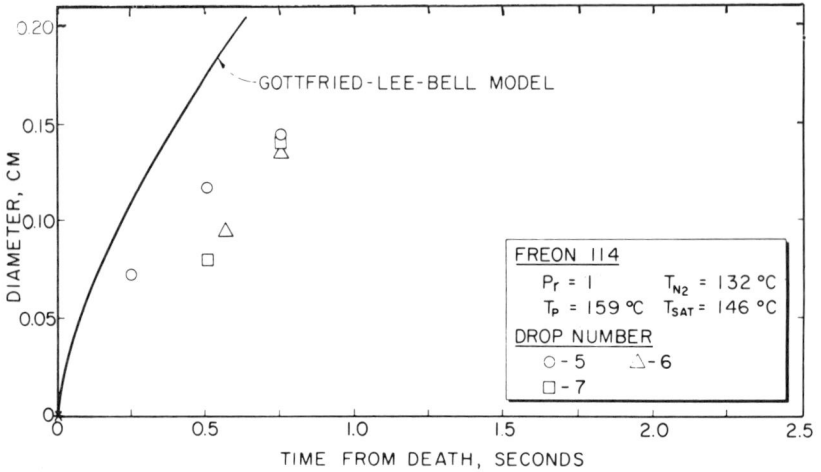

FIG. 4. Freon drop diameter history at $P_r = 1$.

spherical, according to the criterion of Baumeister [2]. The surface temperature of the plate (T_P), the measured temperature of the nitrogen atmosphere above the drop (T_{N2}) and the saturation temperature of the drop at the system pressure are given in the legend block.

From Figs. 2 and 3, it is seen that theory and experiment are in good agreement at least up to the limit of the sphericity assumption. For reduced pressures of 3/4 and 1, however, the agreement is only qualitative, as indicated by Fig. 4. Probably the main cause of the latter discrepancies is that the model is too simple to account for the great changes in physical properties over relatively small temperature ranges in this pressure range. Also, the physical properties of Freon-114 are not well known near the critical.

It is possible to cross-plot the data in [9] to obtain graphs of drop lifetime vs. temperature difference for a given pressure and initial drop diameter, as in

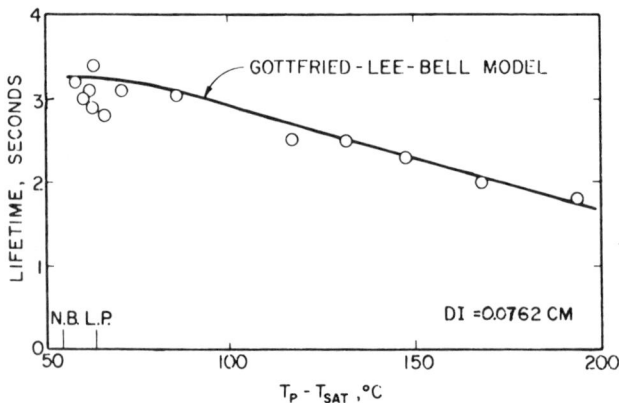

FIG. 5. Freon-114 drop lifetime at $P_r = 1/8$.

Fig. 5. Again, the generally good agreement with theory at P_r = 1/8 is evident. "L. P." denotes the Leidenfrost point, the lowest temperature difference at which film boiling was well established. "N. B." denotes the highest temperature difference at which nucleate boiling existed.

V. CONCLUSIONS

The Gottfried-Lee-Bell model predicts the behavior of Freon-114 drops quantitatively from P_r = 1/8 to P_r = 1/2. At reduced pressures of 3/4 and 1, the agreement is only qualitative.

NOMENCLATURE

Q_c	Total heat conducted through vapor film generated on bottom surface of drop (cal/sec).
Q_{R_1}	Net radiant heat transfer from plate to drop (cal/sec).
Q_{R_2}	Net radiant heat transfer from surroundings (cal/sec).
t	Time (sec).
T_P	Temperature of heated plate (°C).
T_{Sat}	Saturation temperature at cell pressure (°C).
T_{Surr}	Temperature of surroundings (°C).
V	Volume (cm³).
W_1	Rate of mass loss over lower half of drop (g/sec).
W_2	Rate of mass loss over upper half of drop (g/sec).
λ	Latent heat of vaporization (cal/gm).
λ'	Latent heat corrected for superheating (cal/gm).
ρl	Liquid density (gm/cm³).

REFERENCES

1. Adadevoh, J. K.: M.S. Thesis, University of Wisconsin, Madison, 1962.
2. Baumeister, K. J., T. D. Hamill, and G. J. Schoessow: Proceedings Third International Heat Transfer Conference, Chicago, 1966, IV, pp. 66–75.
3. Bell, K. J.: Chemical Engineering Progress Symposium Series, 1967, 63, No. 79, pp. 73–82.
4. Emmerson, G. S.: Int. Jour. Heat Mass Transfer, 1975, 18, pp. 381–386.
5. Froessling, N.: Gerlands Beitr. Geophys., 1938, 52, p. 170.
6. Gottfried, B. S., C. J. Lee, and K. J. Bell: Int. Jour. Heat Mass Transfer, 1966, 9, pp. 1167–1188.
7. Leidenfrost, J. G.: Ibid., 1966, 9, pp. 1153–1156.
8. Nikolayev, G. P., V. V. Bychenkov, and V. P. Skripov, Heat Transfer-Soviet Research, 1974, 6, No. 1, pp. 128–132.
9. Rhodes, T. R.: Ph.D. Thesis in preparation, Oklahoma State University, Stillwater, 1976.
10. Sciance, C. T., and C. P. Colver: Trans. ASME, 1970, 92, pp. 659–661.
11. Temple-Pediani, R. W.: Proc. Instn. Mech. Engrs., 1969–1970. 184, Pt. 1, No. 38, pp. 677–696.

FORCED CONVECTION LAMINAR FILM CONDENSATION ON ARBITRARY HEAT-FLUX SURFACES*

K. N. Seetharamu and M. V. Krishnamurthy
Indian Institute of Technology, Madras, India

INTRODUCTION

Recent engineering development such as nuclear reactors, aerospace applications, etc., require the knowledge of heat transfer in the condensation process under forced flow. Cess [1] considered forced convection film condensation over a flat plate by neglecting inertia forces, body forces, and convected energy in the governing liquid-flow equations. Koh [2] treated the same problem by including the inertia terms in the liquid-layer equation. Sparrow et al. [3] determined the effects of noncondensible gas and interfacial resistance on the forced convection film condensation. Isa and Chen [4] considered the effect of pressure gradients on laminar forced film condensation on liquids of small Prandtl number. All the above analyses assume an isothermal wall condition. The present investigation deals with the analysis of the wall heat flux, assumed to be an arbitrary function of the distance x, from the leading edge. The only assumption is that it is a continuously differentiable function with respect to x.

ANALYSIS

The physical model and coordinate system are shown in Fig. 1. As is customary for condensation problems, the condensation process is assumed to be controlled by the flow within the liquid layer and is not limited by the supply of vapor at the interface. The boundary layer form of the conservation laws for the liquid layer are as follows:

Continuity
$$\frac{\partial u}{\partial x} + \frac{\partial v}{\partial y} = 0 \tag{1}$$

---

*The abstract of this paper appears on page 384.

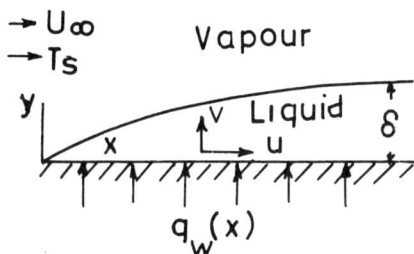

FIG. 1. Physical Model.

Momentum $\qquad\qquad u\dfrac{\partial u}{\partial x} + v\dfrac{\partial u}{\partial y} = \nu\dfrac{\partial^2 T}{\partial y^2}$ $\qquad\qquad$ (2)

Energy $\qquad\qquad u\dfrac{\partial T}{\partial x} + v\dfrac{\partial T}{\partial y} = \dfrac{K}{C_p\rho}\dfrac{\partial^2 T}{\partial y^2}$ $\qquad\qquad$ (3)

Boundary and interface conditions are as follows:

At $\qquad\qquad\qquad\qquad y = 0, \quad u = v = 0$ $\qquad\qquad$ (4)

and $\qquad\qquad\qquad\qquad -K\left(\dfrac{\partial T}{\partial y}\right)_w = q_w(x)$ $\qquad\qquad$ (5)

where $q_w(x)$ is an arbitrary function of x.

At $\qquad\qquad\qquad\qquad y = \delta(x), \quad \dfrac{\partial u}{\partial y} = 0$ $\qquad\qquad$ (6)

$$T = T_s \qquad\qquad (7)$$

$$\left(u\dfrac{d\delta}{dx} - v\right)\rho h_{fg} = k\dfrac{\partial T}{\partial y} \qquad\qquad (8)$$

The following transformations are used for reducing equations (1) to (8):

$$\eta = \dfrac{y}{\delta(x)} = \dfrac{v}{\Delta}\sqrt{\dfrac{U_\infty}{\nu x}} \qquad\qquad (9)$$

Dimensionless stream function

$$F(\eta,\lambda_0,\lambda_1,\ldots,\lambda_n) = \dfrac{\phi(x,y)}{\Delta\sqrt{U_\infty \nu x}} \qquad\qquad (10)$$

Dimensionless temperature

$$\theta = \frac{k}{q_w(x)\Delta} \sqrt{\frac{U_\infty}{vx}}(T - T_s) \tag{11}$$

where $\Delta = \Delta(\{\lambda_n\})$ and $\{\lambda_n(x)\} = \lambda_0(x), \lambda_1(x), \ldots, \lambda_n(x)$

a set of yet undetermined functions of x. The resulting equations still contain the variable x explicitly which is eliminated by the use of λ_n defined as follows.

$$\lambda_n = \frac{x^{n+1/2}}{q_w(x)} \frac{d^{n+1}[q_w(x)x^{1/2}]}{dx^{n+1}} \tag{12}$$

To reduce the resulting equations to ordinary differential equations, the following expansions of F, θ, and Δ are used

$$\begin{matrix} F(\eta,\lambda_n) & P(\eta) & & \\ 0(\eta,\lambda_n) = & H(\eta) + \lambda_0 \\ \Delta & \Gamma & \end{matrix} \begin{Bmatrix} F_0(\eta) \\ \theta_0(\eta) \\ \Delta_0 \end{Bmatrix} + \lambda_1 \begin{Bmatrix} F_1(\eta) \\ \theta_1(\eta) \\ \Delta_1 \end{Bmatrix} + \cdots \tag{13}$$

where P, H, and Γ are isothermal wall solutions and F_0, θ_0, Δ_0, F_1 etc. are perturbed solutions to be determined from the analysis. Substituting Eq. (13) and collecting the coefficients of 1, λ_0, λ_1, etc., a set of infinite equations are obtained, the first few of which were solved on an IBM 370/155 computer, corresponding to the condensation of steam and other refrigerant vapors. The wall temperature variation compared to the isothermal case is given by

$$\frac{T_w - T_s}{(T_w - T_s)_{\text{ISO}}} = 1 + \lambda_0 \left[\left(\frac{\theta_{0w}}{H_w} + \frac{\Delta_0}{\Gamma} \right) \right] + \lambda_1 \left[\left(\frac{\theta_{1w}}{H_w} + \frac{\Delta_1}{\Gamma} \right) \right] + \cdots \tag{14}$$

RESULTS AND DISCUSSION

From the solutions, we get the functions, P, H, Γ, F_0, F_1, O_0, etc., which are universal in nature. Using these, some specific cases for the condensation of steam will be discussed.

1. Power law variation of heat flux given by $q_w(x) = Mx^m$ where M and m are constants. Using Eq. (12), λ_0 and λ_1 are calculated to be $(m + \frac{1}{2})$ and $(m + \frac{1}{2})$ $(m - \frac{1}{2})$. Using these values and the universal functions the

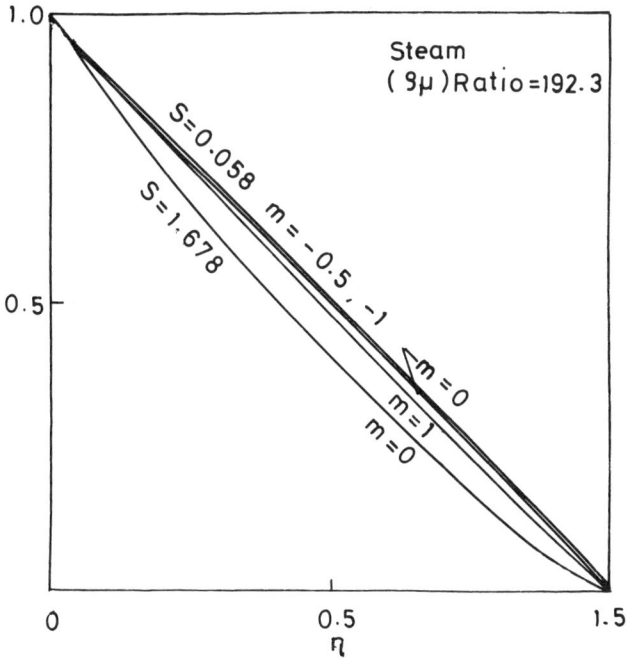

FIG. 2. Temperature distribution in condensate film

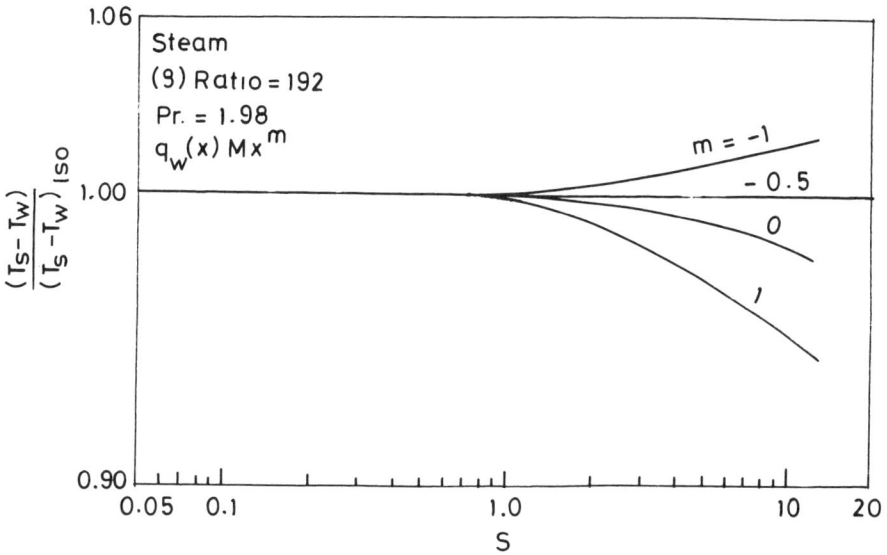

FIG. 3. Wall temperature variation.

temperature profile in the condensate film, Fig. 2, and the wall temperature variation compared to the isothermal case are calculated, Fig. 3. The isothermal case corresponds to $m = -0.5$ and $m = 0$ corresponds to the uniformly heated plate. As the value of m increases, the effect of convection increases in the liquid film as seen from Fig. 2. The same is observed as S increases. It can be observed from Fig. 3 that the wall temperature deviation decreases as S increases compared to the isothermal case so long as m is positive. This is in agreement with the temperature profile shown in Fig. 2; when m is negative it increases. For small values of S (up to 1) the deviation is practically zero. The maximum deviation is about 6.5 percent at $S = 16$. A similar type of behavior is observed for the condensation of refrigerant vapors like F-12, F-22, and ammonia.

2. Exponential variation of heat flux given by $q_w(x) = Me^{mx}$. Using Eq. (12), λ_0 and λ_1 can be calculated. The temperature profile and wall temperature deviation can be calculated as above. The trend observed in case 1 is also seen in this case, however, to a lesser degree and hence not shown here.

NOMENCLATURE

C_p	Specific heat	U_∞	Free stream velocity
h_{fg}	Latent heat	ρ	Density
K	Thermal conductivity	δ	Film thickness
q_w	Heat flux at the wall	η	Similarity variable
Pr	Prandtl number	ν	Kinematic viscosity
S	$C_p(T_s - T_w)/h_{fg}$ Pr	$(\rho\mu)$	Ratio $[(\rho\mu)_L/(\rho\mu)_v]^{1/2}$
T	Temperature		

Subscripts

L	Liquid
S	Saturated
Iso	Isothermal

REFERENCES

1. Cess, R. D. Z. angew. Math. Phys., 1960, **11**, p. 426.
2. Koh, J. C. Y.: Int. J. Heat Mass Transfer, 1962, **5**, p. 941.
3. Sparrow, E. M., W. J. Minkowyez, and M. Saddy: Int. J. Heat and Mass Transfer, 1967, **10**, p. 1829.
4. Isa, I. and Ching-Jenchen: J. Heat Transfer, 1972, **94**, p. 99.

MAXIMUM AND MINIMUM HEAT FLUX IN NEAR CRITICAL POOL BOILING *

E. Hahne
University of Stuttgart, Germany

G. Feurstein
Escher Wyss, Ravensburg, Germany

1. INTRODUCTION

Higher efficiencies in power stations call for higher temperatures and heat fluxes. In steam generation the thermodynamic critical point represents an upper limit for evaporation temperatures while the maximum heat flux marks the upper limit for high heat transfer nucleate boiling and possibly the point of damage with a burnout for the heater. The maximum heat flux density q_{max} characterizes the transition from the nucleate to the film boiling regime. The minimum heat flux q_{min} on the other hand indicates the transition from the film boiling to the nucleate boiling regime characterized by the so-called Leidenfrost point. The technical importance may appear in cryogenics and its applications. Thus q_{min} represents the multitude of effects responsible for a change in boiling regime in the opposite direction as q_{max}. Both values do not coincide. In saturated pool boiling the appearance of a maximum heat flux is considered to be a hydrodynamic phenomenon. The value of the maximum heat flux density q_{max} is determined by a hydrodynamic instability in the removal of vapor from the heater. In power controlled heating systems—as are considered here—q_{max} indicates a local overheating which initiates the formation of an insulating vapor layer across the entire heating surface.

The stochastic character of the phenomenon and the great many parameters involved such as wettability, heat capacity and geometry of the heater, mode of heating [1], fluids, convection flows, and edge effects made theoretical predictions from experimental investigations difficult. Basically the same parameters control the existence of the minimum heat flux. There is, however, a lack of measurements of q_{min} across a larger area of pressures up to the critical. Correlations from literature for both q_{max} and q_{min} in most

---

*The abstract of this paper appears on page 387.

cases give satisfactory agreement to experimental data at atmospheric and moderate pressure conditions.

Effects of heater geometry are usually not taken into account.

It was the intention of this investigation to provide experimental data for three different fluids within a wide pressure range up to the thermodynamic critical point and for four different heater diameters. In order to reduce the influencing parameters and to obtain basic results a comparatively simple experimental setup with a heated wire in a pressure vessel was used.

2. EXPERIMENTAL PROCEDURE

The pressure vessel and measuring equipment are described in detail in [2]. The experiments were performed with horizontal platinum wires of diameters d = 0.5, 0.1, 0.2, and 0.3 mm. These wires were used both as heating elements and as resistance thermometers giving the overall heater temperature. As test fluids were used Carbon dioxide (CO_2), Chlortrifluormethane (R 13), and Trifluormethane (R 23). Changes in the thermodynamic state of the bulk fluid were obtained by heating the entire pressure vessel in a water bath. Glass windows on both ends of the pressure vessel allowed direct observation of the boiling phenomena. The experiments for maximum heat flux were performed in the following way: for a constant thermodynamic state a range of values for maximum heat flux was registered in preliminary experiments. In the final experiments the heat flux was increased in very small increments starting from a value when safely nucleate boiling is obtained, up to the value when a transition to film boiling occurs. This transition was registered both as a temperature increase and a change in flow pattern. The heat fluxes immediately below and above this transition characterized a lower and upper boundary for q_{max}. The value of q_{max} was taken as the arithmetic mean value of these upper and lower boundary values. These measurements were performed for five consecutive tests in intervals of 10 min to restore equal initial positions. Data points given are obtained as average data from these five tests. The maximum scatter was below ± 10%.

The experiments for minimum heat flux were performed in the same way with the heat flux being decreased from a value safely providing film boiling. The maximum scatter in these tests was below ± 20%.

3. RESULTS AND COMPARISON

3.1 Maximum Heat Flux Density

The experimental results are plotted in diagrams giving the heat flux density q_{max} versus the pressure ratio p/p_c. In Figs. 1, 2, and 3 data for CO_2, R 13, and R 23 are presented, respectively. As a parameter in each figure the diameter of the platinum wire is used.

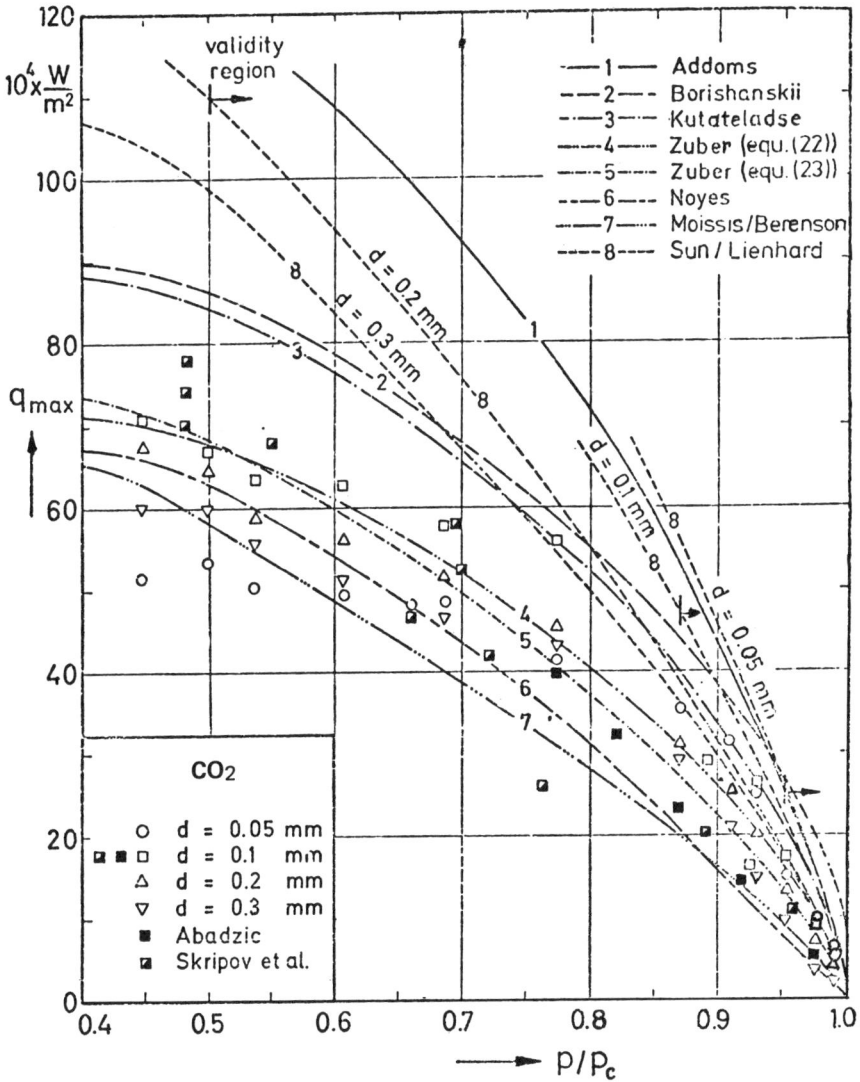

FIG. 1

FIGS. 1, 2, 3. Comparison of experimental data to various predictions from literature for maximum heat flux density vs. reduced pressure and three different fluids: CO_2, R13, R23.

The lines drawn in the diagrams represent various correlations from literature. The line denoted by 1 gives an equation by Addoms [3] which was found from experiments on horizontal platinum wires and tubes in water and on flat plates in organic liquids. It is in the form

$$q_{max} = 2.4 \, h_e \rho_v \sqrt{\frac{\rho_1 - \rho_v}{\rho_v}} \sqrt[3]{\frac{g\lambda_1}{\rho_1 c_{pl}}} \qquad (1)$$

The constant 2.4 in Eq. (1) was determined for moderate pressures only.

Kutateladse [4] using dimensional analysis and a great many experimental results derived the following equation for flat plates

$$q_{max} = 0.16 h_e \sqrt{\rho_v} \sqrt[4]{\sigma g(\rho_1 - \rho_v)} \qquad (2)$$

while Borishanskij [5] extended the model underlying Eq. (2) by taking into account the influence of viscosity to obtain

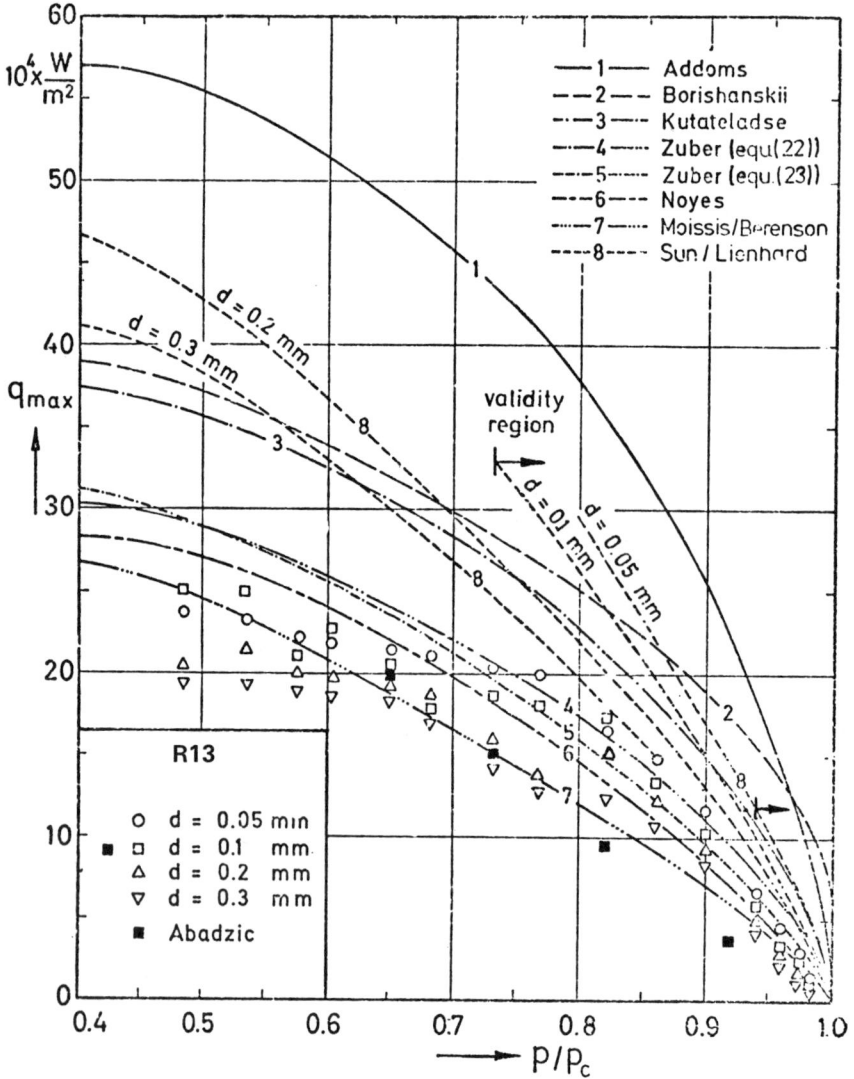

FIG. 2

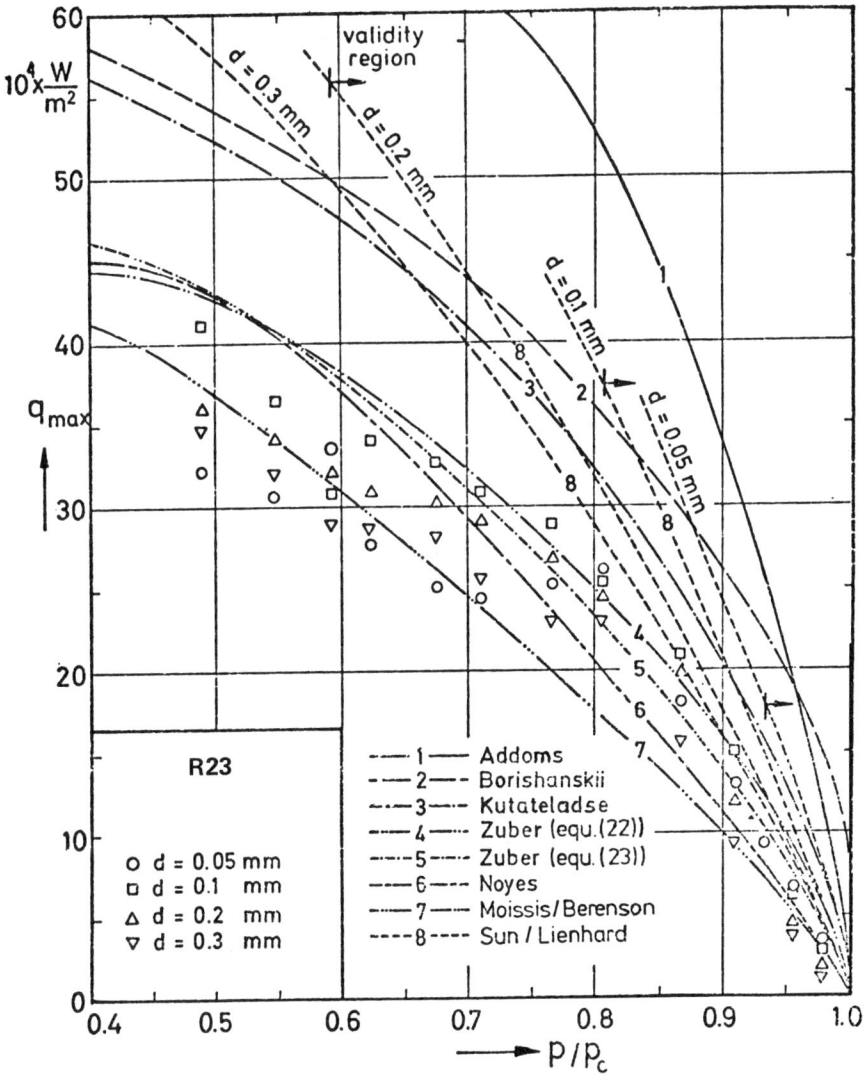

FIG. 3

$$q_{\max} = h_e \sqrt{\rho_v} \sqrt[4]{g(\rho_1 - \rho_v)} \left\{ 0.13 + 4\left[\frac{\eta^2}{\rho_1}\sqrt{\frac{g(\rho_1 - \rho_v)}{\sigma^3}}\right]^{0.4}\right\} \qquad (3)$$

From the theory of hydrodynamic instability Zuber [6] and Zuber/Tribus [7] derived an analytical relation in the following form

$$q_{max} = \frac{\pi}{24} h_e \sqrt{\rho_v} \sqrt[4]{\sigma g(\rho_1 - \rho_v)} \sqrt{\frac{\rho_1}{\rho_1 + \rho_v}} \tag{4}$$

represented as curve No. 4 in Figs. 1, 2, and 3.

This was generalized by Zuber et al. [8] into

$$q_{max} = \frac{\pi}{24} \frac{16 - \pi}{16 - \pi(1 - \rho_v/\rho_1)} \cdot h_e \sqrt{\rho_v} \sqrt[4]{\sigma g(\rho_1 - \rho_v)} \sqrt{\frac{\rho_1 + \rho_v}{\rho_1}} \tag{5}$$

shown as curve No. 5 in Figs. 1, 2, and 3.

Upon Zuber's theory Moissis and Berenson [9] developed the following equation

$$q_{max} = 0.18 \frac{h_e \sqrt{\rho_v} \sqrt[4]{\sigma g(\rho_1 - \rho_v)} \sqrt{(\rho_1 - \rho_v)/\rho_1}}{1 + 2\sqrt{\rho_v/\rho_1} + \rho_v/\rho_1} \tag{6}$$

The constant 0.18 in Eq. (6) has to be determined by experiment and depends on the surface effects such as wettability and roughness.

Noyes [10] found an empirical correlation based on his sodium experiments

$$q_{max} = 0.144 \cdot h_e \cdot \rho_v \sqrt{\frac{\rho_1 - \rho_v}{\rho_v}} \sqrt[4]{\frac{\sigma g}{\rho_1}} \left(\frac{\nu_1}{\lambda_1} \rho_1 c_{pl} \right)^{-0.245} \tag{7}$$

All these correlations only give the effect of thermophysical properties upon q_{max}, while all other effects are combined in the constant factor. These must be determined for each fluid/heater combination separately.

One step toward a further disintegration of influencing parameters was taken when Lienhard/Dhir [11] noticed by photographic observation a pronounced influence of heater geometry upon size and distribution of vapor columns at near critical boiling conditions.

Sun/Lienhard [12] developed a hydrodynamic model based on the former observation and dimensional analysis to obtain the following equation:

$$\frac{q_{max}}{q_{max F}} \approx 0.890 + 2.27 \exp(-2.43\sqrt{D'}) \tag{8}$$

with

$$q_{max F} = 0.131 h_e \sqrt{\rho_v} \sqrt[4]{\sigma g(\rho_1 - \rho_v)}$$

and

$$D' = \frac{d}{\sqrt{\sigma/g(\rho_1 - \rho_v)}}$$

valid in a region $D' > 0.30$.

The comparison of experimental data and correlations in Figs. 1, 2, and 3 show that the general tendency of a decreasing maximum heat flux density with an increasing pressure ratio is correctly given by all equations. At the critical point, at a pressure ratio $p/p_c = 1$ the maximum heat flux density approaches the value zero, meaning that in this region a transition from nucleate into film boiling is obtained at very small heat inputs. This again is easily explained by the effect of surface tension which approaches the value zero and the cessation of buoyancy effects at the critical point.

In absolute values, agreement within the scatter of all data is found for equations (4), (5), (6), and (7).

A distinct influence of wire diameters is not clear taking scatter of data into account. The tendency of a higher maximum heat flux for a smaller diameter wire as given in Eq. (8) is not followed by the very thin wire of $d = 0.05$ mm, but is clearly obtained for the larger wires for all fluids and throughout all pressures under consideration. In absolute values, however, predictions according to Eq. (8) are always considerably higher than experimental results.

For further comparison experimental results for wires of $d = 0.1$ mm obtained by Abadzic [13] and Skripov/Dubrovina [14] are included in the figures, where applicable.

3.2 Minimum Heat Flux Density

Relative few experimental and theoretical investigations are performed on the transition from film boiling to nucleate boiling. A review on mathematical models and correlations is given in [15].

Our experimental results are presented as points in a q_{min} vs. p/p_c plot. In Figs. 4, 5, and 6 these results are shown for CO_2, R 13, and R 23, respectively, and for different diameters $d = 0.05, 0.1, 0.2,$ and 0.3 mm. The lines shown in the figures again represent correlations from literature.

Based on experimental results by Lewis [16] for flat plates and the theory of Taylor instability, Zuber [6] derived the following equation:

$$q_{min} = 0.177 h_e \rho_v \sqrt[4]{\frac{\sigma g(\rho_1 - \rho_v)}{(\rho_1 + \rho_v)^2}} \qquad (9)$$

Berenson [17], neglecting influences of vapor velocity and viscosity, obtained also for the flat plate

$$q_{min} = 0.09 h_e \rho_v \sqrt[4]{\frac{\sigma g(\rho_1 - \rho_v)}{(\rho_1 + \rho_v)^2}} \qquad (10)$$

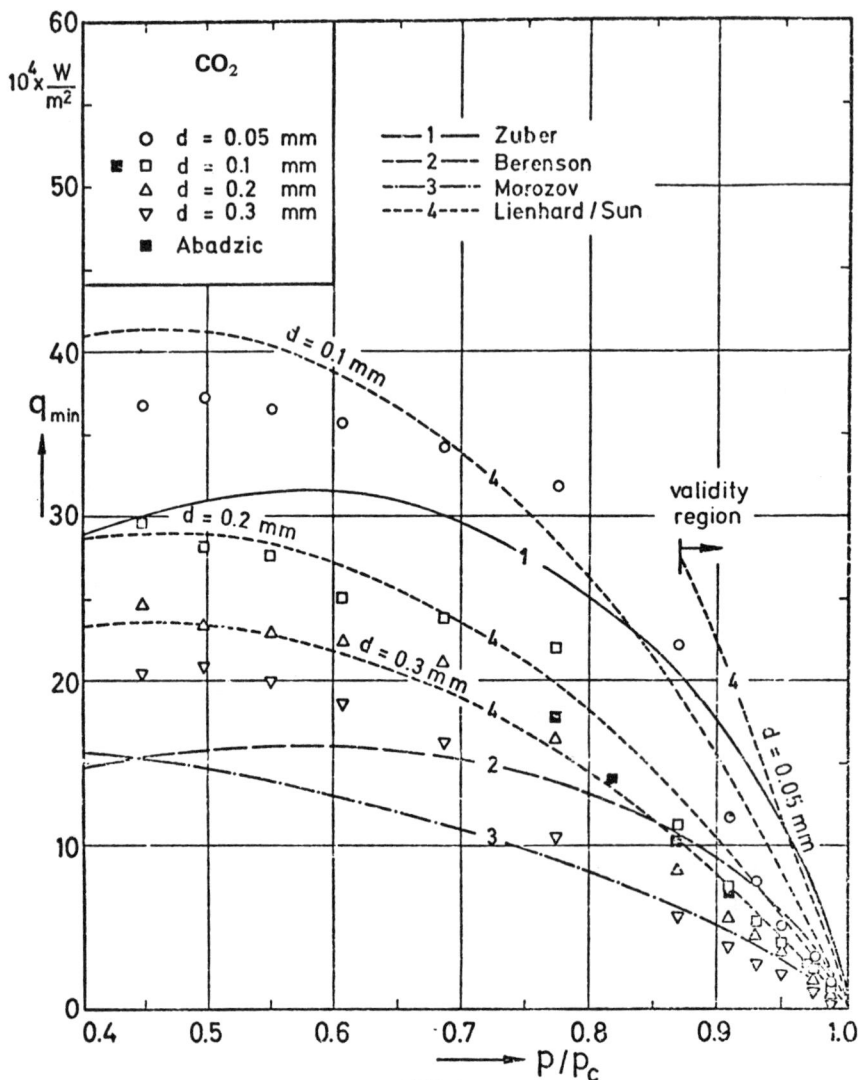

FIG. 4

FIGS. 4, 5, 6. Comparison of experimental data to various predictions from literature for minimum heat flux density vs. reduced pressure and three different fluids: CO_2, R 13, R 23.

By dimensional analysis Morozov [18] obtained the following relation:

$$q_{min} = 0.0267 h_e \rho_v \sqrt[4]{\frac{\sigma g(\rho_1 - \rho_v)}{\rho_v^2}} \qquad (11)$$

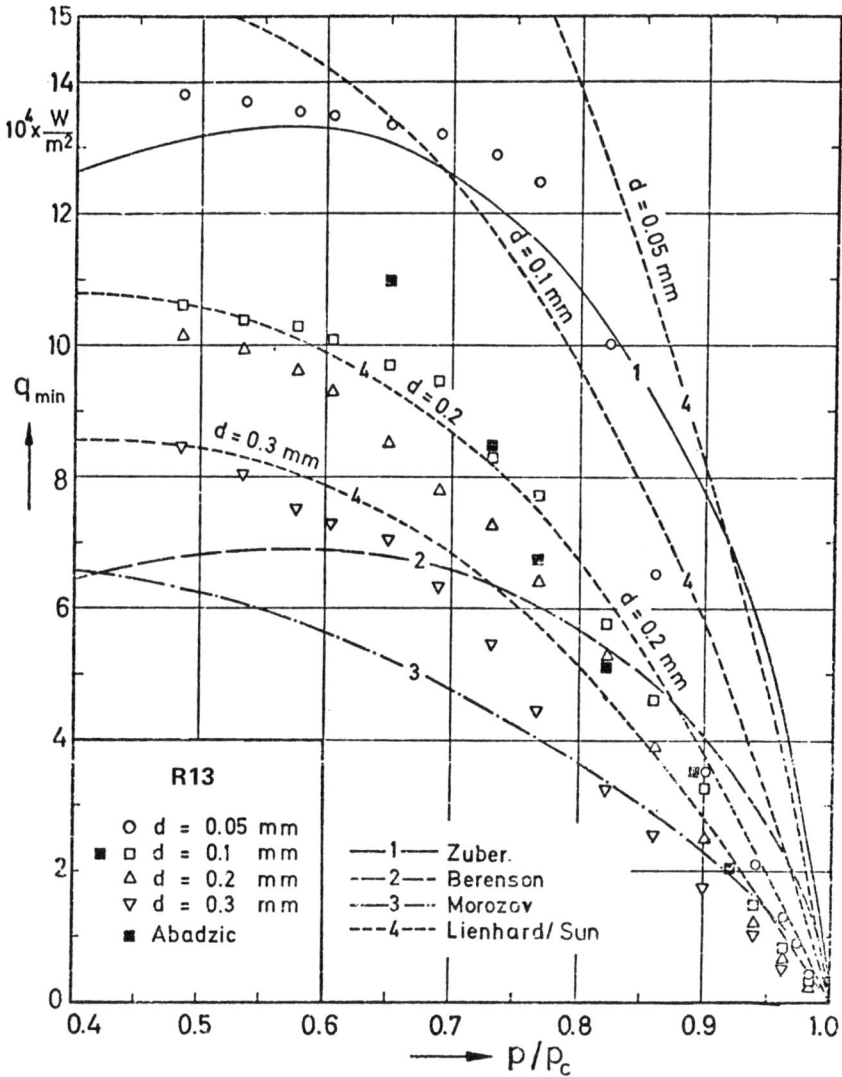

FIG. 5

As for q_{max}-correlations, only Lienhard/Sun [19] took into account an influence of wire diameter for the prediction of q_{min} in their relation

$$q_{min} = 0.03 h_e \rho_v \sqrt[4]{\frac{\sigma g(\rho_1 - \rho_v)}{(\rho_1 + \rho_v)^2}} f(D') \tag{12}$$

with

$$f(D') = \sqrt[4]{\frac{18}{(D'/2)^2 (D'^2/2 + 1)}}$$

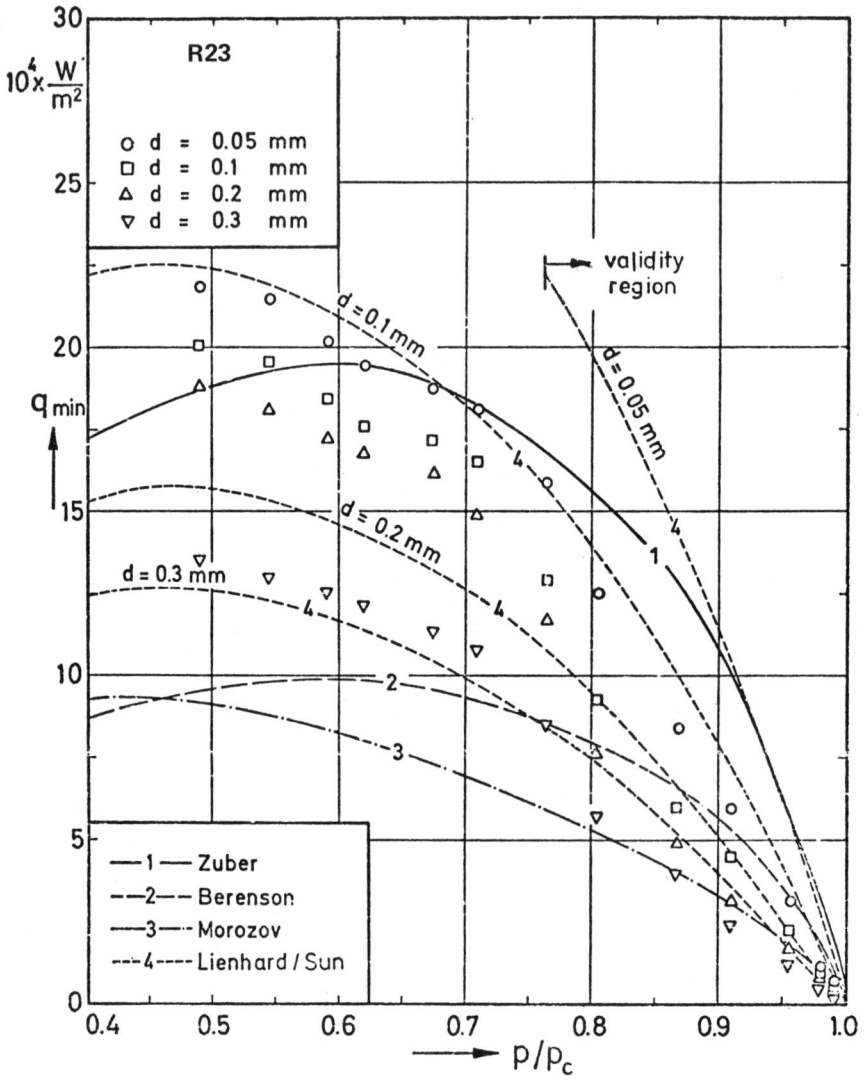

FIG. 6

and
$$D' = \frac{d}{\sqrt{\sigma/g(\rho_1 - \rho_v)}}$$

for $D' > 0.14$.

The general tendency of decreasing values of q_{min} with increasing pressure is represented in all correlations.

As in the case of q_{max}, Q_{min} also approaches zero the the thermodynamic critical point, so that differences between q_{max} and q_{min} become increasingly smaller as this point is approached within the fluid.

A distinct influence of wire diameter is observed in all results and this influence follows in tendency exactly the predictions given by Eq. (12), i.e., the smaller the diameter the higher is the minimum heat flux density q_{min}. In absolute values, however, predictions are considerably higher than experimental results.

4. CONCLUSIONS

4.1 Maximum Heat Flux Density

All correlations taken into consideration give correctly the tendency of decreasing maximum heat flux with increasing pressure. An influence of wire diameter seems to exist. The correlations giving usch an influence appear more appropriate than those which neglect geometry influences; however, the quantitative agreement is not satisfactory yet.

4.2 Minimum Heat flux density

Correlations for minimum heat flux as considered here give correctly the tendency of a decreasing q_{min} with an increasing pressure. An influence of wire diameter is obvious and in tendency agrees well with an existing correlation. The quantitative agreement with our experimental data is not satisfactory, however.

NOMENCLATURE

c_p	isobaric specific heat	η	dynamic viscosity
d	heater wire diameter	λ	thermal conductivity
g	acceleration of gravity	ν	kinematic viscosity
h_e	enthalpy of evaporation	ρ	density
q_{max}	maximum heat flux density	δ	surface tension
q_{min}	minimum heat flux density		

INDICES

l	liquid	v	vapor

ACKNOWLEDGEMENT

The experiments were supported by the Deutsche Forschungsgemeinschaft. This is gratefully acknowledged.

REFERENCES

[1] Stephan, K.: BWK (1965) 17, 571/614.
[2] Feurstein, G.: Dissertation 1973, Techn. Univ. München.
[3] Addoms, J. N.: Sc. D. Thesis, Chem. Eng. MIT (1948).
[4] Kutateladse, S. S.: AEC-tr-3770 (1959).
[5] Borishanskij, V. M.: Zhur. Techk. Fiz. (1956) 25 p. 252.
[6] Zuber, N., Thesis AECV-4439 (1959).
[7] Zuber, N., M. Tribus: UCLA Rept. No. 58-5 (1958).
[8] Zuber, N., M. Tribus, and J. W. Westwater: Int. Heat Transfer Conf. Boulder (1961) Paper No. 27, p. 230–236.
[9] Moissis, R., and P. J. Berenson: J. Heat Transfer, Trans ASME Ser. C. (1963) 85, p. 221–229.
[10] Noyes, R. C.: *Ibid.*, p. 125–131.
[11] Lienhard, J. H. and V. K. Dhir: NASA CR 2270, July (1973).
[12] Sun, K. H. and J. H. Lienhard: Int. J. Heat Mass Transfer (1970) 13, p. 1425–1439.
[13] Abadzic, E.: Diss. TH München, (1967).
[14] Skripov, V. P., and E. N. Dubrovina: Heat Transfer, Soviet Research (1972), 4, No. 2.
[15] Clements, L. D., and C. P. Colver: Ind. Eng. Chem. (1970) 62, p. 26–46.
[16] Lewis, D. J.: Proc. Roy. Soc. London Ser. A-202 (1950) p. 81.
[17] Berenson, P. J.: MIT Heat Transfer Lab. Tech. Rep. No. 17 (1960).
[18] Morozov, V. G.: Insh. Fiz. Zh (1962) 4, 15; Int. Chem. Eng. (1963), 3, p. 48–51.
[19] Lienhard, J. H., and K. H. Sun: Int. J. Heat Transfer Trans. ASME Ser. C. (1970) 92, p. 292–298.

A MATHEMATICAL MODEL OF TWO-PHASE ANNULAR FLOW*

G. F. Hewitt and P. Hutchinson
Thermodynamics Division, AERE Harwell
Didcot, Oxon, England

1. INTRODUCTION

This paper is intended to give a general presentation of developments at the Harwell laboratory in the United Kingdom in the past few years, on the modeling of annular two-phase flow. This regime of two-phase flow is defined as that in which part of the liquid flows on the walls of the channels in the form of a liquid film, and the rest is entrained in the form of droplets in the gas phase which is flowing in the core of the channel. Annular flow is the most important form of two-phase flow, and occurs at qualities (i.e., mass fraction of the vapor) above a few percent in normal systems. During the 1960s, an intensive experimental program was pursued at Harwell to elucidate the mechanisms of annular flow. This and other data are reviewed extensively in the book by Hewitt and Hall–Taylor [1].

It is useful at this stage to mention two important findings of this experimental program at Harwell, which have formed the basis of the more recent theoretical studies and also guided the current experimental program. These findings are:

1. Hydrodynamic equilibrium in annular flow systems is approached only slowly and does, in fact, require hundreds rather than tens of diameters, in contrast to single-phase flow systems. Thus, in nearly all practical applications, the local flow conditions reflect "history effects" resulting from the nature of the upstream conditions and flow development. This is the main reason why general, overall correlations of two-phase flow which take account only the channel cross section and the local phase flow rates (and this applied to practically all correlations used), are bound to be inadequate. To make progress, it is necessary to take account of these

---

*The abstract of this paper appears on page 388.

history effects and this has been the main object of the work described in this paper.

2. A wide range of experiments have been carried out in which liquid film flow rate has been measured in heated channels as a function of heat input, with given inlet conditions. Such experiments have been carried out in a range of geometries and over a range of pressure. Figure 1 shows typical experiments on the variation of film flow rate with power input for water boiling in a vertical round tube at a pressure of 70 bars. The conclusion from all these studies is that burnout (critical heat flux) occurs when the film flow rate proceeds smoothly to zero on increasing power input. Thus, the description of the critical heat flux phenomena is assumed, in what follows, to correspond to a description of the hydro-dynamics of the annular flow in a heated system leading to a situation of zero film flow rate at one point in the channel. Usually, this is at the end of the channel (for uniform heating) but may be displaced from the end of the channel with nonuniform heating.

Thus, the treatment in this present paper is purely hydrodynamic; it might have been expected that heat flux would have some *direct* effect on the local processes in annular flow. However, as we shall see below, experiments reveal

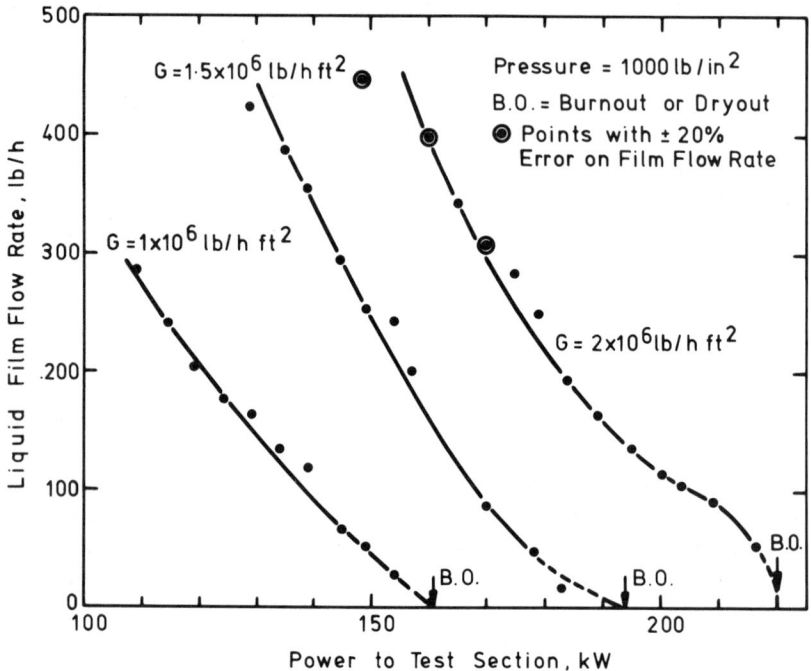

FIG. 1. Variation of liquid film flow rate with power input for evaporation of water in a 12-foot long tube at a pressure of 75 bars [2].

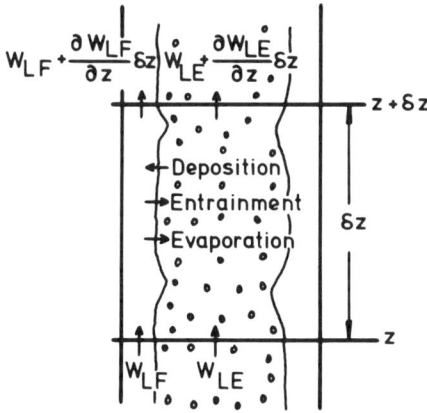

FIG. 2. Parameters for incremental mass balance in annular flow.

that this direct effect is small. There is, nevertheless, an important *indirect* effect of heat flux profile in that it affects the change of quality with distance and this, in turn, has an effect on the hydrodynamics.

2. THE BASES OF THE MODEL

The situation in annular flow in a channel is illustrated schematically in Fig. 2. Considering the change in the flow in the liquid film (W_{LF}) in passing through the element δz, one may write the following mass balance equation:

$$\underbrace{W_{LF} + \pi d_0 D\, \delta z}_{\text{Inflow}} = \underbrace{W_{LF} + \frac{\partial W_{LF}}{\partial z}\, \delta z + \pi d_0 E\, \delta z + \pi d_0 \frac{\phi}{\lambda}\, \delta z}_{\text{Outflow}} + \underbrace{\frac{\partial M_{LF}}{\partial t}\, \delta z}_{\text{Accumulation}} \tag{1}$$

where d_0 is the tube diameter, D is the deposition rate of droplets onto the liquid film (mass flow per unit area per unit time), E is the entrainment rate of droplets from the liquid film (mass flow per unit area of tube wall per unit time), ϕ is the local heat flux, and λ the latent heat of vaporization of the liquid. M_{LF} is the mass of liquid per unit length of the film which, for a circular tube, is given by:

$$M_{LF} = \pi d_0 m \rho_L \tag{2}$$

where m is the film thickness and ρ_L the liquid density. From a conservation equation similar to (1), the change in liquid core mass rate of flow (W_{LE}) is given by:

$$\frac{\partial W_{LE}}{\partial z} = \pi d_0 (E - D) - \frac{\pi d_0^2}{4} \frac{\partial C}{\partial t} \tag{3}$$

where C is the concentration (mass per unit volume) of droplets in the core.

In Sec. 6 below, we will discuss the detailed solution of the above equations for a variety of applications in both steady state and time-varying flows. However, we first discuss, in more detail, the calculation of the terms, D and E, giving the deposition rate and entrainment rate, respectively.

3. DEPOSITION MODEL

The droplets in the gas core are subject to random motion due to their interactions with the gas phase turbulence. This random motion can be represented formally as a diffusion process, with a diffusion constant η. The mass conservation equation for any given volume element of the flow in the gas core is as follows:

$$\int_{\text{volume}} \left[\frac{\partial}{\partial z} \bar{U} n(r,z) - S(r,z) + \frac{\partial}{\partial t} n(r,z) \right] dV - \int_{\text{surface}} \zeta \nabla_r n(r,z)\, ds = 0 \quad (4)$$

where U is the axial velocity, $S(r, z)$ is a source term, $n(r, z)$ is the local mean concentration of particles (particles per unit volume), dV is a volume element and ds is an element of surface area. In Eq. (4), derivatives with respect to the axial direction have been ignored. The first term gives the effect of dilation arising from varying axial velocity (due, for instance, to the expansion of the gas under its own pressure gradient) and the final term accounts for the loss of particles by diffusion in directions perpendicular to the direction of gas flow. Equation (4) can be grossly simplified by assuming steady state flow, considering only the radial motion of the particles (axial velocity of the particles being thus equal by definition to that of the gas) and by transforming the coordinates to a system moving with the axial gas velocity. This leads to the following equation [3]:

$$\frac{\partial}{\partial t} (\bar{U}(t) n(r,t)) - \bar{U}(t) S(r,t) - \zeta \nabla_r^2 (\bar{U}(t) n(r,t)) = 0 \quad (5)$$

The solution of Eq. (5) for a relatively simple case is discussed in Ref. 3. However, the order of complexity of such solutions, in more practical cases, is probably greater than is justified in the present state of knowledge of droplet diffusion. A simpler representation of the deposition rate can be obtained from the classical chemical engineering mass transfer expression:

$$D = k\bar{C} \quad (6)$$

where $\bar{C}$ is the mean concentration in the gas core (usually calculated assuming homogeneous flow in the gas core) and k is a mass transfer

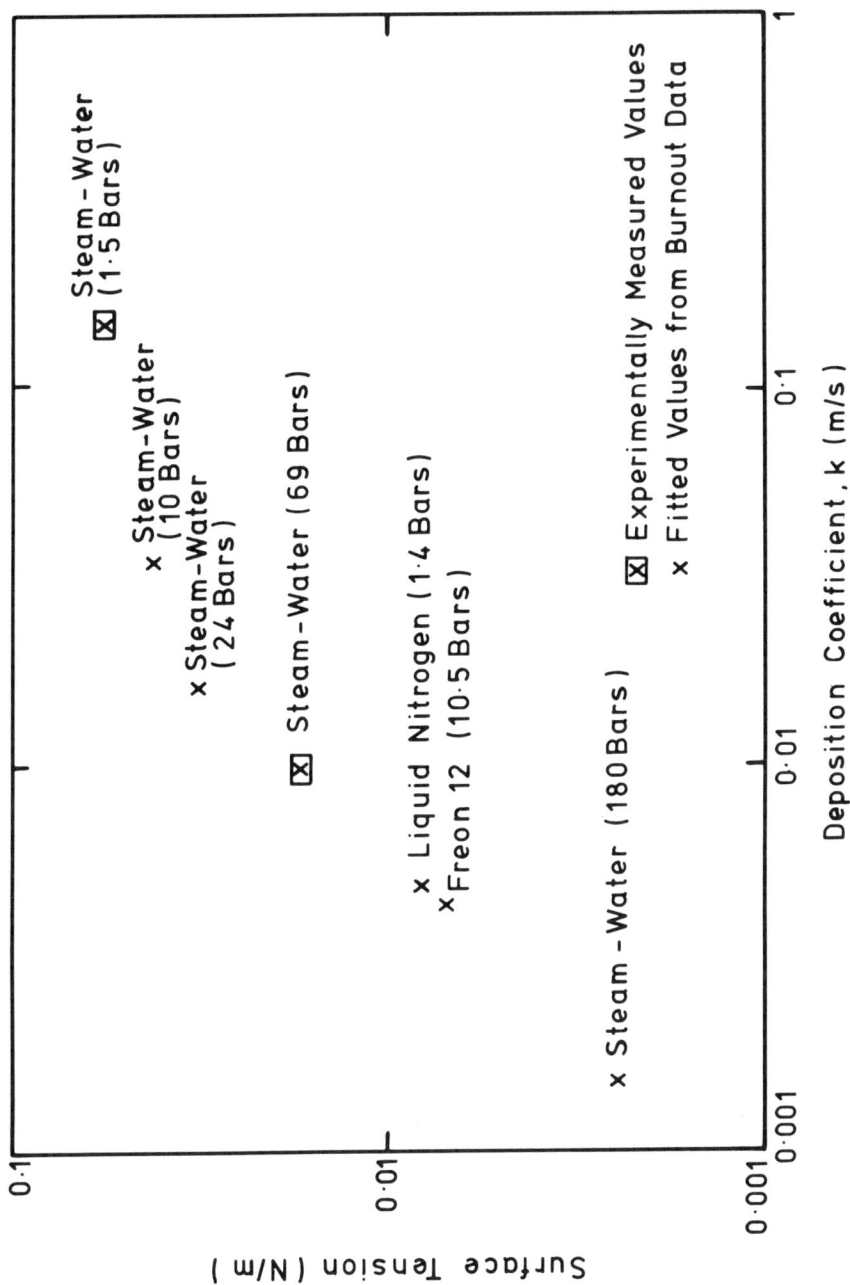

FIG. 3. Relationship between droplet mass transfer coefficient and surface tension (tentative).

175

coefficient. The value of k can be estimated from experimental data on deposition or from the diffusion model. k will vary with length for any given set of nonequilibrium boundary conditions but, for simplicity, we have usually assumed it to be constant at its asymptotic value.

A correlation for k which has been used in many of the predictions described below, is shown in Fig. 3. The relationship between k and σ is surprising and may be partly a reflection of other systematic physical property variations. Work is proceeding to obtain an improved correlation for k but Fig. 3 is shown below to be a reasonable working hypothesis. Note that Fig. 3 includes both experimentally determined values and also values fitted from estimation of burnout fluxes (see below). The consistency of the experimental and fitted values gives some support for the relationships proposed.

4. ENTRAINMENT MODEL

The rate of entrainment from liquid films can be deduced by interpretation of data from hydrodynamic equilibrium annular flow where the rate of entrainment is equal to the rate of deposition. The rate of entrainment can thus be calculated if the above relationships for deposition are applied with a knowledge of the mass transfer coefficient, and is given by:

$$E = k\bar{C}_{\mathrm{E}} \tag{7}$$

where $\bar{C}_E$ is the equilibrium homogeneous gas core concentration. If k is constant, then E is proportional to $\bar{C}_E$. A number of possible correlations have been examined for E, the most promising form being that in which E is related to the group $(\tau_i m/\sigma)$ where τ_i is the interfacial shear stress, m the liquid film thickness and σ the surface tension. In the first version of this correlation [4] $\bar{C}_E$ was plotted against $(\tau_i m/\sigma)$ as shown in Fig. 4. The correlation was found to fit a wide range of data to an acceptable accuracy but contains an obvious inconsistency. If E (calculated from Eq. 7) is plotted rather than $\bar{C}_E$, then the data separate if values of k are used as shown in Fig. 3. It is now our opinion that Fig. 4 taken together with Fig. 3 gives a reasonable combined representation of deposition and entrainment. The inconsistency may possibly reflect a concentration dependence of k. Further discussion of entrainment correlations is given in Ref. 5.

In using the entrainment correlation, it is necessary to know the interfacial shear stress τ_i, and the local film thickness m. These can be calculated provided the film flow rate at a particular point is known by a combination of the following relationships (1):

(1) the "triangular relationship" relating film flow, liquid film thickness and pressure drop via a suitable turbulent velocity profile relationship for the liquid film.

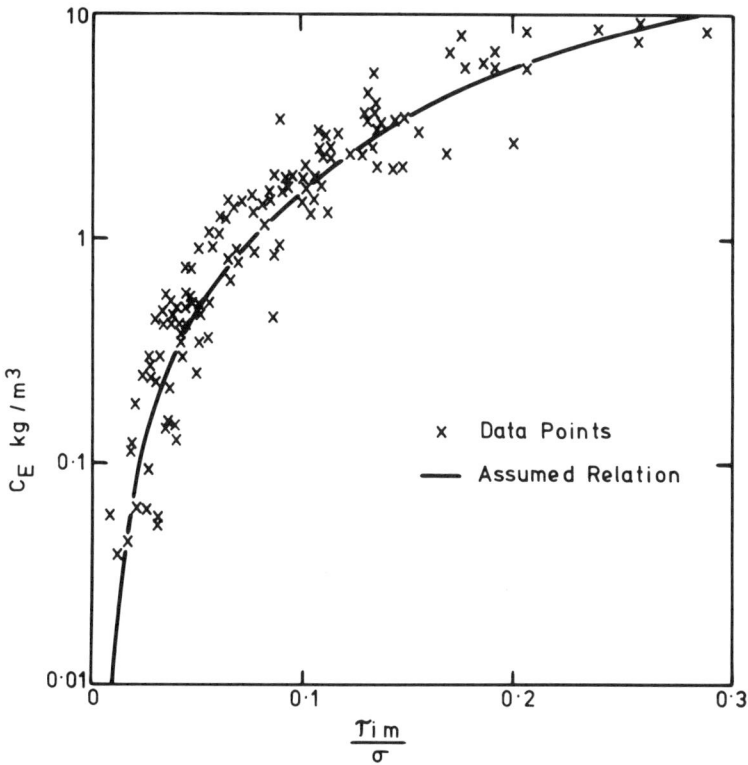

FIG. 4. Correlation for equilibrium concentration (and hence entrainment rate).

(2) the "geometrical simularity" relationship which relates the thickness of the liquid film to the effective roughness presented by the film interface to the gas core.

5. EFFECT OF HEAT FLUX ON ENTRAINMENT AND DEPOSITION

In applying the above relationships for entrainment and deposition, it is assumed that there is no *direct* effect of heat flux. Thus, it is assumed that nucleate boiling within the liquid film, if it occurs, does not cause significant extra entrainment. Also, it is assumed that the vapor flux away from the film surface does not materially inhibit the droplet deposition onto the film. Extensive investigations of heat flux effects are summarized in Ref. 6. Figure 5 illustrates some experiments carried out in which entrained liquid flow was determined as a function of axial distance. The channel used had successive heated and unheated zones, the relative length of which could be varied. It will be seen from Fig. 5 that the rate of change of entrained liquid flow rate with length is not appreciably changed on going from a heated to an unheated zone. Three examples are shown taken respectively, in regions of positive, nil,

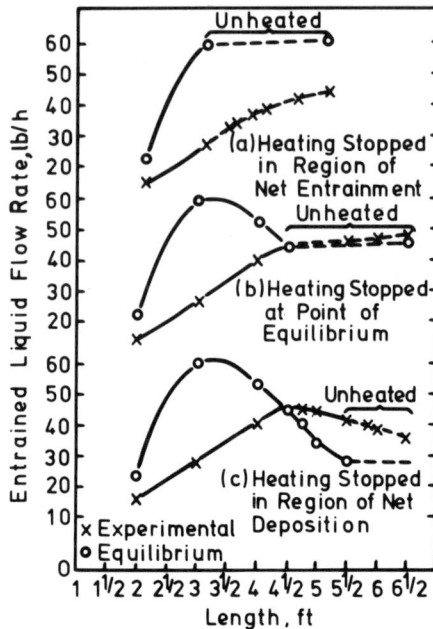

FIG. 5. Effect of length and heat input on entrained liquid flow. Flow rate (inlet pressure 55 psia,; mass flux 2. 19 × 10⁵ lb/ft² h; tube diameter 0.366 in.).

and negative rates of change of entrained liquid flow rate with length. The experiments illustrated in Fig. 5 were carried out at a pressure of about 2.5 bars. Higher pressure systems would be expected to show less effect, as indeed was confirmed in experimental work discussed in Ref. 6.

6. APPLICATIONS OF THE MODEL

A. *Steady Adiabatic Flows.* Annular adiabatic flows reach equilibrium only slowly, the process taking up to several hundred channel diameters. In fact, only a pseudoequilibrium is reached since the gas pressure itself changes with length. Figure 6, from Ref. 7 shows calculations of entrained liquid flow rate as a function of position, made by the diffusion theory of Ref. 3, compared with experiments. Reasonable agreement is obtained between the predictions and experiments, though other experiments taken in the region closer to the entrance show some discrepancies. The probable reason for this is that the interfacial waves, which govern the entrainment rate and also the interfacial shear stress, are not fully developed near the inlet. Means of allowing for wave structure development are discussed in Ref. 3.

B. *Calculation of Burnout Heat Flux in Single Tubes (Steady State).* The burnout (zero film flow) condition can be estimated by integration of Eq.

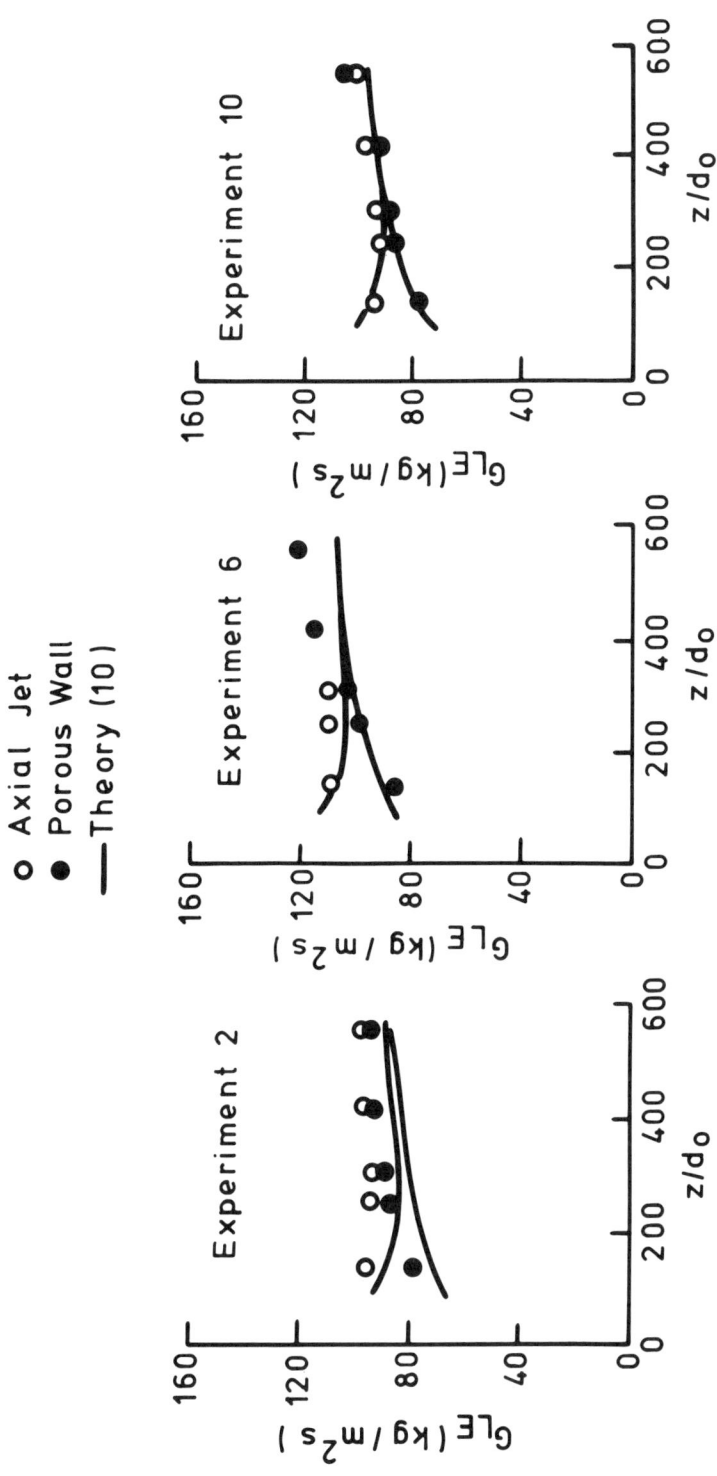

FIG. 6. Comparison of experimental and predicted entrained mass flow rate for air-water flow in a long 31.75 mm diameter tube. Data obtained by injecting water at entrance through a jet and through a porous wall section respectively.

(1) along a heated channel. For this purpose, a boundary condition must be chosen and that selected is 99% entrainment at a quality of 1% for the onset of annular flow. This boundary condition is, of course, somewhat arbitrary, but studies have shown that the prediction of burnout is relatively insensitive to the selection of initial boundary conditions, except for rather short tubes. The length at which burnout occurs can be predicted as a function of the quality and the inlet mass velocity. The results are exemplified by Fig. 7 and more details are given in Ref. 8. Figure 7 is plotted in terms of burnout quality, whereas for most practical purposes it is more convenient to make an iterative calculation to obtain burnout heat flux for a given length of tube. A comparison of calculated and experimental fluxes for evaporation of water is shown in Fig. 8.

C. *Burnout During a Flow Transient.* Equation 1 can be integrated in both time and space, again given satisfactory boundary conditions. The

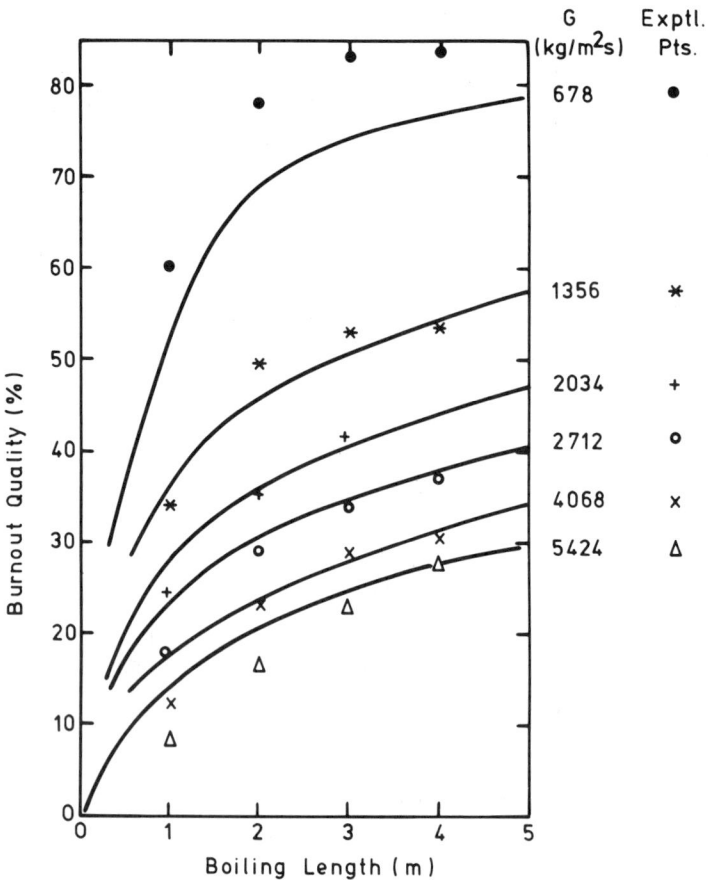

FIG. 7. Burnout quality predictions compared with experimental data for steam water flow at 69 bars in a 12.6 mm diameter tube.

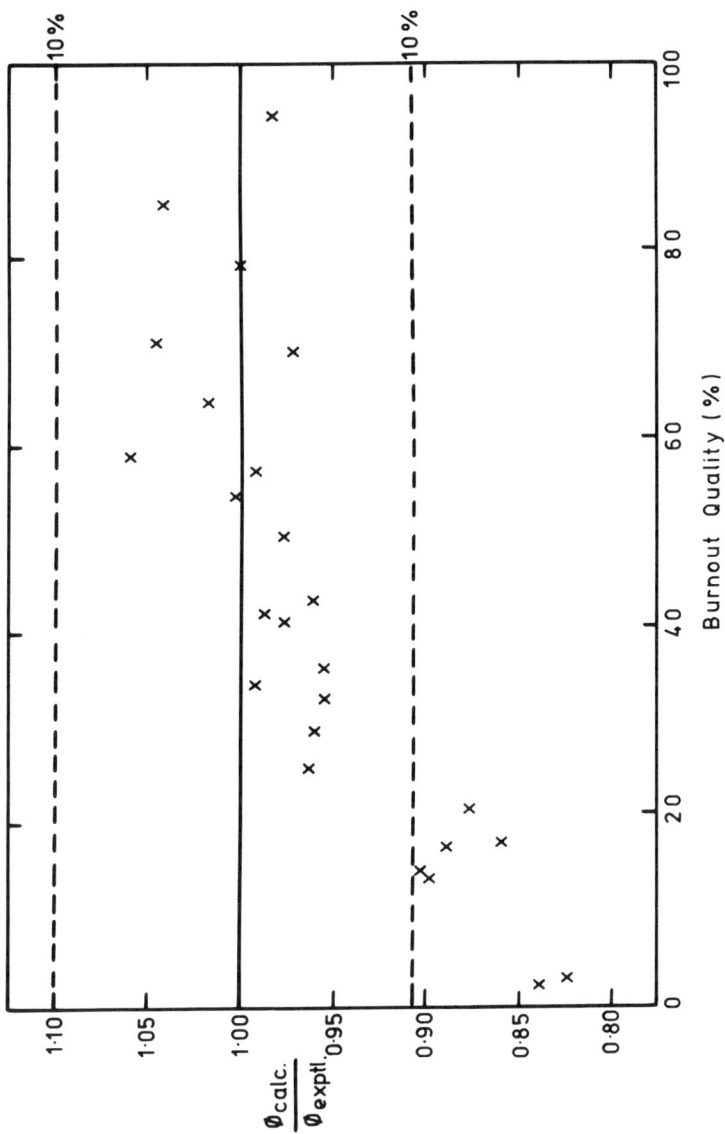

FIG. 8. Comparison of calculated and experimental burnout heat fluxes for steam-water flow at 91 bars.

181

boundary conditions could be, for instance, a known inlet flow transient of the type illustrated in Fig. 9 which is taken the experiments of Moxon and Edwards (9). The integration of Eq. (1) leads to an outlet mass velocity variation as also indicated on Fig. 9.

In addition to predicting total mass flow, local film flow rate is also predicted during the transient and the point where film flow rate goes to zero (i.e., the onset of burnout) can be predicted as a function of time. In uniformly heated channels, with a decreasing inlet mass flow, the maximum channel length which could be wetted can be calculated. In the initial conditions of the Moxon and Edwards experiment, this maximum channel length exceeds the physical length of the channel but decreases with time as illustrated in Fig. 10. After about 0.6 seconds, the maximum wettable length is less than the physical channel length and burnout starts to propagate down the channel as shown.

Further development of the transient calculation methods is proceeding but there is also a need for more sensitive experimental methods for the detection of the onset of burnout and its propagation.

D. *Burnout in Annuli (Steady State).* In an annulus, in annular flow, there are films in both surfaces. Burnout occurs either on the inside surface or the outer surface depending on the absolute and relative magnitudes of the fluxes on the two surfaces.

To model the case of an annulus two equations like Eq. (1) must be integrated simultaneously, one for each surface. Burnout is calculated to occur when the film flow rate on either surface is zero. A difficult problem in describing annuli using the methods described above is that the interfacial shear stress on the wavy interface is affected not only by the film thickness but also by the effective equivalent diameter of the flow adjacent to the surface. This is governed by the position of the surface of zero shear which moves toward the interface of decreasing relative roughness. The above problem is solved by using a simplified turbulent flow model to solve for the interfacial shear at each surface, the details of which are given in Ref. 10. Figure 11 shows calculations of burnout quality as a function of mass velocity and power distribution within the annulus, compared with the data of Jensen and Mannov [11]. Both the qualitative trends of the data (i.e., the attainment of a peak quality at a dryout when the distribution of heat fluxes produces a simultaneous dryout on both surfaces) and the quantitative values are predicted well by the theory.

E. *Burnout in Rod Bundles (Steady State).* The annular flow model is now being applied to the prediction of burnout in parallel flow in bundles of rods. Successful predictions have been obtained with bundles of between 7 and 37 rods. Further details of these calculations will be given in Ref. 12. The procedure is to divide the rod bundle into a series of "rod centered" subchannels as illustrated in Fig. 12. Inventories are made of the liquid

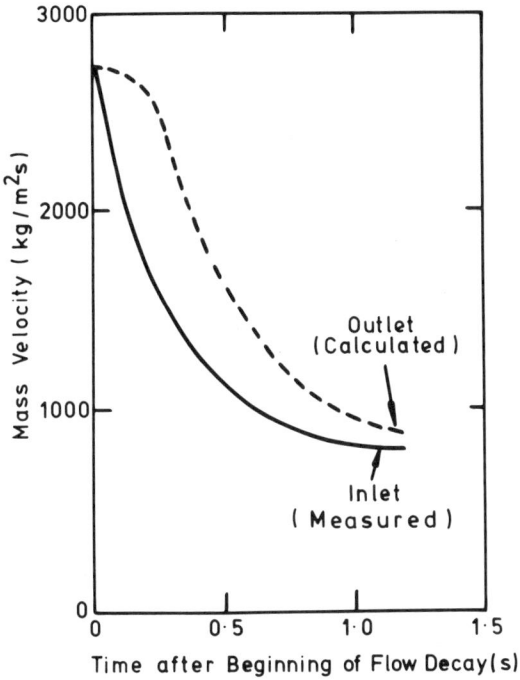

FIG. 9. Inlet and outlet flow variation in a typical flow transient measurement (inlet flow as measured in Ref. 9; outlet flow as calculated from present model.)

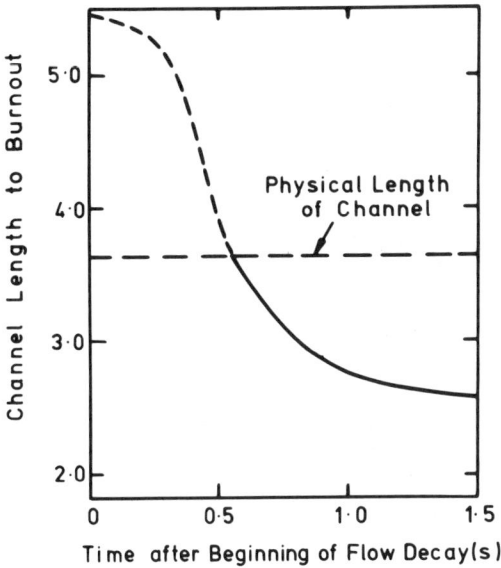

FIG. 10. Calculated wetted length for the inlet flow transient illustrated in Fig. 9.

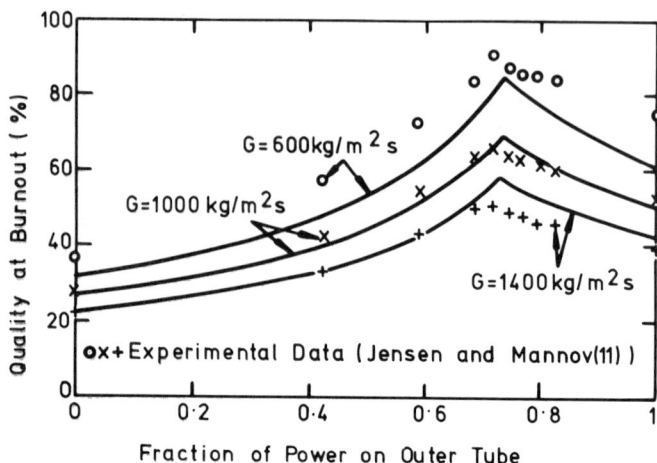

FIG. 11. Comparison of calculated and experimental burnout quality for water evaporation in an 8.5 mm internal diameter, 13 mm external diameter annulus at 70 bars.

films on each rod and those on the outer surface of the containing pressure tube (which is usually unheated). Burnout occurs when the film flow goes to zero on any given rod. The interfacial friction is calculated on an equivalent diameter basis and account is taken of mass transfer between subchannels due to both interchannel droplet diffusion and interchannel bulk mass transfer arising from pressure differentials between the channels.

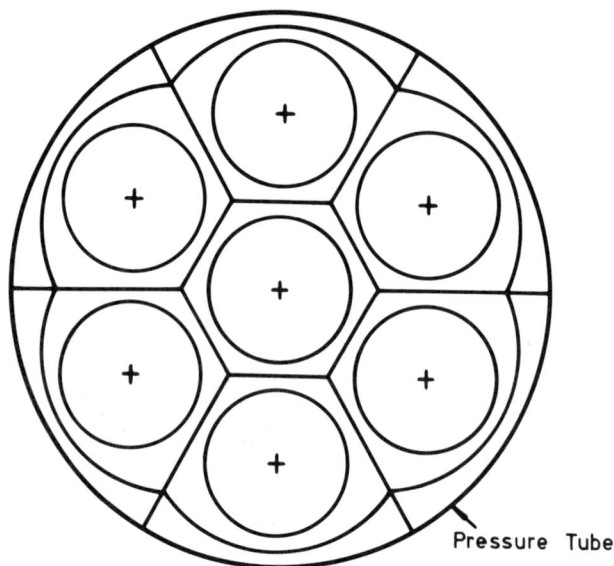

FIG. 12. Rod-centered subchannels illustrated by the case of the 7-rod bundle.

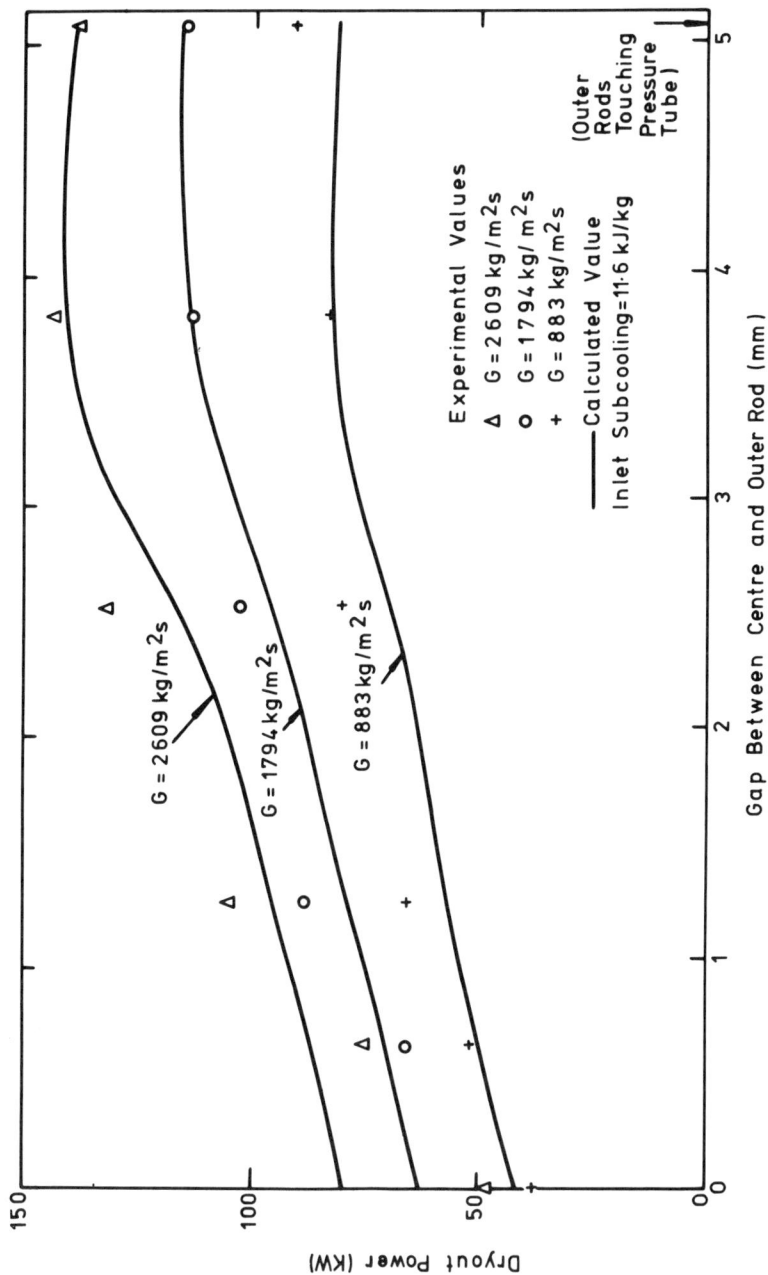

FIG. 13. Comparison of experimental and predicted dryout power for a 7-rod bundle in which the positions of the outer 6 rods are adjusted systematically. Data is for Freon-12 evaporation at 10.7 bars.

185

Both rod-to-rod and axial heat flux variations can be dealt with in the model.

The excellent fit to experimental data obtained for rod bundle predictions is illustrated by an example which can be said to represent a sensitive check on the analysis. This is the case of a Freon-12 flow in a 7-rod bundle where the gap between the center rod was varied between the extremes of contact between the outer rods and the containing tube, and contact between the outer rods and the inner rod. Results are illustrated in Fig. 13. Again both the trends and quantitative data are predicted accurately.

7. CONCLUSION

The use of annular flow modeling for flow and dryout prediction is proving useful in practical assessments. However, it should be recognized that there are many areas in this field which need closer attention. Among these, we would list:

(1) Further data are needed on mass transfer coefficients in annular flow, particularly for high droplet concentrations and low surface tension.
(2) A better understanding of the development of interfacial waves in annular flow is required, particularly where rapid changes are taking place along the channel.
(3) Numerical methods must be further developed to deal with complex transients.
(4) There is a need to link the preburnout calculations described above to calculations in the postburnout region.
(5) A closer evaluation of the rewetting rate of a dry, hot, region following reinstatement of a flow, should be carried out.

Work on all the above topics is proceeding at Harwell and elsewhere.

REFERENCES

1. Hewitt, G. F. and N. S. Hall-Taylor: "Annular Two-Phase flow," Pergamon Press, Oxford, (1970).
2. Bennett, A. W., G. F. Hewitt, H. A. Kearsey, R. K. F. Keeys, and R. A. Stinchombe: "Measurement of Liquid Film Flow Rates at 1000 psia in Upward Stream Water Flow in a Vertical Heated Tube," AERE-R5809, (1969).
3. Hutchinson, P., P. B. Whalley, and G. F. Hewitt: Int. J. Multiphase Flow, 1974, 1, p. 383.
4. Hutchinson, P., and P. B. Whalley: Chem. Eng. Sci., 1973, 28, p. 974.
5. Whalley, P. B., G. F. Hewitt, and P. Hutchinson: Proceedings, Symposium on Multiphase Flow Systems, The Institution of Chemical Engineers, 1974, Symposium Series 38, Paper A1.
6. Hewitt, G. F.: Fourth International Heat Transfer Conference, Versailles 1970, Paper B6.6.

7. Brown, D. J., A. Jensen, and P. B. Whalley: "Non-Equilibrium Effects in Heated and Unheated Annular Two-Phase Flow," AERE-R8154 (1975).
8. Whalley, P. B., P. Hutchinson, and G. F. Hewitt: Fifth International Heat Transfer Conference, Tokyo, 1974, Paper B6.11.
9. Moxon, D., and P. A. Edwards: "Dryout during Flow and Power Transients," AEEW-R553 (1967).
10. Whalley, P. B., P. Hutchinson, and G. F. Hewitt: Paper presented at the European Two-Phase Flow Group Meeting, June 1975.
11. Jensen, A., and G. Mannov: Paper presented at the European Two-Phase Flow Group Meeting, June 1974.
12. Whalley, P. B.: "The Calculation of Dryout in a Rod Bundle," AERE-R8319 (1976).

BUBBLE GROWTH RATES IN AQUEOUS BINARY SYSTEMS AT SUBATMOSPHERIC PRESSURES *

S. J. D. van Stralen, W. Zijl and D. A. de Vries
Laboratory for Fluid Dynamics and Heat Transfer
Eindhoven University of Technology, Eindhoven, The Netherlands

1. SCOPE OF THE INVESTIGATIONS

1.1. Bubble Growth Rates

Experiments

The growth rate of vapor bubbles during nucleate boiling has been investigated up to departure both in: *water* (at pressures between 2.04 and 26.7 kPa, the corresponding Jakob number decreasing from 2689 to 108 [1]) and in the following *"positive" aqueous binary systems* with a more volatile organic component: *water-ethanol* (up to 31 wt% ethanol, at pressures between 4.08 and 6.65 kPa, the Jakob number varying from 1989 to 1075), *water-1-butanol* (up to 2.4 wt% 1-butanol, at pressures between 3.60 and 4.08 kPa, the Jakob number varying from 2760 to 1989), and *water-2-butanone* (up to 15 wt% 2-butanone, at pressure between 7.31 and 9.07 kPa, the Jakob number varying from 1519 to 683).

The investigations are limited to relatively small concentrations of the organic component as advanced isobaric (heat and mass) diffusion-controlled bubble growth is expected to show a minimum in this range. Contrarily, the initial (isothermal) phase of growth is approximately independent of concentration.

Theory

Experimental bubble growth is compared with the Van Stralen et al. theory [2]. This model combines the initially dominating Rayleigh solution with a (heat and mass) diffusion-type solution for advanced growth, which accounts for the contributions due to both the "relaxation microlayer"—around the lower part of the bubble dome—and the "evaporation microlayer"—beneath the bubble.

---

*The abstract of this paper appears on page 389.

1.2. Bubble Oscillations

Numerical solutions are given for the growth and implosion of spherical and rotationally symmetric vapor bubbles in the transition period between the isothermal and isobaric phases. Oscillations in both the bubble radius and the temperature of the bubble boundary are predicted to occur.

Experimental verification follows from previous observations by Van Stralen [3] and from present data including the behavior of imploding free bubbles in water, subjected to a stepwise subcooling.

2. BUBBLE GROWTH RATES DURING ADHERENCE

2.1. Initial Growth

It has been shown previously by several investigators [1, 2], that the initial (isothermal) phase of bubble growth is hydrodynamically controlled: the bubble is blown up, due to an excess pressure resulting from the superheating of the bubble boundary, according to the following solution of the modified Rayleigh equation [8]:

$$R_1(t) = \left[\frac{2\rho_2 l \theta_0 \exp -(t/t_1)^{1/2}}{3\rho_1 T}\right]^{1/2} t = 0.816\left[\frac{c}{T} \exp -\left(\frac{t}{t_1}\right)^{1/2}\right]^{1/2} (\text{Ja})^{-1/2}\theta_0 t$$

$$(1)$$

For the investigated binary mixtures with $x_0 \ll 1$, R_1 is approximately independent of liquid composition.

2.2. Advanced Growth

Advanced (isobaric) bubble growth in binary systems is determined by combined heat and mass diffusion (of the more volatile component) toward the bubble boundary [2]:

$$R_{2,m}^{(t)} \cong \frac{C_{1,m}}{C_{1,p}} \left[\left(1.954b^* + 0.373\frac{C_{1,m}}{C_{1,p}} \text{Pr}^{-1/6}\right)\left\{\exp -\left(\frac{t}{t_1}\right)^{1/2}\right\}\right.$$

$$\left. + 1.954\frac{\Delta\theta_0}{\theta_0}\right] \text{Ja} \, (at)^{1/2}$$

$$(2)$$

The substantial effect of mass diffusion on R_2 is expressed by the ratio $C_{1,m}/C_{1,p}$, Eqs. (4) and (5). The first term between brackets in the right side of Eq. (2) is due to the "relaxation microlayer" accounting for evaporation at the lower part of the bubble boundary. The second term is due to heat transmission through the thin (liquid) "evaporation microlayer," which is formed between the bubble and the heating wall. The third term expresses the effect of the superheating $\Delta\theta_0$ of the bulk liquid. Generally, $\Delta\theta_0/\theta_0$ is the order of magnitude of 10^{-2} to 1. Equation (2) combines the effects of relaxation and evaporation microlayers, which results in an interaction resulting in superposition of the corresponding bubble radii [2].

A spatially uniform vapor temperature and thermodynamic equilibrium at the bubble boundary are assumed in the derivation of Eqs. (1) and (2). The dominating effect of pressure and of the Jakob number Ja = $\rho_1 c\theta_0/\rho_2 1 \sim 1/p$, on bubble growth is obvious. In general one has in case of free bubble in an initially uniformly superheated liquid:

$$R_1 \sim \rho_2^{1/2} \sim p^{1/2} \text{ and } R_2 \sim 1/p.$$

2.3. Transitional Growth

According to Cooper and Vijuk [4], the bubble radius during the transition between initial and asymptotic growth is approximated by the following empirical expression:

$$R(t) = \frac{1}{1/R_1(t) + 1/R_2(t)} = \frac{R_1(t)R_2(t)}{R_1(t) + R_2(t)} \tag{3}$$

The ratio $R_1/R_2 \sim p^{3/2} t^{1/2}$ ($\to 0$ as $p \to 0$ and/or $t \to 0$); therefore, Eq. (3) reduces to the appropriate limits: $R \to R_1$ as $p \to 0$ or $t \to 0$ and $R \to R_2$ as $p \to \infty$ or $t \to \infty$, the latter in case of free bubbles, in which an evaporation microlayer is missing: $t_1 = \infty$, $b^* = 1$, and $\Delta\theta_0 = 0$ [see Eq. (2)].

The former result follows from the concept of time-independent heat removal at the bubble boundary (to satisfy the requirement of latent heat of vaporization), the liquid being a semi-infinite body: $q = \rho_2 1 \dot{R}(t)$ leading to $\rho_2 1 R_1(t)$ = constant as $t \to 0$. The initial decay of the superheating $\theta_R(t)$ of the bubble boundary is thus proportional to $\rho_2^{3/2} \sim p^{3/2}$, and vanishes as $p \to 0$. Consequently, "Rayleigh bubbles," Figs. 1 and 5, have been observed in water boiling at pressures below 4 kPa [1].

Actually, the combined action of R_1 and R_2 results in the occurrence of bubble oscillations, which are due to simultaneous fluctuations $\theta_R(t)$ [see Sec. 5.3].

2.4. Determination of the Bubble Growth Parameter

The numerical value of the bubble growth parameter b^* (which denotes the relative height of the relaxation microlayer around the lower part of the bubble boundary) is determined by substituting $t = t_1$, the experimental departure time, into Eqs. (1)–(3), respectively by substituting the experimental departure radius $R(t_1)$ into Eq. (3). In general, $b^* \sim p$ in a limited range of subatmospheric pressures.

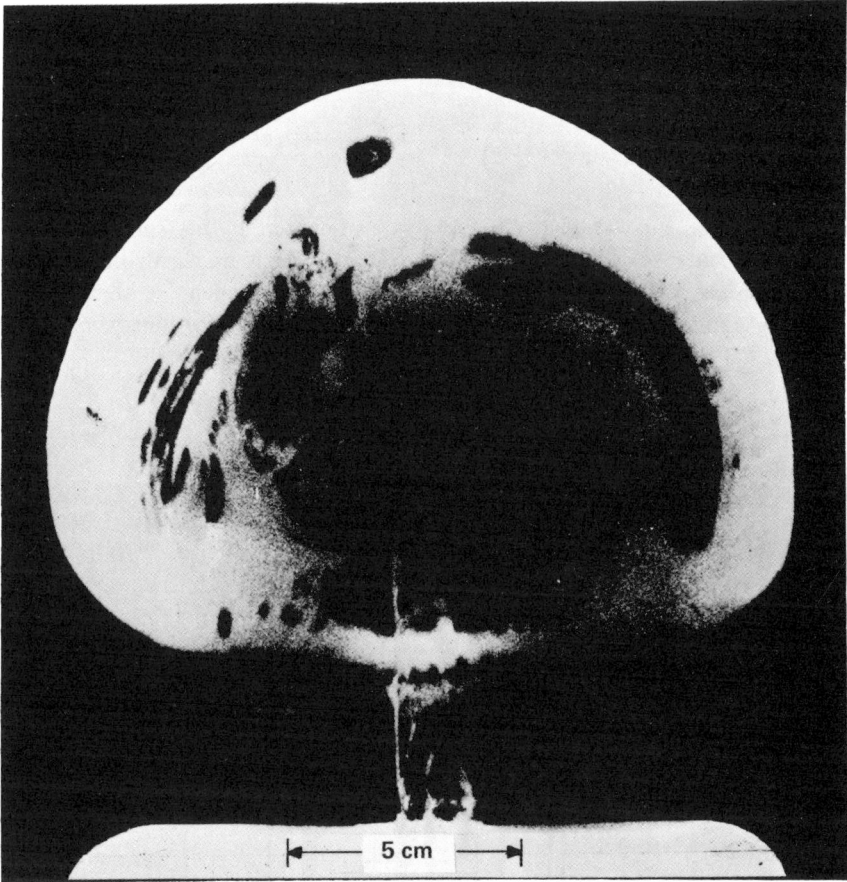

FIG. 1. *Water* boiling at 1.04 kPa (cf. Figs. 5 and 7). A thin secondary vapor column succeeds a large "Reyleigh vapor bubble." Turbulent eddies are visible.

2.5. Slowing-down Effect of Mass Diffusion on Advanced Bubble Growth in Binary Systems

The bubble growth constant $C_{1,m}$ [cf. Eq. (2)] for asymptotic diffusion-controlled growth of a spherically symmetric free bubble in an initially uniformly superheated infinite liquid is given by the following expression [3, 5, 6]:

$$C_{1,m}(x_0) = \left(\frac{12}{\pi}\right)^{1/2} \frac{a^{1/2}}{(\rho_2/\rho_1)[1/c + (a/D)^{1/2}\Delta T/G]} \qquad (4)$$

where ΔT and G both depend on x_0.

In case of a unary (1-component) system and in azeotropic mixtures, the increase in dew-point of the vapor, ΔT, is zero, resulting in $C_{1,m} = C_{1,p}$, where

$$C_{1,p} = \left(\frac{12}{\pi}a\right)^{1/2}\frac{\mathrm{Ja}}{\theta_0} = 1.954\,\frac{(k\rho_1 c)^{1/2}}{\rho_2 l} \tag{5}$$

It follows from Eq. (4), that $C_{1,m}$ shows a minimum at the (small) fraction x_0 of the more volatile component, at which $\Delta T/G$ is maximal; i.e., asymptotic bubble growth is slowed down maximally. Van Stralen [7] derived the following expression for ideal systems (or more generally, for binary mixtures with $x_0 \ll 1$) at constant pressure:

$$\frac{\Delta T}{G} = \frac{T}{\rho_2 l}[K(x_0) - 1]^2 p x_0 \tag{6}$$

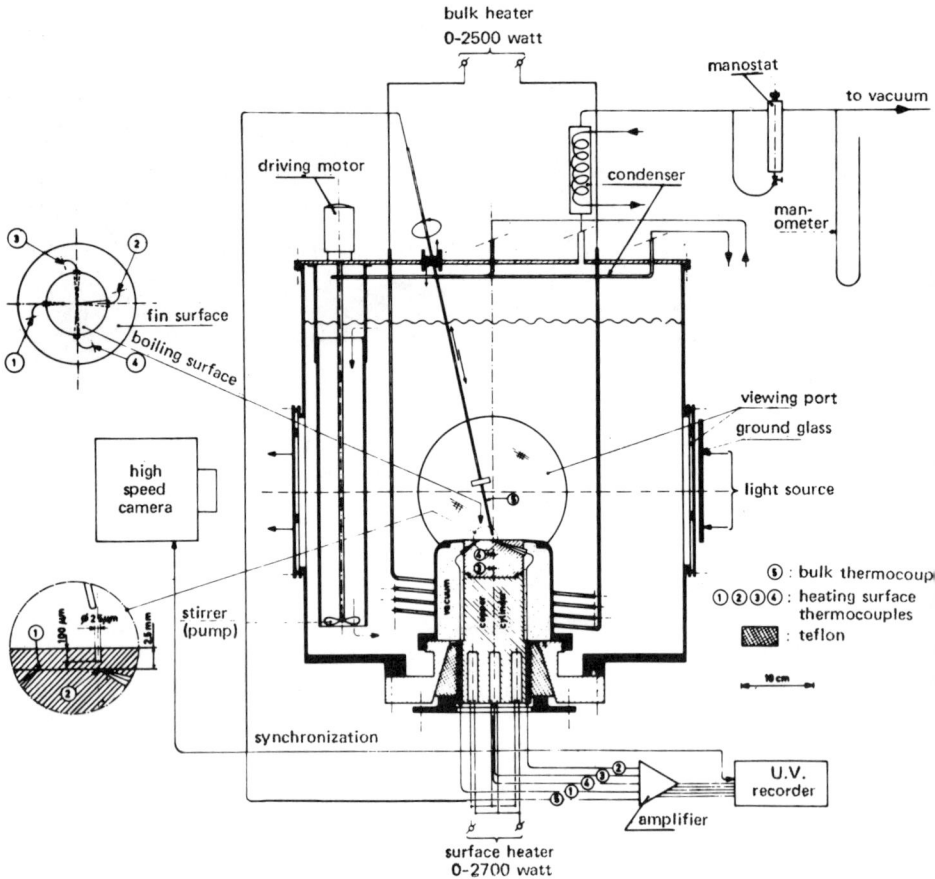

FIG. 2 Diagram of boiling apparatus.

The ratio $\Delta G/T$ is thus (at low pressures) approximately independent of pressure, of $G(\ll 1)$ and of θ_0. Equation (6) expresses $\Delta T/G$ into material properties and is useful to derive $C_{1,m}(x_0)$ for binary systems at subatmospheric pressures, especially when equilibrium data are not given in the literature. As an approximation for those cases, one can make use of the values of $\Delta T/G$, which are derived graphically from the known equilibrium diagram at atmospheric pressure.

3. EXPERIMENTAL SETUP

The experimental setup, Fig. 2, and the application of high-speed cinematographic techniques have been described previously [1].

4. EXPERIMENTAL RESULTS IN COMPARISON TO THEORETICAL PREDICTIONS

4.1. Water

The most important results of the investigations are the following:

(i) The vapor bubbles generated on an artificial nucleation site (with radius R_0 = 12.5 μm) at the horizontal upper surface of the copper heating cylinder (Fig. 2) are hemispherical during an initial period of approximately 10 ms only (Fig. 3). Afterwards, the ratio R_c^*/R of the contactradius R_c^* to the equivalent bubble radius R decreases gradually from an initial value of $2^{1/3}$ = 1.26 to zero (Fig. 3).

(ii) The experimental bubble radius curves $R(t)$ are in the entire investigated range of subatmospheric pressures in quantitative agreement with the Van Stralen et al. theory [2] (Figs. 4 and 5).

The experimental results are also compared with the following theoretical approximations: (a) the model of Cooper and Vijuk [4], which combines the Rayleigh solution [Eqs. (1) and (3)] with a diffusion-type solution accounting for an evaporation microlayer only, leaving the relaxation microlayer out of consideration; (b) the model of Mikic et al. [8], which combines the Rayleigh solution with a relaxation microlayer only; (c) the relaxation microlayer model of Van Stralen [5, 6] (Figs. 4 and 5). The models [2], [4], and [8] all reduce to the Rayleigh solution as $p \to 0$ (Fig. 5).

(iii) The experimental bubble departure time on an artificial nucleation cavity with constant diameter is in good agreement with the theoretical expression for a wall of high thermal conductivity [1]:

$$t_1 = 0.65 \frac{k\rho_1 c}{q_{w,\,co}^2} \left(\frac{\sigma T}{\rho_2 l R_0} \right)^2 \tag{7}$$

It follows from Eq. (7), that $t_1 \sim 1/p^2$; in agreement with this result the departure radius $R(t_1)$ increases substantially with decreasing pressure

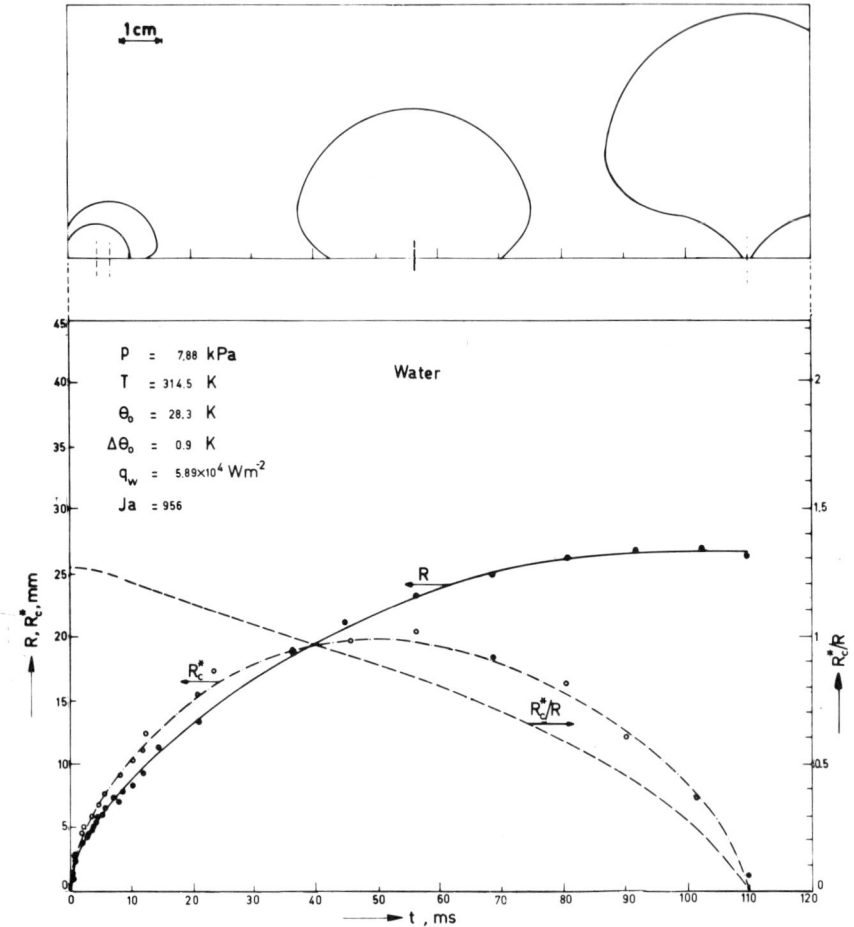

FIG. 3. *Water* boiling at 7.88 kPa (cf. Figs. 4, 6, and 7). Equivalent bubble radius R, contactradius R_c^*, and ratio R_c^*/R, in dependence on time. Corresponding bubble profile and limit of mode of hemispherical growth ($R_c^*/R = 2^{1/3}$) are shown in the upper drawings.

(approximately also $\sim 1/p^2$). Consequently, the bubble frequency on a nucleus decreases considerably with decreasing pressure. Equation (7) has also been verified in the investigated aqueous binary mixtures, in which $\sigma_m < \sigma_p$ at constant pressure (Figs. 3 and 6).

(iv) "Rayleigh bubbles," which are blown up due to an excess pressure, are observed at pressures below 4 kPa (Figs. 1 and 5).

(v) At low subatmospheric pressures, substantial periodic fluctuations (with amplitudes up to 5 K) are observed inside the copper heating cylinder, Fig. 2, at distances of 2–30 mm below the artificial nucleus. The local temperature drops simultaneously during bubble growth and increases during the succeeding waiting time. This result is in good agreement with observations of temperature dips at the heating surface beneath bubbles, which have been made by several investigators. The fluctuations cause material fatigue due to thermal expansions.

(vi) A remarkable bubble cycle occurs at pressures of 2–8 kPa. In general, directly after departure of a large bubble, a vapor column showing a very rapid initial growth rate is generated at the nucleation site (Figs. 1 and 7). This secondary bubble displaces a liquid jet with a high velocity; the jet penetrates into the flattened lower boundary of the preceding large

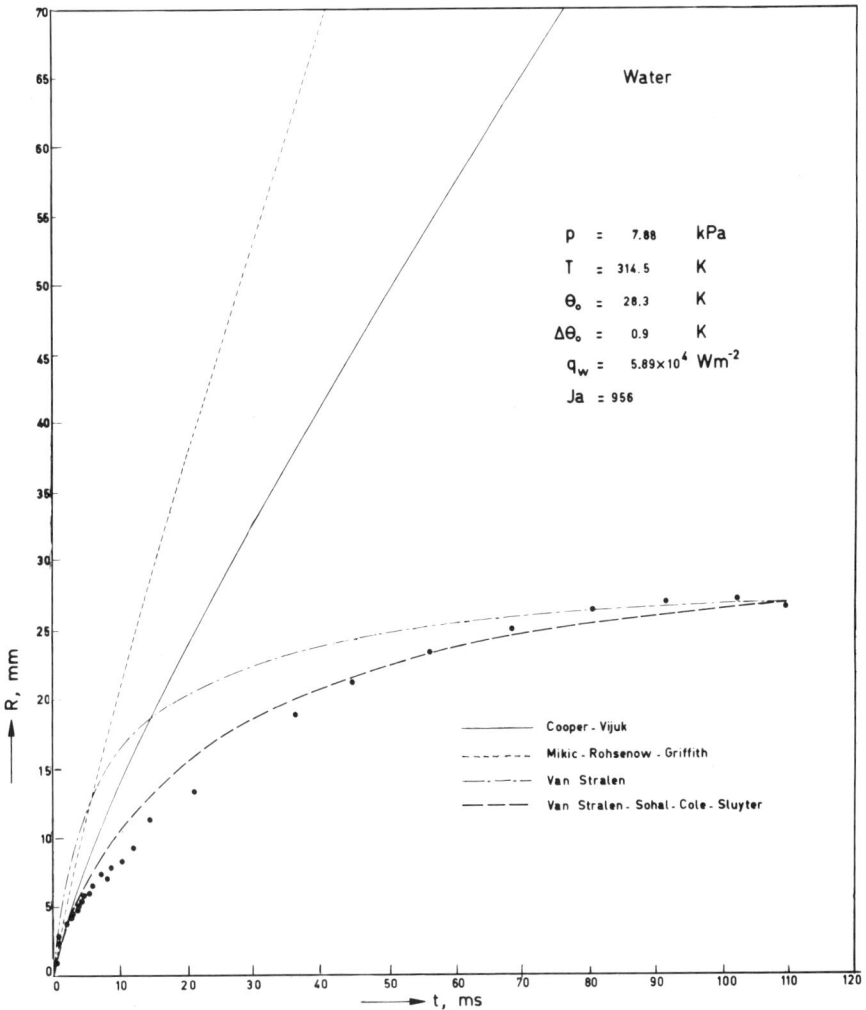

Water

$p = 7.88$ kPa

$T = 314.5$ K

$\theta_o = 28.3$ K

$\Delta\theta_o = 0.9$ K

$q_w = 5.89 \times 10^4$ Wm^{-2}

Ja $= 956$

——————— Cooper - Vijuk

— — — — — Mikic - Rohsenow - Griffith

— · — · — Van Stralen

— — — — Van Stralen - Sohal - Cole - Sluyter

FIG. 4. *Water* boiling at 7.88 kPa (cf. Fig. 3). Experimental bubble growth data up to departure in comparison with theoretical predictions.

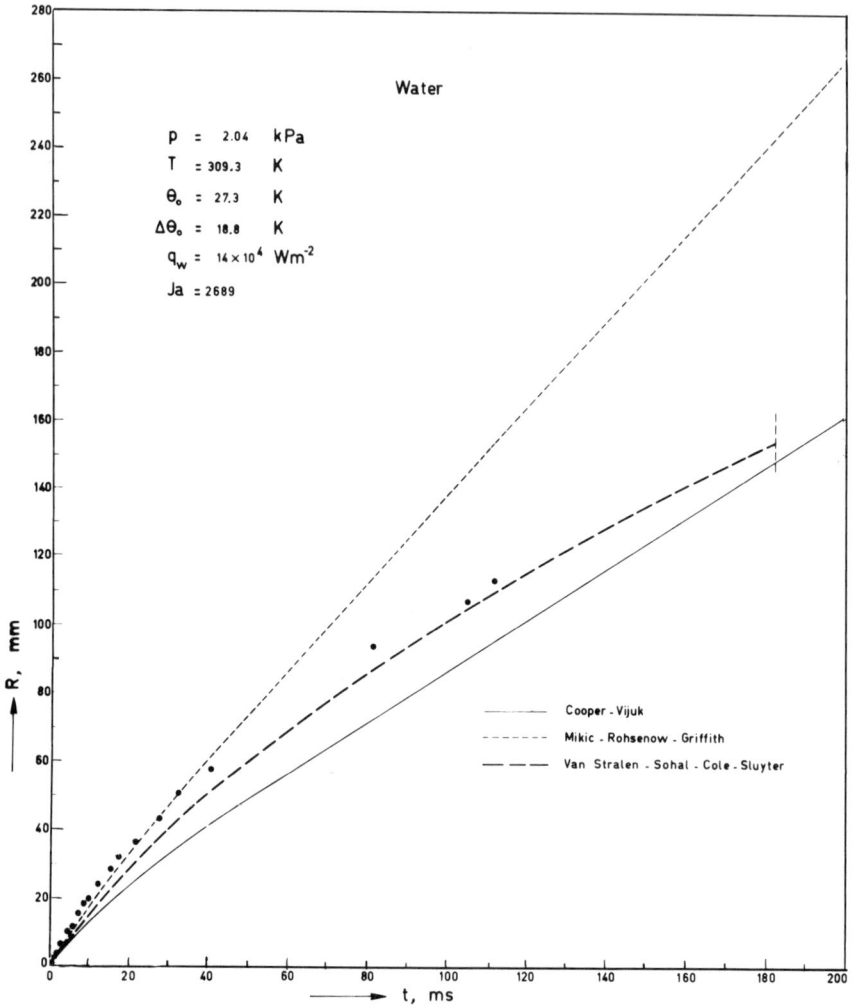

FIG. 5. *Water* boiling at 2.04 kPa (cf. Fig. 1). Experimental bubble growth data up to departure in comparison with theoretical predictions. The bubble growth rate approximates a constant [cf. the Rayleigh solution, Eq. (1)].

bubble. Occasionally, even a penetration through the upper bubble cap occurs, Fig. 7. Similar phenomena have been observed in cavitation, where the high-velocity jets (up to 300 m/s) cause damage to the surface of rotating immersed machine parts such as ship's propellers. The remarkable behavior of the secondary vapor column is attributed to the formation of a hot dry area beneath the center during the growth of the primary bubble.

(vii) According to Eq. (2), the ratio of the contributions to the bubble growth rate due to the evaporation microlayer and the relaxation microlayer is independent of both time and initial wall superheating, and amounts to $0.191 \ Pr^{-1/6}/b^*$. This ratio increases gradually from a value of 0.10 at 26.7 kPa to 0.97 at 4.08 kPa [1]; i.e., the evaporation micro-layer contributes substantially only at low subatmospheric pressures.

4.2. Aqueous Binary Systems

At low concentrations of the more volatile component in binary systems, the dominating influence of mass diffusion is demonstrated by the following effects [2, 7]:

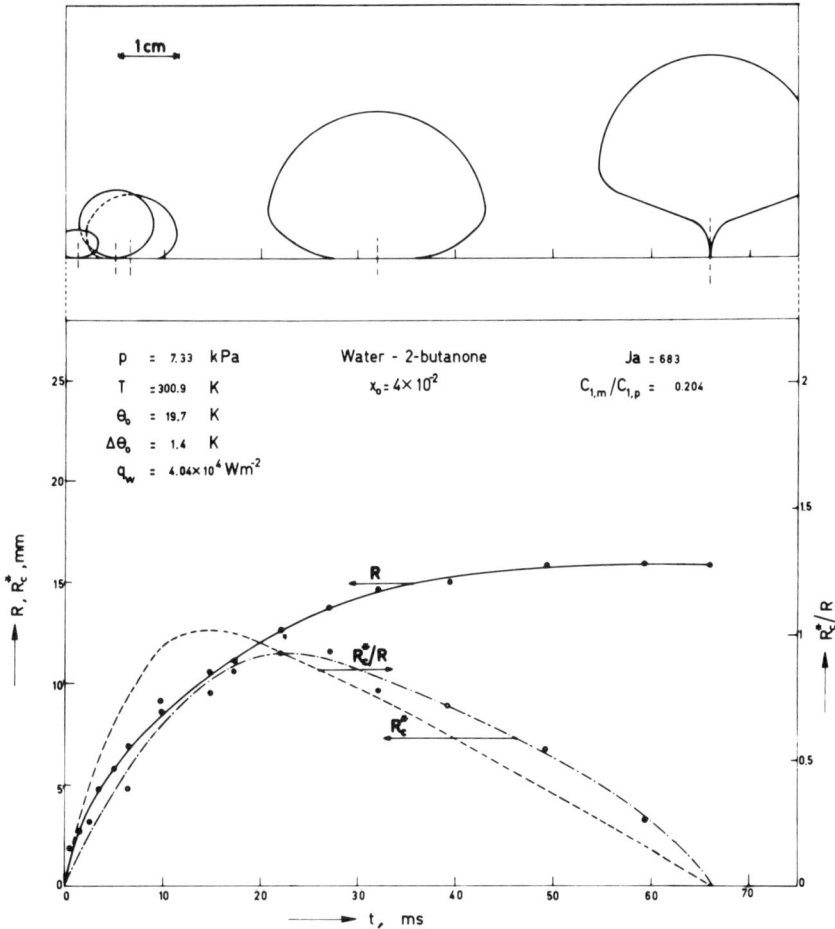

FIG. 6. *Water-2-butanone (4 wt% 2-butanone)* boiling at 7.33 kPa (cf. Fig. 3). Equivalent bubble radius R, contact radius R_c^*, and ratio R_c^*/R, in dependence on time. Corresponding bubble profile is shown in the upper drawings. Initial mode of growth is ellipsoidal: $R_c^*/R \to 0$ as $t \to 0$.

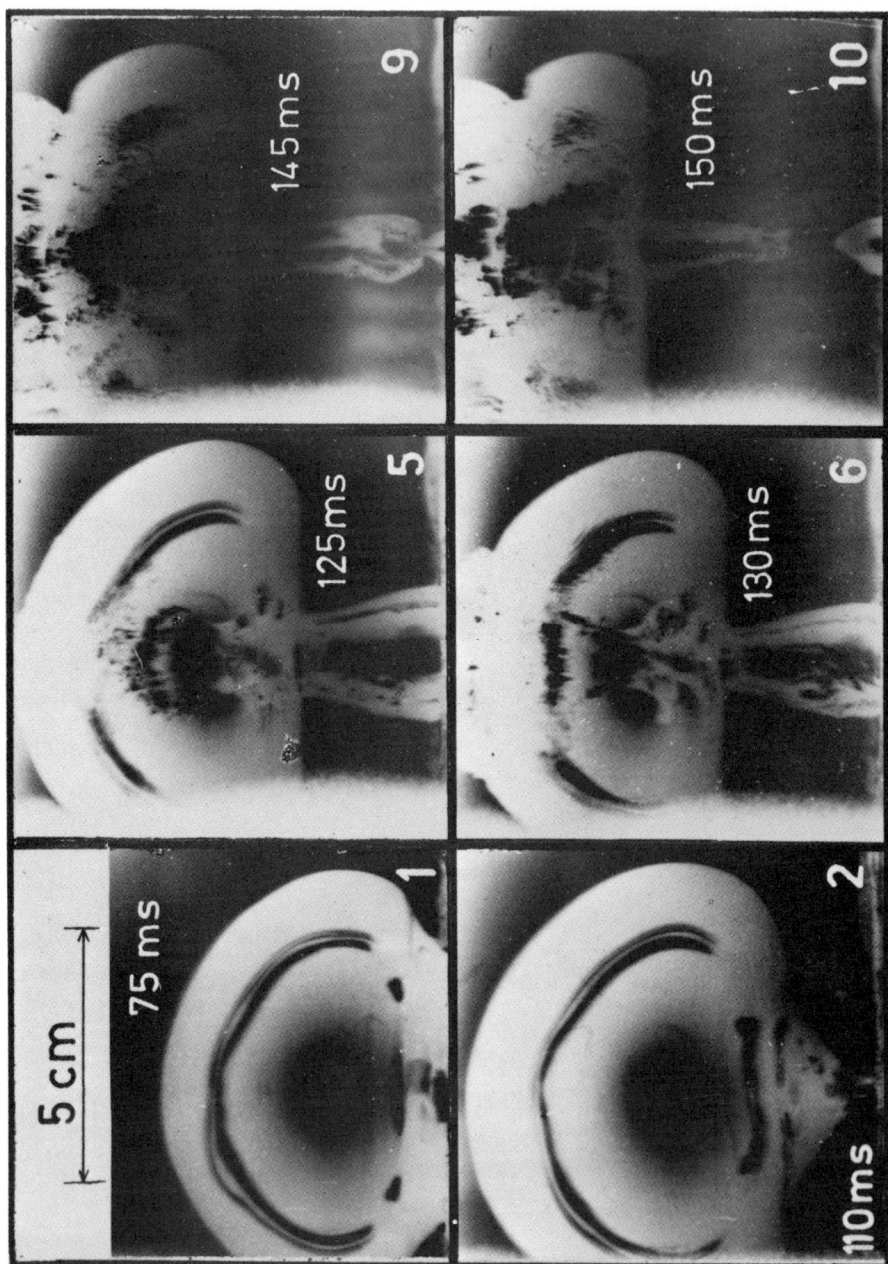

198

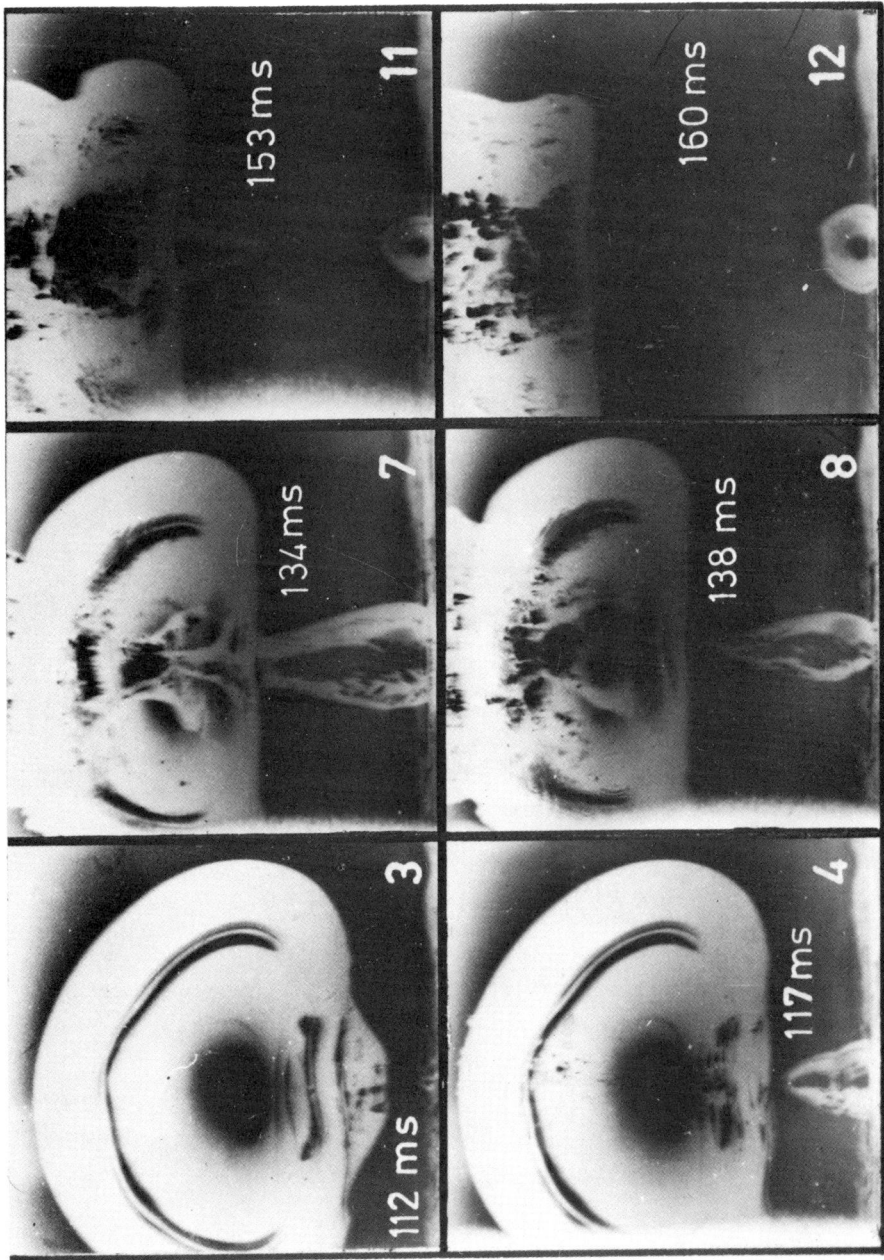

FIG. 7. *Water* boiling at 7.88 kPa (cf. Figs. 1, 3, and 4). A diverging high-velocity (5 m/s) liquid jet followed by a hot vapor column is penetrating through a large preceding vapor bubble.

199

(i) Advanced bubble growth is slowed down substantially, Figs. 3 and 6. For the investigated mixture (4 wt% 2-butanone in water), $C_{1,m} = 0.20\, C_{1,p}$, Eqs. (4) and (5).

(ii) The formation of dry areas beneath growing bubbles is prevented, even at low subatmospheric pressures. As a consequence, the bubble cycle is regular.

(iii) The lower part of the bubble is contracted due to the Marangoni effect, which is based on the occurrence of a surface tension gradient (due to a concentration gradient) along the bubble boundary. Consequently, the bubble has initially the shape of a spheroid, instead of a hemisphere (cf. ratio R_c^*/R in Fig. 6).

(iv) The contribution of the evaporation microlayer to bubble growth will be diminished in comparison to water, Eq. (2).

5. OSCILLATIONS OF VAPOR BUBBLES

5.1. Numerical Approximation

Combination of the continuity, momentum, and linearized Clapeyron equations results in the extended Rayleigh equation for a spherically symmetrical bubble:

$$R\ddot{R} + \tfrac{3}{2}\dot{R}^2 + \frac{2\sigma}{\rho_1 R} - \frac{\rho_2 l}{\rho_1 T}\theta_R + \frac{4\eta \dot{R}}{\rho_1 R} = 0 \qquad (8)$$

The temperature of the surrounding liquid is governed by the heat diffusion equation

$$\frac{\partial \theta}{\partial t} + \dot{R}\left(\frac{R}{r}\right)^2 \frac{\partial \theta}{\partial r} - a\left(\frac{\partial^2 \theta}{\partial r^2} + \frac{2}{r}\frac{\partial \theta}{\partial r}\right) = 0 \qquad (9)$$

The boundary conditions of Eq. (9) at $r = R$ and $r \to \infty$ are given respectively by the heat balance:

$$\left(k\frac{\partial \theta}{\partial r}\right)_{r=R} = \rho_2 l\dot{R} \qquad (10)$$

and
$$\theta(r \to \infty, t) = O_0 \qquad (11)$$

The initial conditions are:

$$\theta(r,0) = \theta_0 \quad \text{and} \quad \dot{R}(0) = \ddot{R}(0) = 0 \quad \text{or} \quad R(0) = R_0 \quad \text{and} \quad \theta_R(0) = \theta_0 \qquad (12)$$

It can be understood qualitatively that $R(t)$ may oscillate. During bubble

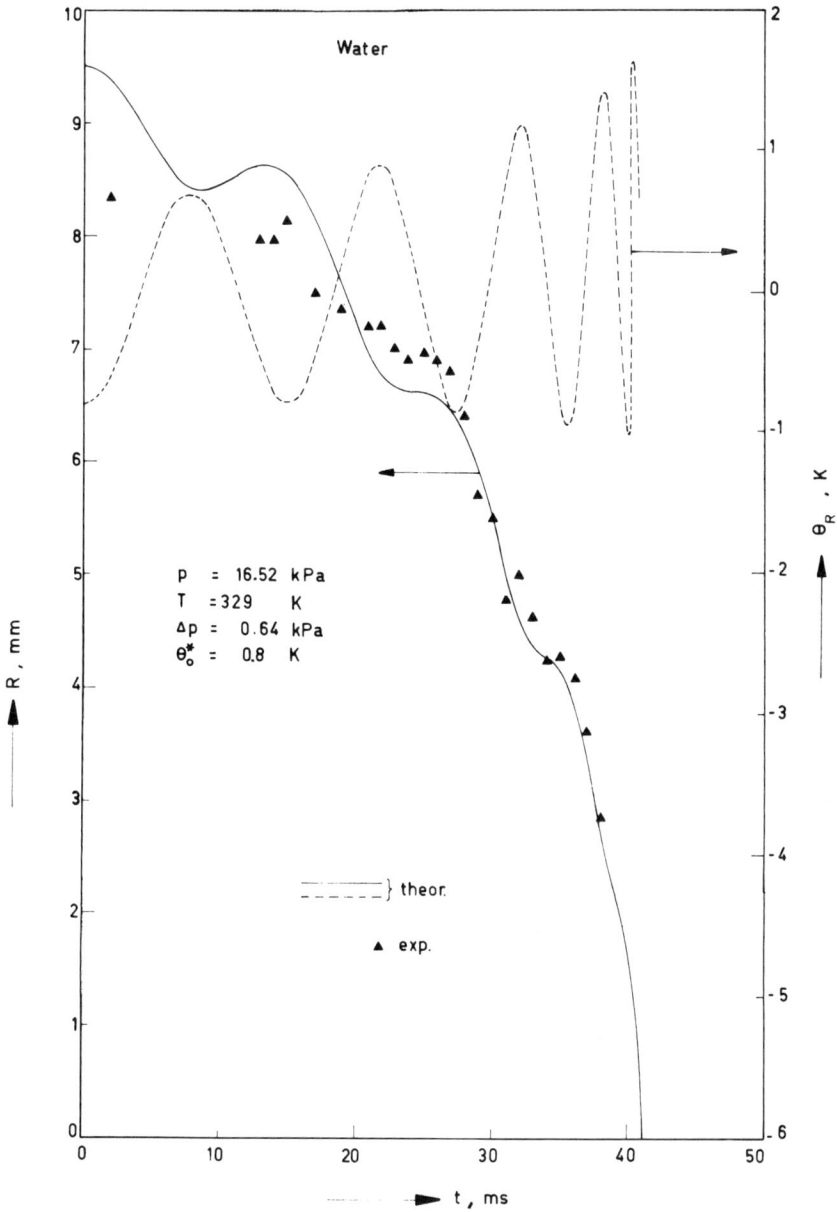

FIG. 8 *Water* boiling at 16.52 kPa. Experimental (▲) and theoretical (−) implosion of an oscillating free vapor bubble due to a spatially uniform subcooling. Instead of Eq. (13), a polynomial of the third degree in $1/r$ has been inserted for the trial function.

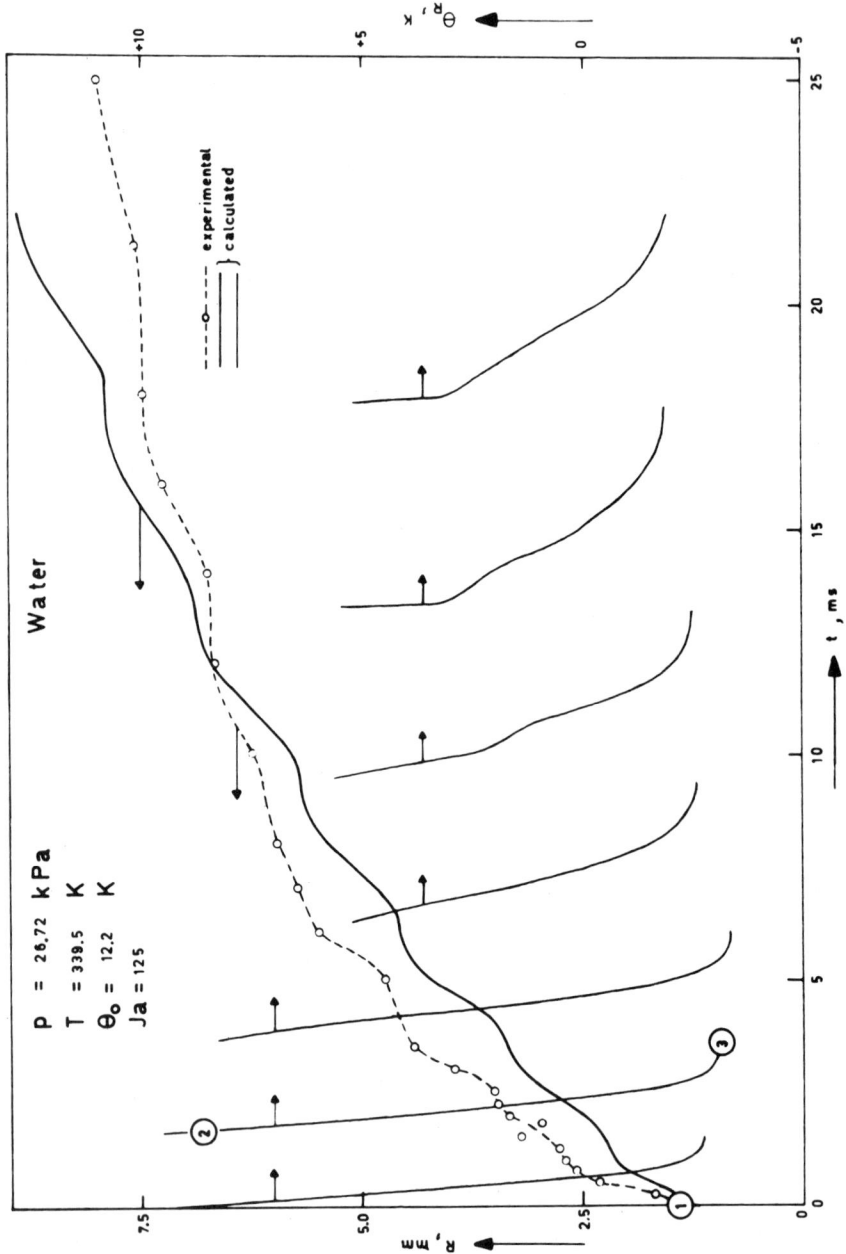

FIG. 9. *Water* boiling at 26.72 kPa (cf. Fig. 10). Comparison between experimental (—○—) and numerical (———) equivalent bubble radius. Both $R(t)$ and $\theta_R(t)$ are oscillating. The initial temperature distribution in the liquid is assumed to be cylindrically symmetric; the boundary conditions are specified in the legend. Calculations ignore an evaporation microlayer. Numbers 1–3 correspond with Fig. 10.

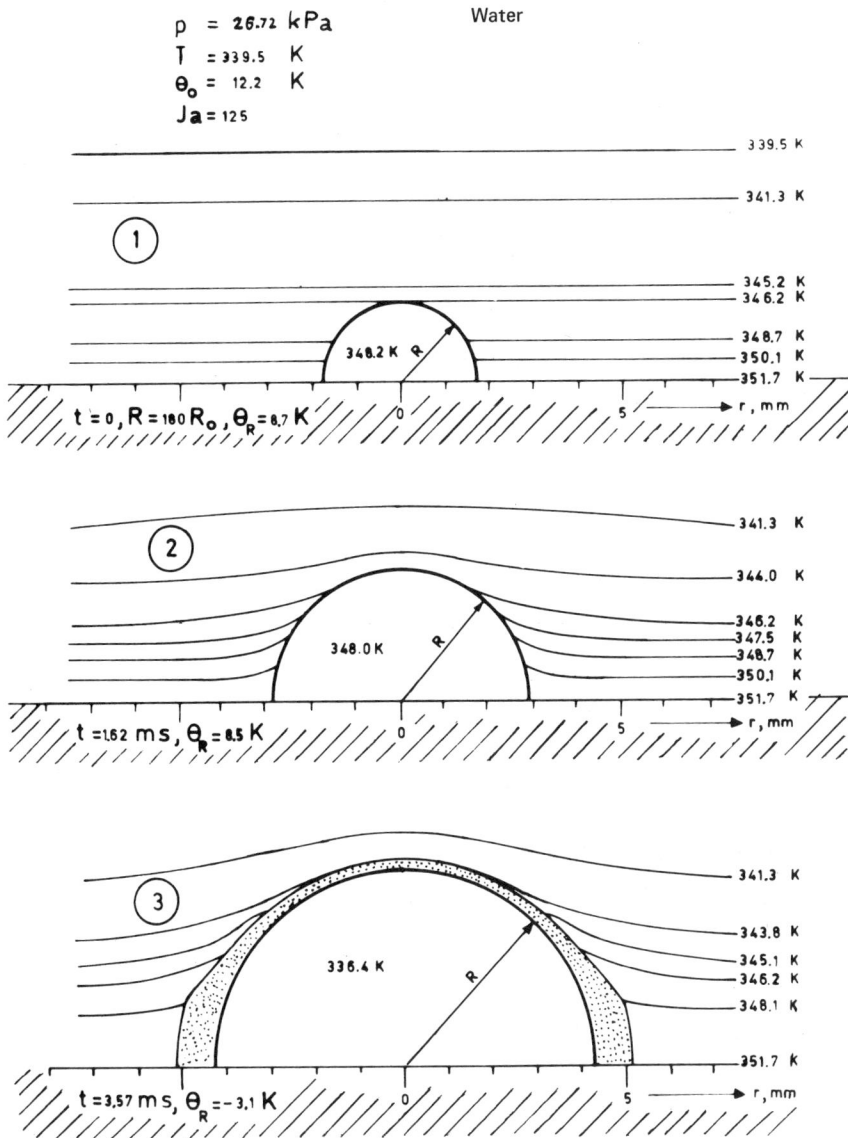

FIG. 10 *Water* boiling at 26.72 kPa. Numerically calculated isotherms around the vapor bubble shown in Fig. 9. The thermal boundary layer has been dotted. Numbers 1–3 correspond with Fig. 9.

growth, the superheating $\theta_R(t)$ decreases due to the required heat of vaporization, Eq. (10). This in turn leads to a decrease in the bubble growth rate—in accordance with Eq. (8), and continues until $\dot{R} < 0$, i.e., the bubble implodes during a short period of time. Condensation of vapor during the implosion leads to an increase in θ_R and consequently in $\dot{R}$, etc. To solve the problem under consideration quantitatively, a simple version of the collocation method [9] has been applied in order to transform the partial differential system into a system of ordinary differential equations. For this purpose $\theta(r, t)$ is expanded into:

$$\theta(r, t) = \theta_0 + [\theta_R(t) - \theta_0] \exp\left[\frac{R(t) - r}{\delta(t)}\right] \qquad (13)$$

The expansion coefficients $\theta_R(t)$ and $\delta(t)$ are determined by substitution of Eq. (13) into Eq. (10) and into Eq. (9) only on the bubble wall. We thus find:

$$\frac{d}{dt}\left(\frac{\theta_R - O_0}{O_0}\right)^2 = \frac{2}{a}\frac{\dot{R}^2}{Ja} \qquad (14)$$

In determining the signs after taking the square root of the solution of Eq. (14), one has for physical reasons:

$$\theta_R(t)\begin{cases} < \theta_0 & \text{if } \dot{R}(t) > 0 \quad \text{(growth)} \\ = \theta_0 & \text{if } \dot{R}(t) = 0 \quad \text{(adiabatic bubble wall)} \\ > \theta_0 & \text{if } \dot{R}(t) < 0 \quad \text{(implosion)} \end{cases} \qquad (15)$$

5.2. Specific Results in Comparison with Experimental Data

The system of ordinary differential Equations (8–12) can easily be solved numerically by standard methods, e.g., Euler-Cauchy or Runge-Kutta [10]. An application is given in Fig. 8.

Zwick and Plesset [11] presented a solution of this problem too, using a combined perturbation and iteration method. However, the method presented here has the advantage, that it can easily be extended to nonspherically symmetric flow—and temperature fields, Figs. 9 and 10, Eq. (8) being replaced by the Bernoulli equation, combined with the Laplace equation for the velocity potential.

ACKNOWLEDGEMENTS

The authors are indebted to W. M. Sluyter for cooperation, and to Ir. L. J. Bour and J. G. H. Joosten for carrying out experiments and numerical calculations.

Footnote

High-speed motion picture. A high-speed motion picture (taken at a rate of 4000 fps) illustrates bubble growth in water during nucleate boiling at gradually decreasing subatmospheric pressure (from 13 to 4 kPa). As an extreme case, at a pressure of 4 kPa, large "Rayleigh bubbles" are observed during the entire adherence time.

Also, a remarkable bubble cycle is observed at the lowest investigated pressures, which is attributed to the combined action of a high-velocity liquid jet, originating in the wake following a large primary bubble, and a succeeding secondary vapor column, generated at the adjacent dry spot at the heating wall beneath the primary bubble.

Projection of the motion picture at the usual rate of 20 fps results in a slowing down with a factor of 200.

NOMENCLATURE

a	liquid thermal diffusivity	m^2/s
b^*	dimensionless bubble growth parameter during adherence	
c	liquid specific heat at constant pressure	$J/kg \cdot K$
C_1	bubble growth constant	$m/s^{1/2} \cdot K$
D	mass diffusivity of more volatile component in less volatile component	m^2/s
G	vaporized mass fraction	
Ja	$(= \rho_1 c \Theta_0 / \rho_2 1)$ Jakob number	
k	liquid thermal conductivity	$W/m \cdot K$
K	equilibrium constant of more volatile component in binary mixture (ratio of mass fractions of vapor and liquid)	
l	latent heat of vaporization	J/kg
p	ambient pressure	$Pa = N/m^2$
Δp	pressure step causing a subcooling Θ_0^*	Pa
Pr	liquid Prandtl number	
q	heat flow density	W/m^2
r	radial distance from bubble center	m
R	equivalent spherical bubble radius	m
R_0	$(= 2\sigma T/\rho_2 1 \Theta_0)$ radius of nucleation site	m
$R(t_1)$	equivalent spherical radius at bubble departure	m
R_1	equivalent bubble radius according to modified Rayleigh solution	m
R_2	equivalent bubble radius according to total diffusion (combined evaporation and relaxation microlayer) solution	m
R^*	radius of hemispherical bubble	m
R_c^*	radius of contact area between bubble and heating surface	m
t	time during bubble growth or implosion	s
t_1	bubble departure time	s
T	absolute boiling temperature	K
ΔT	increase in temperature of liquid mixture at bubble boundary with respect to original liquid	K
x_0	mass fraction of more volatile component in original binary mixture	
z	normal distance from heating surface	m

Greek Symbols

δ	instantaneous local thermal boundary layer thickness around bubble	m
η	liquid dynamic viscosity	$kg/m \cdot s$
θ	instantaneous local liquid superheating	K

θ_0	superheating of heating surface or initial uniform liquid superheating	K
$\theta_0{}^*$	uniform liquid subcooling due to pressure step Δp	K
θ_R	superheating of bubble boundary	K
$\overline{\theta_R}$	mean superheating of rotationally symmetric bubble	K
$\Delta\theta_0$	superheating of bulk liquid	K
ν	liquid kinematic viscosity	m^2/s
ρ_1	liquid density	kg/m^3
ρ_2	saturated vapor density	kg/m^3
σ	surface tension	kg/s^2

Subscripts

co	convective heat flow density
m	binary mixture
p	pure liquid
w	heating surface

REFERENCES

1. Van Stralen, S. J. D., R. Cole, W. M. Sluyter, and M. S. Sohal: *Int. J. Heat Mass Transfer,* **18**, 655, 1975.
2. Van Stralen, S. J. D., M. S. Sohal, R. Cole, and W. M. Sluyter: *Ibid.,* **18**, 453, 1975.
3. Van Stralen, S. J. D.: *Ibid.,* **11**, 1467 and 1491, 1968.
4. Cooper, M. G. and R. M. Vijuk: "Proc. 4th Int. Heat Transfer Conf., Paris-Versailles," **V**, p. B2.1. Elsevier, Amsterdam (1970.
5. Van Stralen, S. J. D.;*Int. J. Heat Mass Transfer,* **9**, 995, 1021, 1967; **10**, 1469, 1485, 1968.
6. Van Stralen, S. J. D.: "Abstracts of Papers and communications of the third All-Union Heat and Mass Transfer Conference," Minsk (1968) 67, 256; Proceedings (1968), **2**, 219; Van Stralen, S. J. D.: *Chem. Engng. Sci.,* (1970), **25**, 149.
7. Van Stralen, S. J. D., W. M. Sluyter, and R. Cole: submitted for publication to Int. J. Heat Mass Transfer (1976) **19**.
8. Mikic, B. B., W. M. Rohsenow, and P. Griffith: *Int. J. Heat Mass Transfer,* **13**, 657, 1970.
9. Finlayson, B. A.: "The Method of Weighted Residuals and Variational Principles," Academic Press, New York/London (1972).
10. Demidowitch, B. P., I. A. Maron, and E. S. Schuwalowa: "Numerische Methoden der Analysis," V. E. B. Verlag, Berlin (1968).
11. Zwick, S. A. and M. S. Plesset: *J. Math. Phys.,* **33**, 308, 1957.

INTERFEROMETRIC INVESTIGATION OF LIQUID SURFACE EVAPORATION AND BOILING DUE TO DEPRESSURIZATION*

S. S. Grewal, C. Shih, and M. M. El-Walil
University of Wisconsin, Madison, Wisconsin

INTRODUCTION

The phenomena of liquid evaporation, boiling, voidage, and expulsion, under sudden depressurization is of importance in the safety analysis of such systems as water- and liquid-metal-cooled nuclear reactors, high-pressure heat exchangers, steam generators, etc. Events such as surface evaporation; bubble nucleation, growth and detachment; slug formation, growth and motion; and fluid expulsion, are under study by various means. Many of these fail to provide detailed field quantitative data, or continuous records of an evolving sequence.

The present investigation uses Mach–Zehnder interferometry, high-speed photography and associated equipment to study the phenomena. The interferometer, normally used to study heat transfer boundary layers in gases (mostly air), is adapted for the study of temperature fields in a liquid. This technique has advantages over such methods that use thermocouples and anemometers in that it produces a two-dimensional field rather than point data; it can, with the proper analysis, produce three-dimensional data; it does not interfere physically with the field of study; and its response is instantaneous and continuous. Liquid systems have been studied by interferometry [1, 2], though quantitative and boiling data are lacking. The present investigation yields quantitative results that explain some of the observed phenomena.

In this study, liquid Freon-11, originally subcooled, is suddenly subjected to a loss of pressure under which it undergoes phase change. Two basic phenomena were observed:

---

*The abstract of this paper appears on page 389.

1. Free-surface evaporation and boiling occurring at the top surface of the liquid and resulting in the expulsion of a two-phase mixture.
2. Bubble formation, growth and detachment, originating at nucleation sites on the bottom unheated surface of the cell containing the liquid similar to pool-boiling in a heated system.

This paper deals with the first of these two phenomena.

THE EXPERIMENTAL APPARATUS

The fluid used is Freon-11 (Trichlorofluoromethane). It has convenient optical qualities and a normal boiling point (22.2°C) which allows boiling studies without subjecting the optically-flat glass plates (below) to thermal stresses.

The liquid is placed in a specially constructed cell, Fig. 1, which is placed in the test path of the interferometer. Optically flat glass plates, placed perpendicular to the light path, are used to enclose the liquid. The plate surfaces are totally free of nucleation sites. The aluminum spacer is polished to a fine finish and provides no nucleation at up to 70 cm Hg vacuum. An

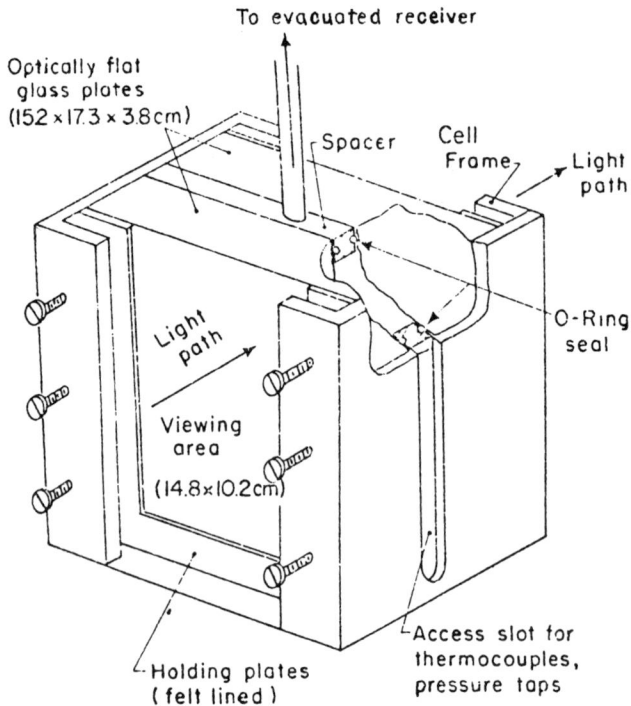

FIG. 1. Test cell.

O-ring seal was found to contain a large number of cavities and was replaced by a bead of locally available molding and potting compound. The cell is connected to an evacuated reservoir through a quick opening solenoid valve with an opening time on the order of 10 msecs. This supplies the desired quick depressurization.

An 8 in. Mach–Zehnder interferometer [3] is used with a xexon flash tube as a light source. The flash tube is controlled by a stroboscope and a programmable pulse generator. Each pulse had a 4.8 μsec duration and 0.27 Mwatts power. A 5330 Å filter with a 70 Å full width at half maximum was used to monochromatise the light. The interferograms are recorded on 13.3 cm wide aerial photography film rated at 400 ASA, mounted in 206 cm lengths on a drum camera. The number and frequency of flashes are synchronized with the speed of the drum camera and the length of film used. The upper limit of operation for the system is 24, 1/2 frame (8 cm × 13.3 cm) interferograms at intervals of 5 msecs.

EXPERIMENTAL RESULTS

Runs are typically recorded during a period of 2.4 seconds after depressurization with interferograms taken every 0.1 seconds. Samples from a typical run at moderate superheats are shown by the infinite-fringe interferograms of Fig. 2. The conditions for this run are a temperature of 21°C and a pressure of 1 atmosphere depressurized to 20.3 cm Hg vacuum, corresponding to a saturation temperature of 15.3°C and a fluid superheat of 5.7°C.

The first effect is observed at 0.3 sec when two small hemispherical indentations appear symetrically below the outlet tube to the vacuum reservoir. At 0.5 sec two additional indentations appear on either side of these. A closer look reveals that the appearance of these indentations marks a lowering of the liquid surface. The next interferogram shows the disappearance of the indentations and the appearance of interference fringes of extremely close spacing, indicating high rates of heat transfer upward. This is interpreted as indicating the supply of the latent heat of evaporation and a cooling of the liquid below the surface. The cooled liquid layer travels downward because of its increased density and is simultaneously heated by the warmer fluid below it.

The fringe spacing becomes larger with time, indicating a reduction in heat transfer rates. The fringe pattern changes, showing discrete volumes or drops of subcooled liquid to which heat from the adjacent liquid bulk is transferred. These "drops" of relatively cold liquid assumed cylindrically symmetric shapes which distort slowly with time. This is thought to be due to their downward motion, expansion due to heating, and convective effects surrounding them. A typical drop that has departed from symmetry is shown in the enlargement in

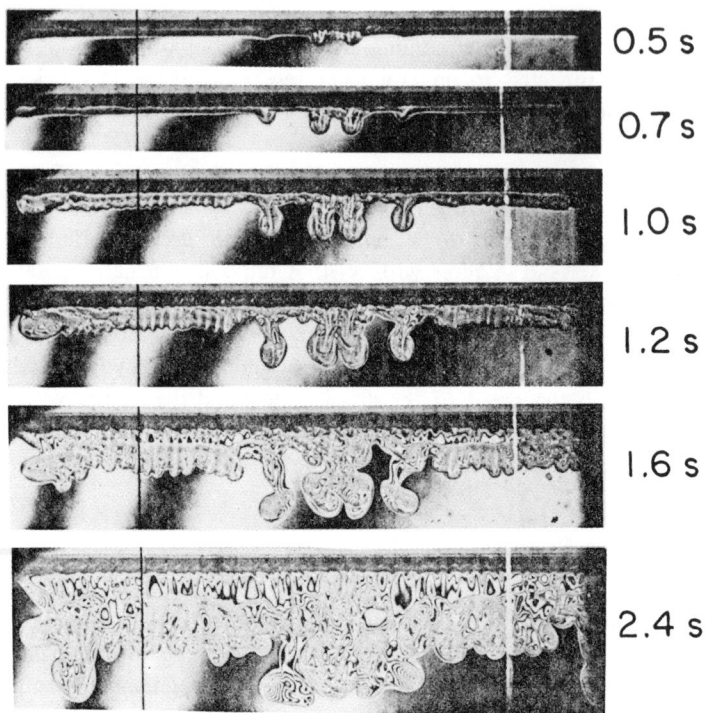

FIG. 2. Surface boiling interferograms; Freon–11, 5.7 initial superheat; at 0.5, 0.7, 1.0, 1.2, 1.6, 2.4 sec after depressurization.

Fig. 3, which is part of a sequence as shown in Fig. 2, taken at 1.7 sec after depressurization.

INTERFEROGRAM ANALYSIS

The early interferograms contain extremely closely-spaced fringes. These are not sufficiently identifiable to permit accurate analysis. The first and last analyses are done at 0.8 and 1.4 sec after depressurization. In the analysis of the "drops" the three-dimensional nature was taken into account by a modification of the method in [4].

Figure 4 shows the resulting temperature profiles on a horizontal axis across the middle of the drop of Fig. 3. The minimum temperature was found to be 9.4°C at the middle of the drop at 0.8 sec. It is possible that this may

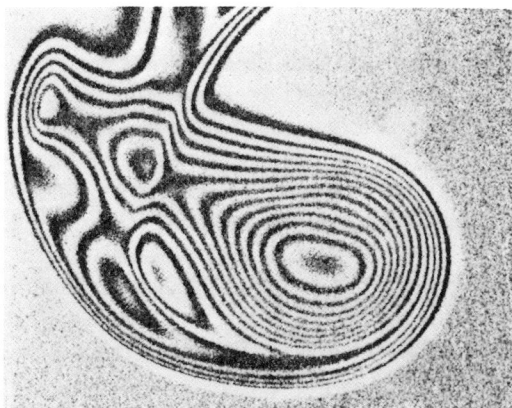

FIG. 3. Enlarged interferogram of analyzed drop at 1.7 sec.

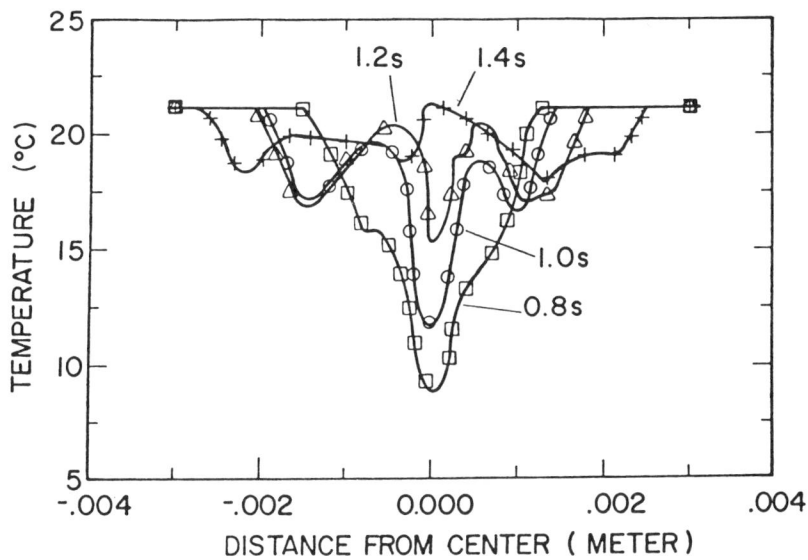

FIG. 4. Temperature profiles in liquid drop at 0.8, 1.0, 1.2, 1.4 sec.

have been lower at an earlier time, when the interferograms could not be analyzed. This temperature is 5.9°C below the saturation temperature of 15.3°C corresponding to 20.3 cm Hg vacuum. It can thus be concluded that there is a large subcooling during flash boiling from the surface. The degree of subcooling, however, decreases with time. At 1.4 secs, the minimum temperature climbs to 17.8°C, which is above the saturation temperature, indicating that the drop has become superheated again.

Analysis of drops beyond 1.4 sec was deemed unreliable because of the extreme asymmetry of the drop and the difficulty of assigning an accurate axis of revolution. This is the case, for example, of the drop as shown in Fig. 3.

Figure 5 shows a plot of the heat flux vs. distance from the center of the drop at 0.8 and 1.4 sec, the first and last interferograms analyzed. The plot shows the expected reduction in heat flux with time, as well as the wavy nature of the flux.

It is also of interest to show the temperature profiles at the liquid top surface. This is done at 1.7 sec at three arbitrarily chosen positions, using a two-dimensional analysis (Fig. 6). It is believed, however, that this analysis is less accurate than that for the drop, above.

CONCLUSIONS

The measured subcooling and superheat may now be used to explain the two-phase expulsion phenomena, also observed by other experimenters [5]. At and just below the surface is a heterogeneous mixture of subcooled, saturated, and superheated liquid. Evaporation and boiling thus take place with boiling seeming to occur at more-or-less regularly spaced intervals along

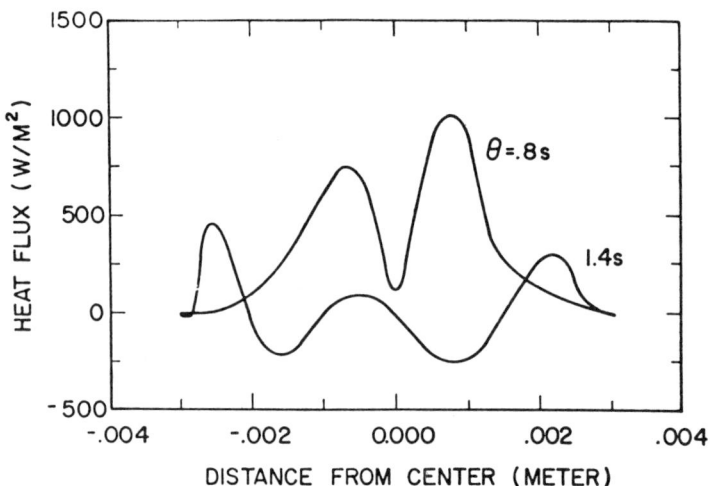

FIG. 5. Heat flux profiles in liquid drop at 0.8, 1.4 sec.

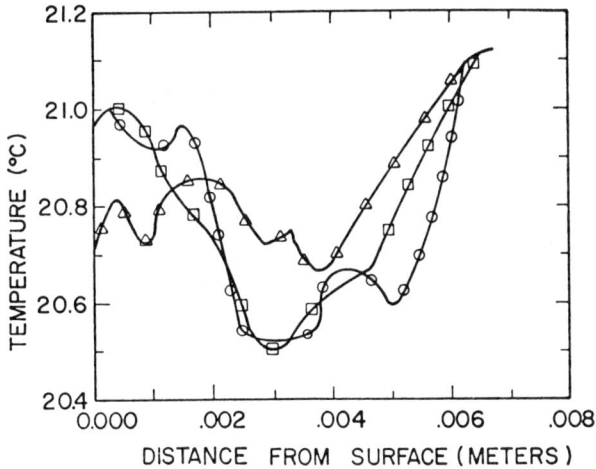

FIG. 6. Liquid temperatures measured from the top at 3 vertical axes, using a two-dimensional analysis, at 1.7 sec.

the surface. The resulting vapor ejects upwards carrying with it some of the subcooled liquid. The result is a two-phase expulsion from the free surface of the liquid.

REFERENCES

1. Hauf, W. and U. Grigull: "Optical Methods in Heat Transfer," Advances in Heat Transfer, **6, p.** 133, 1970, Academic Press.
2. Mayinger, F. and W. Panknin: "Holography in Heat and Mass Transfer," 5th International Heat Transfer Conference, Tokyo, 1974, **VI,** pp. 28, (IL3).
3. Gloudemans, J. R.: "The Transition from Forced to Free-Convective Heat Transfer From a Heated Vertical Plate in Air with Buoyant and Inertial Forces Opposed," Ph.D. Thesis, University of Wisconsin, 1973.
4. Ladenburg, R., J. Winkler, and C. C. Van Voorhis: "Interferometric Study of Faster than Sound Phenomena, Part I," *Phys. Rev.,* **73,** 1359, (1948).
5. Grolmes, M. A. and H. K. Fauske: "Axial Propagation of Free Surface Boiling into Superheated Liquids in a Vertical Tube," 5th International Heat Transfer Conference, Tokyo, 1974, **IV,** p. 80, Paper B1.7.

AN APPROXIMATE SOLUTION OF THE GENERALIZED STEFAN PROBLEM IN A POROUS MEDIUM WITH VARIABLE THERMAL PROPERTIES*

I. J. Kumar and L. N. Gupta
Scientific Analysis Group, R and D Organization, New Delhi

INTRODUCTION

Luikov [1] has studied steady state heat and mass transfer between a capillary porous medium and an external gas stream during drying. The same problem was studied by Morgan and Yerazuins [2] for laminar and turbulent gas flows with variable porous surface temperatures. It is well known that the moving evaporation front divides the porous system into two regions; while the moisture in one region is in vapor form only, in the other region it is in a mixed (vapor and liquid) form. The problem reduces to the simultaneous solution of the pure heat conduction problem in one region and solution of a coupled problem of heat and mass transfer in the other region, the two regions being separated by a variable boundary whose position is to be determined from the solution of the problem. We have called this problem a "Generalized Stefan Problem."

In an earlier paper [3] we have formulated the above problem and obtained its solution with the assumptions that the thermal diffusion factor is negligible and the thermal properties are constant.

In the present paper, we have obtained a solution of the "Generalized Stefan Problem" with variable thermal properties and taking thermal diffusion into account. As distinct from similar available works, the mechanism of the deepening of the evaporation surface is described with the aid of Stefan's problem.

It is shown that the effect of the deepening of the evaporation front on unsteady heat and mass transfer in a porous body is characterized by a parameter v, the nondimensional heat of vaporization. The case of the generalized Stefan problem with constant thermal properties reduces to a particular case. Moreover, it is shown that with the increase in the criterion Pn

---

*The abstract of this paper appears on page 405

(Posnov number), there is a reduction in the nondimensional mass transfer potential and the rate of motion of the evaporation surface. Results have been exhibited graphically.

STATEMENT OF THE PROBLEM

Consider the flow of heat and moisture through a porous half-space ($x >$ 0) in which the surface $x = 0$ is at temperature t_s where t_s is greater than the vaporizing temperature of the liquid. Initially, the whole body is at temperature t_0 and moisture potential θ_0. The evaporation front divides the porous system into two regions; in the region $0 < x < s$, moisture is in vapor form only and there is no moisture gradient, while in the region $s < x < \infty$ moisture is in mixed form.

In an earlier paper [3] we have reported the solution of the problem with the assumption that the thermal diffusion factor is negligible and the thermal properties of the porous body are constant. However, in situations where wide temperature ranges are involved, the assumption of constant thermal properties is not justified and we cannot neglect the thermal diffusion factor. Here we take

$$a_q = a_{q0}(1 + \lambda_1 T + \lambda_2 \bar{\theta}) \tag{2.1}$$
$$a_m = a_{m0}(1 + \mu_1 T + \mu_2 \bar{\theta}) \tag{2.2}$$

where λ_1, λ_2, μ_1, and μ_2 are constant parameters. Following [3] the problem can be stated mathematically with the help of nondimensional equations as

$$\left. \begin{aligned} \frac{\partial T_1}{\partial \text{Fo}} &= \frac{\partial^2 T_1}{\partial x^2} \\ \bar{\theta}_0 &= \bar{\theta}_v \end{aligned} \right\} \quad 0 < x < S \tag{2.3}$$
$$\tag{2.4}$$

$$\left. \begin{aligned} \frac{\partial T_2}{\partial \text{Fo}} &= \frac{\partial}{\partial x}\left[(1 + \lambda_1 T_2 + \lambda_2 \bar{\theta}_2)\frac{\partial T_2}{\partial x}\right] - \epsilon\,\text{Ko}\,\frac{\partial \bar{\theta}_2}{\partial \text{Fo}} \\ \frac{\partial \bar{\theta}_2}{\partial \text{Fo}} &= \frac{\partial}{\partial x}\left[\text{Lu}_0(1 + \mu_1 T_2 + \mu_2 \bar{\theta}_2)\frac{\partial \bar{\theta}_2}{\partial x}\right] \\ &\quad - \text{Lu}_0\text{Pn}\frac{\partial}{\partial x}\left[(1 + \mu_1 T_2 + \mu_2 \bar{\theta}_2)\frac{\partial T_2}{\partial x}\right] \end{aligned} \right\} \quad S < x < \infty \tag{2.5}$$
$$\tag{2.6}$$

The interface conditions are

$$\left.\begin{array}{r}
\dfrac{\partial T_1}{\partial x} - (1 + \lambda_1 T_2 + \lambda_2 \bar{\theta}_2)K_{21}\dfrac{\partial T_2}{\partial x} = -v_s \dfrac{\partial S}{\partial \mathrm{Fo}} \\[2ex]
T_1 = T_2 = T_v \\[1ex]
\bar{\theta}_1 = \bar{\theta}_2 = \theta_v = 1
\end{array}\right\} \quad x = S$$

$$\tag{2.7}$$
$$\tag{2.8}$$
$$\tag{2.9}$$

The initial and boundary conditions are

$$T(x,0) = 0 \qquad x > 0 \tag{2.10}$$

$$\bar{\theta}(x,0) = 0 \qquad x > 0 \tag{2.11}$$

$$T(0,\mathrm{Fo}) = 1 \qquad \mathrm{Fo} > 0 \tag{2.12}$$

SOLUTION OF THE PROBLEM

An approximate solution of the problem can be obtained by a local potential technique [4]. The problem reduces to the solution of a pure heat conduction problem in one region and a coupled problem of heat and mass transfer with variable thermal properties in the other region.

(a) Solution for the Region $0 < X < S$

For the temperature distribution in the region $0 < X < S$ the exact solution for a semi-infinite body is used [3].

$$\left.\begin{array}{l}
T_1 = \dfrac{T_v - 1}{\mathrm{erf}(\lambda)}\,\mathrm{erf}\!\left(\dfrac{x}{2\sqrt{\mathrm{Fo}}}\right) + 1 \\[2ex]
\bar{\theta}_1 = \bar{\theta}_v
\end{array}\right\} \quad 0 < x < S$$

$$\tag{3.1}$$
$$\tag{3.2}$$

where $$S = 2\lambda\sqrt{\mathrm{Fo}} \tag{3.3}$$

(b) Solution for the Region $S < X < \infty$

In the region $S < X < \infty$, a coupled problem of heat and mass transfer with variable thermal properties and boundary conditions of the first kind is to be solved. The solution of the problem is obtained by a boundary layer approach in local potential [4]. Let $q(\mathrm{Fo})$ and $Q(\mathrm{Fo})$ be nondimensional thermal and mass penetration distances, respectively. We can assume parabolic profiles for the temperature and moisture distributions as

$$T_2 = T_v\left(1 - \frac{x - S}{q}\right)^2 \tag{3.4}$$

$$\bar{\theta}_2 = \left(1 - \frac{x - S}{Q}\right)^2 \tag{3.5}$$

Kumar [5] has established the form of local potential for coupled problem of heat and mass transfer in a porous medium. For a one-dimensional case it can be written as:

$$J = \int \left[\frac{1}{2} \left(\frac{\partial T_2}{\partial x} \right)^2 + T_2 \frac{\partial T_2{}^0}{\partial \text{Fo}} + \epsilon \, \text{Ko} \, T_2 \frac{\partial \bar{\theta}_2{}^0}{\partial \text{Fo}} + \frac{\text{Lu}_0}{2} \left(\frac{\partial \bar{\theta}_2}{\partial x} \right)^2 + \bar{\theta} \frac{\partial \bar{\theta}_2{}^0}{\partial \text{Fo}} \right] dx \quad (3.6)$$

Here, the variation is to be taken independently and exclusively over T and $\bar{\theta}$ where

$$T = T^0 + \delta T \tag{3.7}$$
$$\bar{\theta} = \bar{\theta}^0 + \delta \bar{\theta} \tag{3.8}$$

The unknown parameters q and Q in the profiles (3.4) and (3.5) are to be determined from

$$\frac{\partial J}{\partial q} = 0 \tag{3.9}$$

$$\frac{\partial J}{\partial Q} = 0 \tag{3.10}$$

with the self-consistency conditions

$$q = q^0 \tag{3.11}$$
$$Q = Q^0 \tag{3.12}$$

Now, the problem reduces to obtaining S(Fo) and parameters q and Q from the set of equations (2.7), (3.9), and (3.10). Goodman [6], Biet [7], and Kumar and Narang [8] have shown that the penetration distance varies as $A\sqrt{\text{Fo}}$ where A is some constant to be determined from the equations. Therefore, we assume

$$q = 2\beta \sqrt{\text{Fo}} \tag{3.13}$$
$$Q = 2\alpha \sqrt{\text{Fo}} \tag{3.14}$$

when $q < Q$ the set of equations (2.7), (3.9), and (3.10) can be written as

$$(4\beta^2 - 10 + 10\beta\lambda) + \frac{60\epsilon \text{Ko}\beta^3 \bar{\theta}_s}{\alpha T_s} \left(\frac{1}{6} - \frac{1}{10} \frac{\beta}{\alpha} \right) - 9\lambda_1 T_s$$

$$+ 30\lambda_2 \bar{\theta}_s \left(-\frac{1}{3} + \frac{1}{30} \frac{\beta^2}{\alpha^2} \right) + 30\epsilon \text{Ko}\bar{\theta}_s \cdot \frac{2\lambda\beta^2}{T_s\alpha} \left(\frac{1}{3} - \frac{\beta}{6\alpha} \right) = 0 \quad (3.15)$$

$$(4\alpha^2 - 10\,Lu_0 + 10\lambda\alpha) - \frac{60\,Lu_0 P_n T_s}{\bar{\theta}_s}\left(-\frac{1}{2} + \frac{1}{3}\frac{\beta}{\alpha}\right)$$

$$+ 30\,Lu_0\mu_1 T_s\left(-\frac{2\beta}{3\alpha} - \frac{2\beta^3}{15\alpha^3} + \frac{1\beta^2}{2\alpha^2}\right)$$

$$- 9\,Lu_0\mu_2\bar{\theta}_s - 60\,Lu_0 P_n\mu_1 T_s^2\left(-\frac{1}{4} + \frac{1\beta}{10\alpha}\right)\bar{\theta}_s$$

$$- 120\,Lu_0 P_n\mu_2 T_s\left(-\frac{1}{2} + \frac{2\beta}{3\alpha} - \frac{5\beta^2}{12\alpha^2} - \frac{1\beta^3}{10\alpha^3}\right) = 0 \quad (3.16)$$

$$e^{-\lambda^2}|\sqrt{\pi}\ \text{erf}\,(\lambda) - (1 + \lambda_1 T_s + \lambda_2\bar{\theta}_s)p|\beta = \upsilon\lambda \tag{3.17}$$

and when $q > Q$ it can be replaced with (3.16), (3.17), and the equation

$$(4\beta^2 - 10 + 10\beta\lambda) + 60\epsilon Ko\beta^3\bar{\theta}_s\left(\frac{1}{6} - \frac{\alpha}{10\beta}\right)|\alpha T_s - q\lambda_1 T_s$$

$$+ 30\lambda_2\bar{\theta}_s\left(-\frac{1}{3} + \frac{\beta^2}{30\alpha^2}\right) + 30\epsilon Ko\bar{\theta}_s\,|\,T_s\,\frac{2\lambda\beta^2}{\alpha}\left(\frac{1}{3} - \frac{1}{6}\frac{\alpha}{\beta}\right) = 0 \quad (3.18)$$

Therefore, the set of equations (3.1) to (3.5) with these determined values of α, β, and λ gives us the temperature and moisture profiles.

SOME PARTICULAR CASES OF THE PROBLEM

We get particular cases of the problem by giving suitable values to the constant parameters λ_1, λ_2, μ_1, μ_2, Pn and λ

Case 1. when $\lambda = 0$

When $\lambda = 0$, Eq. (3.3) implies that $S = 0$ for all values of Fo. This is possible only when the initial and boundary conditions are such that the evaporation front is fixed at the surface. From the statement of the problem it is obvious that λ can be zero only when the temperature at the surface is less than or equal to the vaporizing temperature of the liquid; therefore, complete vaporization cannot take place inside the porous system. Consequently, the problem reduces to a coupled heat and mass transfer problem in a porous half-space with boundary conditions of the first kind, variable thermal properties, and taking the thermal duffusion factor into account. Its solution can be obtained by taking $\lambda = 0$ in the set of equations (3.4), (3.5), and (3.15) to (3.18).

Case 2. when $\lambda_1 = 0$, $\lambda_2 = 0$, $\mu_1 = 0$, $\mu_2 = 0$, and Pn = 0

In this case the problem reduces to the generalized Stefan problem with constant thermal properties [3]. The solution of the problem is obtained by taking these values in the set of equations (3.4), (3.5), and (3.15) to (3.18). The solution derived from Eqs. (3.4), (3.5), and (3.15) to (3.18) is identical to the one obtained in [3].

Case 3. when $\lambda_1 = 0$, $\lambda_2 = 0$, $\mu_1 = 0$, $\mu_2 = 0$, Pn $\neq 0$

This will represent the solution of the generalized Stefan problem with constant thermal properties and taking thermal diffusion terms into account.

The solution can be obtained by taking $\lambda_1 = 0$, $\lambda_2 = 0$, $\mu_1 = 0$, and $\mu_2 = 0$ in the set of equations (3.15) to (3.18).

NUMERICAL RESULTS AND DISCUSSION

The values of the parameters chosen for the numerical work of the above problems are

$$L_{u_0} = 0.5, \epsilon = 0.5, k_0 = 1.2, k_{21} = 1.0, p = 1.0, \nu = 10.0,$$
$$\lambda_1 = 0.5, \lambda_2 = 0.0, \mu_1 = 0.0, \mu_2 = 0.5$$

The results of numerical calculations for various values of Pn and fixed values of the parameters υ, λ_1, λ_2, μ_1, and μ_2 are depicted in Figs. 1 and 2. In Fig. 1, the nondimensional rate of motion of evaporation surface $\dot{S}$ is plotted against $\log_{10}$ (Fo) for values of Pn = 0, 0.5, 0.8. It is seen that $\dot{S}$ decreases as Pn increases.

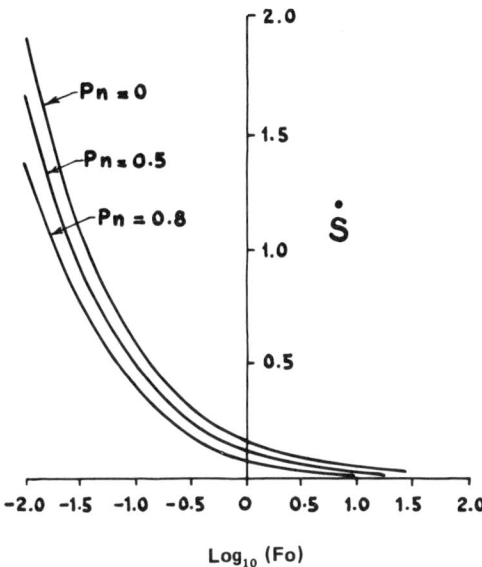

FIG. 1. Effect of Posnov number on the rate of motion of evaporation surface.

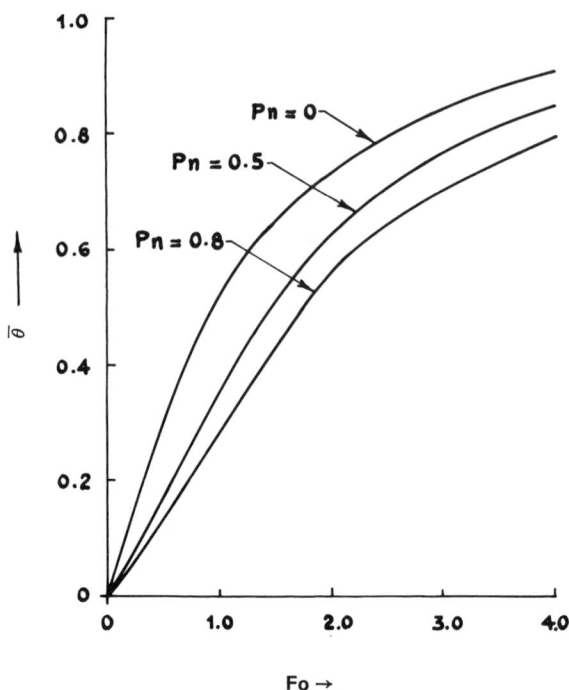

FIG. 2. Effect of Posnov number on the mass transfer potential.

In Fig. 2, the nondimensional mass transfer potential $\bar{\theta}$ has been plotted against $\log(\text{Fo})_{10}$ for values of Pn = 0, 0.5, 0.8. It is noted that with the increase in the criterion Pn, the nondimensional mass transfer potential and rate of motion of the evaporation surface diminishes. In other words, the greater the criterion Pn, the more slowly the material dries.

CONCLUSION

In this paper an approximate solution of the generalized Stefan problem with variable thermal properties and taking thermal diffusion into account has been obtained. As distinct from similar available studies, the mechanism of the deepening of the evaporation surface is described with the aid of the Stefan problem. The solution of the generalized Stefan problem with constant thermal properties follows as a particular case. It is shown that with an increase in the criterion Pn (Posnov number) there is a reduction in the nondimensional mass transfer potential and the rate of motion of the evaporation surface.

NOMENCLATURE

a_q thermal diffusivity
a_m moisture diffusivity

c_m	specific mass capacity
c_q	specific heat capacity
Fo	Fourier number, $a_q \tau / l^2$
K	thermal conductivity
Ko	Kossovitch number, $L c_m \Delta\theta / c_q \Delta t_s$
l	the characteristic length
L	latent heat of vaporization of liquid per unit mass
Lu	Luikov number, a_m / a_q
p	k_{21} $(t_v - t_0)/(t_s - t_v)$
$s(\tau)$	position of evaporation front
$S(\text{Fo})$	nondimensional position of evaporation front
t	temperature
t_s	temperature at surface $x = 0$
T	nondimensional temperature $(t - t_0)/(ts - t_0)$
x	length coordinate
X	nondimensional, x/l

Greek Symbols

α, β, λ	Nondimensional constants defined by Eqs. (3.14), (3.13), and (3.3)
δ	Soret coefficient
$\Delta\theta$	$\theta_0 - \theta_v$
Δt_v	$t_s - t_v$
Δt_s	$t_s - t_0$
ϵ	coefficient of internal evaporation
θ	mass transfer potential
$\bar{\theta}$	nondimensional mass transfer potential $(\theta_0 - \theta)/(\theta_0 - \theta_v)$
μ	$\epsilon \text{Ko}/T_v$
v	nondimensional latent heat of vaporization of liquid $\left[(1 - \epsilon)p_{mq} \cdot L\right]/(c_q \cdot \Delta t_v)$
ρ_m	density of moisture per unit volume
ρ_q	density of porous medium per unit volume
ρ_{mq}	nondimensional density of liquid, ρ_m / ρ_q
τ	time

Subscripts

ϑ	vaporizing state
1	first region, $0 < x < S$
2	second region $S < X < \infty$
21	ratio of properties of region 2 to 1
0	initial state

REFERENCES

1. Luikov, A. V.: Advances in Heat Transfer, **1**, edited by T. F. Irvine and J. P. Hartnett, Academic Press, New York, 1964.
2. Morgan, R. P. and S. Yerazunis: AIChE J., 1967, **13**, 132.
3. Gupta, L. N.: *Int. J. Heat Transfer,* 1974, **17**, 313.
4. Kumar, I. J. and L. N. Gupta: Int. J. Pure Appl. Math, 1971, **2**, 692.
5. Kumar, I. J.: *Int. J. Heat Mass Transfer,* 1971, **13**, 1759.
6. Goodman, T. R.: Advances in Heat Transfer, **1**, edited by T. F. Irvine and J. P. Hartnett, Academic Press, New York, 1964.
7. Biot, M. A.: *J. Aeronaut. Sci.,* 1957, **24**, 857.
8. Kumar, I. J. and H. N. Narang: *Int. J. Heat Mass Transfer,* 1967, **10**, 1095.

EXPERIMENTAL STUDY OF HEAT TRANSFER TO EMBEDDED TUBE AND COIL SURFACES IN A FLUIDIZED BED COMBUSTOR*

V. N. Vedamurthy
College of Engineering, Madras, India

V. M. K. Sastri
Indian Institute of Technology, Madras, India

INTRODUCTION

In many countries, a major portion of the coal mined is used for steam generation, either for production of power or process heating. The mode of combustion in all these units had hitherto been either in a fixed bed of coal spread over stokers or in a pseudogaseous state by injecting pulverized coal particles into an air stream and transferring the released energy to the working fluid by convection and radiation. The inherent difficulties faced while burning large ash content, low grade coal by both these methods have necessitated the need for a breakthrough in the technology of coal combustion. The recent energy crisis has added a new dimension to this situation and calls for a radical departure from the conventional modes of release and transfer of energy from low grade solid fuels.

Fluidized combustion of coal, enabling it to be burned in a pseudoliquid state, is now widely regarded as providing this long awaited breakthrough and extensive research on laboratory and pilot plant scales is in progress in a number of countries.

A great deal of work had been done investigating the heat transfer in fluidized beds operating at low temperatures and an extensive literature spans the effect of various physical parameters such as fluidizing velocity, particle size, bed height, etc. Since the technique of fluidized bed combustion is currently under development, the literature available in this field is relatively meagre. So it is expected that the present investigation will, to some extent, provide an insight into the heat transfer characteristics of embedded surfaces in a fluidized combustor.

---

*The abstract of this paper appears on page 415.

EXPERIMENTAL APPARATUS

The experimental investigation was carried out in a 250 mm diameter, shell type, water cooled combustor shown schematically in Fig. 1. To facilitate the variation of the bed height, it was made of sections of 100 mm height. During the present investigations, three sections were used giving a total test section height of 300 mm, excluding the thickness of the packings.

A steel pipe of 25 mm diameter and 150 mm long was fixed to the top test section with its axis inclined at 45° to that of the combustor. A 30 mm diameter metallic hose was attached to the outer end of the pipe and this combination served as the overflow passage for the bed materials. An 18 mm steel pipe was housed centrally in the bottom test section and a helical coil of 25 mm pitch, and made of 12 mm steel pipe was fixed centrally to the middle test section. The test sections were joined together with a 6 mm thick asbestos packing sheet and the bed height during the investigations was maintained at 265 mm by ensuring a slight overflow of the bed materials through the overflow tube.

A perforated plate type of air distributor with 7 percent of its area open for airflow was provided at the bottom of the test section and air was supplied from a 3 HP blower through a conical diffuser as shown in Fig. 1. The combustor test section was covered by an uptake tube of the same

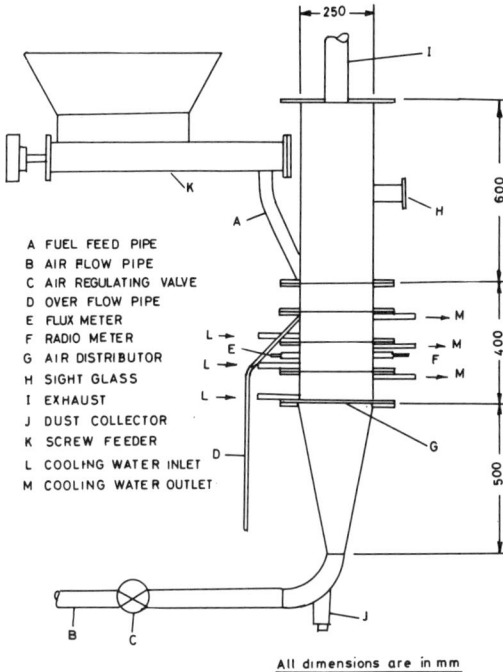

A   FUEL FEED PIPE
B   AIR FLOW PIPE
C   AIR REGULATING VALVE
D   OVER FLOW PIPE
E   FLUX METER
F   RADIO METER
G   AIR DISTRIBUTOR
H   SIGHT GLASS
I   EXHAUST
J   DUST COLLECTOR
K   SCREW FEEDER
L   COOLING WATER INLET
M   COOLING WATER OUTLET

All dimensions are in mm

FIG. 1. Experimental Apparatus

diameter. The pressure of the gases at different bed heights was measured by static pressure probes located at the wall surface and connected to a multilegged manometer. The bed temperature was measured by an unshielded chromel-alumel thermocouple located inside a 3 mm diameter stainless steel tube sheathing. The bed side surface temperatures of the straight tube were measured by a total of 12 thermocouples. Four thermocouples, 90° apart and coinciding with the top, bottom, and sides of the tube were located in each of the quarter and middle sections of the tube. Four thermocouples were used to measure the bedside surface temperature of the coil. Two of them were located on the upstream side and the other two on the downstream side.

Crushed particles of LECO, briquetted lignite marketed by the Neyveli Lignite Corporation, India, and sieved to sizes of −3.15, −4, −5, and −6.3 mm were fed from the top and burned in the combustor at fluidizing velocities ranging from 0.20 to 0.45 m/s at bed temperatures varying from 800-1000°C.

RESULTS AND DISCUSSION

The overall heat transfer coefficients for the tube, coil, and bed wall surface [1] for different fluidizing velocities at a bed temperature of 900°C and with different particle sizes are shown in Figs. 2 to 5. The results of the analysis assuming the emulsion packet to radiate as a black body [2] and those assuming it to absorb and emit radiation with absorption and extinction coefficients (K) of 0.5 and 2.0 [3] are also shown on the graphs. It may be seen that both the analytical and experimental values exhibit a maximum at a fluidizing velocity of about 0.38 m/s for a particle size of −3.15 mm. As the particle size increases, there is a tendency for this point to move slightly to higher velocities and for a particle size of −6.3 mm, it occurs at a velocity of

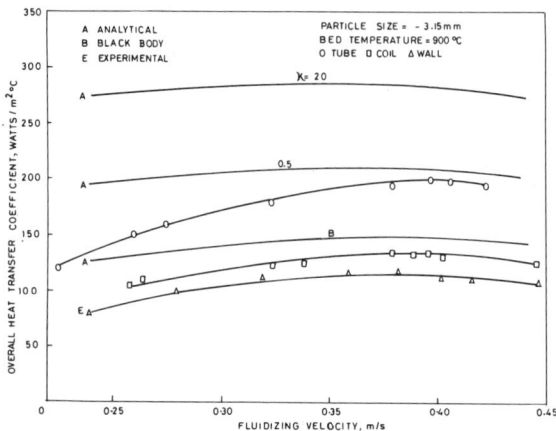

FIG. 2. Variation of overall heat transfer coefficient with fluidizing velocity.

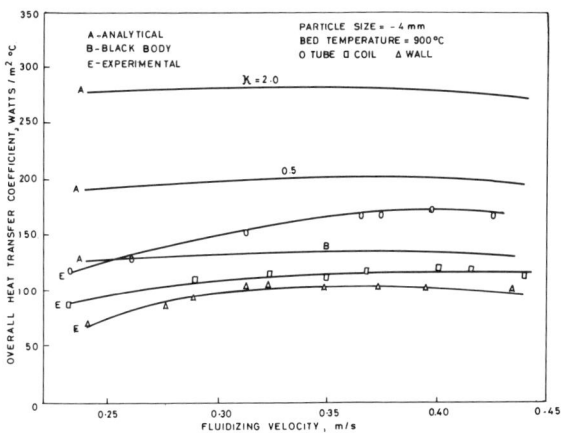

FIG. 3. Variation of overall heat transfer coefficient with fluidizing velocity.

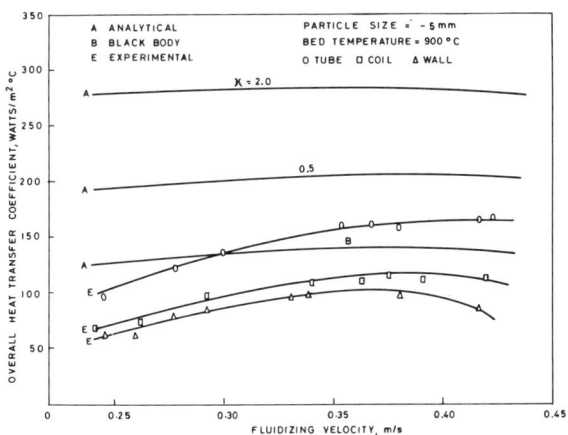

FIG. 4. Variation of overall heat transfer coefficient with fluidizing velocity.

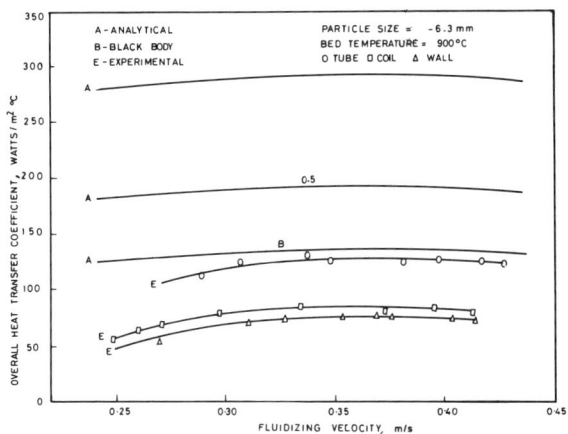

FIG. 5. Variation of overall heat transfer coefficient with fluidizing velocity.

225

0.4 m/s. This might be because of the higher minimum fluidizing velocity for larger particle sizes. The experimental overall heat transfer coefficients are greater for the immersed surfaces as compared to the combustor wall. This may be attributed to the fact that the film effect will be reduced in their lower half due to the impingement of the fluidized solids as well as the upward flowing gas stream. The lower coefficient for the coil as compared to the tube may be due to the complete fluidization of the portion of the bed enclosed by the coil. It may be further seen that the values for the wall and coil are theoretically approximated by the black body assumption whereas the values for the tube are approximated by the assumption of an absorbing and emitting medium with absorption and extinction coefficients of 0.5. This could be due to the black body assumption underestimating the temperature of the emulsion packet as being of the same order as the effects due to increased gas film thickness actually present with those surfaces.

Figure 6 shows the variation of the experimental and analytical values of the overall heat transfer coefficients with particle sizes at a bed temperature of 900°C and for a fluidizing velocity of 0.3 m/s. It can be seen that they decrease with increase in particle size. This may be explained by the fact that for the same fluidizing velocity, the void ratio will be less for larger particles, increasing the area exposed to the emulsion phase and consequently decreasing the direct radiant interchange of heat between the core of the bed and the transfer surfaces. It may also be noted that the decrease is more rapid for the experimental values. This is understandable, since the analysis is based on a uniform particle size whereas, in an actual bed, nonuniform particle size is unavoidable, resulting in a reduced void ratio.

The variation of the experimental and analytical values of the overall heat transfer coefficients with bed temperatures for a particle size of −3.15 mm and fluidizing velocity of 0.3 m/s is shown in Fig. 7. The difference between

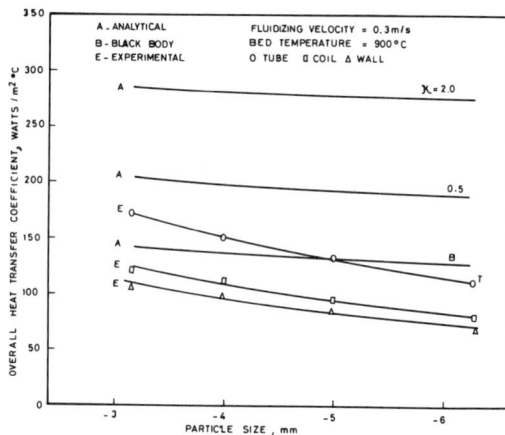

FIG. 6. Variation of overall heat transfer coefficient with particle size.

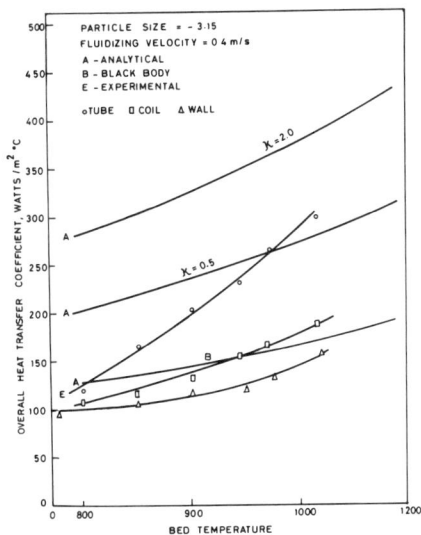

FIG. 7. Variation of overall heat transfer coefficient with bed temperature

the experimental values of the overall coefficients for the wall, coil, and tube increases with increase in temperature. This can be attributed to the effect of reduced gas film thickness on the tube, resulting in an increase in temperature around it, and to the effect of incomplete fluidization in the portion of the bed contained inside the coil, resulting in lower local temperatures on the surface of the coil. It can be seen that the heat transfer to the wall surface can be approximated by the black body radiation. The difference between the analytical values for an absorption coefficient of 0.5 and the experimental values are higher at bed temperatures above $965°C$. This may also be attributed to the film resistance for the tube, causing an increase in conductive heat transfer, and also due to the increased radiative action of the gas film at higher temperatures.

CONCLUSION

From the experimental investigations reported in the foregoing and their comparison with the earlier analytical studies of the authors, the following major conclusions are drawn:

1. The results obtained by the assumption of black body conditions for the emulsion packet underestimate the heat transfer rates. However, these results can be used for estimating the heat transferred to combustor wall surface.
2. The inclusion of the absorbing and emitting characteristics of the emulsion packet in the analysis improves the results considerably and so can be used to predict the heat transferred to embedded surfaces.

REFERENCES

1. Vedamurthy, V. N., and V. M. K. Sastri: "An experimental study of the influence of bed parameters on heat transfer in a fluidized bed combustor," International seminar on future energy production—heat and mass transfer problems, International Center for heat and mass transfer, Dubrovnik, August 1975. Hemisphere Press Inc., Washington, (1976).
2. *Idem:* "An analysis of the conductive and radiative heat transfer to the walls of fluidized bed combustors," *Int. J. Heat Mass Transfer,* 17, 1, pp. 1–9 (1974).
3. *Idem:* "An analytical study of the effect of radiative properties of the emulsion phase on heat transfer in a fluidized bed combustor," Third National Heat and Mass Transfer Conference, Bombay, December 1975.
4. Vedamurthy, V. N.: "Analytical and experimental investigation of heat transfer in a fluidized bed combustor," Ph.D. Thesis, Indian Institute of Technology, Madras, June 1974.

INFLUENCE OF VELOCITY AND PARTICLE MASS ON THE HEAT EXCHANGE IN A "FLUIDIZED" BED AIR COOLER*

S. P. Dichev

Higher Institute of Food Industry, Plovdiv, Bulgaria

INTRODUCTION

The utilization of "fluidized" beds for the intensification of heat exchange in a refrigeration apparatus has not yet found many applications. To date, fluidization as a physical method is utilized in refrigeration techniques only for the freezing of foodstuffs in bulk [1, 2]. Possibilities for a greater utilization of fluidization for heat-transfer intensification in freezing equipment are discussed in [3]. The results obtained during preliminary studies on the application of a "fluidized" bed of solid particles for heat-transfer intensification in air coolers are discussed in [4, 5, 6, 7].

The factors influencing the heat exchange between fluidized bed particles and a given heat exchange surface are numerous and various, namely: the mass velocity of the fluidizing air flow, the geometrical and the thermophysical parameters of the solid particles, the geometrical shape, and the position of the heat exchange surface in the fluidized bed, etc.

PURPOSE OF THE STUDY

The present paper reports on an investigation of the influence of the fluidizing flow velocity and the bulk mass of the layer of solid particles on the heat exchange in a vertical multiple tube air cooler.

EXPERIMENTAL STAND

A model of a vertical multiple tube heat exchanger (air cooler) is shown in Fig. 1. The apparatus is composed of two ring-shaped headers (1) made of copper tubes of 16 mm in diameter and connected with eighteen vertical

---

*The abstract of this paper appears on page 416.

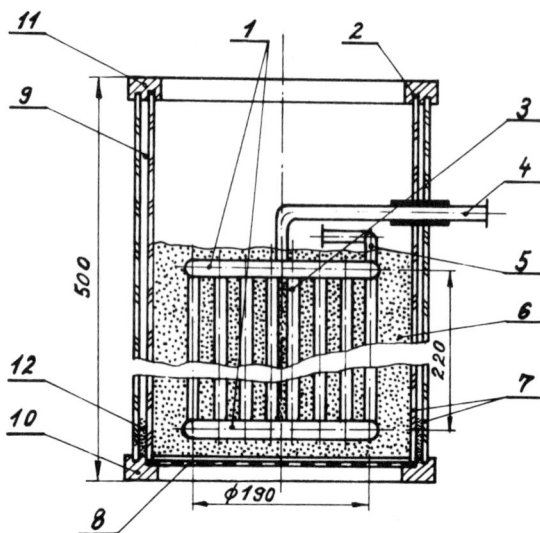

FIG. 1. Model of the vertical multiple tube fluidized bed air cooler.

copper tubes 12 mm in diameter. The fluid, Freon–12, is supplied to the lower header by a feed tube (4) connected at the lower end with the header by means of three 6-mm diameter copper tubes radially disposed through 120°. The flow of vapor is carried out through the upper header by a tube connection (5). The heat-transfer surface of the air cooler is immersed in a bed of solid particles of 3.77 mm in equivalent diameter. The heat exchange surface and the bed are placed in a transparent double wall cylinder of organic glass. A perforated metal plate (8) is located at the lower end of the cylinder and is covered by the solid particles. The transparent cylinder walls give the possibility to observe visually the fluidization process. The air space (9) between the two cylinders furnishes thermal insulation which is so selected that the temperature of the external cylinder is higher than the dew point in order to prevent moisture condensation from the surrounding air. The air space insulation is spaced by flanges (10) and (11). The air in this space is maintained dry by means of a filled layer of silica gel (12). The main geometrical dimensions of the heat transfer surface are as follows: total length of the tubes L = 4.97 m; total external heat exchange surface F_0 = 0.189 m^2; working height H = 220 mm and mean header diameter D_c = 190 mm.

The heat transfer investigation in the abovementioned model of a vertical multiple tube fluidized bed air cooler was carried out in a closed duct (Fig. 2). The air flow caused by a centrifugal fan (35) passes through the experimental section (the air cooler shown in Fig. 1) vertically upward and creates a uniform fluidization of the layer of solid particles. The air cooler is connected with a separate hermetic freon unit (33). A finned air cooler (24) connected with two hermetic refrigerating units (32 and 32′) is built in the

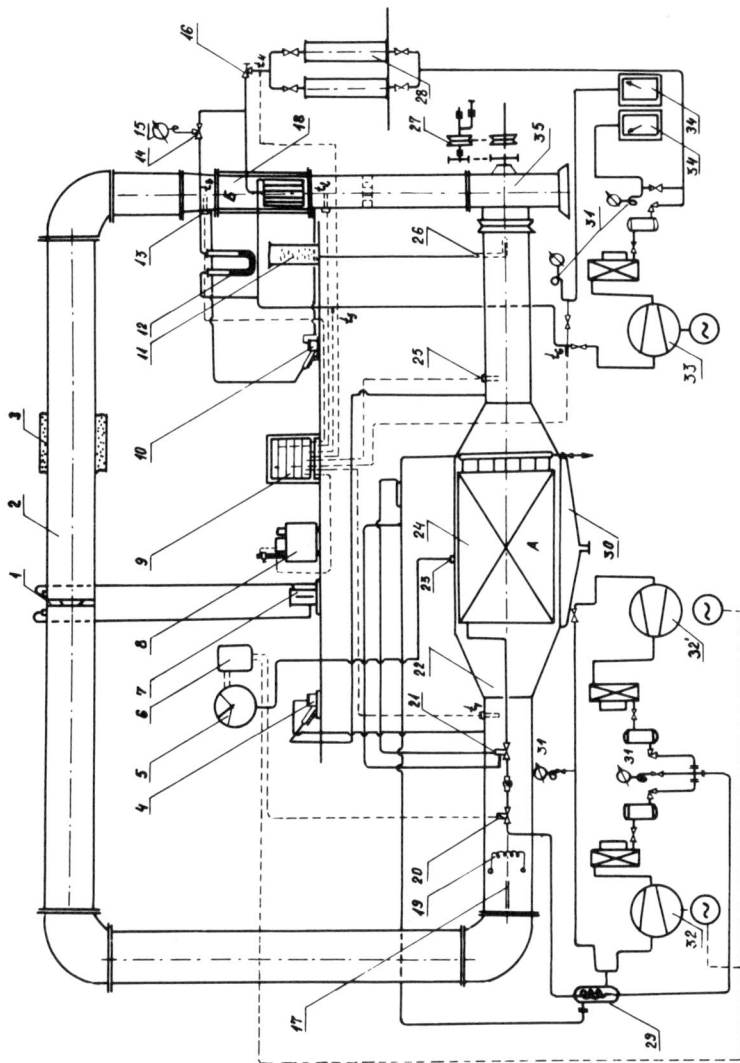

FIG. 2. Experimental apparatus.

231

streamline duct contour in order to maintain fixed temperature conditions of the air flow. The fluidizing air flow velocity can be controlled over a large range by means of a continuous variator (27) and a gate valve (17). The flow rate of the refrigerant entering the evaporator is measured by glass graduated cylinders (28) and the air flow rate is determined with the help of a diaphragm (1) and a micromanometer (7). The temperature of the refrigerant air flow and the characteristic points of the cycle are registered by thermocouples (copper-constantan) and a self-recording multipoint potentiometer type MKVT (9). The evaporation and condensation pressure is measured by means of model manometers type MO 1213 (31) and a self-recording manometer type MTC-711 (34).

METHODS OF INVESTIGATION

We utilize the heat balance equation in order to determine the heat transfer coefficient of the fluidized bed air cooler:

$$Q_0 = K_0 F_0 \Delta t_m \qquad w \qquad (1)$$

$$Q_0 = G_a q_0 \qquad w \qquad (2)$$

or
$$Q_0 = k_0 F_0 \Delta t_m = G_a q_0 \qquad w \qquad (3)$$

We obtain from Eq. (3):

$$k_0 = \frac{Q_0}{F_0 \Delta t_m} \qquad w/m^2k \qquad (4)$$

where: Q_0—cold production of the air cooler, w; G_a—mass flow rate of the refrigerant, kg/s; g_0—specific cold production, kJ/kg; k_0—heat-transfer coefficient related to the external heat transfer surface (F_0), w/m^{2}K; ΔT_m—average logarithmic temperature difference, °C.

The mass flow rate of the refrigerant is determined by graduated cylinders and the specific cold production is measured from the enthalpy difference of the refrigerant at the air cooler inlet and outlet with the help of a log p–i diagram for Freon-12. The average logarithmic temperature difference is determined from the expression:

$$\Delta t_m = \frac{(t_1 - t_{OI}) - (t_2 - t^*)}{\ln\left[(t_1 - t_{OI})/(t_2 - t^*)\right]} \qquad °C \qquad (5)$$

where: t_1, t_2—air flow temperature before and after the air cooler, respectively, °C; t_{01}—evaporation temperature, °C; t^*—average temperature

HEAT EXCHANGE IN A "FLUIDIZED" BED AIR COOLER   233

between t_{01} and the temperature of the vapor at the air cooler outlet t', °C.

$$t^* = \frac{t_{0I} + t'}{2} \quad °C \tag{6}$$

The following relationships have been experimentally established (Figs. 3–6):

$$k_0 = f_1(w\rho) \quad w/m^2k \tag{7}$$

$$k_{max} = f_2\left(\frac{G_p}{F_0}\right) \quad w/m^2k \tag{8}$$

where: G_p–mass of the solid particles, kg; k_{max}–maximum value of the heat-transfer coefficient, $w/m^2 \cdot k$.

The value of the heat-transfer coefficient can be also referred to the total heat-transfer surface including that of the evaporator tubes F_0 and the surface of the layer of solid particles F_p. The specific surface of the layer particles is S_p = 1.73m²/kg. The total surface of the particles is:

$$F_p = G_p \cdot S_p \quad m^2 \tag{9}$$

The experiments were carried out in a range of air flow mass velocity from 2 to 4 kg/m²s at three values of the mass of the particles: I–G_p = 2600 g; II–G_p = 2700 g; III–G_p = 3300 g.

RESULTS

The experimental results obtained for the influence of the air flow mass velocity ($w\rho$) on the heat-transfer coefficient k_0 at various values of G_p are

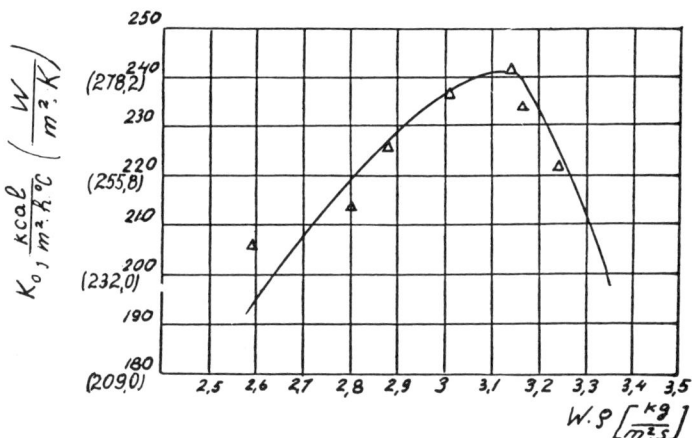

FIG. 3. $k_0 = f_1(w\rho)$ for G_p = 2600 g.

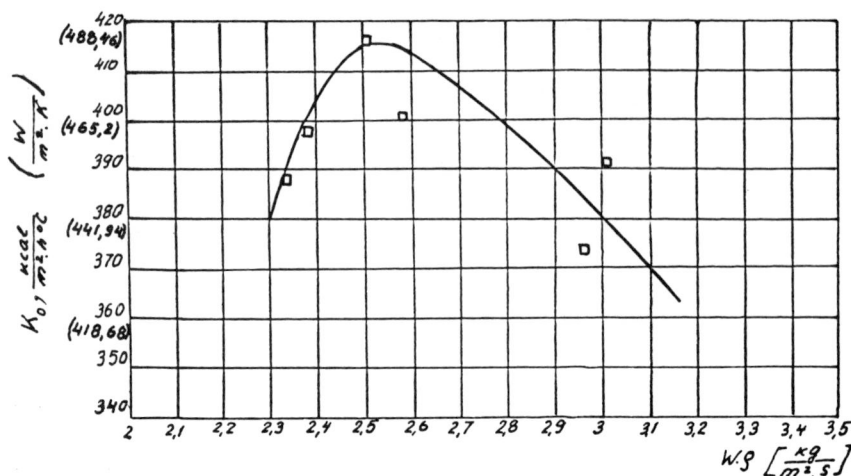

FIG. 4. $k_0 = f_1(w\rho)$ for $G_p = 2700$ g.

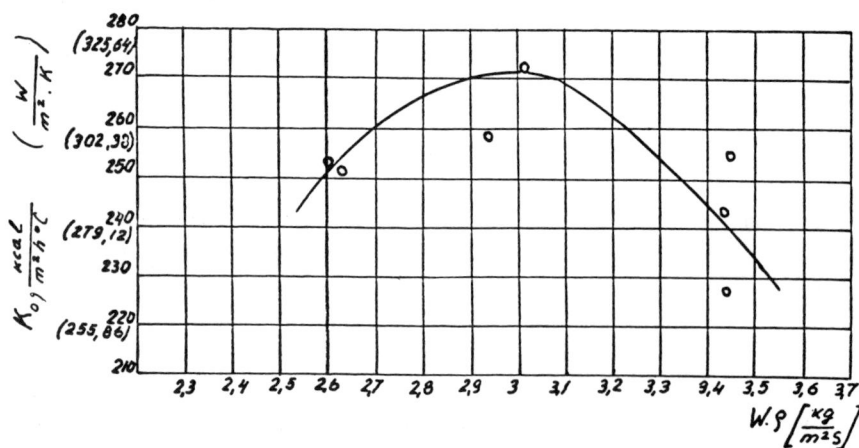

FIG. 5. $k_0 = f_1(w\rho)$ for $G_p = 3300$ g.

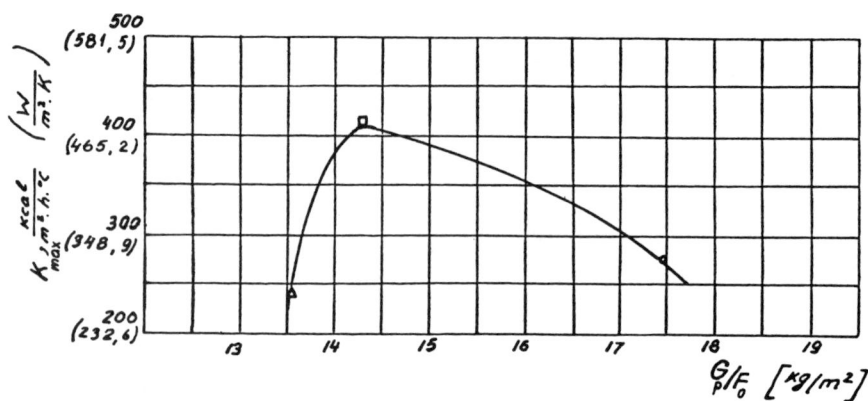

FIG. 6. $k_{max} = f_2(G_p/F_0)$.

shown in Figs. 3, 4, and 5. It is evident from these graphs that the dependence (7) has a pronounced maximum. The dependence (8) has also a marked maximum (Fig. 6).

CONCLUSIONS

The following conclusions can be drawn on the basis of the experimental results:

I. The dependence $k_0 = f_1(w\rho)$ is marked by the presence of a maximum. The increase of k_0 up to k_{max} with increasing the mass velocity of the air flow $(w\rho)$ is connected with the turbulization action of the particles during the flow-around of the tube surface and the decrease of k_0 below k_{max} with the further increasing $(w\rho)$ is associated with the increase of the layer porosity.

II. An optimum velocity of the air flow is one that corresponds to k_{max}.

III. The dependence $k_{max} = f_2(G_p/F_0)$ is marked by the presence of a maximum. This maximum determines the optimum value of the mass of the layer solid particles.

REFERENCES

1. Dichev, S. P.: Hydrodynamics and Heat Transfer During Fluidization Freezing, dissertation, Higher Institute of Food Industry, Plovdiv, 1973. (Bulgaria).
2. Persson, P. O.: Fluidizing technique in food freezing, *ASHRAE J.*, 1967, 6.
3. Dichev, S. P.: Utilization of a "Fluidized" Bed for Heat Transfer in a Freezing Apparatus. Jubilee session of the Higher Institute of Food Industry, Plovdiv, November 8-10, 1973. (Bulgaria).
4. Dichev, S. P.: A Study on the Possibilities for the Application of Fluidization for Heat Transfer Intensification in Air Coolers, *Journal Khranitelna Promishlenost*, 1974, 5. (Bulgaria).
5. Dichev, S. P.: Studies on the Hydrodynamics and Heat Transfer in a Fluidized Bed Air Cooler, Scientific Works of the Higher Institute of Food Industry, **XX**, pt. II, Plovdiv, 1974. (Bulgaria).
6. Dichev, S. P.: Effectiveness of the Fluidized Bed Air Coolers, Transactions of the I National conference on freezing technology, refrigerators, and freezing equipment, November 12-14, 1974, Higher Institute of Food Industry, Plovdiv. (Bulgaria).
7. Dichev, S. P.: Investigation of the Possibilities for the Application of a New Type of "Fluidized" Bed Air Cooler, XIV Congress, IIR, Moscow, 1975. (Eng.).

EXPERIMENTAL FRICTION FACTORS FOR FULLY DEVELOPED FLOW OF DILUTE AQUEOUS POLYETHYLENE-OXIDE SOLUTIONS IN SMOOTH WALL TRIANGULAR DUCTS*

William J. Leonhardt and Thomas F. Irvine, Jr.
State University of New York, Stony Brook, New York

I. INTRODUCTION

Because of wide applications in engineering, especially in the design of compact heat exchangers, much effort has been spent in determining the pressure drop and heat transfer characteristics of noncircular ducts. There exists a special class of noncircular ducts, namely those with a triangular cross section, in which the transition process from laminar to turbulent flow occurs in a special way. Eckert and Irvine [1] have established the simultaneous existence of laminar and turbulent flow in narrow triangular ducts during this transition process, with the narrow apex region containing laminar flow while the flow in the remainder of the duct is turbulent. This existence of an exaggerated laminar sublayer in turbulent flow provides a valuable tool in the study of the transport processes between a duct wall and a moving fluid.

Of late, much interest has been generated in the use of dilute polymer solutions (known as viscoelastic fluids) to reduce frictional drag in both the internal and external flow of liquids. Sometimes known as Tom's Phenomenon, this reduction of fluid drag is the subject of much current research. As of this writing, there exists, among investigators, a lack of agreement on the physical mechanisms involved. Most notably are the effects on normal transition from laminar to turbulent flow, the dependence on polymer concentration of the Reynolds number at which the onset of drag reduction occurs, the concentration of a particular polymer which yields the largest friction reduction and the effect of polymer addition on the velocity profile near the wall. It is felt that this latter phenomenon is amenable to being investigated in triangular ducts due to the magnification of the laminar sublayer by the geometry in the narrow apex region. Virk [12] provides the most current review of work in this field.

---

*The abstract of this paper appears on page 428.

With few exceptions, all the research concerned with pipe flow of drag reducing polymers has been limited to circular tubes. It is the intent of this work to provide a better understanding of the friction reduction encountered in the flow of solutions of polyethylene-oxide (PEO) and water in triangular ducts having apex angles of 10, 15, 20, and 60° and to acquire this basic friction data as a prelude to subsequent investigations of the velocity and temperature fields.

The data are presented in the form of friction factor f as a function of Reynolds number. The friction factor utilized here is the Darcy friction factor defined by the following equation:

$$\frac{dp}{dx} = \frac{\rho \bar{u}^2}{2} \frac{f}{d_h} = \frac{\dot{m}^2}{2\rho A^2} \frac{f}{d_h} \qquad (1)$$

The Reynolds number uses the hydraulic diameter as its characteristic length in the following manner:

$$\text{Re}_{d_h} = \frac{\rho \bar{u} d_h}{\mu} = \frac{\dot{m} d_h}{\mu A} \qquad (2)$$

The hydraulic diameter is defined as

$$d_h = \frac{4A}{P} \qquad (3)$$

which reduces to the actual diameter for a circle. All the friction factors considered are for fully developed flow in smooth walled tubes with constant fluid properties.

The procedure for determining the theoretical value of laminar flow friction factors in triangular ducts is well known and is reported by Shah and London [3]. The relation used is $f \cdot \text{Re}_{d_h} = C_L$, where C_L is a constant dependent on apex angle only. Although a satisfactory theoretical solution for turbulent flow in a triangular duct does not exist, the experimental results of Carlson and Irvine [4] have been used with good agreement by other investigators. Their results yield $f \cdot \text{Re}_{d_h}^{0.25} = C_T$, where C_T, in a manner similar to C_L, is again a constant dependent on apex angle only.

In this work, Newtonian friction factors are measured and compared to the solutions of Shah and London [3] and Carlson and Irvine [4] to provide a check on the experimental apparatus and procedure. In addition, polymer concentrations of 10, 20, 50, and 100 wppm (parts per million by weight) are also studied. The results are compared with the findings of investigators of round tubes who utilized the same polymer additive and concentrations.

The polymer additive chosen was polyethylene-oxide (PEO) manufactured by "Union Carbide Corp." under the trade name of Polyox WSR-301. Polyox was chosen because it has been widely reported in the literature; therefore, many experimental data were available for comparison. The triangular tube data are compared with that of the following investigators of round tubes: Virk, Merrill, Mickley, and Smith [5]; Goren and Norbury [6]; Paterson and Abernathy [7]; Gadd [8]; and van der Meulen [9].

II. EXPERIMENTAL APPARATUS AND PROCEDURE

A schematic of the experimental apparatus is found in Fig. 1. Solutions of 0, 10, 20, 50, and 100 wppm PEO were prepared using an "Ainsworth" type 1C analytical balance to weigh the polymer additive and a "Fairbanks" model 41-1000 FA scale to weigh the solvent (ordinary water). The solutions were prepared following the method described by van der Meulen [9], namely, the polymer was sifted into the solvent and this was allowed to stand for a minimum of 16 hours. In addition, the solutions were stirred for 15 minutes using a length of 8 mm plastic tubing which was rotated by a small electric motor at approximately 100 rpm. The stirring was accomplished just prior to transferring the solutions to the constant head tank. Each batch of solution was prepared in three 120 liter containers. The estimated error in concentration for the most dilute solution (worst case) is 0.36 percent.

After mixing was complete, the solutions were transferred, by gravitational methods to avoid degradation, to the constant head tank and were allowed to stand until they reached room temperature. During experimental runs, regulated high pressure air was used to maintain a constant pressure head in the tank and to assure that the error in flow rate incurred by the change in liquid level was negligible. A "Heise" pressure gauge, graduated in 345 Pa increments, was used to monitor the tank pressure. In this way, verification that the pressure head did not vary during an experimental run was accomplished. The solutions were allowed to flow out of the bottom of the constant head tank and the flow rates were controlled by a gate valve and/or the air pressure on top of the solutions. In this manner, the Reynolds numbers through the test sections could be varied.

A flow straightener was made from a piece of steel pipe 20 cm long and 5.08 cm in diameter. The pipe was packed with plastic straws of the same length and located just upstream of the test section. The flow straightener was used to dampen out swirls and eddies from the preceding gate valve and pipe bend.

Next, the solutions flowed through a test section. In all, four different test sections were used, each having a different apex angle. The angles used were 10, 15, 20, and 60 degrees. A typical cross section is shown in Fig. 2 where the principal dimensions of each test section are tabulated. A more complete description of the design and construction of the test sections is given by Leonhardt [10]. Each test section was fitted with flanges at its ends, allowing it to be mated with the 5.08 cm circular piping used in the rest of the system.

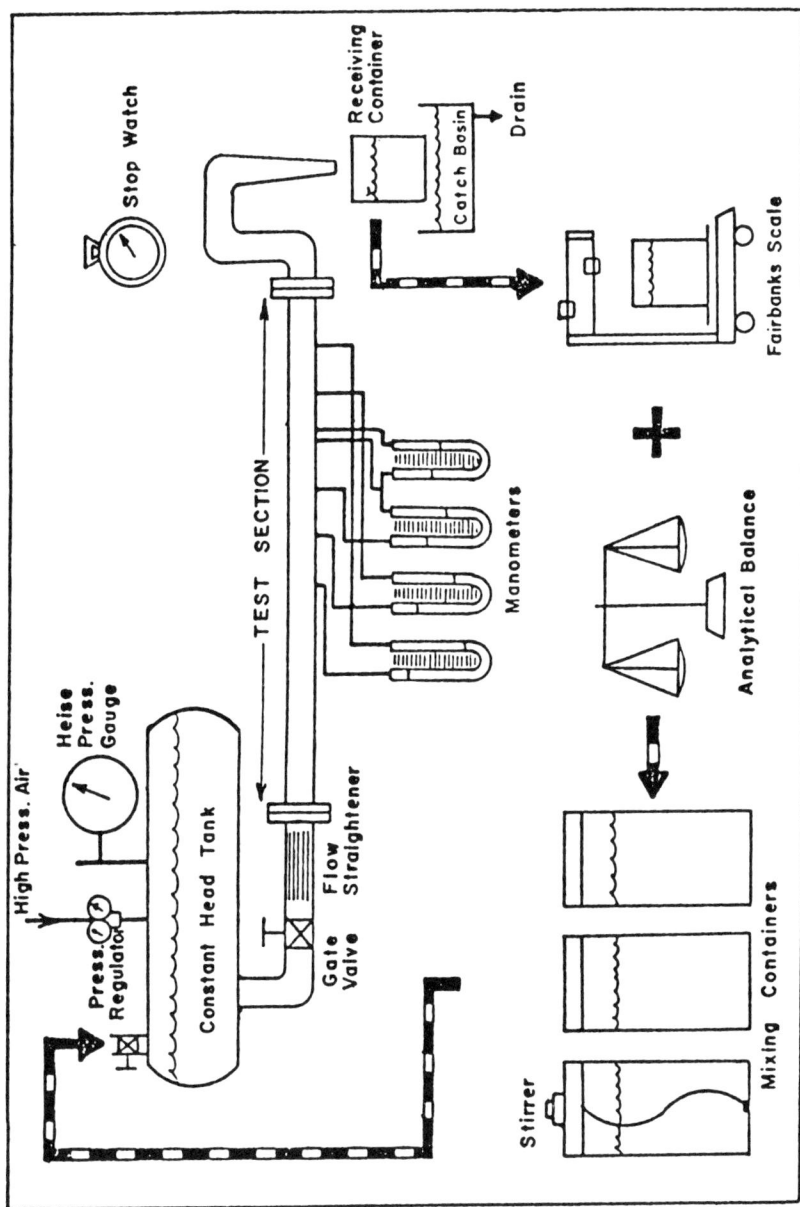

FIG. 1. Schematic of experimental apparatus.

240

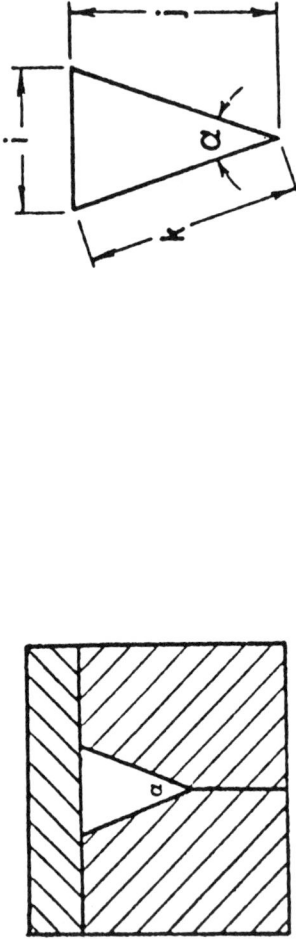

α Nominal	α Actual	i mm	j mm	k mm	d_h mm	% Error in f	% Error in Re_{d_h}	Length total	Length to 1st tap
10°	10.2°	4.70	26.16	26.26	4.29	6.8%	1.6%	$852d_h$	$365d_h$
15°	14.4°	4.60	18.21	18.34	4.06	4.4%	1.3%	$900d_h$	$385d_h$
20°	19.7°	6.12	17.65	17.91	5.18	2.9%	1.6%	$706d_h$	$302d_h$
60°	60.0°	10.95	9.47	10.95	6.32	2.8%	1.2%	$578d_h$	$247d_h$

FIG. 2. Principle dimensions of test sections and errors in the measurement of f and Re_{d_h}.

The entrances to the test sections were made abrupt so that fully developed flow would occur in the shortest possible downstream distance. The four test sections had seven pressure taps, which were connected to four "Refienry Supply Co." 100 cm water manometers. During test runs, the manometers were used to determine the pressure drop per unit length of duct. The distances between the pressure taps are 195.6, 117.4, 39.1, and 9.8 cm for all ducts. By reading two or more manometers during a test run, and then comparing the calculated friction factor from each, entrance effects, if present, could be detected. The manometer whose pressure taps were the farthest apart was deemed most accurate; however the manometers can only read a maximum pressure difference of 90 cm of water and, once this difference was exceeded, the manometer was shut off and the next most accurate one utilized.

After leaving the test section, the solutions flowed through three $90°$ elbows arranged to form an inverted U to insure the test section be completely full even at the lowest Reynolds numbers. After this vertical bend, the solutions ran into a large catch basin and, finally, into a floor drain.

During the test runs, the system was first allowed to reach steady state (constant flow rate). When steady state was attained (indicated by no fluctuations in both the "Heise" gauge and the manometers), a receiving container was inserted into the flow just before the catch basin, and, simultaneously a stop watch was started. After a sufficient amount of fluid was collected, the receiving container was removed and the stop watch was, again simultaneously, stopped. The receiving container was weighed before and after insertion into the flow using the same "Fairbanks" model 41-1000 FA scale as in the solution preparation. A thermometer was placed in the receiving container and the solution temperature was placed in the receiving container and the solution temperature noted so the values of density and dynamic viscosity used in the calculations could be specified.

The measured quantities were elapsed time, mass change of the receiving container, solution temperature, and heights of liquid in the manometers. This raw data was fed into an IBM 370 computer which performed the calculations that yielded the values of Reynolds number and friction factor, their errors, and the least squares fit of the data. The error in friction factor and Reynolds number is included in the data of Fig. 2.

III. PRESENTATION OF RESULTS

The Newtonian data are presented in standard form (f vs. Re_{d_h}) in Figs. 3, 4, 5, and 6. The laminar theoretical solution for an isosceles triangular duct is given by Shah and London [3] and a least squares fit of the laminar data was employed to theoretically calculate the value of the hydraulic diameter for

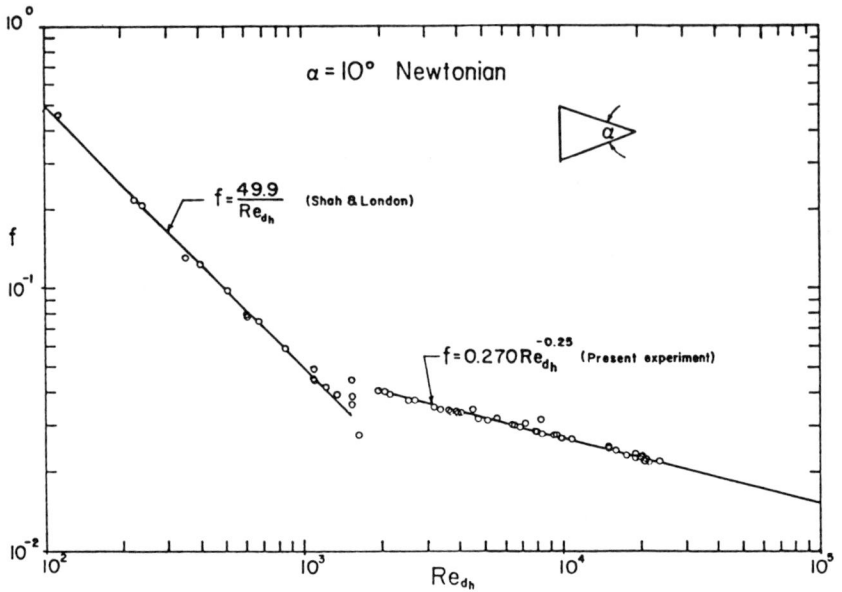

FIG. 3. Presentation of Newtonian data, $\alpha = 10°$. Note—d_h determined through best fit of laminar data to theoretical solution of Shah and London [3].

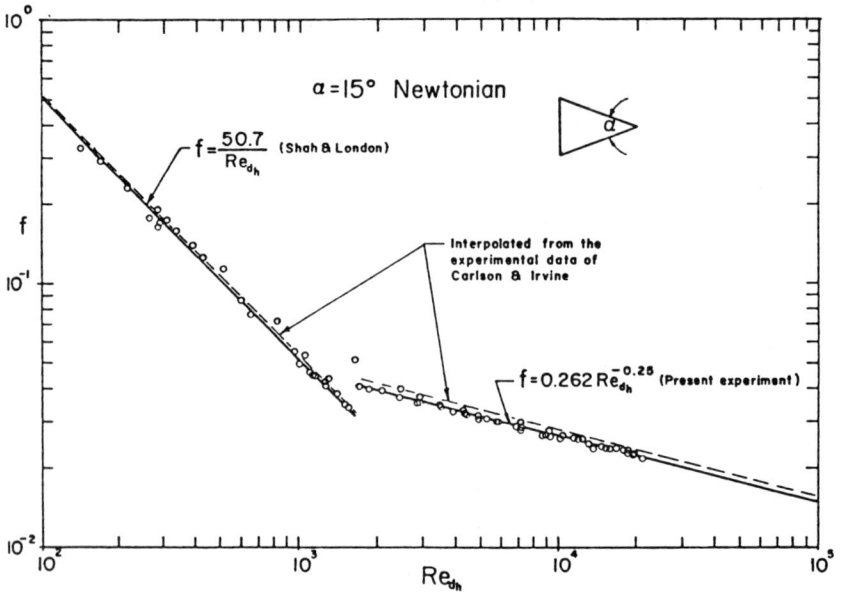

FIG. 4. Presentation of Newtonian data, $\alpha = 15°$, and comparison with experimental results of Carlson and Irvine [4]. Note—d_h determined through best fit of laminar data to theoretical solution of Shah and London [3].

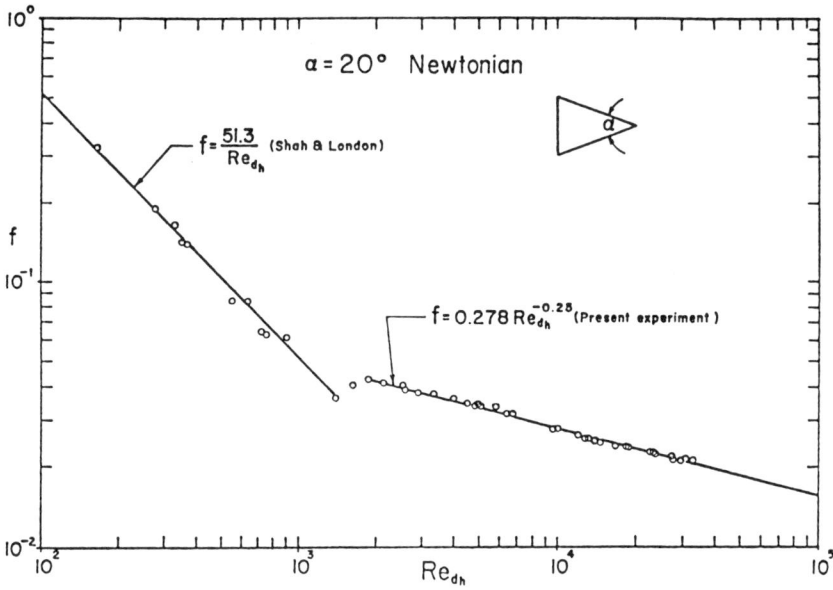

FIG. 5. Presentation of Newtonian data, $\alpha = 20°$. Note—d_h determined through best fit of laminar data to theoretical solution of Shah and London [3].

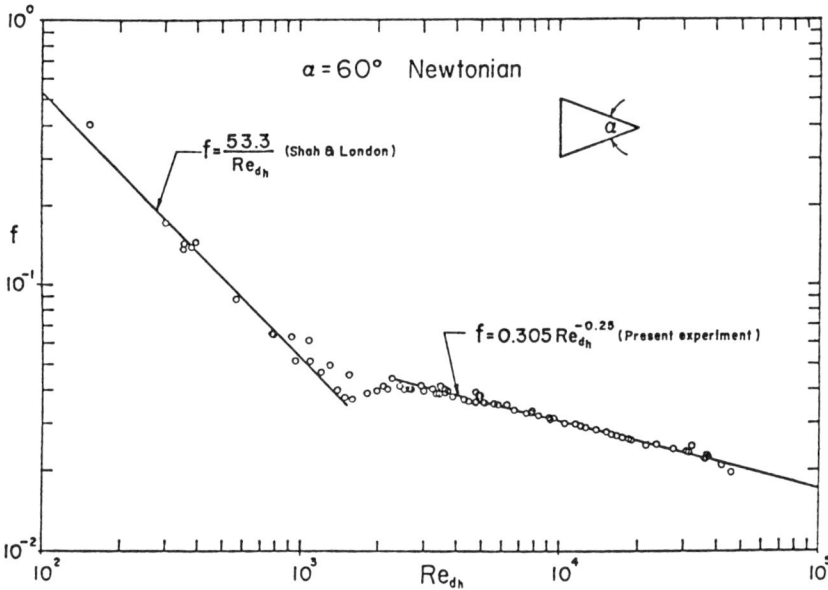

FIG. 6. Presentation of Newtonian data, $\alpha = 60°$. Note—d_h determined through best fit of laminar data to theoretical solution of Shah and London [3].

each duct. These values are found to deviate less than 1.5 percent (in the worst case) from the physically measured values. Upon comparing the two methods of determining the hydraulic diameters, theoretical and physical measurement, it was decided that the theoretically determined hydraulic diameters were more accurate and, therefore, these values will henceforth be used. The primary shortcoming of the physical measurements is that the accuracy of the measuring instrument and techniques is poor considering the size of the hydraulic diameters. A least squares fit is also utilized in drawing straight lines through the turbulent data, assuming a $-\frac{1}{4}$ slope on log Re_{d_h}. It should be noted that the Reynolds number is based on the solvent (water) viscosity, which was determined using a capillary tube viscometer.

All of the Newtonian data are compared with the data of Carlson and Irvine [4], whose ducts had hydraulic diameters several times larger than the present work, and who used air as the working fluid rather than water. The interpolated data of Carlson and Irvine [4] for a 15° duct are shown plotted in Fig. 4 along with the present work. A better comparison of the turbulent data is shown in Fig. 7 by plotting $f \cdot Re_{d_h}{}^{0.25}$ vs. apex angle α. Error bars drawn in show good agreement between both sets of data. The laminar data of Carlson and Irvine [4] are not presented but are between 0.8 percent and 1.9 percent higher than the present work.

Figures 8, 9, 10, and 11 present the data for the polymer solutions in standard form. The Newtonian lines, both laminar and turbulent, are drawn

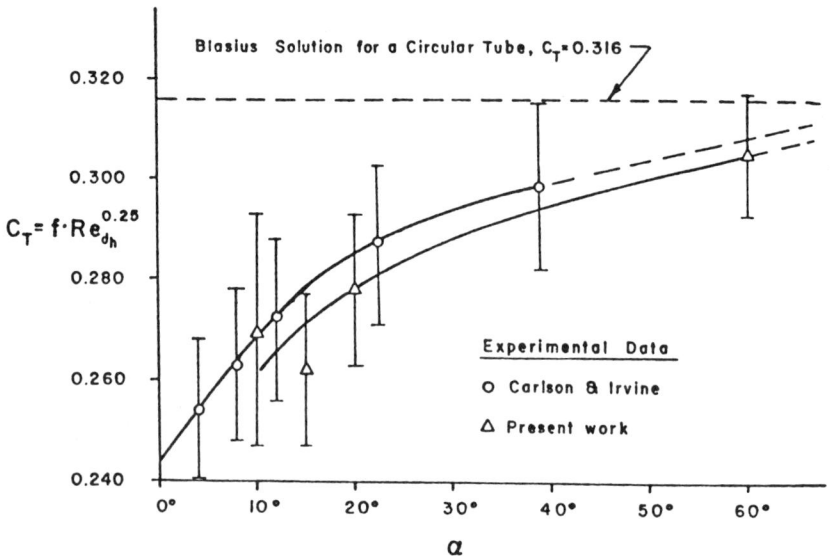

FIG. 7. Comparison of Carlson and Irvine [4] experimentally determined values of C_T with present work.

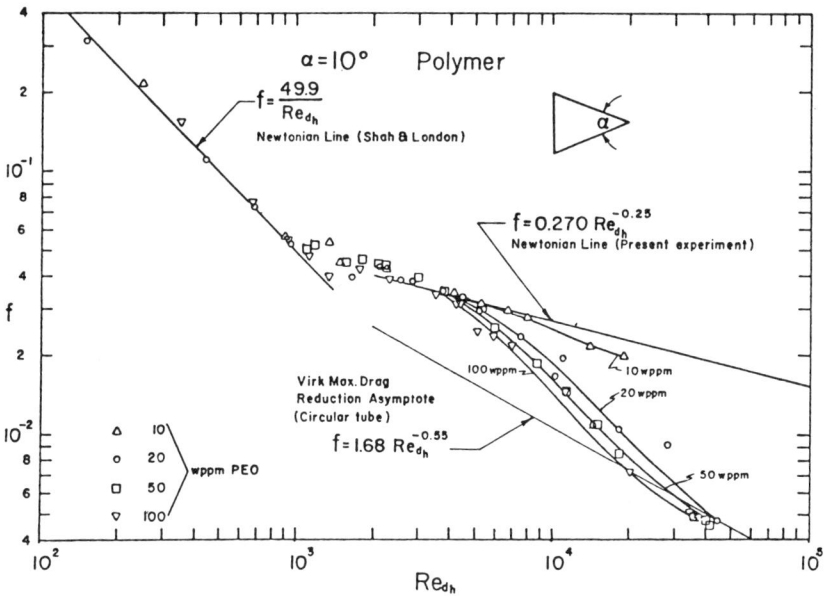

FIG. 8. Presentation of drag reduction data for concentrations of 10, 20, 50, and 100 wppm PEO, $\alpha = 10°$. Comparisons with: Shah and London [3] and Virk, et al. [5].

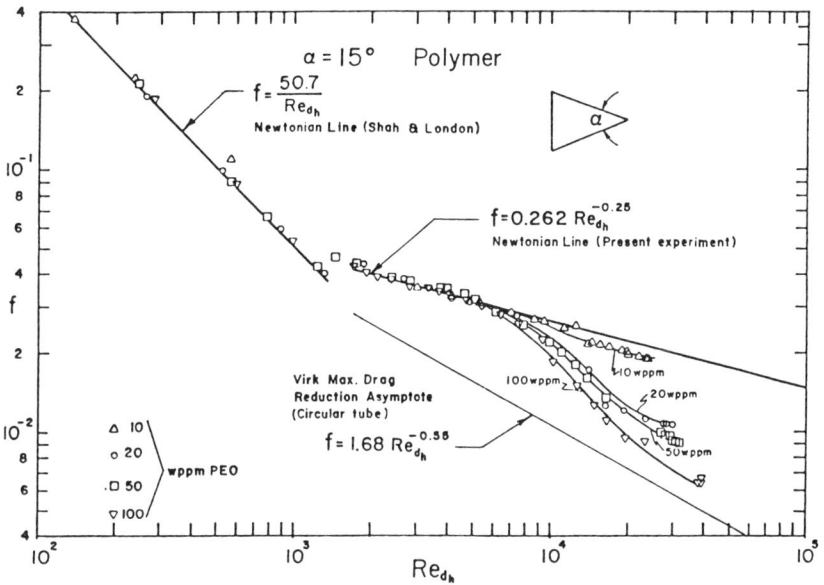

FIG. 9. Presentation of drag reduction data for concentrations of 10, 20, 50, and 100 wppm PEO, $\alpha = 15°$. Comparisons with: Shah and London [3] and Virk, et al. [5].

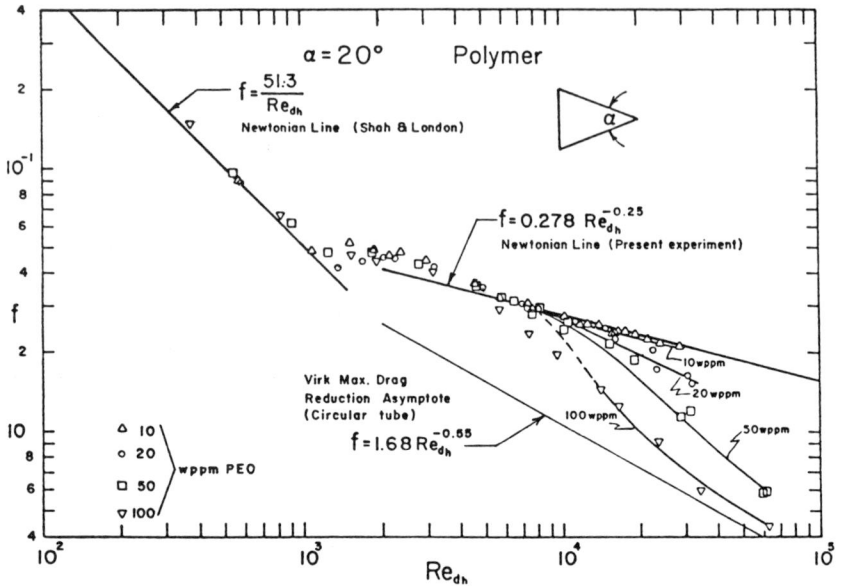

FIG. 10. Presentation of drag reduction data for concentrations of 10, 20, 50, and 100 wppm PEO, $\alpha = 20°$. Comparisons with: Shah and London [3] and Virk, et al. [5].

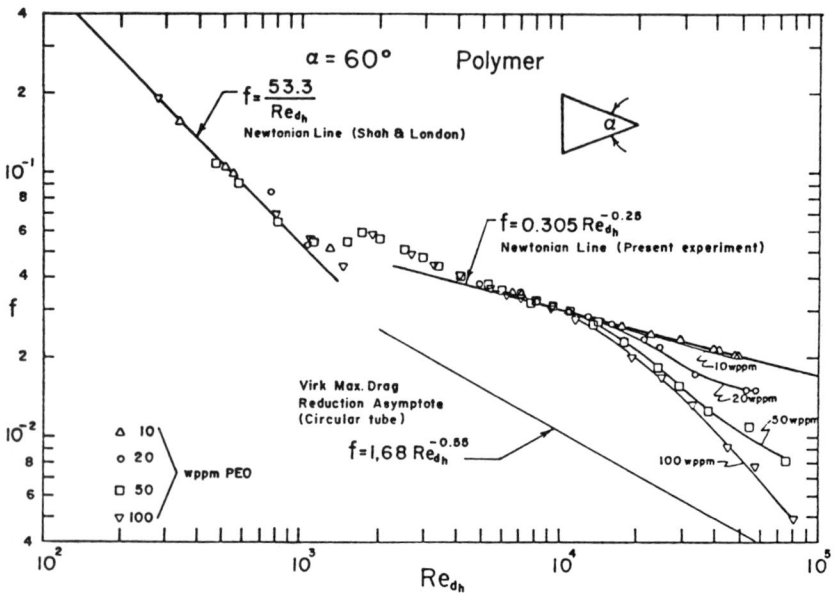

FIG. 11. Presentation of drag reduction data for concentrations of 10, 20, 50, and 100 wppm PEO, $\alpha = 60°$. Comparisons with: Shah and London [3] and Virk, et al. [5].

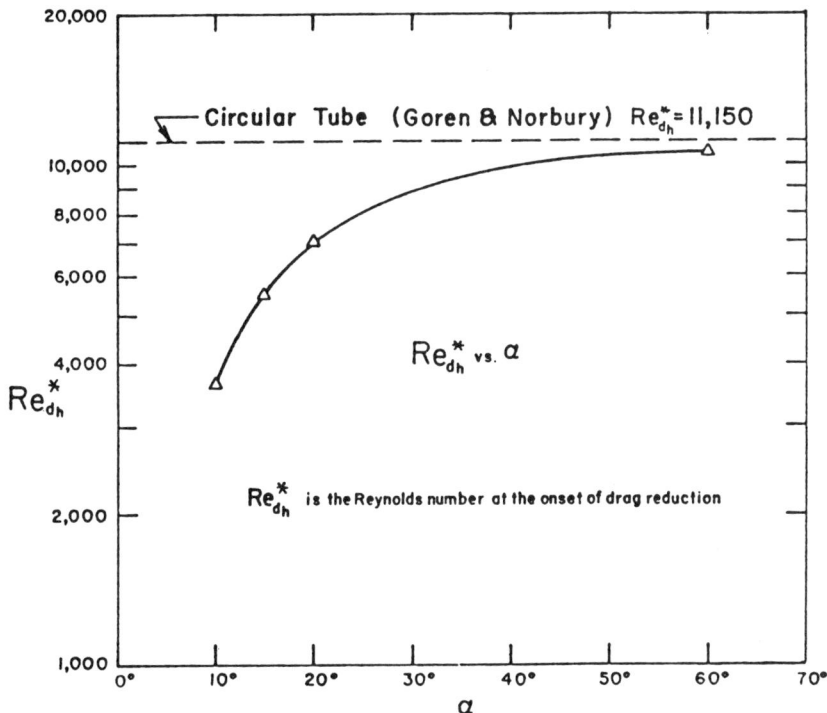

FIG. 12. Comparison of the onset Reynolds number as a function of apex angle with circular tube data of Goren and Norbury [6].

and so labeled. Visual best fit lines are drawn through points of the same concentration in an effort to indicate the trend of the drag reduction and to extrapolate to a point of "onset" of drag reduction. Figure 10 contains a dashed line which indicates how the 100 wppm line would continue if it followed the trend of the other ducts. The "Virk maximum drag reduction asymptote" for circular ducts ($f = 1.68 \, Re^{-0.55}$ as given in [5]) is drawn in on each figure to indicate the maximum drag reduction that could be expected if the ducts were of circular cross section.

In Fig. 12, $Re_{d_h}^*$ is plotted against apex angle α. $Re_{d_h}^*$ is defined as the Reynolds number at which the onset of drag reduction occurs. The values for each duct are, again, obtained by visual means. Data for a circular tube from Goren and Norbury [6] are also included for comparison.

IV. DISCUSSION OF RESULTS

The turbulent Newtonian data of Figs. 3, 4, 5, and 6 agree well with the previously established results of Carlson and Irvine [4]. The greatest deviation encountered was with the $15°$ duct, but, as shown in Fig. 7, this deviation is well within the combined experimental accuracies of both sets of data. The Carlson and Irvine data are also compared with the present measurements in Fig. 4 and show good agreement. The transition from laminar to turbulent flow occurs at approximately the same Reynolds numbers ($1500 \leqslant \mathrm{Re}_{d_h} \leqslant 2000$) for both the present work and the data of Carlson and Irvine [4].

Comparison of the Newtonian data (Figs. 3, 4, 5, and 6) with the polymer data (Figs. 8, 9, 10, and 11) shows that, in all ducts and at all concentrations, the polymer additives had no detectable effect on normal transition from laminar to turbulent flow. This is consistent with the findings of Virk, et al. [5], Goren and Norbury [6], Paterson and Abernathy [7], and Gadd [8].

The findings of other investigators seem to differ considerably when the dependence of the onset Reynolds number ($\mathrm{Re}_{d_h}^*$) on polymer concentration is examined. In the present work, $\mathrm{Re}_{d_h}^*$ exhibited essentially no dependence on polymer concentration for any of the four ducts tested. This agrees well with the findings of Virk, et al. [5] and Goren and Norbury [6], but Paterson and Abernathy [7] and Gadd [8] show $\mathrm{Re}_{d_h}^*$ to be a function of concentration, where $\mathrm{Re}_{d_h}^*$ decreases as concentration increases.

The present work does show that $\mathrm{Re}_{d_h}^*$ decreases with decreasing apex angle. This is graphically shown in Fig. 12. The data of Goren and Norbury [6] are included to show a relationship between the circular and triangular tube results. It must be pointed out that the circular $\mathrm{Re}_{d_h}^*$ is for a two in. diameter pipe and lower $\mathrm{Re}_{d_h}^*$ are reported when using capillary tubes. A comparison of figs. 7 and 12 shows a similar functional dependence on apex angle of C_T and $\mathrm{Re}_{d_h}^*$.

The "Virk maximum drag reduction asymptote" was exceeded in all concentrations except 10 wppm in the $10°$ duct. This does not seem unreasonable since the Virk line is only for circular ducts and previous measurements have shown that Newtonian friction factors for triangular ducts are less than friction factors for circular ducts with the same hydraulic diameter and Reynolds number.

In essentially all cases, the largest drag reduction was produced by the highest concentration (100 wppm). Careful examination of Fig. 8 ($10°$ duct) shows, however, that the largest friction reduction obtained (75.9%) is at Re_{d_h} = 41,200 with a concentration of 50 wppm. Note that at this point, the friction reductions for 100 wppm and for 50 wppm are seen to be very close and, considering the experimental errors, the 100 wppm might acutally yield a larger friction reduction than the 50 wppm. This best performance of 100 wppm concentration is supported by Gadd [8] but does not agree with the findings of Goren and Norbury [6], Paterson and Abernathy [7], and van der Meulen [9], who report the largest reduction in friction at concentrations of 10 or 20 wppm. Gadd [8] also reported that the slopes of the drag reduction lines become sharper with increasing concentration. This is also demonstrated by all the polymer data of the present work.

V. SUMMARY

Friction factor vs. Reynolds number data are presented for both a Newtonian fluid (water) and a viscoelastic polymer solution (polyethyleneoxide). The Newtonian results agree well with existing experimental data. The polymer data show that the Reynolds number at which the onset of drag reduction occurs is independent of polymer concentration and decreases with decreasing apex angle. The normal transition from laminar to turbulent flow is unaffected by the polymer additive in the concentration range studied and the amount of friction reduction increases with increasing concentration.

ACKNOWLEDGEMENT

The authors wish to express their appreciation to the National Science Foundation for supporting this work under a program of "Cooperative Research between the USSR and the USA," and to Professor J. P. Hartnett of the University of Illinois, Chicago Circle, USA, who performed the viscosity measurements.

LIST OF SYMBOLS

A	Cross sectional area
C_L	Laminar coefficient, $C_L = f \cdot \mathrm{Re}_{d_h}$
C_T	Turbulent coefficient, $C_T = f \cdot \mathrm{Re}_{d_h}^{0.25}$
d_h	Hydraulic diameter as defined in Eq. (3)
f	Darcy friction factor as defined in Eq. (1)
$\left.\begin{array}{l} i \\ j \\ k \end{array}\right\}$	Duct interior dimensions (Fig. 2)
$\dot{m}$	Mass flux
p	Pressure
P	Wetted perimeter
PEO	Polyethylene–oxide
Re_{d_h}	Reynolds number as defined in Eq. (2)
$\mathrm{Re}^*_{d_h}$	Reynolds number at onset of drag reduction
$\bar{u}$	Average velocity in x direction
wppm	Weight parts per million
x	Distance parallel to tube axis
μ	Dynamic viscosity
ρ	Density

REFERENCES

1. Eckert, E. R. G., and T. F. Irvine, Jr.: "Flow in Corners of Passages with Noncircular Cross-Sections," Trans. ASME, 78, 1956, pp. 709–718.
2. Virk, P. S.: "Drag Reduction Fundamentals," AIChE J., 21, No. 4, July 1975, pp. 625–656.
3. Shah, R. K., and A. L. London: "Laminar Flow Forced Convection Heat Transfer and Flow Friction in Straight and Curved Ducts–A Summary of Analytical Solutions," Tech Report 75, Nonr 225 (91), (NR-090-342), Office of Naval Research, Nov. 1971, pp. 134–139.

250   W. J. LEONHARDT AND T. F. IRVINE

4. Carlson, L. W. and T. F. Irvine, Jr.: "Fully Developed Pressure Drop in Triangular Shaped Ducts," Trans. ASME, J. Heat Transfer, 83, 1960, pp. 441-444.
5. Virk, P. S., E. W. Merrill, H. S. Mickley and K. A. Smith: "The Toms Phenomenon: Turbulent Pipe Flow of Dilute Polymer Solutions," J. Fluid Mech., 30, part 2, 1967, pp. 305-328.
6. Goren, Y. and J. F. Norbury: "Turbulent Flow of Dilute Aqueous Polymer Solutions," Trans. ASME, J. Basic Engineering, 89, series D, No. 4, Dec. 1967, pp. 814-822.
7. Paterson, R. W. and F. H. Abernathy: "Turbulent Flow Drag Reduction and Degradation with Dilute Polymer Solutions," J. Fluid Mech., 43, part 4, 1970, pp. 689-710.
8. Gadd, G. E., "Reduction of Turbulent Drag in Liquids," Nature Phy. Sci., 230, Mar. 1971, pp. 29-31.
9. Van der Meulen, J. H. J.: "Friction Reduction and Degradation in Turbulent Flow of Dilute Polymer Solutions," Appl. Sci. Rev., 29, Jun. 1974, pp. 161-174.
10. Leonhardt, W. J.: "Experimental Friction Factors for Fully Developed Flow of Dilute Aqueous Polyethylene-Oxide Solutions in Smooth Wall Triangular Ducts," M.S. Thesis, Department of Mechanical Engineering, State University of New York, Stony Brook, New York, Dec. 1975.

INFLUENCE OF DIAMETER ON HEAT TRANSFER AND FRICTION FACTOR FOR NON-NEWTONIAN FLUIDS IN TURBULENT PIPE FLOW*

J. P. Hartnett, K. S. Ng, and J. E. Rasson
University of Illinois at Chicago Circle, Chicago

INTRODUCTION

Non-Newtonian fluids are those for which the shear stress is not directly proportional to the shear rate. Dodge and Metzner [1] reported one of the earliest extensive studies of turbulent pipe flow of non-Newtonian fluids. They found that friction factor of one of the fluids, an aqueous solution of carboxymethylcellulose (CMC), behaved anomalously. They speculated that this abnormal behavior might be due to an elastic effect, so they excluded the CMC from the data analysis. Subsequently Metzner [2] classified the subclass of non-Newtonian fluids having negligible elastic effects as purely viscous fluids. In this paper, only these purely viscous fluids will be considered.

Previous experimenters [3, 4, 5, 6] have reported that the shear stress and shear rate relationship of a majority of these purely viscous non-Newtonian fluids may be represented over a wide range of shear rate by a two-constant power function of the form

$$\tau = K\left(\frac{du}{dy}\right)^n \tag{1}$$

Skelland [7] has shown that the shear stress at the wall of a steady laminar flow of these fluids in a capillary tube viscometer can be written as

$$\tau_w = K'\left(\frac{8V}{D}\right)^{n'} \tag{2}$$

The quantities in Eqs. (1) and (2) are related as follows [7]:

---

*The abstract of this paper appears on page 433.

$$n' = n \quad K' = K\left(\frac{3n + 1}{4n}\right)^n \tag{3}$$

For a power law fluid, the two viscometric parameters n' and K' are constant. The Fanning friction factor is defined as

$$f = \frac{\tau_w}{\frac{1}{2}\rho V^2} \tag{4}$$

Substituting for τ_w from Eq. (2) and rearranging gives

$$f = \frac{16}{\rho V^{2 - n'} D^{n'} / K' 8^{n' - 1}} \tag{5}$$

Metzner and Reed [5] defined the denominator in Eq. (5) as the generalized Reynolds number

$$\text{Re}' = \frac{\rho V^{2 - n'} D^{n'}}{K' 8^{n' - 1}} \tag{6}$$

Thus Eq. (5) becomes

$$f = \frac{16}{\text{Re}'} \tag{7}$$

This may be compared with the well-known result for the laminar flow of a Newtonian fluid

$$f = \frac{16}{\text{Re}} \tag{8}$$

Comparison of Eqs. (7) and (8) reveals that all non-Newtonian fluids which are time independent obey Eq. (7), provided that the generalized Reynolds number in Eq. (6) is used. When the fluid is Newtonian, n' is equal to one and the generalized Reynolds number reduces to the conventional Reynolds number with K' equal to the Newtonian viscosity.

Using the generalized Reynolds number and the definition of Peclet number, Pe, a generalized Prandtl number, Pr', can be defined as

$$\text{Pr}' = \frac{C_p K'}{k_f}\left(\frac{8V}{D}\right)^{n' - 1} \tag{9}$$

Dodge and Metzner [1] based their theoretical analysis on a power law constitutive equation, and concluded that the relationship between the friction factor and the generalized Reynolds number in turbulent pipe flow has the form

$$\sqrt{\frac{1}{f}} = A_n \log (\text{Re}'(f)^{1 - (n'/2)}) + C_n \tag{10}$$

where A_n and C_n are constants to be determined empirically.

They used extensive data collected from purely viscous power law fluids and found that

$$A_n = \frac{4.0}{(n')^{0.75}} \qquad C_n = \frac{-0.40}{(n')^{1.2}} \tag{11}$$

Thus Eq. (10) becomes

$$\frac{1}{\sqrt{f}} = \frac{4.0}{(n')^{0.75}} \log \text{Re}' f^{1 - (n'/2)} - \frac{0.40}{(n')^{1.2}} \tag{12}$$

Equation (12) reduces to the well-known expression given by Nikuradse for the Newtonian case. Dodge and Metzner [1] also concluded that Eq. (12) is applicable to nonpower law fluids provided that the parameters (K' and n') are evaluated at the existing wall shear stress under the proposed turbulent condition.

Metzner and Friend [8] extended the analogy between heat and momentum transfer for Newtonian fluids to include purely viscous non-Newtonian fluids in turbulent pipe flow and obtained the relationship

$$\text{St} = \frac{f/2}{1.20 + C_1 \sqrt{f/2}} \tag{13}$$

where C_1 is to be determined empirically. Studies of the heat transfer performances of aqueous Carbopol, corn syrup, and Attagel clay [4, 9, 10] led Metzner and Friend [8] to a determination of C_1 and established the final form of Eq. (13) as

$$\text{St} = \frac{f/2}{1.20 + 11.8(f/2)^{0.5}(\text{Pr}_w - 1)(\text{Pr}_w)^{-1/3}} \tag{14}$$

where Pr_w is the Prandtl number using the apparent viscosity evaluated at the wall shear stress ($\tau_w/(du/dy)$). Metzner and Friend [8] also showed that Eq. (14) is only applicable when

$$\frac{Pr'Re'}{(n')^{0.25}}\sqrt{\frac{f}{2}} > 5000 \tag{15}$$

Clapp [11] extended the Karman–Martinelli [12, 13] analogy to purely viscous non-Newtonian fluids and suggested an empirical correlation which covered only a very limited range of the flow behavior index. Therefore, this analysis shall not be reported here.

Yoo and Hartnett [14] studied Carbopol and Attagel solutions in turbulent pipe flow and presented the algebraic expression

$$Nu = \frac{0.0118}{n'}(Re')^{0.9}(Pr')^{0.3} \tag{16}$$

which correlated all available data above a generalized Reynolds number of 10,000 to within ± 20 percent, with few exceptions.

Although abundant experimental data have been reported, it is difficult to determine whether the differences between reported results are due to differences in tube diameter or in experimental technique. To check the possibility of diameter effects, the same fluid must be used in tubes of differing diameter. The present experimental setup fulfills this requirement.

EXPERIMENT AND CALIBRATIONS

A schematic diagram of the flow loop is shown in Fig. 1. A detailed description of the apparatus may be found in Yoo [15].

A positive displacement pump discharges the test fluid into one of the two horizontal tubes of diameters .51 in. (1.30 cm) and .87 in. (2.21 cm) as desired. The fluid is then recycled back to the entrance of the pump. Each stainless steel tube consists of four sections: an initial calming section, a hydrodynamic test section (2nd stage calming section), a heat transfer test section, and a mixing section. To assure fully developed hydrodynamic conditions, the heat transfer sections of the .51 in. and .87 in. diameter tubes are preceded respectively, by 96 in. (244 cm) and 137 in. (384 cm) lengths of the same size tube. Six pressure taps are installed in the hydrodynamic section of the smaller tube and fourteen pressure taps are installed in the larger diameter tube. Thirty-six thermocouples made from 30 gage copper constantan are placed along the length of each heat transfer test section to determine the local heat transfer coefficient.

Pressure drop measurements were taken by using a set of parallel 6 ft manometers with an adjustable air pressure head to balance the pressure in the tube. Thermocouple readings were taken with a six-dial precision potentiometer. Flow rates were measured by a calibrated flow meter, and in some instances by direct weighing. The voltage drop over the length of each tube was measured for each run using a digital voltmeter.

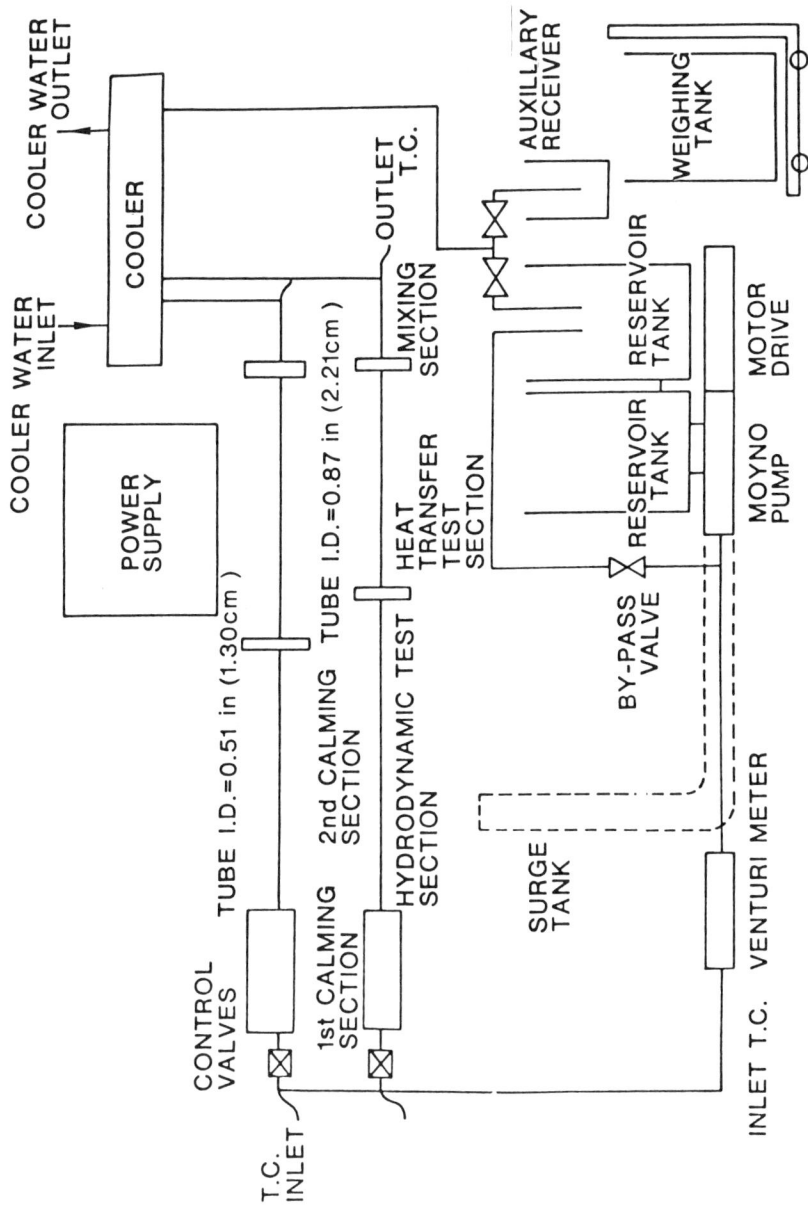

FIG. 1. Schematic diagram of the flow loop.

255

Rheological data were obtained from two .1172 cm diameter capillary tubes in the capillary tube viscometer. One capillary tube has an L/D ratio of 190, the other 390. The capillary tubes were calibrated with double distilled water and were used interchangeably to assure that correct rheological data were obtained. After each measurement was completed, the capillary tubes were flushed clean with water, then checked again.

Two typical flow curves, one for Carbopol and one for Attagel, are shown in Fig. 2. The flow curves of three concentrations of Carbopol 934 were found to have the form

$$\ln \tau_w = a_1 \ln \frac{8V}{D} + a_2 \tag{17}$$

whereas the flow curve of Attagel 40 was not linear for the shear stress–shear rate on a log-log plot. The following form was found to describe the flow curve satisfactorily:

$$\ln \tau_w = b_1 + b_2 \ln \frac{8V}{D} + b_3 \left(\ln \frac{8V}{D} \right)^2 \tag{18}$$

All other fluid properties except specific heat were measured in the laboratory with specially built apparatus [15]. Thermocouples were calibrated as described by Yoo [15]. Thermal conductivity and resistivity of the stainless steel tubes were measured using specially built instruments.

The properties of the fluids studied are tabulated in Tables 1 and 2, along with the range of variables covered in the experiments.

Water runs were made initially for each tube to check the reliability of the system. Results for pressure drop and heat transfer are shown in Fig. 3. After

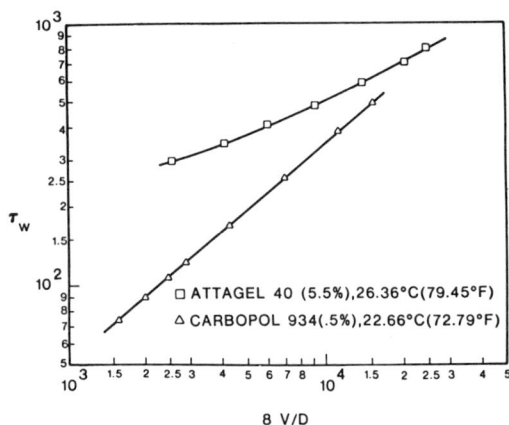

FIG. 2. Flow curves for Carbopol 934 (1.0%) and Attagel 40 (5.5%).

TABLE 1. Properties of Fluids Studied

Fluid	Conc. (%)	n'	$K' \times 10^1$	ρ	C_p*	$k_f \times 10^3$
Carbopol						
A	.20	.91	.42	.998–1.00	0.998	1.45–1.46
B	.50	.84	1.5	1.00	0.998	1.45–1.46
C	1.0	.71–.73	11.–7.7	1.00	0.998	1.44–1.47
Attagel						
D	5.5	.40–.60	12.–1.8	1.03	0.950	1.49–1.50

*From Ref. 15.

each set of non-Newtonian fluid measurements were completed, the system was completely disassembled and thoroughly cleaned. Then water runs were performed to ensure system reliability. Four sets of data taken in this manner at different times throughout the experimental program are included in Fig. 3. It is clear that these data are in excellent agreement with the well-established correlations.

EXPERIMENTAL RESULTS AND DISCUSSIONS

The fluids selected for this investigation were three concentrations of Carbopol 934 and one concentration of Attagel 40, in aqueous solutions. All four may be classified as purely viscous non-Newtonian fluids.

The friction factor measured in the fully developed hydrodynamic region is shown in Fig. 4. No visible diameter effect can be seen in the fully developed turbulent region. However, for the .51 in. (1.30 cm) tube, the transition region of the 1% Carbopol solution occurs at a higher generalized Reynolds number than in the case of the 0.87 in. (2.21 cm) tube. In Dodge and Metzner's experiments [1] there is additional evidence to suggest that the transition

TABLE 2. Ranges of Variables Studied

Fluid	Conc. (%)	n'	$Re' \times 10^{-3}$	Pr'	Nu	Pr_w
Carbopol						
A	.20	.91	6.65–52.3	18.1–14.3	80.9–536	17.6–11.0
B	.50	.84	8.97–27.8	34.7–26.9	132–413	25.1–17.5
C	1.0	.71–.73	5.09–11.9	83.9–65.3	98.5–278	56.5–36.0
Attagel						
D	5.5	.40–.60	5.65–15.0	100–41.6	178–326	30.1–12.9

*From Ref. 15.

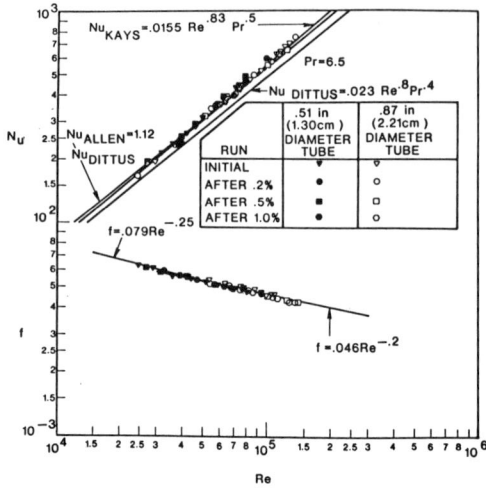

FIG. 3. Experimental Nusselt number–Reynolds number and Fanning friction factor–Reynolds for Newtonian fluid (system run, water).

Reynolds number is dependent on the tube diameter. Friction factor data obtained in the fully developed flow region are plotted in Fig. 5 against Metzner's predicted values (Eq. 12). The percent standard deviation is found to be 5.33, and 50 of the 79 data points fall within this region. Their

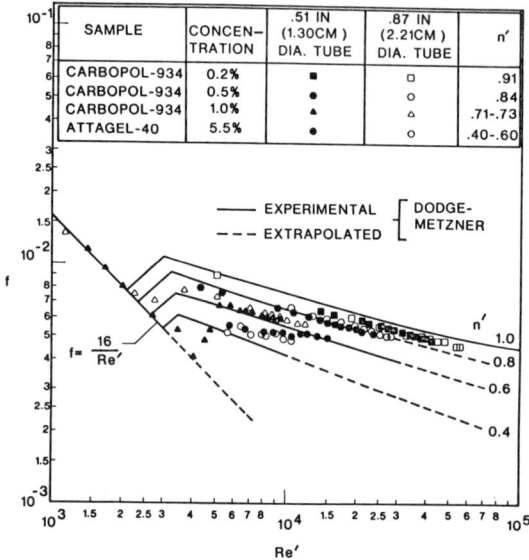

FIG. 4. Experimental pressure drop results for purely viscous non-Newtonian fluids, Fanning friction factor-generalized Reynolds number.

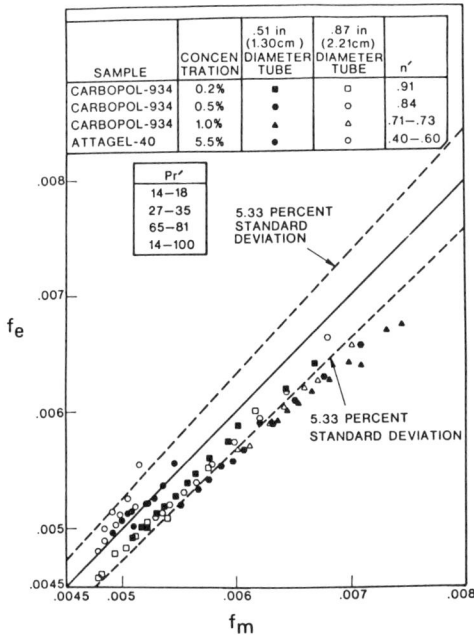

FIG. 5. Comparison of experimental Fanning friction factors with those predicted by Dodge and Metzner (Eq. 12).

prediction is found to be good within 8 percent of the present experimental data in the flow behavior index range above 0.4 and at a generalized Reynolds number above 5,400.

To obtain the experimental Nusselt values, heat fluxes were measured by the enthalpy rise of the fluid and the potential differences across the heat transfer test section. The two values were found to agree within 4 percent. This gave additional assurance of the accuracy of the measurements. The heat transfer coefficients as reported were based on the enthalpy rise of the fluids.

The temperature difference between the inner tube wall and the bulk fluid were kept small throughout the experiment to minimize radial variations of temperature and fluid properties. The maximum temperature difference in all the runs was less than 25°F (14°C). Rheological measurements made on Carbopol solutions at 20°F higher than the fluid bulk temperature revealed no appreciable difference in the flow curves. The same result was found for Attagel solutions measured at 12°F apart.

The experimental fully developed turbulent heat transfer results, interpreted as Nusselt numbers, are shown in Fig. 6. No apparent diameter effect can be seen.

The present heat transfer data were used to test the validity of the two proposed correlations by Metzner and Friend (Eq. 14), and Yoo and Hartnett (Eq. 16).

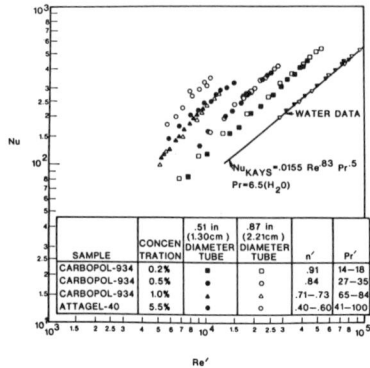

FIG. 6. Experimental heat transfer results for purely viscous non-Newtonian fluids, Nusselt number-generalized Reynolds number.

In Fig. 7, experimental Nusselt numbers are plotted against the predicted values of Metzner and Friend for the generalized Reynolds number range of 5,400–53,000. The percent standard deviation is calculated to be 11.7, with 64 of the 82 points flaling within this region. Metzner and Friend's prediction gives a lower Nusselt number for n' greater than .85 and a higher Nusselt number for n' less than .85.

In Fig. 8, the same set of experimental data is plotted against Yoo and Hartnett's predicted values. The percent standard deviation is 13.8, with 69 of the 82 data points falling within this region. When compared with the present experimental values, their prediction gives a good approximation for n' values between .4 and 1.0. For generalized Reynolds numbers below 8,000, their predicted values deviate gradually from the experimental values, generally being higher. This gives additional support to Yoo and Hartnett's [14] claim

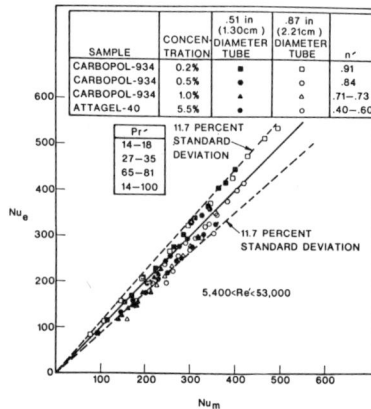

FIG. 7. Comparison of experimental Nusselt numbers with those predicted by Metzner and Friend (Eq. 14).

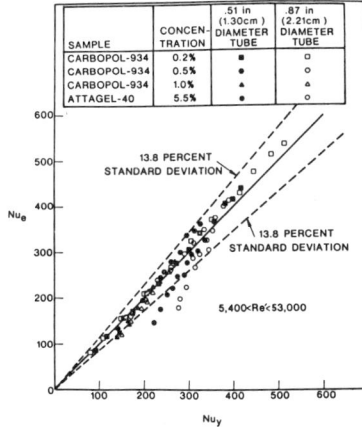

FIG. 8. Comparison of experimental Nusselt numbers with those predicted by Yoo and Hartnett (Eq. (16))–above a generalized Reynolds number of 5,400.

that there may exist an n' dependence for generalized Reynolds numbers below 10,000.

Yoo and Hartnett [14] recommend that their prediction be used above a generalized Reynolds number of 10,000, and the present experimental values show that their prediction starts to fail gradually as the generalized Reynolds number falls below 8,000. This is shown in Fig. 9, where all data points above the generalized Reynolds number of 8,000 are plotted against the predicted values. The predicted and experimental values are in good agreement, with the percent standard deviation being 6.73% with 50 out of 65 points falling within this region.

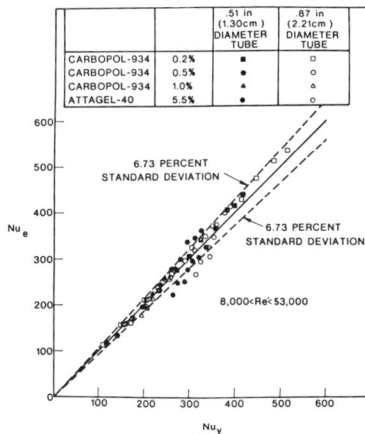

FIG. 9. Comparison of experimental Nusselt numbers with those predicted by Yoo and Hartnett (Eq. (16))–above a generalized Reynolds number of 8,000.

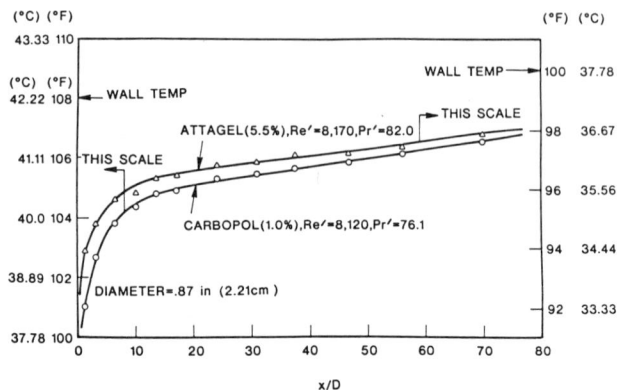

FIG. 10. Temperature profiles for Carbopol 934 (1.0% and Attagel 40 (5.5%) in the .97 in. (2.21 cm) diameter tube.

The hydrodynamic entrance length for the larger tube is of the same magnitude as that of Newtonian fluids, about 10 to 15 pipe diameters. In Figs. 10 and 11 the thermal entrance lengths in both tubes are seen to have the same magnitude as Newtonian fluids, about 10 to 15 pipe diameters, which confirms the earlier result reported by Yoo and Hartnett [14].

CONCLUSIONS

The results of this study may be summarized as follows:

1. The fully developed friction factor and Nusselt numbers are independent of pipe diameter above a generalized Reynolds number of 5,400.

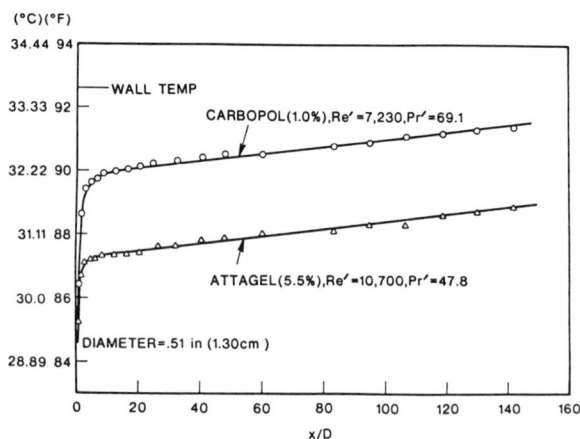

FIG. 11. Temperature profiles for Carbopol 934 (1.0% and Attagel 40 (5.5%) in the .51 in. (1.30 cm) diameter tube.

2. Dodge and Metzner's prediction of the friction factor (Eq. 12) is within 8% of the experimental values for purely viscous fluids for values of the flow behavior index range above .4.
3. For generalized Reynolds numbers from 5,400 to 8,000, Metzner and Friend's prediction (Eq. 14) of the Nusselt number for purely viscous non-Newtonian fluids in the flow behavior range of .4 to 1.0 is preferred to Yoo and Hartnett's (Eq. 16).
4. Above a generalized Reynolds number of 8,000, in the flow behavior index range of .4 to 1.0, Yoo and Hartnett's prediction (Eq. 16) for the Nusselt number of purely viscous non-Newtonian fluids is recommended. It is more accurate than Metzner and Friend's and simpler to use because a prior knowledge of the friction factor is not required.
5. It appears that the transition region is sensitive to the flow behavior index and tube diameter.

ACKNOWLEDGMENT

The authors acknowledge the financial support of the National Science Foundation under its Grant No. 75 02992. In addition, the authors wish to express appreciation for the assistance of Armando Argumedo and Mohamad Farooqui in carrying out the experimental program. Finally the critical comments, suggestions and advice offered by Dr. Thomas Tung in the preparation of the paper have been most helpful.

NOMENCLATURE

a_1, a_2	coefficients of straight line
A_n	dimensionless function of flow behavior index
b_1, b_2, b_3	coefficients of the second order polynomial
C_n	dimensionless function of flow behavior index
C_p	specific heat, Cal/g°K
D	tube diameter, cm
f	Fanning friction factor, dimensionless, $f = 2\,\tau_w/\rho V^2$
h	convective heat transfer coefficient, Cal/cm² s °K
k_f	thermal conductivity, Cal/cm s °K
K	fluid consistency index, g s$^{n'-2}$/cm
K'	defined by Eq. (2), $K' = K(3n + 1/4n)^n$ for power law model
n	flow behavior index, dimensionless
n'	defined by Eq. (2), $n' = n$ for power law model, dimensionless
Nu	Nusselt number, dimensionless, Nu = hD/k_f
Pr′	generalized Prandtl number, dimensionless, Pr′ = $(C_p K'/k_f)(8V/D)^{n'-1}$
Pr	Prandtl number, dimensionless, $\mu C_p/k_f$
St	Stanton number, dimensionless, St = Nu/RePr = Nu/Re′Pr′
V	average velocity, cm/s
ρ	density of fluid, g/cm³
τ	shear stress, dyne/cm²

Subscripts

e	obtained by present experiments
m	obtained by Metzner and his coworkers
w	evaluated at the wall
y	obtained by Yoo and Hartnett

REFERENCES

1. Dodge, D. W., and A. B. Metzner: "Turbulent Flow of Non-Newtonian Systems," *AIChE Journal,* **5**, 189 (1959).
2. Metzner, A. B.: "Heat Transfer in Non-Newtonian Fluids," *Advances in Heat Transfer,* **2**, Edited by J. P. Hartnett and T. F. Irvine, Jr., Academic Press, New York (1965).
3. Ward, H. C.: "Co-Current Turbulent–Turbulent Flow of Air and Water-Clay Suspensions in Horizontal Pipes," Ph.D. Thesis, Georgia Institute of Technology, Atlanta, Georgia (1953).
4. Friend, P. S.: "Turbulent Non-Newtonian Heat Transfer," M.Ch.E. Thesis, University of Delaware, Newark, Delaware (1958).
5. Metzner, A. B., and J. C. Reed: "Flow of Non-Newtonian Fluids–Correlation of the Laminar, Transition, and Turbulent-Flow Regions," *AIChE Journal,* **1**, 434 (1955).
6. Weltman, R. N.: "Friction Factors for Flow of Non-Newtonian Materials in Pipelines," *Ind. Eng. Chem.,* **48**, 386 (1956).
7. Skelland, A. H. P.: *Non-Newtonian Flow and Heat Transfer,* John Wiley and Sons, New York (1967).
8. Metzner, A. B., and P. S. Friend: "Heat Transfer to Turbulent Non-Newtonian Fluids," *Ind. Eng. Chem.,* **51**, 879 (1959).
9. Haines, R. C.: "Heat Transfer to Pseudo-Plastic Solution in Turbulent Flow," B.Ch.E. Thesis, University of Delaware, Newark, Delaware (1957).
10. Raniere, F. D.: "Heat Transfer to Pseudo-Plastic Suspensions in Turbulent Flow," B.Ch.E. Thesis, University of Delaware, Newark, Delaware (1958).
11. Clapp, R. M.: "Turbulent Heat Transfer in Pseudo-Plastic Non-Newtonian Fluids," *International Developments in Heat Transfer,* **652**, D211, ASME, New York (1963).
12. Eckert, E. R. G., and R. M. Drake, Jr.: *Heat and Mass Transfer,* McGraw-Hill Book Co., New York (1959).
13. Kays, W. M.: *Convective Heat and Mass Transfer,* McGraw-Hill Book Co., New York (1966).
14. Yoo, S. S., and J. P. Hartnett: "Heat Transfer and Friction Factors for Purely Viscous Non-Newtonian Fluids in Turbulent Pipe Flow," *Int'l. Heat and Mass Transfer Conference,* **II**, 218, Tokyo, Japan (1974).
15. Yoo, S. S.: "Heat Transfer and Friction Factors for Non-Newtonian Fluids in Turbulent Pipe Flow," Ph.D. Thesis, University of Illinois, Chicago, Illinois (1974).

INTENSITIES IN THE ν_3-FUNDAMENTALS
OF CO_2 AND N_2O*

Prasad Varanasi
State University of New York, Stony Brook, New York

The recently published measurements by Tubbs and Williams [1, 2] of the line strengths and collision-broadened half-widths at low temperatures in the ν_3-fundamentals of CO_2 (Ref. 1) and N_2O (Ref. 2) are of significant value to planetary spectroscopists. The data of Tubbs and Williams appear to be the only measurements that have been performed in a laboratory at low temperatures which are relevant to the atmospheres of the terrestrial planets. Therefore, it is essential that their validity be ascertained, especially, since the line strengths measured by Tubbs and Williams [1, 2], which the authors used in determining the half-widths of lines from their curves-of-growth, seem to lead to the values for the absolute intensities of the bands that are significantly lower than previously published measurements by others [3]. The discrepancies between the results obtained in Refs. 1 and 2 and earlier measurements, insofar as the band intensities are concerned, are, in our view, beyond commonly acceptable bounds of experimental error. Such a view has prompted us to remeasure the absolute intensities of the ν_3-fundamentals of CO_2 and N_2O in our laboratory using the Wilson-Wells-Penner-Weber technique [4].

The results of our measurements are shown in Fig. 1 and Table 1. The ν_3-fundamental of CO_2 was measured using a commercial, double-beam, grating spectrophotometer with a spectral resolution of 1 cm^{-1} and broadening pressures between 15 and 22 atmospheres. Measurements on the N_2O band were performed on the abovementioned double-beam instrument and with higher resolution (0.15 cm^{-1}) on a commercial, Ebert-mounted grating specrometer, which, being nonevacuable, could not be employed in our studies on the CO_2 band. Broadening pressures between 8 and 12 atmospheres were found to be adequate to smear out any rotational fine structure in the N_2O band. Calibrated gas mixtures were obtained from a commercial supplier. The

---

*The abstract of this paper appear on page 436.

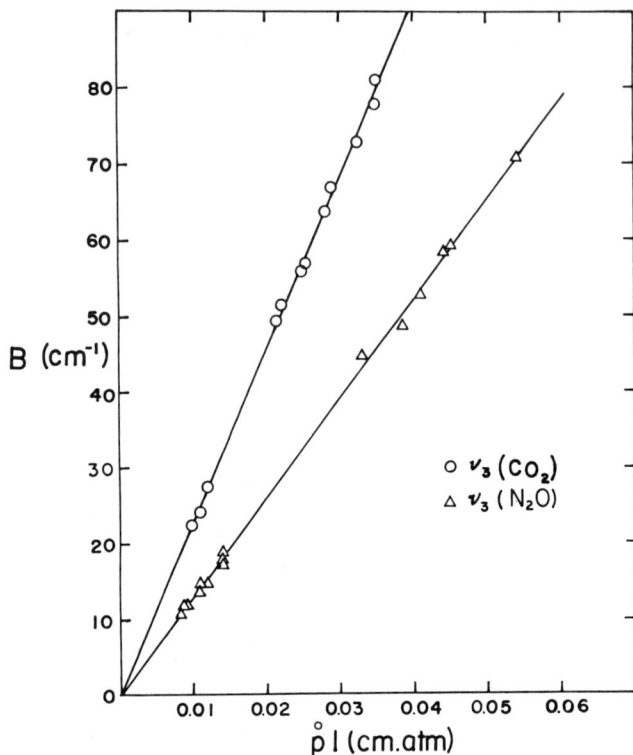

FIG. 1. Plots of the quantity

$$B \equiv \int_{\nu_1}^{\nu_2} \ln\left(1/\tau_\nu{}^a\right) d\nu$$

versus $p°l$ for the ν_3 fundamentals of CO_2 and N_2O at $296°K$. τ_ν^a is the apparent fractional transmittance measured between limiting wave numbers ν_1 and ν_2 for the amount $p°l$ of the absorbing gas, $p°$ derived using van der Waals constants for N_2.

mixture concentrations used were 330 ppm of CO_2 in N_2 and 1095 ppm of N_2O in N_2. Our best estimates obtained from Fig. 1 for the absolute intensities at $296°K$ are 2452 ± 72 cm^{-2}(atm^{-1})$_{STP}$ for the CO_2 band and 1411 ± 54 cm^{-2}(atm^{-1})$_{STP}$ for the N_2O band. These values include contributions from the hot band ($01'0 \rightarrow 01'1$) and the isotopic bands. Our estimate for the combined intensity of the CO_2 bands is in perfect agreement with the value 2448 cm^{-2}(atm^{-1})$_{STP}$ obtained at $298°K$ by Tubbs and Williams [1]. The value 1323 cm^{-2}(atm^{-1})$_{STP}$ reported by Tubbs and Williams [2] for the combined intensity of the N_2O bands is lower by only 6.5% of our result.

By using the relation $k_\nu \simeq S_J/\delta_J$ applicable at sufficiently high pressures, at which the rotational fine structure in a band is smeared out completely, line

TABLE 1. Measured data at 296°K for the spectral absorption coefficient, k_ν (cm^{-1}atm^{-1}), line spacing δ_J (cm^{-1}) and line intensity S_J (cm^{-2}atm^{-1}) in the ν_3 fundamentals of N_2O and CO_2.

Molecule	Line	δ_J*	k_ν	$S_J = k_\nu \delta_J$
N_2O	R(28)	0.645	22.05	14.20
	R(27)	0.623	22.58	14.07
	R(26)	0.637	24.56	15.64
	R(25)	0.665	25.77	17.14
	R(24)	0.651	26.96	17.55
	R(23)	0.664	27.21	18.07
	R(22)	0.670	29.46	19.74
	R(21)	0.682	30.60	20.87
	R(20)	0.674	31.22	21.04
	R(19)	0.697	31.91	22.24
	R(18)	0.707	32.36	22.88
	R(17)	0.707	32.29	22.83
	R(16)	0.698	32.22	22.49
	R(15)	0.739	32.06	23.69
	R(14)	0.699	31.82	22.24
	R(13)	0.747	31.56	23.58
	R(12)	0.738	30.59	22.58
	R(11)	0.753	30.25	22.78
	R(10)	0.736	29.06	21.39
	R(9)	0.784	27.54	21.59
	R(8)	0.759	25.56	19.40
CO_2	R(28)	1.182	35.60	42.10
	R(26)	1.210	40.40	48.90
	R(24)	1.232	44.60	55.00
	R(22)	1.257	49.50	62.20
	R(20)	1.283	52.30	67.10
	R(18)	1.307	54.40	71.10
	R(16)	1.331	55.60	74.00
	R(14)	1.353	57.00	77.10
	R(12)	1.385	54.80	75.90
	R(10)	1.400	52.20	73.20

*δ_J are obtained from the following papers: for CO_2, T. K. McCubbin, Jr., J. Pliva, R. Pulfrey, W. Telafir, and T. Todd, *J. Mol. Spectrosc.*, **49**, 136 (1974); for N_2O, J. Pliva, *J. Mol. Spectrosc.*, **12**, 360 (1964).

intensities S_J may be determined from measured spectral absorption coefficients k_ν and known local line spacings δ_J. Line intensity data obtained using this method were shown [5, 6] to be within 8 percent of the high-resolution data for the 00°0-02°1 and 10°0-00°1 bands of N_2O (Ref. 5) and for the CO fundamental [6]. Application of this procedure to lines in the R-branches of the ν_3-fundamentals of CO_2 and N_2O measured in the present experimental study have yielded the line intensity data presented in Table 1.

These data are in good agreement with the high-resolution line intensity measurements of Tubbs and Williams (see Table 1 of Ref. 1 and Fig. 2 of Ref. 2).

ACKNOWLEDGMENT

Supported by NASA under Grant No. NGR 33-015-139 and NSF under Grant No. ENG75-02986.

REFERENCES

1. L. D. Tubbs, and D. Williams: *J. Opt. Soc. Am.,* **62,** 284 (1972).
2. Tubbs, L. D., and D. Williams: *Ibid.,* **63,** 859 (1973).
3. For references on earlier measurements see Refs. 1 and 2.
4. S. S. Penner: *Quantitative Molecular Spectroscopy and Gas Emissivities,* Addison-Wesley, Reading, Massachusetts, 1959.
5. Varanasi, P., and B. R. P. Bangaru: *JQSRT,* **14,** 1253 (1974).
6. Varanasi, P., and S. Sarangi, *Ibid.,* **15,** 473 (1975).

THERMAL BEHAVIOR OF CONFINED ARCS WITH LOCAL FLUID CONSTRICTION

C. J. Cremers and H. S. Hsia
University of Kentucky, Lexington

INTRODUCTION

The electric arc at atmospheric pressure is an efficient mechanism for transformation of electrical energy to thermal energy of a gas at elevated temperatures. Such a system in a more developed state offers the possibility of employment in devices for plasma processing. Of particular current interest are schemes for producing certain fertilizers and plastic feedstocks from coal. One plasma device which appears to be of value for such applications is the channel arc with local fluid constriction (LFC).

The concept of LFC originated in 1937 with the work of Kirschstein and Koppelmann [1, 2]. They attempted to extinguish a dc arc between pointed electrodes with an inwardly directed circumferential jet of air. In 1964 Elliott and Gomez [3] carried out similar tests on the local fluid constriction of an otherwise free-burning electric arc. In their tests an arc was passed through a hole in two brass discs sandwiched together to form a circumferential slit through which a high velocity gas could be blown radially into the discharge column. Later, Mahan and Cremers [4], and most recently Hsia [5], have studied the effect of LFC upon an arc already confined by cooled circumferential walls.

LFC causes a local constriction of the arc column near the axial position of blowing by means of radial injection of a coolant gas stream. The arc column then redevelops downstream of this location. The constriction causes an increase in the local electric-field strength and consequently there is a greater energy density in this region caused by increased dissipation. Downstream, where the arc is redeveloping, a cool layer of gas along the wall causes a reduction of the local heat flux there.

It appears that the concept of LFC should be quite applicable to plasma

---

*The abstract of this paper appears on page 440.

processing problems for which conversion of electrical to thermal energy at high efficiency is desired. It also appears that LFC offers the opportunity for enhanced mixing of a powdered feedstock with the arc column. The object of the present investigation is to study, in the absence of chemical reactions, some of the fluid mechanics and heat transfer phenomena occurring in such a system.

EXPERIMENTAL RESULTS

The experiments were carried out with a plasma reactor described in detail by Hsia [5] and shown schematically in Fig. 1. It consists of a water-cooled ring anode, a pin cathode, a cylindrical channel made up of electrically isolated and individually water-cooled segments, and a radial gas-injection section. The inside diameter of the cylindrical channel is 9.5 mm. The distances between the cathode or anode and the injection section may be varied by the addition or removal of channel segments.

The channel is made up of electrically and thermally isolated segments that can be used both as potential probes and calorimeters. In addition, each segment is provided with a static pressure tap so that the local static pressure in the channel can be determined during operation. Electrical and thermal isolation is provided by rubber seals which space the segments apart slightly while serving to align the segments and seal the channel against leaks. The thickness of each segment, including the seal, is 5 mm.

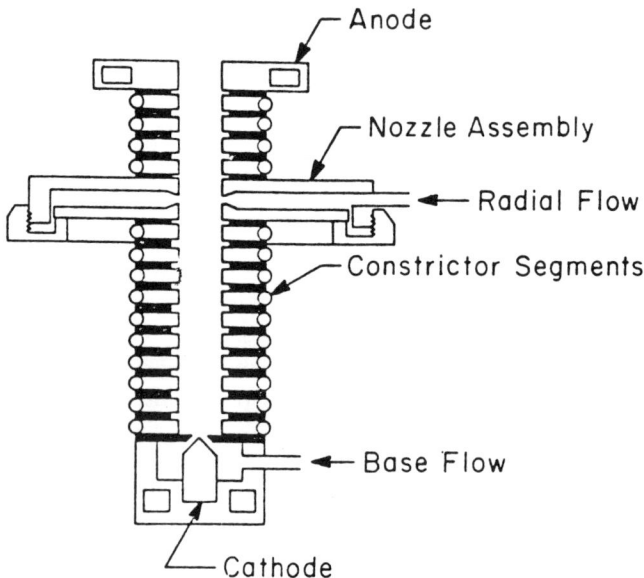

FIG. 1. Plasma generator with local fluid constriction.

The radial injection nozzle is formed by two adjacent oversized channel segments. They are isolated electrically, held apart by springs, and may be adjusted so that the injection slit width may be varied.

Part of the working fluid, here designated as the base mass flux $\dot{m}_B$, is injected at the cathode end. Enough channel segments are included upstream of the cathode so that the arc can become fully developed before the injection slot location. At the injection slot the remainder of the working fluid is injected with no axial component of velocity. The mass flux at this location is designated $\dot{m}_R$. In the present study argon was used exclusively so that chemical reactions could be avoided. More appropriate gases for plasma processing applications will be studied later.

The character of the flow in the injection region depends on Reynolds number in much the same way as for a low temperature flow. Flow visualization studies have been carried out in conjuction with the present experiment in which a water model is used for simplicity. Back and Roschke [6] showed earlier that a water model gives a good qualitative description for geometrically similar plasma problems. The present studies indicate that if the upstream Reynolds number is to the laminar range, then flow is as shown in Fig. 2. There is a large region of boundary layer separation just downstream of the injection slot and the central streamlines are closest together just downstream of the slot location.

Not shown on Fig. 2 is a considerably smaller separation region, which also

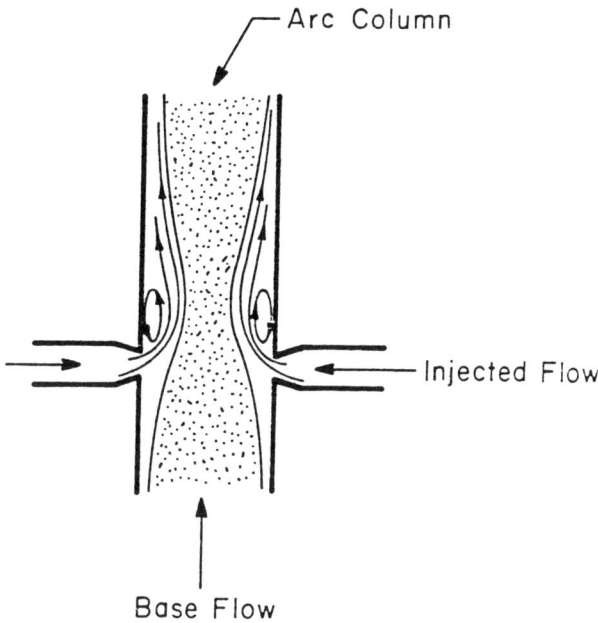

FIG. 2. Flow pattern in injection region.

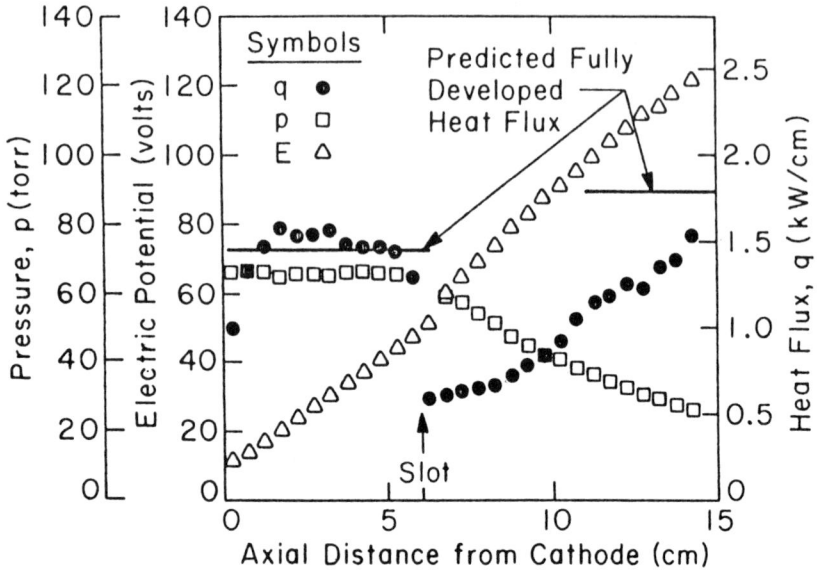

FIG. 3. Axial distribution of flow variables.

occurs just upstream of the slot. Detailed measurements of the arc voltage in the region of the slot indicate that the electric field actually is negative just upstream and also just downstream of the slot, while elsewhere in the arc column it is positive. This indicates that the flow separation has greatly distorted conducting paths near the wall. Therefore, the electric field as measured by way of the wall potential in this region is not representative of the electric field in the arc column at the same axial location.

Typical distributions of electric field strength, wall heat flux, and static pressure (measured with respect to atmospheric pressure) are given in Fig. 3. These are for a current of 215 amps, a base mass flux of 0.169 g/s and an injection rate of 0.244 g/s.

Each point, as measured along the abscissa of the graph, represents a reading taken at one of the channel segments of either pressure, heat flux or electric potential with respect to the cathode. Probably because of the influence of the changing channel cross-section near the cathode, flow development is not as rapid as it might otherwise be. The heat flux shows a rapid increase toward the fully developed value. However, it appears to overshoot this.

There are two possibilities for this apparently anomalous increase in heat flux. One may be that the radiation heat flux from the cathode tip is highest for these segments. More probably, there is another flow separation region here caused by the rapid expansion in flow cross-sectional area in the region of the cathode tip. In the region downstream of the cathode, then, one would expect hot gasses to be swept into the wall region in the direction opposite to the flow. The rapid drop in heat flux for the segment just upstream of injection is somewhat misleading because this segment makes up half of the injection slot and it is not only water-cooled for the calorimetric measurements, but it also has a heat loss to the cold gas being injected into the mainstream. Downstream of the injection flow, development is much slower because of the increased Reynolds number of this flow.

The system being described in this paper is a candidate device in which to carry out high temperature endothermic reactions. Therefore, the energy dissipation per unit of volume is of interest as it has a strong influence on overall thermal efficiency. Consequently, the effect of injection on the electric field is of direct interest. In order to measure the detailed field in the nozzle region, a second injection section was used in which the channel wall was made up of thin copper segments 0.91 mm thick separated by insulators 0.33 mm thick. The upstream and downstream sides of the injection nozzle itself were also copper and were 0.9 mm thick. If the electric field in the slot region is defined as the difference in potential across the slot divided by the distance between the centers of the segments upstream and downstream of the slot, then its magnitude is on the order of 10–40 V/cm for the range of flow rates investigated in this experiment. However, as mentioned previously, there seem to be some rather severe effects of flow separation on the wall potential readings in the slot region. That is, it may well be that the potentials measured do not reflect the conditions in the arc column at the same axial location.

Our experiments show that the results are rather independent of slot width. Therefore, either injection velocity or mass flow rate can be used to characterize the flow. Figure 4, for instance, shows how the electric field intensity, as defined above, varies with radial injection velocity for several currents at a fixed base mass flow rate. The magnitude of the current does not seem to be important and at these rather low velocities it appears that the resistance of the arc column is almost proportional to the blowing rate. It appears that the cross-sectional area of the arc column will simply decrease as the injection velocity increases.

The same kind of information, only for a large range of injection flow rates, is shown in Fig. 5. The measured electric field strength at the slot location seems to reach a maximum and then decrease again. It appears that this is caused by the mixing between the two streams being more effectively carried out at higher velocities so that there is no longer a well defined arc column for which the radius is reduced by the radial injection. We have noted that the temperature profiles for the higher rates of injection flow appear broader at the higher velocities, indicating that there is probably this kind of mixing occurring.

Spectroscopic measurements were also taken in a plasma generator similar to that shown in Fig. 1, only with thin cylindrical quartz windows between the segments. In that way temperature profiles could be measured at 5 mm

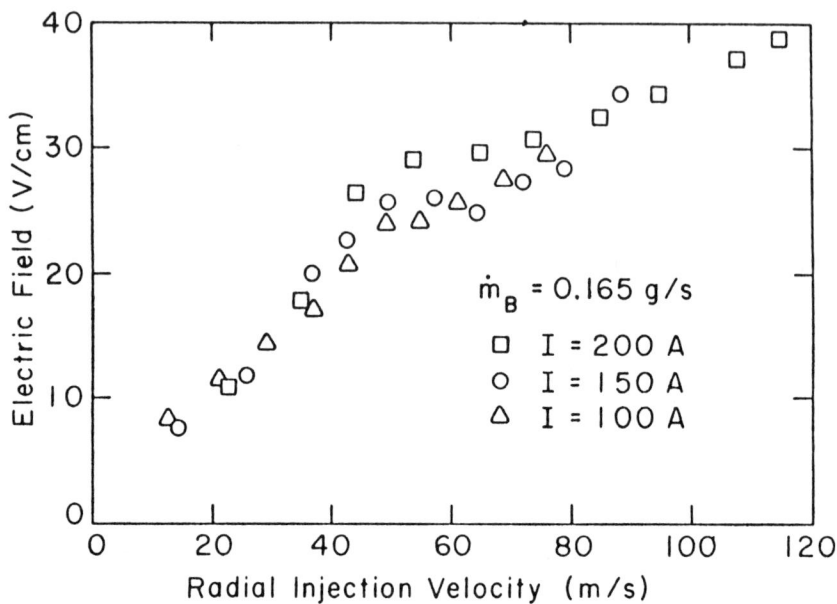

FIG. 4   Effect of blowing velocity on electric field.

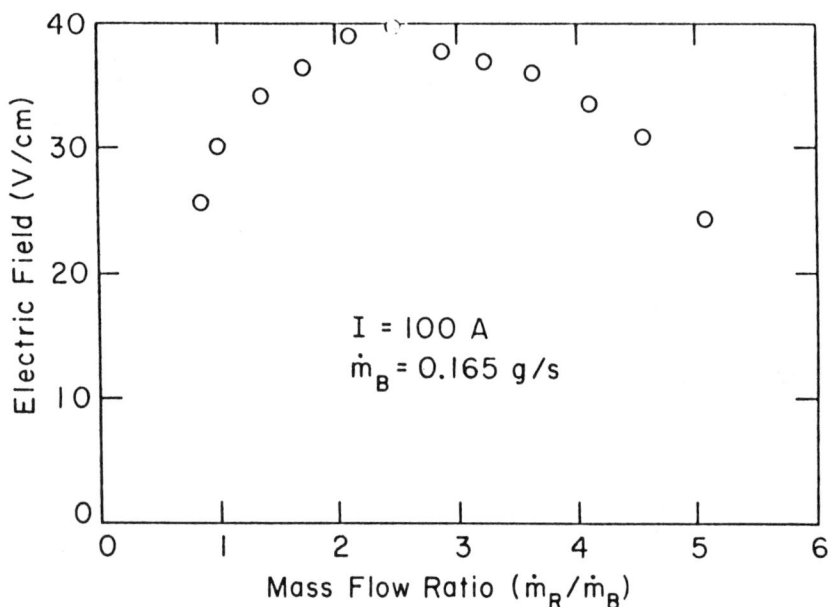

FIG. 5. Effect of blowing rate on electric field.

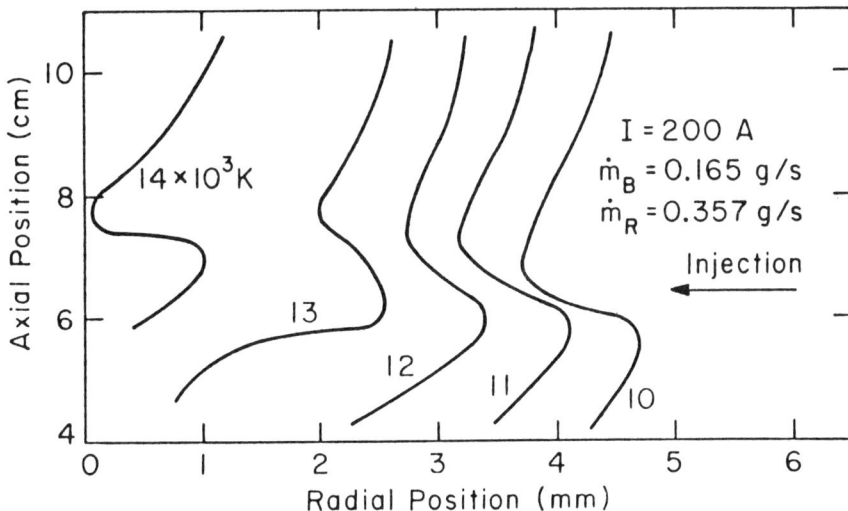

FIG. 6. Isotherms in blowing region.

intervals both upstream and downstream of the injection location. Figure 6 shows a representative set of isotherms that were constructed based on data taken at a number of axial locations. Note that this is for a current of 200 amps and a radial injection rate of approximately twice the base flow rate. There does not seem to be as much of an effect as one might expect. The isotherms show a definite inward curvature caused by the blowing, but this has been frequently found in a number of other studies of interactions between arc columns and fluid flows. The arc column seems to have a remarkable capacity for resisting deformation by fluid effects.

REFERENCES

1. Kirschstein, G., and Koppelmann, F., Wiss. Veroff, Siemens-Werken, 1937, Vol. XVI, No. 1, pp. 51–71.
2. Kirschstein, G., and Koppelmann, F., Wiss. Veroff, Siemens-Werken, 1937, Vol. XVI, No. 3, pp. 26–55.
3. Elliott, H. F., and Gomez, R. V., *An Experimental Study of the Energy Exchange in the Column of an Electric Arc with a Radial Gas Flow Interaction,* GE/EE 64-6, Air Force Institute of Technology, Wright-Patterson Air Force Base, Ohio, 1964.
4. Mahan, J. R., and Cremers, C. J., J. Basic Eng. Trans. ASME Ser. C., 1972, Vol. 94, No. 4, pp. 818–824.
5. Hsia, H. S., *Confined Electric Arc with Local Fluid Constriction,* Ph.D. Dissertation, University of Kentucky, Lexington, Kentucky, 1975.
6. Back, L. H., and Roschke, E. J., Applied Mech. Trans. ASME Ser. E., 1972, Vol. 94, No. 2, pp. 677–681.

SIMULTANEOUS RADIATIVE, CONVECTIVE AND CONDUCTIVE HEAT TRANSFER IN AN EXTENDED SURFACE

Takeshi Kunitomo
Kyoto University, Kyoto, Japan

Sadayuki Tanaka
Fukui Technical College, Fukui, Japan

INTRODUCTION

Extended surfaces with various geometrical configurations have been used to increase the heat flux from a surface. However, information on the heat transfer characteristics of the extended surfaces when radiation dominates is still insufficient. Analyses on longitudinally finned plane surfaces were published by a few authors [1], [2], [3], but the results were insufficient for design and there remain other extended surfaces which have not yet been studied in detail. This paper treats the simultaneous radiative, convective, and conductive heat transfer of longitudinally finned plane surfaces (system A) and cylinders (system B) and a circumferentially finned cylinder (system C) shown in Fig. 1, making allowance for the radiative property of the surface and applying the Monte Carlo method of analysis.

2. ANALYSIS

2.1 Assumptions

1. The convective fluid does not absorb radiation.
2. The convective heat transfer coefficient is constant on the whole surface.
3. The thermal conductivity of fin material is independent of temperature.
4. There is one-dimensional heat conduction through the fins.
5. The emission of radiation follows Lambert's law and the reflection is both specular and diffuse.

2.2 Energy Equation and Boundary Conditions

The temperature distribution in each fin of the array is assumed to be the same and two adjacent fins, as shown in Fig. 1, are considered. The energy

---

*The abstract of this paper appears on page 440.

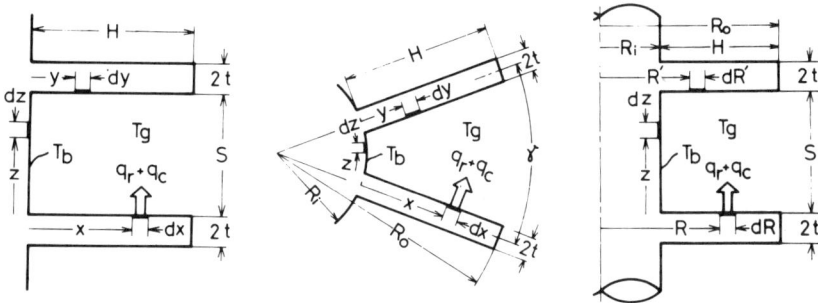

FIG. 1. Extended surfaces.

equations are obtained by applying the energy conservation principle to an infinitesimal fin element as follows:

$$kt \cdot \frac{d^2 T(x)}{dx^2} = q_{rf}(x) + q_{cf}(x) \qquad \text{for } A \text{ and } B$$

$$\frac{kt}{R} \cdot d\, R \frac{dT(R)}{dR}\, dR = q_{rf}(R) + q_{cf}(R) \quad \text{for } C \tag{1}$$

The boundary conditions are:

at $x = 0$, $T = T_b$

at $x = H$ $\quad k\left(\dfrac{dT}{dx}\right) + \epsilon\sigma\,[T^4(H) - T_c^4] + h\,[T(H) - T_g] = 0$

for A and B

and at $R = R_i$, $T = T_b$

at $R = R_0$ $\quad k\left(\dfrac{dT}{dR}\right) + \epsilon\sigma[T^4(R_0) - T_c^4] + h[T(R_0) - T_g] = 0$ for C $\tag{2}$

Using the absorption factor, $B_{dA_\xi - dA_x}$, the radiative heat flux $q_{rf}(x)$ is expressed as follows.

$$dA_x \cdot q_{rf}(x) = \epsilon\sigma T^4(x)\, dA_x - \int dA_\xi \epsilon\sigma T^4(\xi) B_{dA_\xi - dA_x} \tag{3}$$

Introducing the nondimensional parameters N_c and N_h and applying the reciprocity relation, $dA_\xi B_{dA_\xi - dA_x} = dA_x B_{dA_x - dA_\xi}$, the above equations may be expressed as follows:

$$\frac{d^2 T^*(X^*)}{dX^{*2}} = N_c \frac{\phi_{rf}(X^*)}{\epsilon} + N_h\,[T^*(X^*) - T_g^*]$$

$$\frac{\phi_{rf}(X^*)}{\epsilon} = T^{*4}(X^*) - \int T^{*4}(\xi^*) B_{dA_{X^*} - dA_{\xi^*}} \qquad \text{for } A \text{ and } B$$

$$\frac{d^2 T^*(R^*)}{dR^{*2}} + \frac{dT^*(R^*)/dR^*}{R^* + R_i(R_0 - R_i)} = \frac{N_c \phi_{rf}(R^*)}{\epsilon} + N_h [T^*(R^*) - T_g^*] \left.\right\} \quad (4)$$

$$\frac{\phi_{rf}(R^*)}{\epsilon} = T^{*4}(R^*) - \int T^{*4}(\xi^*) B_{dA_{R^*} - dA_{\xi^*}} \quad \text{for } C$$

$$X^* = 0 \quad T^* = 1$$

$$X^* = 1 \quad \frac{dT^*}{dX^*} + \frac{\{N_c[T^{*4}(1) - T_c^{*4}] + N_h[T^*(1) - T_g^*]\}t}{H} = 0$$

$$\text{for } A \text{ and } B \quad (5)$$

$$R^* = 0 \quad T^* = 1$$

$$R^* = 1 \quad \frac{dT^*}{dR^*} + \frac{\{N_c[T^{*4}(1) - T_c^{*4}] + N_h[T^*(1) - T_g^*]\}t}{H} = 0 \quad \text{for } C$$

2.3 Procedure of Numerical Analysis

For system A and B, Eq. (4) is integrated twice using boundary condition (5) and the temperature distribution is obtained by the following equations:

$$T^*(X^*) = 1 - C_0 X^* - X^* \int_X^1 \Phi[T^*(X'^*), X'^*] dX'^*$$

$$- \int_0^{X^*} X'^* \Phi[T^*(X'^*), X'^*] dX'^*$$

$$\text{where } C_0 = \frac{\{N_c[T^{*4}(1) - T_c^{*4}] + N_h[T^*(1) - T_g^*]\}t}{H}$$

$$\Phi[T^*(X^*), X^*] = \frac{N_c \phi_{rf}(X^*)}{\epsilon} + N_h[T^*(X^*) - T_g^*]$$

$$(6)$$

$$\text{and} \quad \frac{\phi_{rf}(X^*)}{\epsilon} = T^{*4}(X^*) - \sum_{\xi^*} T^{*4}(\xi^*) B_{dA_{X^*} - dA_{\xi^*}}$$

The range of ξ^* is the closed surface which is divided into $(2M + 2)$ elements of the fin surface elements ΔX_j^*, $\Delta Y_j^*(j = 1 \sim M)$, the base surface D and the opening C, as shown in Fig. 2. Applying the Monte Carlo method [4], the absorption factor $B_{\Delta X_{i^*} - \Delta \xi_{j^*}}$ is obtained and the local heat flux is calculated by

$$\frac{\phi_{rf}(X_i^*)}{\epsilon} = T^{*4}(X_i^*) - \sum_{\xi_j^*} T^{*4}(\xi_j^*) B_{\Delta X_j^* - \Delta \xi_j^*} \quad (7)$$

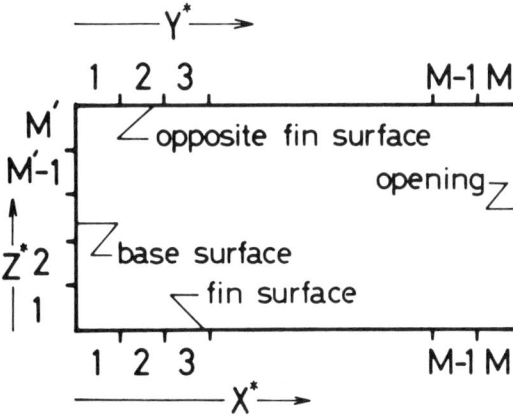

FIG. 2. Subdivision of two adjacent fin surfaces (system A).

A numerical interative procedure is employed to obtain the solution of Eq. (6). Another possible method of obtaining the numerical solution is the usual finite difference approximation of Eq. (6). The latter is applied to the case of the large values of N_c and N_h. For system C, a similar method is adopted.

2.4 Heat Transfer Characteristics

The heat transfer is evaluated by the following three nondimensional quantities.

(1) Nondimensional total heat transfer rate Q*

The radiative heat transfer from the base surface must be calculated to obtain Q^*. Dividing the base surface in to M' equal elements (ΔZ_i^*, $i = 1 \sim M'$) and applying the Monte Carlo method, $B_{\Delta Z_i^* - \Delta \xi_j^*}$ is obtained. The local radiative heat flux is calculated from

$$\frac{\phi_{rb}(Z_i^*)}{\epsilon} = 1 - \sum_{\xi_j^*} T^{*4}(\xi_j^*) B_{\Delta Z_i^* - \Delta \xi_j^*} \qquad (8)$$

The equation is similar for system C. Q^*'s are calculated by the following equations:

$$
\begin{aligned}
Q^* &= \frac{Q_t}{(S + 2t)\sigma T_b^4} \quad \text{for } A \\[2mm]
&= \frac{Q_t}{R_i \gamma \sigma T_b^4} \quad \text{for } B \\[2mm]
&= \frac{Q_t}{2\pi R_i (S + 2t)\sigma T_b^4} \quad \text{for } C
\end{aligned}
\right\} \qquad (9)
$$

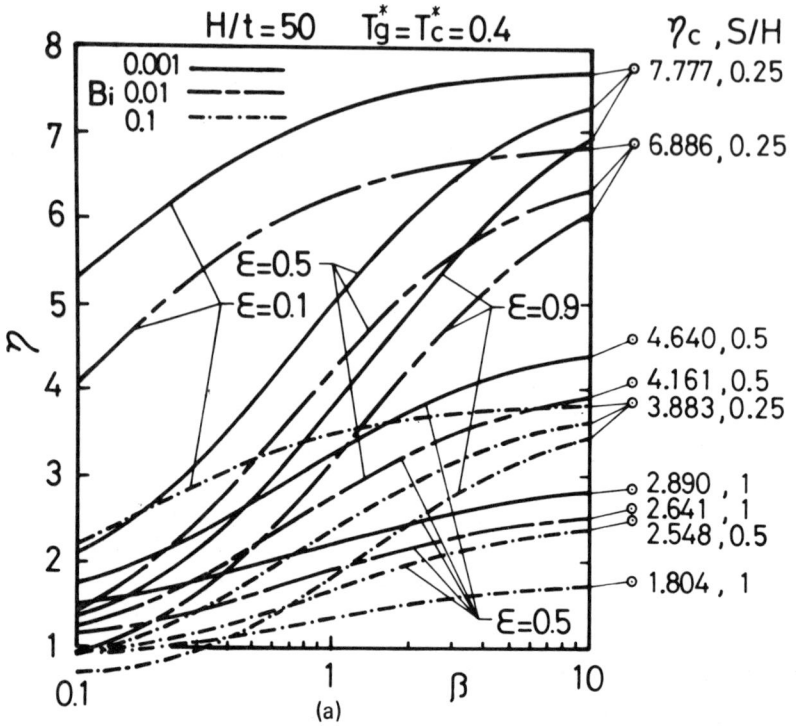

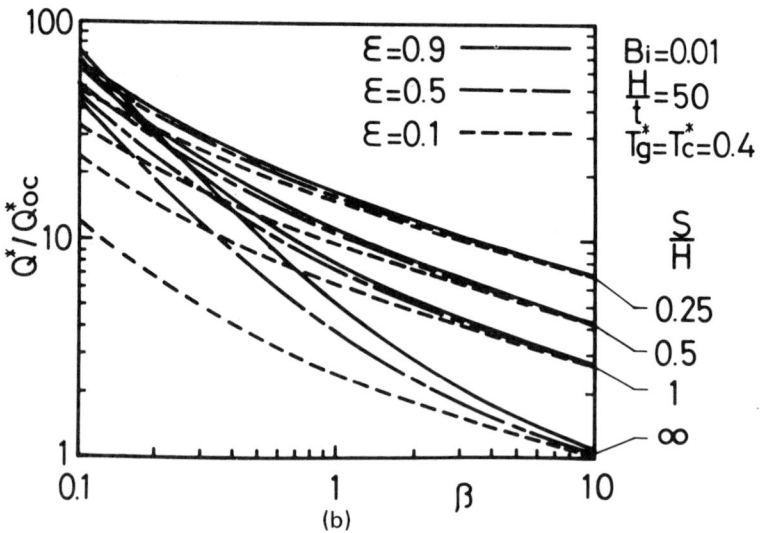

FIG. 3A. The effect of β on η and Q^* (Diffuse reflection).

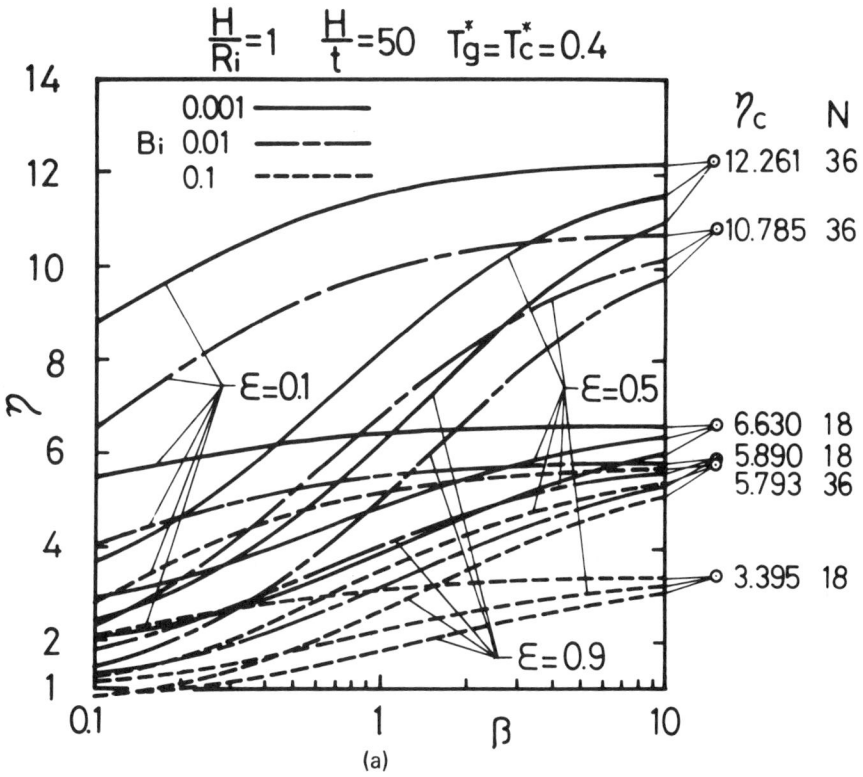

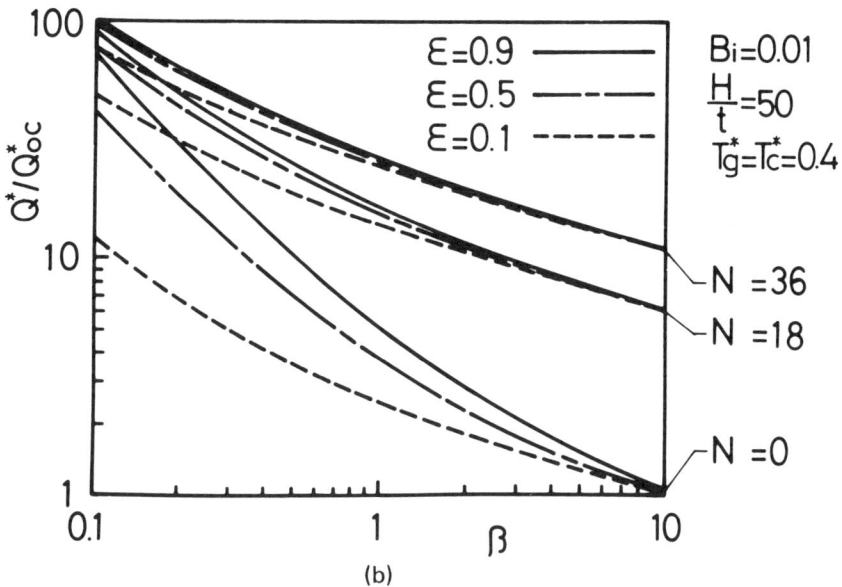

(a)

(b)

FIG. 3B.

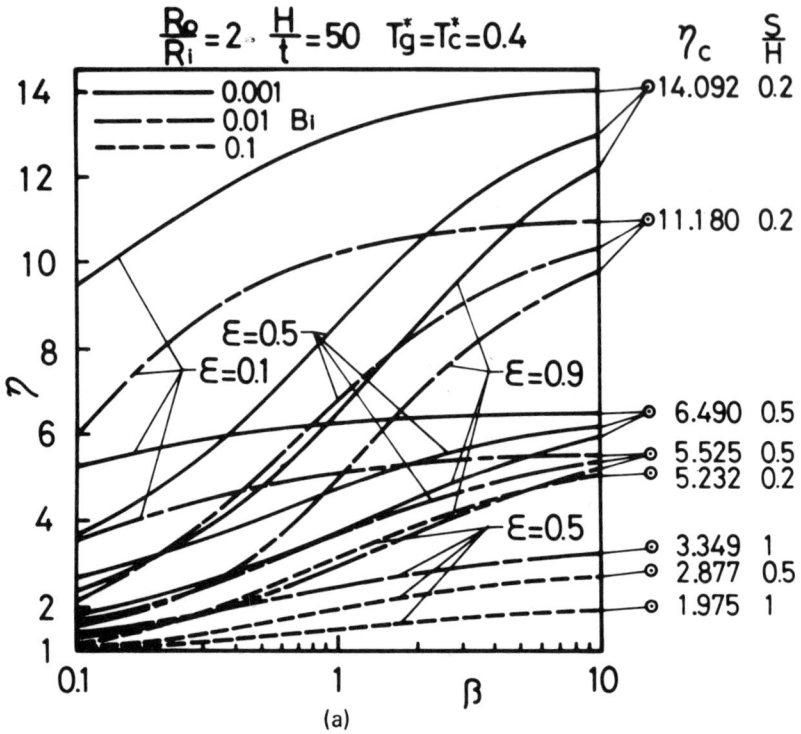

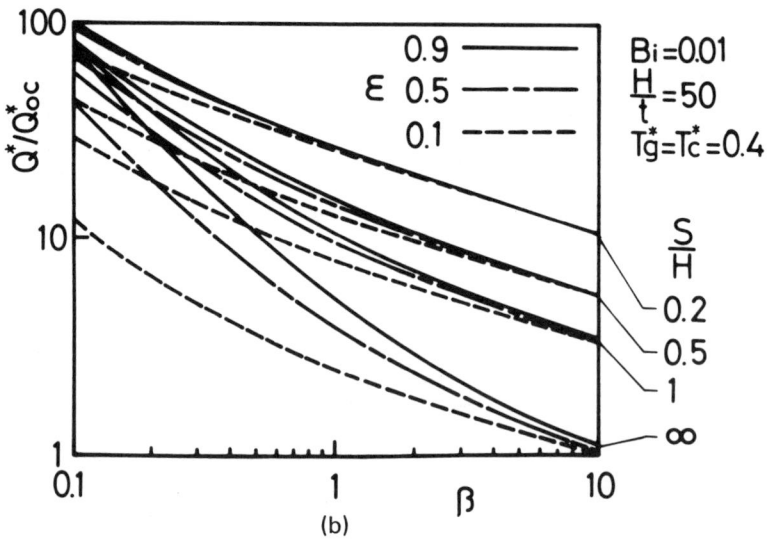

FIG. 3C.

(2) *The system effectiveness* η

The system effectiveness is defined by

$$\eta = \frac{Q^*}{Q_0^*} = \frac{Q^*}{\epsilon(1 - T_c^{*4}) + \beta(1 - T_g^*)} \tag{10}$$

where $Q_0{}^*$ is the nondimensional total heat transfer without fins.

(3) The nondimensional heat transfer increase per unit fin volume $Q_V{}^*$

$Q_V{}^*$ is defined by

$$Q_V^* = \frac{Q^* - Q_0^*}{V^*} \tag{11}$$

where V^*'s are defined as follows.

$$
\left.
\begin{aligned}
V^* &= \frac{2Ht}{(S + 2t)^2} && \text{for } A \\[2mm]
&= \frac{2Ht}{(R\gamma)^2} && \text{for } B \\[2mm]
&= \frac{2\pi t (R_0{}^2 - R_i{}^2)}{\pi R_i{}^2(S + 2t)} && \text{for } C
\end{aligned}
\right\} \tag{12}
$$

3. RESULTS AND DISCUSSIONS

3.1 Reliability of the Solution

The reliability of the present solutions by the Monte Carlo method is confirmed by comparing them with the previous results obtained by Sparrow et al. [5], [6], [7], and Donovan et al. [3]. Good agreement was obtained for all these cases.

3.2 Calculated Results

Although N_c and N_h are generally used as parameters to express the results, in this study these are divided into two heat transfer parameters (R_n, B_i) and one geometrical parameter (H/t) to examine the geometrical effect separately.

(a) Diffuse reflection $(r_{sp} = 0)$

(1) Relation between (η, Q^*) and (β, ϵ, B_i)

Typical examples of calculated results are shown in Fig. 3A–C. The value of η_c is the system effectiveness for the case that only convection exists. For each B_i, the effect of radiation on η is very weak, when T_b is low and convection is predominant. Accordingly as T_b increases and β decreases, η decreases rapidly. This tendency is remarkable when B_i is small, ϵ is large and the spacing is small. In some cases, η becomes smaller than 1.0. For system C, finning is always effective. The values of η increase in the order of system A, B, and C for similar geometrical configurations, because of the effect of the opening.

An example of the effect of B_i at constant R_n (= 0.1) is shown in Fig. 4 for system A. At constant ϵ, Q^* increases monotonically as B_i increases. But, the increase rate of Q^* is very small at small B_i, i.e., at the region where radiation is predominant and the increase rate becomes large at large B_i, i.e., at the region where convection is predominant. For each ϵ, the curve of η has a maximum value at the different B_i which shifts to the larger value as ϵ increases, since the value of B_i where convection becomes predominant increases as ϵ increases.

The effect of ϵ on Q^*, η and Q_V^*, in the case when radiation is predominant, is shown in Fig. 5. Q^*, which determines the scale of the heat exchanger, increases rapidly with an increase of ϵ at the region of small ϵ and slowly at the region of large ϵ. And, the smaller the S/H, the smaller the increasing rate and the effect of emissivity, except at the extremely small region close to 0 of ϵ. For system B and C, Q_0^* does not exceed Q^* and the effect of finning is maintained even when the strong radiative interaction

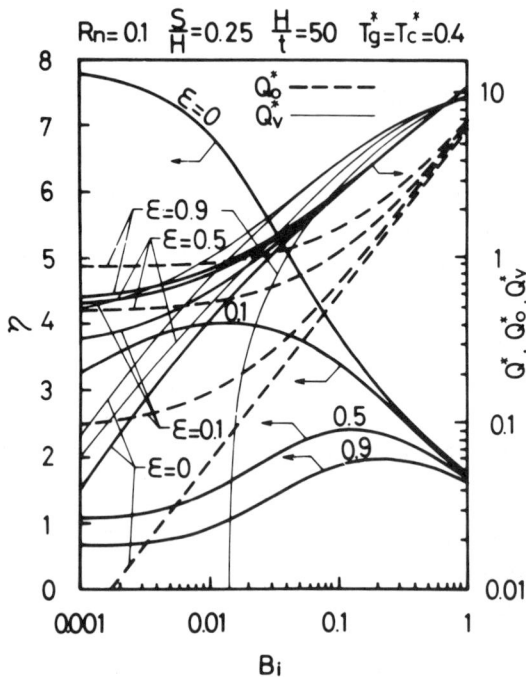

FIG. 4. The effect of B_i on η (System A, Diffuse reflection).

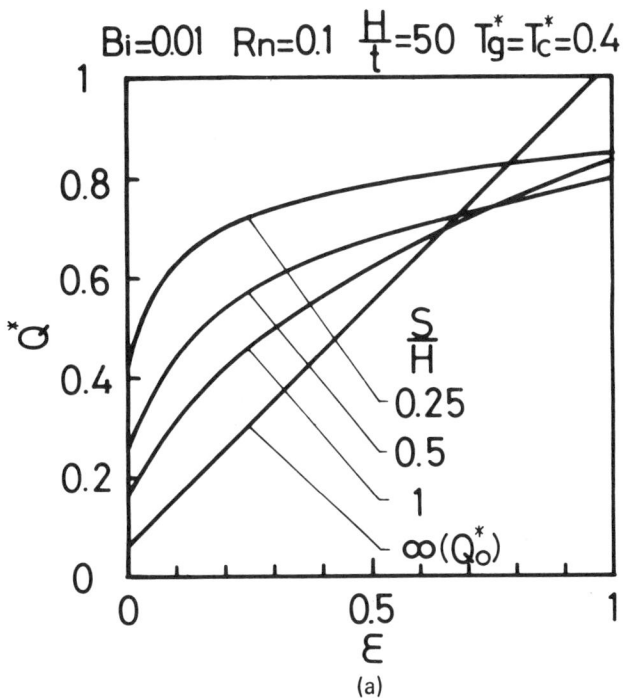

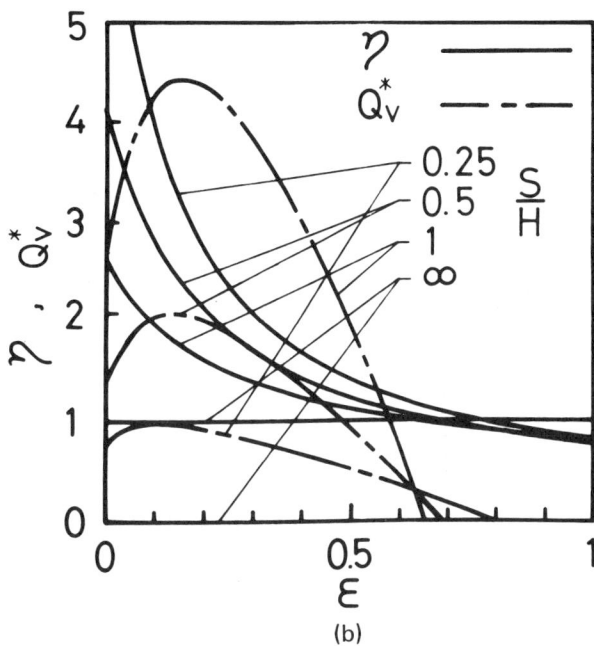

FIG. 5A. The effect of ϵ on η, Q^*, and Q_{V^*} (Diffuse reflection).

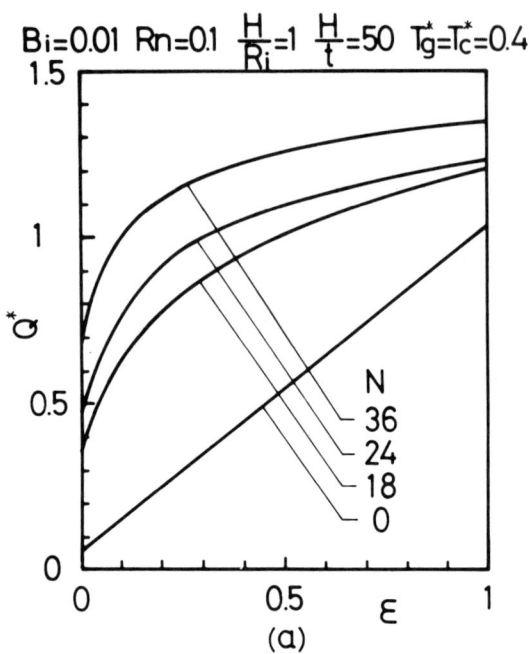

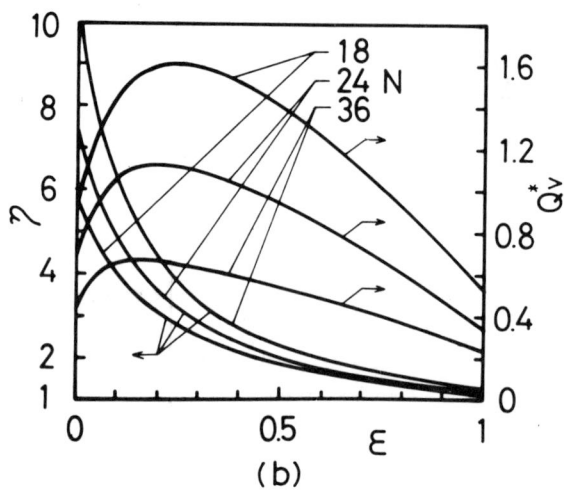

$B_i=0.01$ $Rn=0.1$ $\dfrac{H}{R_i}=1$ $\dfrac{H}{t}=50$ $\vec{T_g}=\vec{T_c}=0.4$

N
36
24
18
0

(a)

(b)

FIG. 5B.

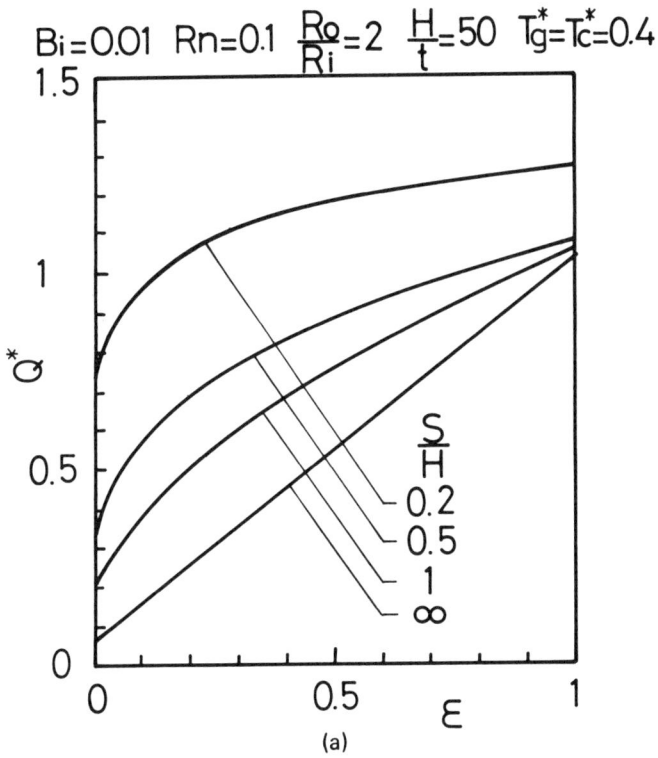

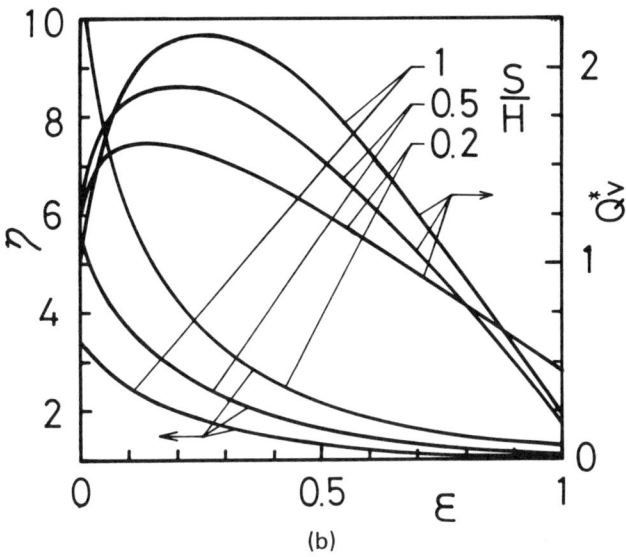

FIG. 5C.

exists. For system A, however, $Q_0{}^*$ exceeds Q^* and η becomes smaller than 1.0 at the region of large ϵ. The heat transfer increase per unit fin volume, which allows an evaluation of the fin cost to increase the heat flux, and has a maximum point for each geometrical condition so one can choose the optimum combination of ϵ and the geometrical condition for fixed values of the other parameters.

(2) The relation between η and (S/H or N, H/t)

Typical examples are shown in Fig. 6. For systems A and C, the curve of η with constant S/H has a maximum at a specific value of H/t or V^*. And, at constant H/t, η is rapidly improved with a decrease of S/H. However, further improvement may not be expected because assumption (2) cannot be applied to the case of extremely narrow fin spacing. For system B, maximum points are not found in the curves of η since N is adopted instead of S/H in this figure.

(3) The effects of $T_g{}^*$ and $T_c{}^*$ on η

The effects of $T_g{}^*$ and $T_c{}^*$ on η are shown in Fig. 7 only for system A. For systems B and C, practically the same results are obtained. In the case of $\beta = 0.1$ i.e., strong radiative interaction, the effect of $T_g{}^*$ is very weak. In the case of $\beta = 1$ i.e., weak radiative interaction, the effect is very strong. The effect

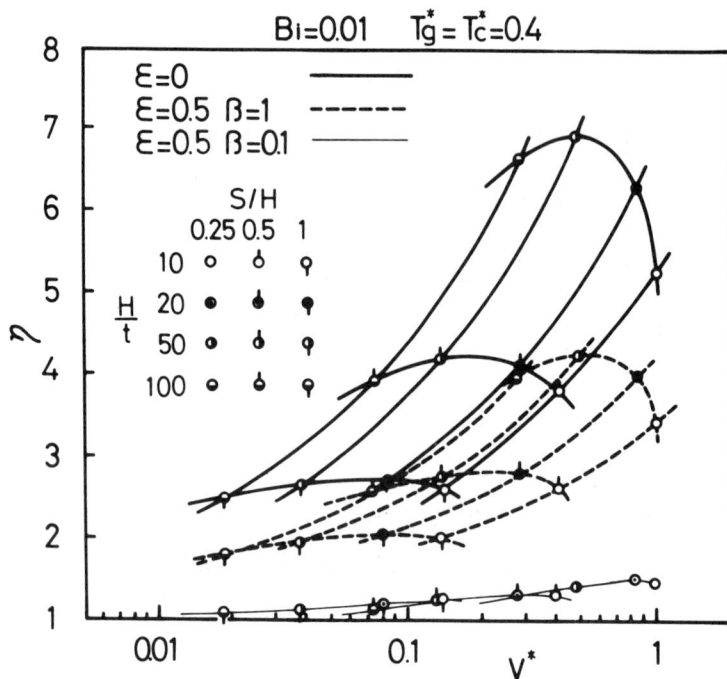

FIG. 6A. The effect of geometrical parameters on η (Diffuse reflection).

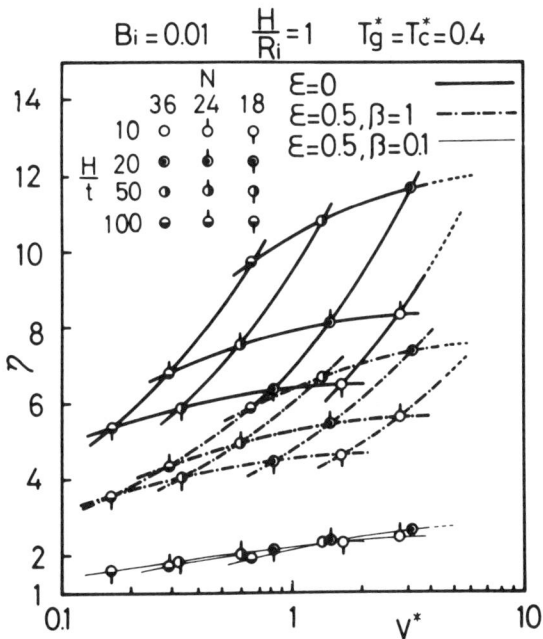

FIG. 6B.

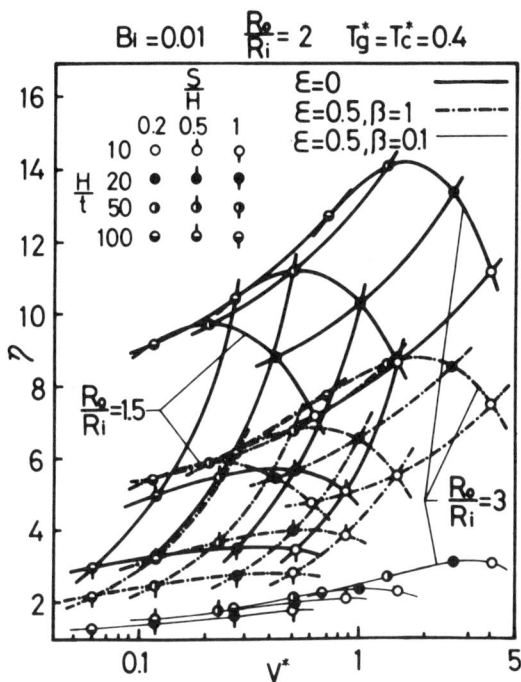

FIG. 6C.

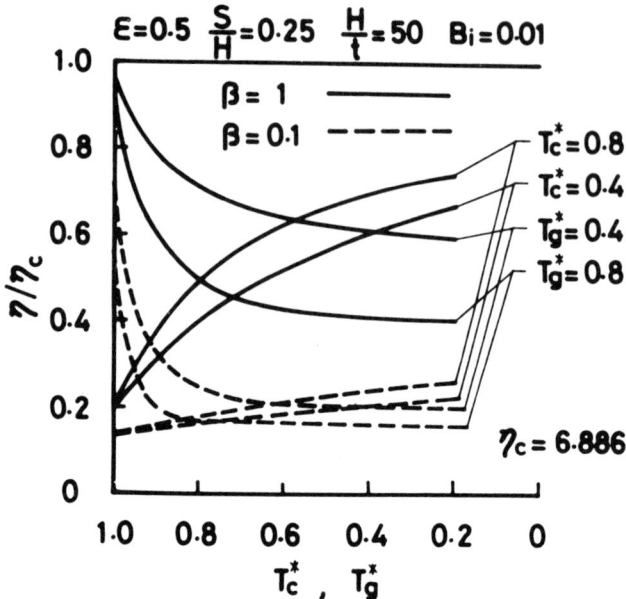

FIG. 7. The effect of $T_c{}^*$ and $T_g{}^*$ on η (System A, Diffuse reflection).

of $T_c{}^*$ is especially strong in the region of $T_c{}^* = 0.8 \sim 1.0$ both for $\beta = 1$ and $\beta = 0.1$.

(b) Nondiffuse reflection ($r_{sp} = 0$)

Typical results for nondiffuse reflection are shown in Fig. 8 at conditions of $r_{sp} = 0.2, 0.5, 0.8$, and 1.0 (specular reflection), and comparing them with the results of diffuse reflection. Heat is more effectively transferred in the case of specular reflection than in the case of diffuse reflection. The difference between the combined reflection and the diffuse reflection increases with an increase of r_{sp} and with a decrease of β for each r_{sp}. When ϵ and β are kept constant, the difference increases with an increase of B_i. The effect of r_{sp} increases with a decrease of S/H. The effect of ϵ becomes strong at the region a little less than 0.5 of ϵ for small S/H and at the middle value of ϵ for large S/H. This result can be understood by the following explanation. At extremely small value of ϵ, the reflective property may affect the radiative transfer strongly because of multireflection. However, the radiative heat transfer rate is small compared with the total heat transfer rate. Thus, the effect of r_{sp} is weak. At extremely large value of ϵ, r_{sp} does not affect radiation because of low reflection. In the middle region of ϵ, the radiative heat transfer rate becomes relatively large and the effect of reflection becomes stronger. Thus, the maximum effect of r_{sp} is found in this region. As shown in Fig. 5, the effect of ϵ on η and Q^* is weak at the region of ϵ greater than 0.5 for small S/H. So the maximum effect of r_{sp} is found at the region a little

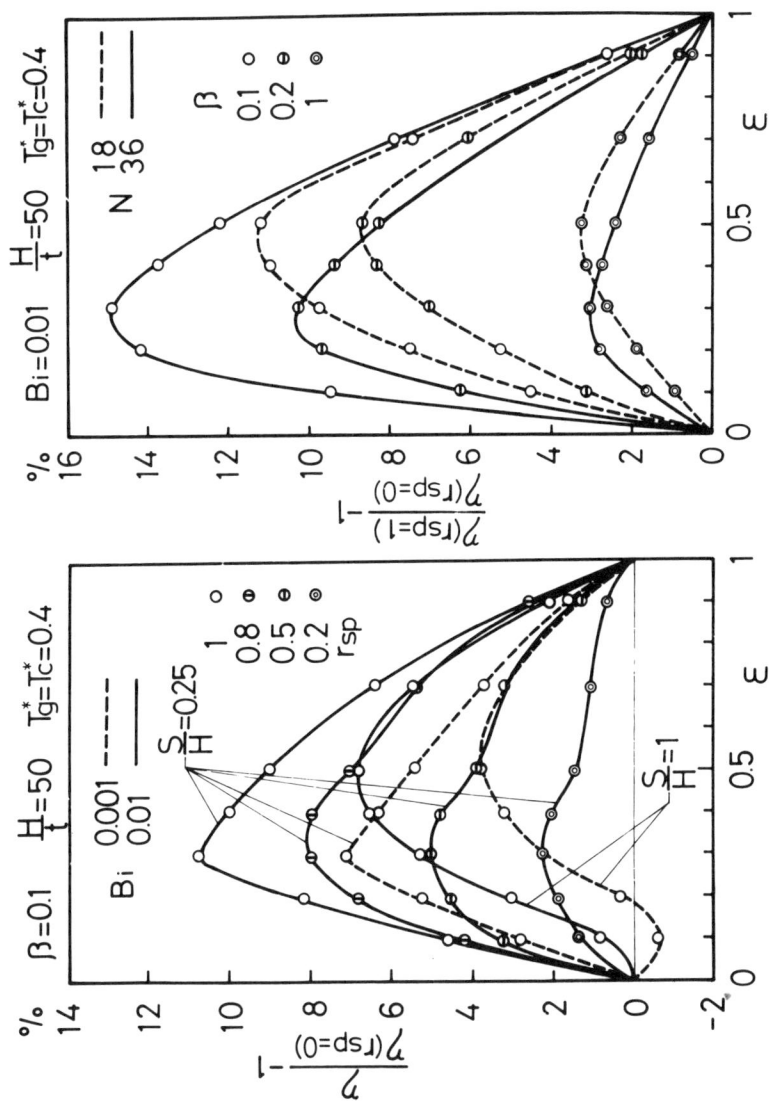

FIG. 8. The effect of r_{sp} on η.

less than 0.5 of ϵ. For large S/H, the effect of ϵ on η and Q^* is still strong even at the middle value of ϵ. Thus, the maximum effect of r_{sp} is found just around the middle value of ϵ.

CONCLUDING REMARKS

From the present analysis on simultaneous radiative, convective, and conductive heat transfer with an extended surface, the effects of the heat transfer parameters and the geometrical parameters on the system effectiveness, the total heat transfer rate and the heat transfer increase per per unit fin volume can be seen. In addition, the optimum combinations of these parameters for the design of a high temperature heat exchanger may be estimated.

NOMENCLATURE

$B_{dA_x - dA_\beta}$	adsorption factor
B_i	Biot number hH/k
H	fin height
h	convective heat transfer coefficient
k	thermal conductivity
M, M'	number of subdivision of fin surface and fin height
N	number of fins of system B
N_c	radiation-conduction parameter $\epsilon\sigma T_b^3 H^2/kt (\equiv R_n \cdot H/t)$
N_h	convection-conduction parameter $hH^2/kt (\equiv B_i \cdot H/t)$
Q^*	nondimensional heat transfer rate
Q_t	total heat transfer rate
Q_v^*	see Eq. (11)
q_c	convective heat flux
q_r	radiative heat flux
R_n	radiation number $\epsilon\sigma T_b^3 H/k$
r_{sp}	ρ_s/ρ
S	fin spacing
T	absolute temperature
T_b	base surface temperature
T_c	equivalent blackbody temperature of opening
T_g	convective fluid temperature
T^*, T_c^*, T_g^*	$T/T_b, T_c/T_b, T_g/T_b$
t	half of fin thickness
V^*	nondimensional fin volume
X^*, Y^*, Z^*	$x/H, y/H, z/H$
x, y, z	coordinates on the fin surface, on the opposite fin surface, and on the base surface
β	convection-radiation parameter $h/\sigma T_b^3$
ϵ	emissivity
η	system effectiveness
η_c	pure convective system effectiveness
ρ	reflectivity $\rho \equiv 1 - \epsilon \equiv \rho_s + \rho_d$
ρ_d, ρ_s	diffuse and specular component of ρ

σ	Stefan Boltzmann constant
Φ	see Eq. (6)
ϕ	nondimensional radiative heat flux

Subscripts

b base surface c convective heat transfer
f fin surface r radiative heat transfer

REFERENCES

1. Okamoto Y.: *Trans. JSME,* 1965, **31**, 226, p. 964.
2. Frost, W. and A. H. Eraslan: Proc. of 1968 Heat Trans. and Fluid Mech. Inst., 1968, 206.
3. Donovan, R. C. and W. M. Rohrer: Trans. ASME, Ser. C, 1971, **93**, 2, 41.
4. Howell, J. R.: Advances in Heat Transfer V, Edited by J. P. Hartnett and T. F. Irvine, Jr., Academic Press, N.Y., 1968.
5. Eckert, E. R. G. and E. M. Sparrow: *Int. J. Heat and Mass Transfer,* 1961, **3**, 42.
6. Sparrow, E. M. and J. L. Gregg: Trans. ASME, Ser. C., 1962, **84**, 8, 270.
7. Sparrow, E. M., G. B. Miller, and V. K. Jonson: *J. Aerospace Sciences,* 1962, Nov., 1291.

THERMAL RADIATION MODELS FOR THE ATMOSPHERES OF JUPITER AND SATURN*

Robert D. Cess

State University of New York, Stony Brook, New York

INTRODUCTION

The atmospheres of Jupiter and Saturn are composed primarily of hydrogen and the main thermal opacity results from the pressure-induced far infrared spectrum of hydrogen, including enhancement due to mixing with helium. Model atmospheres for Jupiter and Saturn, as well as for Uranus and Neptune, have been constructed by Trafton [1], and further discussion is given by Trafton and Münch [2]. More recently brightness temperature observations by Gillett et al. [3], in addition to limb brightening observations by Gillett and Westphal [4], suggest the existence of a temperature inversion within the upper stratosphere of Jupiter, and this is possible due to solar absorption by the 3.3μ fundamental band of methane, with methane being a minor atmospheric constitutent.

Methane opacity has been included in the model atmosphere calculations for Jupiter by Hogan et al. [5], and they do predict a temperature inversion. As pointed out by Encrenaz [6], however, comparison of calculated brightness temperatures for ammonia lines [7] with the observations of Aitken and Jones [8] implies that the inversion layer of [5] is at too low an altitude. Cess and Khetan [9] suggest that this is most likely due to use of a line-averaged absorption coefficient for the methane bands, which overestimates the methane opacity, whereas inclusion of line structure yields a much higher inversion layer [9].

The purpose of the present paper is to review certain aspects of the model atmosphere calculations of Cess and Khetan [9] for Jupiter and Saturn, and in particular to use these results to suggest possible models for the upper cloud layers of these planets. Such clouds are presumably composed of ammonia ice.

---

*The abstract of this paper appears on page 444.

RADIATIVE TRANSFER

Let us now briefly consider radiative transfer due to the pressure-induced spectrum of hydrogen, and this is discussed in more detail by Cess and Khetan [9]. The basic procedure for formulating the net radiative flux is quite analogous to the exponential kernel approximation for a gray gas, and this consists of fitting an exponential function to the pressure-induced hydrogen emissivity. Since the kernel function for the integral formulation of the radiative flux may be expressed in terms of the hydrogen emissivity, then the exponential approximation to the emissivity allows the integral formulation for the radiative flux to be converted to a second-order differential equation.

Consider now a hydrogen atmosphere in radiative-convective equilibrium. This consists of an atmospheric region in radiative equilibrium (stratosphere), below which lies the convective troposphere. In that altitude is an inconvenient variable for both Jupiter and Saturn, we consider temperature as a function of pressure, and from Cess and Khetan [9] the vertical temperature structure is described by

$$T = T_e\left(0.5 + \frac{3a_1 HP^2}{8}\right)^{1/4} \qquad P \leqslant P_1 \tag{1a}$$

$$T = T_1\left(\frac{P}{P_1}\right)^{R/C_p} \qquad P \geqslant P_1 \tag{1b}$$

where P is atmospheric pressure, $H = RT_e/g$ the atmospheric scale height, T_e the effective temperature of the planet, a_1 appears in the exponential fit to the emissivity, R and C_p are the gas constant and specific heat at constant pressure, respectively, and T_1 and P_1 denote temperature and pressure at the top of the convective troposphere (tropopause). The tropopause location has been evaluated [9] by employing overall conservation of energy. Results for both Jupiter and Saturn are summarized in Table 1.

Hydrogen is not the only opacity source within the atmospheres of Jupiter and Saturn, and for present purposes we take the mixing ratios of other

TABLE 1. Model atmosphere quantities for Jupiter and Saturn

	Jupiter	Saturn
T_e, °K	135	97
H, km	21	38
a_1, km^{-1} atm^{-2}	0.68	0.68
P_1, atm	0.36	0.29
T_1, °K	141	104
C_p/R	3.05	2.87

constituents to be as follows:

$$\frac{He}{H_2} = 0$$

$$\frac{CH_4}{H_2} = 10^{-3}$$

$$\frac{NH_3}{H_2} = 2 \times 10^{-4}$$

Helium should be the second most abundant gas within both atmospheres. It has no opacity of its own, although it does collisionally enhance the hydrogen opacity. As shown by Cess and Khetan [9], however, this effect may be ignored providing one interprets P as the hydrogen partial pressure rather than the total atmospheric pressure.

The methane and ammonia mixing ratios correspond roughly to cosmic abundances for carbon and nitrogen and are consistent with abundance observations for Jupiter [10]. Uniform mixing should occur for CH_4, but this is not the case for NH_3 due to condensation, and the NH_3 mixing ratio refers to atmospheric levels where NH_3 is not saturated.

Both CH_4 and NH_3 possess a conventional infrared spectrum consisting of vibration-rotation bands. The abundance of NH_3 within the radiatively controlled region of the atmospheres, however, is limited by condensation, such that NH_3 is not a significant opacity source. Condensation of methane does not occur, and since the methane opacity varies as a linear function of pressure while that of hydrogen goes as the square of pressure, it follows that methane will become an increasingly important opacity source, relative to hydrogen, as one progresses to higher levels (lower pressures) within the atmospheres. The methane opacity gives rise to two separate effects. The 7.7μ fundamental band transmits infrared radiation, but since this lies in the short wavelength tail of Planck's function, it is a relatively inefficient transmitter of radiative energy. The 3.3μ fundamental band, which lies in the long wavelength tail of the solar spectrum, is a source of solar absorption and is a possible cause for the previously discussed temperature inversion.

The procedure for including solar absorption and infrared transmission by methane is discussed by Cess and Khetan [9]. The methane contribution becomes important only at sufficiently high altitudes for which the hydrogen opacity is optically thin. This in turn allows the methane opacity to be incorporated by asymptotic matching to the atmospheric thermal structure given by Eq. (1). Typical atmospheric temperature profiles for the combined H_2-CH_4 opacity are illustrated in the next section.

MODEL ATMOSPHERES

Atmospheric temperature profiles for Jupiter and Saturn are shown in Figs. 1 and 2, and these correspond to a diurnal average at the equator. The effect

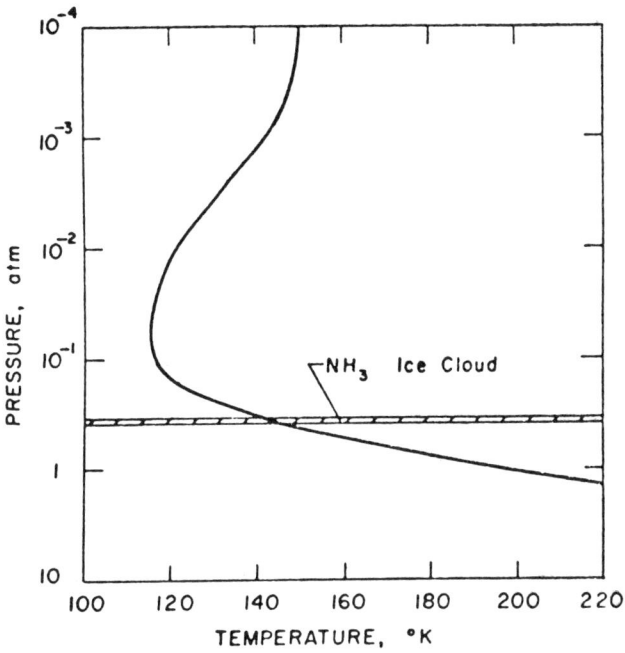

FIG. 1. Model atmosphere results for Jupiter.

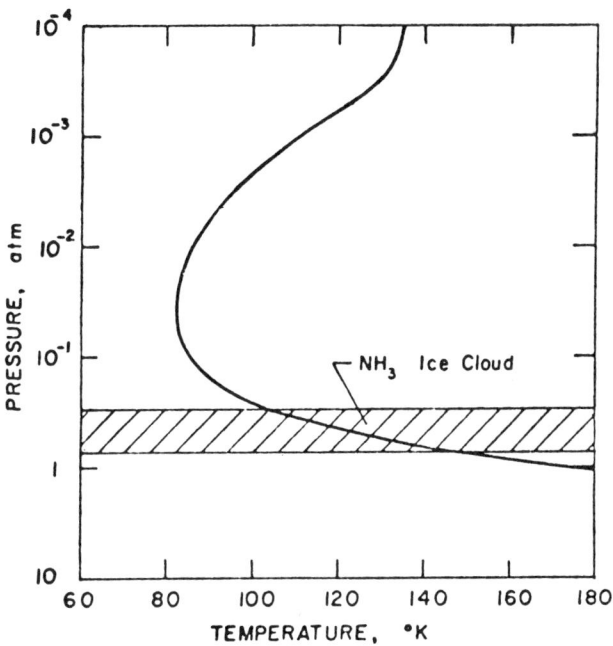

FIG. 2. Model atmosphere results for Saturn.

of latitude is quite slight and enters solely through the formulation for solar absorption by the 3.3μ CH$_4$ band. Considering Jupiter, for example, the temperature asymptotically approaches 151°K in the upper stratosphere for conditions at the equator, whereas the corresponding global mean value is 150°K. Figures 1 and 2 clearly illustrate the methane inversion layer, and if methane were not included, the inversion region would instead be isothermal at a temperature corresponding roughly to the minimum temperature in Figs. 1 and 2.

The main point is that the methane opacity does not influence the atmospheric thermal structure within the lower stratosphere and troposphere, and for this reason Eq. (1) may be employed for interpreting models for ammonia ice clouds within both atmospheres. Cloud models for Jupiter have been presented by Lewis [11], with the uppermost cloud being composed of ammonia ice, while a lower cloud consists of NH$_4$SH. An upper cloud of NH$_3$ ice should also exist on Saturn, and it is the possible thickness of these NH$_3$ ice clouds which we wish to consider in view of the model atmosphere calculations.

The formation of an ammonia ice cloud is of course due to condensation of gaseous ammonia and, as is conventional, we take the location of the cloud bottom to correspond to the location at which condensation of ammonia occurs. Letting P_N denote the partial pressure of ammonia, the saturation vapor pressure is given by [12]

$$P_N \text{ (atm)} = (1.31 \times 10^7) \exp\left(-\frac{3754}{T}\right) \tag{2}$$

Employing this together with Eq. (1) and the assumed unsaturated mixing ratio for ammonia, $P_N/P = 2 \times 10^{-4}$, the cloud bottom locations, expressed in terms of temperature, are listed in Table 2. These are virtually identical to those of Trafton [1].

Trafton [1] has suggested that the cloud top be located at the top of the convective zone (tropopause), although for Jupiter the cloud top is conventionally located at a much higher level corresponding to a temperature of 115°K (e.g., Newburn and Gulkis [10]). From dynamical considerations,

TABLE 2. Cloud model characteristics for Jupiter and Saturn

	Jupiter	Saturn
T, Cloud Top	141°K	104°K
T, Cloud Bottom	145°K	150°K
Cloud Thickness	2 km	51 km

however, Trafton's suggestion seems the most plausible, since it is vertical convection which supports the cloud particles as well as supplying gaseous ammonia to form the particles. Following Trafton [1], and employing the tropopause locations as determined by Cess and Khetan [9], the cloud top temperatures are listed in Table 2 together with the corresponding cloud thickness. The clouds are also schematically illustrated in Figs. 1 and 2.

It should be emphasized that a radiative-convective model atmosphere is a crude representation of reality, and the present cloud models should be regarded only as qualitative. What is important is that the Saturn cloud is much thicker than that of Jupiter. Furthermore, the cloud density should be considerably greater for Saturn than for Jupiter, since the lower portion of Saturn's cloud lies in a region of strong convection, and it is convection which supplies gaseous ammonia to form the cloud.

DISCUSSION OF RESULTS

There is observational evidence for a thin cloud on Jupiter versus a thick cloud on Saturn. A summary of evidence for an optically thin NH_3 ice cloud on Jupiter is given by Newburn and Gulkis [10]. In addition, near infrared observations of ammonia bands (e.g., Newburn and Gulkis [10]) indicate an ammonia abundance of roughly 13 meter amagat (i.e., meter atm at NTP), while assuming that the partial pressure of gaseous ammonia equals its saturation vapor pressure above the cloud bottom, the present model atmosphere yields only 0.4 m amagat of ammonia above the cloud bottom. Thus the ammonia abundance observations are evidently "seeing" through the NH_3 ice cloud.

Aside from the implications for an optically thin cloud on Jupiter, there is evidence that it is also physically thin. Carleton and Traub [13] have observed center-to-limb variations in the equivalent widths of H_2 quadrupole lines within the Jovian atmosphere and compare their results with theoretical calculations by Hunt [14] and Hunt and Margolis [15]. These calculations assume a cloud bottom at $146°K$ (nearly identical with the present $145°K$), but place the cloud top at the previously mentioned $115°K$, which corresponds to a cloud thickness of 15 km. From their observations Carleton and Traub [13] suggest that the upper cloud should be physically thin rather than the relatively deep 15 km cloud, and this is at least qualitatively consistent with the present estimated thickness of only 2 km.

Consider now Saturn, for which the present estimate for the cloud thickness is 51 km and, as previously discussed, the cloud density should be much greater than that for Jupiter. Thus, in contrast to the Jovian cloud, it would appear that for Saturn the NH^3 ice cloud is both optically and physically thick, and there is evidence for this. For the sake of brevity we again refer to the summary by Newburn and Gulkis [10], in which they conclude that "there is no evidence in any of the optical measurements for

TABLE 3. Ammonia abundances for Saturn

Temperature, °K	NH$_3$ Abundance, cm amagat
120	0.3
130	4.0
138	22
140	33
150 (cloud bottom)	220

definite penetration of radiation from below a single ammonia cirrus cloud layer."

The existence of an optically thick NH$_3$ cloud offers a possible explanation for the apparent lack of ammonia within Saturn's atmosphere from observations of near infrared bands [10]. If such observations are "seeing" only partially into the NH$_3$ cloud, then it is quite possible that there is insufficient gaseous ammonia above the "level" of observation to be detected by optical means. To illustrate this, NH$_3$ abundances above given temperature levels within the NH$_3$ cloud, assuming that ammonia is saturated, are listed in Table 3.

Cruikshank [16] has set an upper limit of 20 cm amagats of ammonia based on the 1.5μ ammonia band observations of Kuiper, et al. [17]. From Table 3 this would correspond to a temperature level of 138°K, which is consistent with Trafton's [18] observed rotational temperature of 132°K and Bergstralh's [19] range of rotational temperatures from 122 to 142°K. Of course, the 20 cm amagat abundance is only an upper limit, and the realities of events within a scattering cloud are certainly far more complicated than a simple single-level interpretation.

In conclusion it is again emphasized that the present cloud models are based on an extremely simplified dynamical interpretation, and there are a variety of phenomena which could alter the cloud thicknesses. Ramanathan [20], for example, has illustrated that convective overshoot can raise the tropopause location, resulting in thicker clouds for both planets. Furthermore, the present thermal structure models neglect cloud opacity, an effect which might be significant for Saturn, and inclusion of cloud opacity for Saturn would raise the tropopause location, resulting in even a thicker cloud. The primary conclusion, of course, is that Saturn's cloud should be much thicker than that for Jupiter, and thus one should not expect the atmosphere of Saturn to simply be a colder analog of the Jovian atmosphere.

ACKNOWLEDGEMENT

This work was supported by the US–USSR Cooperative Research Program through the US National Science Foundation.

REFERENCES

1. Trafton, L. M.: *Astrophys. J.,* **147**, 765 (1967).
2. Trafton, L. M. and G. Münch: *J. Atmos. Sci.,* **26**, 813 (1969).
3. Gillett, F. C., F. J. Low, and W. A. Stein, *Astrophys. J.,* **157**, 925 (1969).
4. Gillett, F. C. and J. A. Westphal: *Astrophys. J.,* (Letters) **179**, L153 (1973.)
5. Hogan, J. S., S. I. Rasool, and Th. Encrenaz: *J. Atmos. Sci.,* **26**, 898 (1969).
6. Encrenaz, Th.: private communication.
7. Encrenaz, Th.: *Astr. Astrophys.,* **16**, 237 (1972).
8. Aitken, D. K. and B. Jones: *Nature,* **240**, 230 (1972).
9. Cess, R. D. and S. Khetan: *J. Quant. Spectrosc. Radiat. Transfer,* **13**, 995 (1973).
10. Newburn, R. L. and S. Gulkis: *Space Science Reviews,* **14**, 179 (1973).
11. Lewis, J. S.: *Icarus,* **10**, 365 (1969).
12. International Critical Tables (1928).
13. Carleton, N. P. and W. A. Traub: "Observations of Spatial and Temporal Variations of the Jovian H_2 Quadrupole Lines," paper presented at the Copernicus Symposium IV, IAU Symposium No. 65, Torun, Poland, September 1973.
14. Hunt, G. E.: *Icarus,* **18**, 637 (1973).
15. Hunt, G. E. and J. S. Margolis: *J. Quant. Spectrosc. Radiat. Transfer,* **13**, 417 (1973).
16. Cruikshank, D. P.: *Bull. Am. Astron. Soc.,* **3**, 282 (1971).
17. Kuiper, G. P., D. P. Cruikshank, and U. Fink: *Bull. Am. Astron. Soc.,* **2**, 235 (1970).
18. Trafton, L. M.: *Bull. Am. Astron. Soc.,* **3**, 282 (1971).
19. Bergstralh, J. T.: *Icarus,* **18**, 605 (1973).
20. Ramanathan, V.: private communication.

HEAT TRANSFER AT THE SOLID-GAS INTERFACE: THERMAL ACCOMMODATION COEFFICIENTS FOR HELIUM ON GAS COVERED TUNGSTEN*

B. J. Jody and S. C. Saxena
University of Illinois at Chicago Circle, Chicago

INTRODUCTION

We have developed a hot wire column instrument capable of determining several thermal properties of gases and metal wires from steady state heat transfer measurements. The instrument and the related experimental details in various stages of development are described in previous papers [1-3]. In this paper, the thermal accommodation coefficient α values for He on gas covered tungsten are given for the temperature range 700-2300 K, Table 1.

Two sets of α values are generated since the analysis of the heat transfer data is carried out by the constant power as well as by the constant temperature difference methods [4]. In the first method, the α values are determined directly against T_e while in the second method, the α values are determined against T_H. The determination of $\alpha(T_H)$ involves the approximation that $T_e \simeq T_H \cdot T_H - T_e$ is proportional to $Q_H/2L$ and to $(2 - \alpha)/2\alpha$. For a gas like helium which has high thermal conductivity and a low α the difference in the two temperatures is large and so appreciable error is likely to creep in $\alpha(T_H)$ values. Therefore, greater reliance is attached to those values determined by the first method, i.e., $\alpha(T_e)$.

In this work, the experimental α values are utilized in conjunction with the knowledge of α for He on clean tungsten and with the available information on the adsorption of gases on tungsten, to characterize the tungsten surface. The measured α values thus correspond to a known gas-solid interface.

---

*The abstract of this paper appears on page 451.

TABLE 1. Experimental values of thermal accommodation coefficients, α, as a function of temperature, T_e or T_H.

T, K	$\alpha(T_e)^a$	$\alpha(T_H)^b$
700	0.0274	0.0163
800	0.0305	0.0190
900	0.0340	0.0220
1000	0.0390	0.0250
1100	0.0438	0.0281
1200	0.0486	0.0312
1300	0.0522	0.0344
1400		0.0376
1500		0.0415
1700		0.0484
1900		0.0540
2100		0.0595
2300		0.0620

[a]The constant power method.
[b]The constant temperature difference method.

DISCUSSION OF α VALUES AND CHARACTERIZATION OF GAS–METAL INTERFACE

During the annealing of tungsten wire in vacuum, 2×10^{-5} mm of mercury, almost all of the gases adsorbed on the wire will desorb. The interaction of the residual oxygen present in the annulus with the carbon impurities in the tungsten at such high temperatures results in partial decarbonization of the wire. The prolonged heating also evaporates the impurities that have lower melting points than tungsten, like silica, which is usually present in tungsten samples [5]. Therefore, at the completion of the annealing process the wire may be regarded as clean and also almost stress free. The gaseous material in the annulus is made up mainly of oxygen, nitrogen, and carbon monoxide which is a product of the decarbonization process. No attempt is made to clean the other cold surfaces, i.e., glass walls and other metal mounting parts of the conductivity column or to purify the helium test gas. The cold surfaces are initially saturated with gas adsorbates at room temperature and one atmosphere pressure of air. The evacuation of the column leads to the desorption of some of the adsorbates like hydrogen [6] and carbon monoxide [7] but does not influence, for instance, the already adsorbed oxygen [7].

The α values given in Table 1 and Fig. 1 are greater than those determined for helium on clean tungsten [8] and predicted on the basis of the theory of Goodman and Wachman [9]. This disparity is attributed to the presence of a gas film adsorbed on the wire and to the contamination of the test gas by the

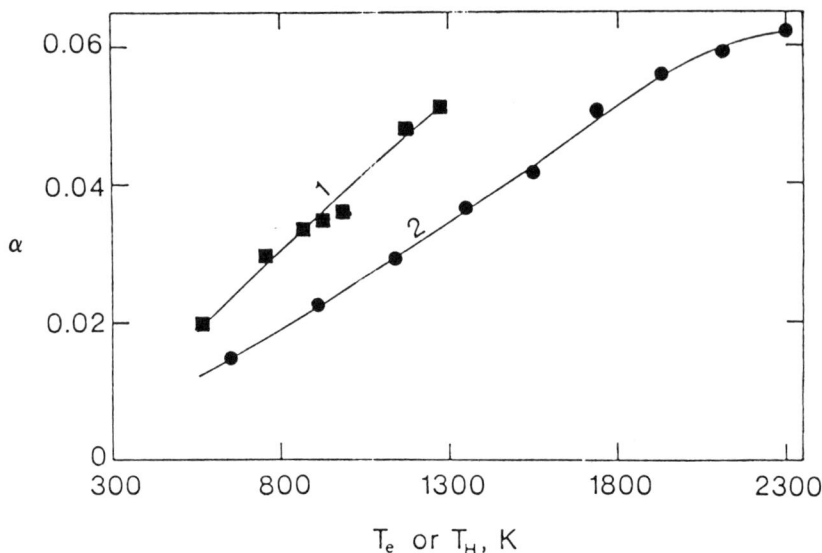

FIG. 1. Variation of thermal accommodation coefficient with temperature for He–W system. Curve 1: Constant power method, values based on the extrapolated gas temperature (T_e). Curve 2: Constant temperature difference method, values based on the tungsten wire temperature (T_H).

desorbing gases from the cold parts of the system. A detailed discussion of the adsorption–desorption of gases on the tungsten surface is given below.

As the annealed tungsten wire is cooled to the cold wall temperature of the column, it will adsorb some of the gaseous molecules present in the environment, mostly oxygen, nitrogen, and carbon monoxide. Oxygen [10, 11] and nitrogen [12] will be adsorbed mostly in the atomic state, while carbon monoxide retains its molecular form upon adsorption [13, 14], and an equilibrium is reached between the adsorbed gases and their partial pressures at room temperature. However, since the available number of molecules of adsorbates is considerably smaller that what is needed for a monolayer coverage, only a small fraction of the wire will be covered. Heating leads to the desorption of some of the adsorbates including atomic oxygen. Desorbing atomic oxygen has the capability to knock out adsorbed gaseous material including oxygen [15] from the glass and other cold parts of the column. In addition, desorption from the cold surfaces will occur as a result of interaction with hot gas molecules. The knocked out gas particles cause an increase in the partial pressures of the corresponding gases in the annulus and thereby increase adsorption on the tungsten. Again, as the wire is heated desorption from the filament takes place and the desorbing oxygen atoms liberate more gas adsorbates from the cold surfaces, leading to an increase in the partial pressure of the adsorbates and therefore more adsorption on the tungsten. This process is repeated every time the temperature of the tungsten wire is increased until a temperature is reached at and above which the sticking coefficients of the adsorbates are so small that no significant adsorption on tungsten is possible. The sticking coefficient is inversely proportional to the temperature as well as to the coverage [16, 17] and it also depends on the crystallographic face of tungsten on which the adsorption takes place [18, 19]. Beyond this temperature, further heating will lead to the desorption of the adsorbates from the tungsten but no readsorption as the partial pressures of the gases increase due to gas particles being liberated from the cold surfaces. Consequently, the desorption from the filament and the cold surfaces will end up in the gaseous phase as contaminants to the test gas.

   The cold surfaces referred to above consist of the glass walls, the copper disks, copper rods, and the steel weights. In addition, the system contains several stopcocks with silicone grease at room temperature. Chang [20] found that the main contaminants on unannealed silicon samples that were chemically etched are oxygen and carbon. The oxygen was about evenly spread over the surface with concentrations in the range 2-5×10^{15} atoms/cm^2. This was interpreted to show that the surface oxygen is adsorbed from the gas phase. On the other hand, the carbon concentration on the surface varied from region to region in the range of 2×10^{14} atoms/cm^2 to 2×10^{15} atoms/cm^2. This suggested that most of the carbon contamination is coming from the bulk silicon by outward diffusion. Interestingly, Chang [20] observed that covering clean silicon with two layers of adsorbed oxygen or even simple exposure to air reduced the surface carbon content. This indicated that adsorbed oxygen blocks the adsorption of carbon-containing gases, like hydrocarbons. Since the glass walls in the column were exposed to air at one atmosphere for long periods of time, it is expected that the walls are covered with multilayers of oxygen and silicon oxides [21-24], and that oxygen whose adsorption is irreversible with pressure [7] is the dominant adsorbate. Further, Chang [20] reported that temperatures of about 1273 K are needed to remove the surface oxygen while temperatures of about 1573 K are needed to reduce the surface carbon content appreciably. Therefore, oxygen is expected to be knocked out from the glass walls much easily than carbon.

   The copper disks and rods are also saturated with adsorbed gases at room temperature from their exposure to atmospheric air. Simmons et al. [25] found that oxygen remained as the major adsorbate on copper surfaces even after prolonged heating at 1023 K and 2×10^{-10} mm of mercury pressure. Pope and Olson [26] investigated the degassing of commercial copper wires over the temperature range 1008 to 1218 K. The desorbing gases were oxygen, carbon monoxide, carbon dioxide, hydrogen, and water vapor. They [26] pointed out that the originally adsorbed species are oxygen and carbon monoxide on the copper and hydrogen which was adsorbed on the metallic impurities in the copper, since copper itself does not adsorb hydrogen [27-33]. Carbon dioxide and water vapor are the results of the reaction of the three originally adsorbed gases on the copper surface. Therefore, only oxygen and carbon monoxide are adsorbed on commercial copper to any appreciable extent. Nitrogen is found not to adsorb on copper [26, 27, 34].

Lee and Farnsworth [27] suggested that at exposures above about 1 X 10^{-4} Torr. min., an amorphous oxygen layer is weakly held on the top of the chemisorbed oxygen structures and that this amorphous layer can be desorbed by heating to 473 K. Dell et al. [33] observed that at 273 K the limited adsorption of 20-25 percent of a monolayer of oxygen is followed by rapid oxidation which can continue until several oxide layers are formed and on the top of which additional adsorption of oxygen occurs. In view of this adsorption mechanism and our experimental conditions, the formation of several oxide layers and a weakly held amorphous top oxygen layer is anticipated. Amorphously adsorbed oxygen will be desorbed upon collision with hot gas particles. Smith and Quets [35] found that at room temperature the adsorption of carbon monoxide on copper is irreversible with pressure, but desorbs at about 473 K. Pope and Olson [26] suggested that the desorbing carbon monoxide reacts with the surface cuprous oxide giving carbon dioxide and oxygen. The liberated oxygen may be adsorbed on the tungsten wire and carbon dioxide will contaminate the test gas.

The weights used in the experimental cell are made of cold roll steel, type 1018. They contain in addition to iron about 0.2 percent carbon, 0.75 percent manganese, 0.05 percent sulfur, and 0.04 percent phosphorous. Iron surfaces have been found to adsorb most of the gases present in the air viz., oxygen, hydrogen, carbon monoxide, carbon dioxide, and nitrogen to various degrees [5, 36-42]. However, oxygen is found to be the dominant adsorbate. Roberts [40] found that iron films heated in the presence of oxygen up to about 296 K were covered with about 8.8 to 10.0 oxide layers. Pignocco and Pellissier [42] observed that even annealing at 873 K in a 10^{-10} mm of mercury pressure did not desorb the oxides and the adsorbed oxygen was irreversible with pressure down to about 10^{-13} mm of mercury.

The adsorption of nitrogen on clean iron at room temperature has been found to proceed to about 0.2 of a monolayer only [38, 39]. Similarly the adsorption of hydrogen at room temperature does not proceed to any significant coverage, specially when its partial pressure is very low. Carbon monoxide does not appreciably adsorb on iron in the presence of oxygen since, as pointed out by Porter and Tompkins [43], oxygen rapidly displaces the adsorbed carbon monoxide. Therefore, the dominant adsorbate on the surface of the weights is expected to be oxygen.

In view of the above comments, multilayer adsorption, specially of oxygen, is anticipated on the cold surfaces of the column where adsorption occurred at 1 atmosphere of air pressure and room temperature. Further, while oxygen can adsorb on all the sites in tungsten [10] forming a complete monolayer, nitrogen covers about 50 percent of a monolayer only when it is in the pure state [44]. Oxygen also replaces some of the nitrogen adsorbed on the tungsten and drastically reduces the adsorption of carbon monoxide on tungsten [44–46]. Thus, it is apparent that adsorbed oxygen is the dominant species on the wire surface. Due to the complexity of the problem of considering more than one adsorbate and in the light of the above comments, we will assume that the adsorbed layer on the tungsten surface is made up of oxygen only in the development of the following semiquantitative analysis.

At a pressure of 2×10^{-5} mm of mercury, the total number of gas molecules present in the annulus in the gas phase N_g, assuming ideal gas behavior, is

$$N_g \simeq 3.72 \times 10^{13} \text{ molecules}$$

The total geometrical surface area of the tungsten wire A is

$$A \simeq 7.66 \text{ cm}^2$$

The tungsten surface has about 10^{15} adsorption sites per cm^2 of the geometrical surface area [5, 10]. Therefore, the total number of gas molecules present in the gaseous phase after pumping and annealing is enough to cover about 0.5 percent of the total surface area of the wire or about 0.04 cm^2 only and since all the molecules present cannot get adsorbed, the wire surface can be considered as fairly clean at this point.

After the wire is cooled the test gas, helium is introduced up to a maximum pressure of about 8.92 cm of mercury. According to the manufacturer, the latter contains impurities of neon (5 ppm) and 1 ppm of each of argon, oxygen, nitrogen, and hydrogen. Thus, with each 1 cm of mercury of helium gas are introduced about 1.68×10^{14} molecules of impurities in the column which will contain about 1.86×10^{13} molecules of each of nitrogen and oxygen. At the highest operating pressure the latter number will be about 1.66×10^{14} molecules. The total number of molecules of oxygen and nitrogen present in the column are 1.74×10^{14} and 1.95×10^{14}, respectively. The maximum coverage of the tungsten wire possible on the basis of these two gases is about 10 percent. Consequently, even at the highest pressure of the test gas about 90 percent of the wire is bare or clean at the ambient temperature.

In the above calculations, no mention is made of the adsorption of hydrogen, neon, and argon, which has only a negligible effect. Hydrogen is effectively replaced by nitrogen [6] and oxygen [47] on tungsten. Neon and argon [48] are physically adsorbed on tungsten below 100 K only. Chemisorption of these two gases is unlikely since they have no tendency to share their valence electrons as their valence orbits are full and will not be disturbed up to temperatures much higher than those reached in the column method. Consequently, it is concluded that neon and argon are not adsorbed on the tungsten in our experiments and their effect, if of any importance, is limited to the contamination of the test gas sample. The tungsten surface is, therefore, to be regarded as made up of bare and gas covered surfaces and the test gas is treated here as a mixture of nonreacting gases. Thus, the experimentally determined α is attributed to all the gases present in the column on all the surfaces at the wire. This α value more appropriately represents an apparent value for the gas-solid system and hereafter is denoted by α_{app}.

A general formula for α_{app} referring to i gases and j surfaces may be written as follows:

$$\alpha_{app} = \sum_{i=1}^{\infty} \sum_{j=1}^{\infty} X_i \alpha_i^j A^j \tag{1}$$

α_i^j is the value of the thermal accommodation coefficient for gas i on a tungsten surface covered with gas j, A^j is the fraction of the tungsten surface covered with species j, and the rest of the symbols are as defined in nomenclature. Assuming no interaction among adsorbed particles,

$$j \leqslant i + 1 \tag{2}$$

For the present system under discussion, we have

$$i = He, O_2, N_2, Ar, H_2, \text{ and Ne}, \tag{3}$$

and $j = B(bare), O(oxygen\ covered), \text{ and } N(nitrogen\ covered)$ $\tag{4}$

Initially, the mole fractions of the gases present are about 1.2×10^{-6} for oxygen, nitrogen, and hydrogen, 1×10^{-6} for argon, 5×10^{-6} for neon and 1 for helium. Except for oxygen and nitrogen, the number of molecules of each of these species in the gas phase stay almost constant throughout the experiment. For simplification, it is assumed that all the nitrogen present is covering a fraction of 0.057 of the surface area and this coverage will stay about constant [12] up to about 1900 K. At high temperatures, the oxygen will knock out some of the nitrogen and will replace it, but because of the small nitrogen coverage we will continue to assume it to be constant. Assuming further that all of the initial oxygen is also adsorbed, the fraction of the surface area covered is about 0.046. Therefore, initially, the fraction of the surface that is bare is at least 0.897.

The oxygen coverage at any temperature is given by

$$A^O(T) = 0.046 + \Delta A^O(T) \tag{5}$$

where $\Delta A^O(T)$ is the change in the fraction of the surface covered by oxygen. Equation (1) for the present system then assumes the following form:

$$\alpha_{\mathrm{app}}(T) = [\alpha_{\mathrm{He}}{}^{B}(T)A^{B}(T) + \alpha_{\mathrm{He}}{}^{O}(T)A^{O}(T) + \alpha_{\mathrm{He}}{}^{N}(T)A^{N}(T)]X_{\mathrm{He}}$$
$$+ [\alpha_{\mathrm{Ne}}{}^{B}(T)A^{B}(T) + \alpha_{\mathrm{Ne}}{}^{O}(T)A^{O}(T) + \alpha_{\mathrm{Ne}}{}^{N}(T)A^{N}(T)]X_{\mathrm{Ne}}$$
$$+ [\alpha_{\mathrm{Ar}}{}^{B}(T)A^{B}(T) + \alpha_{\mathrm{Ar}}{}^{O}(T)A^{O}(T) + \alpha_{\mathrm{Ar}}{}^{N}(T)A^{N}(T)]X_{\mathrm{Ar}}$$
$$+ [\alpha_{\mathrm{H}_2}{}^{B}(T)A^{B}(T) + \alpha_{\mathrm{H}_2}{}^{O}(T)A^{O}(T) + \alpha_{\mathrm{H}_2}{}^{N}(T)A^{N}(T)]X_{\mathrm{H}_2}$$
$$+ [\alpha_{\mathrm{N}_2}{}^{B}(T)A^{B}(T) + \alpha_{\mathrm{N}_2}{}^{O}(T)A^{O}(T) + \alpha_{\mathrm{N}_2}{}^{N}(T)A^{N}(T)]X_{\mathrm{N}_2} \quad (6)$$
$$+ [\alpha_{\mathrm{O}_2}{}^{B}(T)A^{B}(T) + \alpha_{\mathrm{O}_2}{}^{O}(T)A^{O}(T) + \alpha_{\mathrm{O}_2}{}^{N}(T)A^{N}(T)]X_{\mathrm{O}_2}$$

Because of the smallness of the mole fractions of Ne, Ar, H_2, N_2, and also of oxygen at low temperatures while it is still adsorbing on the tungsten surface, Eq. (6) may be truncated so that

$$\alpha_{\mathrm{app}}(T) = \alpha_{\mathrm{He}}{}^{B}(T)A^{B}(T) + \alpha_{\mathrm{He}}{}^{O}(T)A^{O}(T) + \alpha_{\mathrm{He}}{}^{N}(T)A^{N}(T) \quad (7)$$

Goodman and Wachman [9] developed the following expression for the thermal accommodation coefficient for helium on bare tungsten surface:

$$\alpha_{\mathrm{He}}{}^{B}(T) = 1 - \exp\left(-\frac{T_0}{T}\right) + \alpha(\infty)\tanh\frac{(MT)^{1/2}a}{\alpha(\infty)\lambda}\exp\left(-\frac{T_0}{T}\right) \quad (8)$$

where $T_0 = 0.2$, $\alpha(\infty) = 0.05$, $\lambda = 3200$, and $a = 1.64$. Wachman [49] experimentally determined the value of α for helium on atomically chemisorbed oxygen layer on tungsten at 305 K and found that α_{He}^{O} (305 K) = 0.065. He also found α_{He}^{N} (305 K) = 0.040. No values for α_{He}^{O} and α_{He}^{N} are available at higher temperatures to the best of our knowledge.

The rate of desorption of oxygen from the tungsten is very rapid above 1800 K. Singleton [10] found that while A^O is about 0.565 at 1800 K, it drops to about 0.240 at 2100 K and to about 0.072 at 2300 K at an oxygen pressure of about 5×10^{-7} mm of mercury. His values are in good agreement with those of Mazumdar and Wassmuth [50]. To gain some understanding of the temperature dependence of α_{He}^{O}, we employed Eq. (7) without the last term. This is reasonable as the contribution of this term is less than 4 percent. Now at 1800 K, $A^O = 0.565$ and the simplified version of Eq. (7) in conjunction with the experimental values of α_{app} and α_{He}^{B} from Eq. (8) leads to a value of 0.063 for α_{He}^{O} (1800 K). This value is in good agreement (about 3 percent) with the experimental value at 305 K of Wachman [49]. Therefore, it is concluded that α_{He}^{O} is independent of temperature and 0.065 will be adopted as its value for the entire temperature range. Likewise α_{He}^{N} will be considered as temperature independent with a value of 0.040.

The oxygen contamination of the test gas sample though negligible under 1800 K, is sufficient to be accounted for at still higher temperatures. Consequently for temperatures above 1800 K, we employ Eq. (6) in the following form instead of Eq. (7):

$$\alpha_{app}(T) = [\alpha_{He}{}^B(T)A^B(T) + \alpha_{He}{}^O A^O(T) + \alpha_{He}{}^N A^N(T)] X_{He}$$
$$+ [\alpha_{O_2}{}^B(T)A^B(T) + \alpha_{O_2}{}^O A^O(T) + \alpha_{O_2}{}^N A^N(T)] X_{O_2} \qquad (9)$$

We employ the values of A^O as given by Singleton [10], since the pressure in our experiments is close to that existed in his work [10].

Employing Eq. (7), the two sets of experimental thermal accommodation coefficient values, $\alpha_{He}^O = 0.065$, $\alpha_{He}^N = 0.040$, $A^N(T) = 0.057$, $\alpha_{He}^B(T)$ as given by Eq. (8), we get A^O values given in columns 4 and 7 of Table 2. As pointed out earlier, the maximum coverage due to the adsorption of all the oxygen present at room temperature would be equivalent to $A^O = 0.046$. Since all the oxygen can not get adsorbed the true A^O should be smaller and the value at 700 K obtained from α_{app} based on constant power method procedure is consistent with this contention. However, the results obtained from α_{app} values based on constant temperature difference method are not clear at first sight. These α_{app} values are less accurate than those given in column 2 of Table 2 as explained earlier. The α_{app} values are smaller than those predicted by Eq. (8) at temperatures below 1300 K and this is the reason for negative A^O values. Generally, these α_{app} values are smaller than those of column 2 and hence the calculated values of A^O are also smaller.

At temperatures greater than 1800 K, the desorption from the tungsten filament occurs at a faster rate, resulting in enhanced impingement of atomic oxygen on the glass. This results in increased liberation of molecular oxygen from cold surfaces and larger contamination of the test gas. At 2300 K, the wire may be regarded as practically clean and Eq. (9) further simplifies to the following:

$$\alpha_{app}(T) = \alpha_{He}{}^B(T) X_{He} + \alpha_{O_2}{}^B(T) X_{O_2} \qquad (10)$$

and
$$X_{He} + X_{O_2} = 1 \qquad (11)$$

TABLE 2. Computed Values of $A^O(T)$ from Eq. (7)

T, K	α_{app}^a	α_{He}^B	A^O	α_{app}^b	α_{He}^B	A^O
700	0.0274	0.0250	0.003	0.0163	0.0250	−0.275
900	0.0340	0.0275	0.113	0.0220	0.0275	−0.207
1100	0.0438	0.0297	0.335	0.0281	0.0297	−0.110
1300	0.0522	0.0316	0.549	0.0344	0.0316	+0.016
1500				0.0415	0.0333	0.187
1700				0.0484	0.0345	0.381
1800				0.0512	0.0353	0.459

[a]Experimental values based on the constant power method procedure.
[b]Experimental values based on the constant temperature difference method procedure.

Amdur and Guildner [51] report a value of 0.905 for thermal accommodation coefficient of oxygen on gas covered tungsten. We assume $\alpha_{O_2}^B(T) = 0.90$. This corresponds to $X_{O_2} = 0.028$, which implies a pressure of 1 to 2 mm of mercury in the helium pressure range of 4.3 to 8.9 cm of mercury. This contamination is consistent with our column design as elaborated below. The surface area of the glass walls is about 290 cm² and that of the mounting metal parts (made of steel and copper) is about 120 cm². Thus, we have a total surface area of about 410 cm² and the greased joints all saturated with multilayers of adsorbed gases. Two mm of mercury pressure of oxygen in the column corresponds to about 7×10^{-7} moles or about 4.2×10^{17} molecules. Following the work of Chang [20], desorption of about 57 percent of the adsorbed oxygen on the glass walls will provide enough oxygen to account for the required impurities in the test gas discussed above. The analysis completely substantiates the qualitative appropriateness of the proposed model as described by Eq. (1).

In the temperature range 1800 K $< T <$ 2200 K, some oxygen will be in the gas phase and some still adsorbed on the tungsten wire. Thus Eq. (1) simplifies to:

$$\alpha_{app}(T) = [\alpha_{He}^B(T)A^B(T) + \alpha_{He}^O(T)A^O(T)]X_{He} \\ + [\alpha_{O_2}^B(T)A^B(T) + \alpha_{O_2}^O(T)A^O(T)]X_{O_2} \tag{12}$$

Assuming A^O values are given by Singleton [10] in this temperature range, Eq. (12) in conjunction with experimental values of $\alpha_{app}(T)$, $\alpha_O^B = 0.90$, and $\alpha_{O_2}^O = 0.905$ leads to the X_{O_2} values given in the last column of Table 3. The A^O values employed here may be smaller than their actual values because the partial pressure of oxygen in Singleton's work [10] was smaller than that existed in our column during measurements. This implies that actual X_{O_2} will be somewhat smaller than the listed values in Table 3.

The three sets of A^O values, as determined from the α values of the present work, and those of Singleton [10], are plotted in Fig. 2. Also shown in this

TABLE 3. Computed Values of X_{O_2} in the Test Gas from Eq. (12)

T, K	α_{app}^b	α_{He}^B	A^O	X_{O_2}
1900	0.0540	0.0358	0.43	0.006
2000	0.0570	0.0365	0.34	0.013
2100	0.0595	0.0370	0.24	0.018
2200	0.0607	0.0375	0.15	0.023

^{b}The constant temperature difference method values.

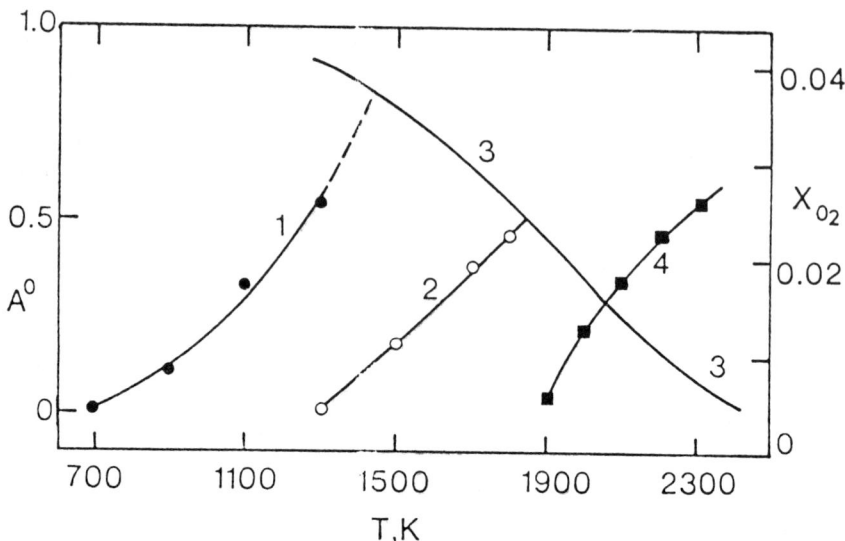

FIG. 2. Dependence of A^O and X_{O_2} on temperature. Curves 1, 2, and 3 are based on $\alpha(T_e)$, $\alpha(T_H)$, and Singleton's work [10]. Curve 4 is the oxygen mole fraction in the gas phase and is based on $\alpha(T_H)$.

figure is a plot of X_{O_2} against temperature based on the α values obtained from the constant temperature difference method.

NOMENCLATURE

A total geometrical surface area of the tungsten wire [cm²]
A^j fraction of the wire surface covered with gas j
a constant in Eq. (8) and is a parameter of the Morse potential [10^{-8} cm]
M molecular weight [g/g mole]
N_g total number of gas molecules present in the column
Q_H power conducted through the gas [W/cm]
T temperature [deg K]
T_e linearly extrapolated temperature of the gas on the surface of the hot wire [deg K]
T_H temperature of the hot wire [deg K]
T_0 parameter in Eq. (8) [deg K]
X_i mole fraction of gas i in the column

Greek Letters

α thermal accommodation coefficients
α_{app} the experimental value of the accommodation coefficient referring to the mixture of gases present in the column on all the surfaces of the hot wire
α_i^j accommodation coefficient of gas i on a surface covered with gas j
$\alpha(\infty)$ constant in Eq. (8)
λ constant in Eq. (8) and its value depends upon the solid surface [10^{-8} g½K½/g mole cm²]

Subscript

i component of the gas mixture

Superscripts

B	bare surface
j	surface covered with gas j
N	surface covered with nitrogen atoms
O	surface covered with oxygen atoms

ACKNOWLEDGMENT

This work is supported by the United States National Science Foundation (Grant No. ENG 75-02992) under a USA–USSR cooperative research program.

REFERENCES

1. Chen, S. H. P. and S. C. Saxena: *High Temp. Sci.*, 1973, **5**, 206.
2. Jody, B. J. and S. C. Saxena: *Phys. Fluids*, 1975, **18**, 20.
3. Jody, B. J., P. C. Jain, and S. C. Saxena: *J. Heat Transfer, Trans. ASME, Series C*, 1975, **97**, 605.
4. Chen, S. H. P. and S. C. Saxena: *Int. J. Heat Mass Transfer*, 1974, **17**, 185.
5. Hayward, D. A. and B. M. W. Trapnell: B. M. W., Chemisorption, Butterworths, London, 1964.
6. Rigby, L. J.: *Canad. J. Phys.*, 1965, **43**, 1020.
7. Schlier, R. E.: *J. Appl. Phys.*, 1958, **29**, 1162.
8. Watt, W. and R. Moreton: 1964, Royal Aircraft Establishment RAE-TN-CPM-80, 21 pp.
9. Goodman, F. O. and H. Y. Wachman: *J. Chem. Phys.*, 1967, **46**, 2376.
10. Singleton, J.: *Ibid.*, 1967, **47**, 73.
11. McCarroll, B.: *Ibid.*: 1967, **46**, 863.
12. Ehrlich, G.: *Ibid.*, 1961, **34**, 29.
13. Ehrlich, G.: *Ibid.*, 1961, **34**, 39.
14. Ustinov, Yu. K., V. N. Ageev, and N. I. Ionov: *Soviet Phys. Tech. Phys.*, 1965, **10**, 851 [*Zh. Tekhn. Fiz.*, 1965, **35**, 1106].
15. Ageev, V. N. and N. I. Ionov: *Progress in Surface Sci.*, 1974, **5**, 118.
16. Ageiken, V. S. and Yu. G. Ptushinskii: *Ukr. Fiz. Zh.*, 1967, **12**, 1483.
17. Ageiken, V. S. and Yu. G. Ptushinskii: *Ibid.*, 1968, **13**, 779.
18. Becker, J. A. and R. G. Brandes: *J. Chem. Phys.*, 1955, **23**, 1323.
19. Müller, E. W.: *Z. Electrochemie*, 1955, **59**, 372.
20. Chang, C. C.: *Surf. Sci.*, 1970, **23**, 283.
21. Law, J. T.: *J. Phys. Chem. Solids*, 1958, **4**, 91.
22. Wolsky, S. P.: *Ibid.*, 1959, **8**, 114.
23. Carosella, C. A. and J. Comas: *Surf. Sci.*, 1969, **15**, 303.
24. Meyer, F. and J. J. Vrakking: *Ibid.*, 1973, **38**, 275.
25. Simmons, G. W., D. F. Mitchell, and K. R. Lawless: *Ibid.*, 1967, **8**, 130.
26. Pope, L. E. and F. A. Olson: *Ibid.*, 1972, **32**, 16.
27. Lee, R. N. and H. E. Farnsworth: *Ibid.*, 1965, **3**, 461.
28. Ponec, V., Z. Knor, and S. Cerny': *J. Catalysis*, 1965, **4**, 485.
29. Pitchard, J. and F. C. Thompkins: *Trans. Faraday Soc.*, 1960, **56**, 540.
30. Pitchard, J.: *Ibid.*, 1963, **59**, 437.
31. Trapnell, B. M. W.: "Chemisorption," Academic Press, 1955.
32. Blodgett, K.: *J. Chem. Phys.*, 1958, **29**, 39.
33. Dell, R. M., F. S. Stone, and P. F. Tiley: *Trans. Faraday Soc.*, 1953, **49**, 195.
34. Trapnell, B. M. W.: *Proc. Roy. Soc. (London)*, 1953, **A218**, 566.
35. Smith, A. W. and J. M. Quets: *J. Catalysis*, 1965, **4**, 163.
36. Brennan, D., D. O. Hayward, and B. M. W. Trapnell: *Proc. Roy. Soc.*, 1960, **A256**, 81.

37. Baker, F. S., A. M. Bradshaw, J. Britchard, and K. W. Sykes: *Surf. Sci.*, 1968, **12**, 426.
38. Anderson, J. R.: "Chemisorption and Reactions on Metallic Films," Vol. 1, Academic Press, 1971.
39. Bagg, J. and F. C. Tompkins: *Trans. Faraday Soc.*, 1955, **51**, 1071.
40. Roberts, M. W.: *Ibid.*, 1961, **57**, 99.
41. Quin, C. M. and M. W. Roberts: *Ibid.*, 1964, **60**, 899.
42. Pignocco, A. J. and G. E. Pellisier: *Surf. Sci.*, 1967, **7**, 261.
43. Porter, A. S. and F. C. Tompkins: *Proc. Roy. Soc., (London)*, 1953, **A217**, 544.
44. Rigby, L. J.: *Canad. J. Phys.*, 1964, **42**, 1256.
45. Ageiken, V. S. and Yu. G. Ptushinskii: *Soviet Phys. Solid State*, 1969, **10**, 1698 [*Fiz. Tverd. Tela*, 1968, **10**, 2168].
46. Ageiken, V. S., Yu. G. Ptushinskii, and B. P. Polozov: *Ibid.*, 1970, **12**, 176 [*Ibid.*, 1970, **12**, 176].
47. Roberts, J. K.: *Proc. Roy. Soc. (London)*, 1935, **A152**, 477.
48. Ehrlich, G. and F. G. Hudda: *J. Chem. Phys.*, 1959, **30**, 493.
49. Wachman, H. Y.: Ph.D. Thesis, University of Missouri, Columbia and Rolla, Missouri, 1957.
50. Mazumdar, A. K. and H. W. Wassmuth: *Surface Sci.*, 1972, **30**, 617.
51. Amdur, I. and L. A. Guildner: *J. Am. Chem. Soc.*, 1957, **79**, 311.

THREE NEW DEVICES FOR MEASURING
HEAT TRANSFER IN BUILDINGS*

Ivo Augusta
R&D Institute of the Prague Building Trust
Prague, Czechoslovakia

INTRODUCTION

The thermotechnical laboratories of the Research and Development Institute of the Prague Building Trust have been concerned with the introduction of thin and light outer building walls during recent years. Special attention has also been given to measuring and defining the room microclimate, with respect to modern concepts of maintaining biological comfort.

In order to obtain more experimental data on the behavior and functioning of modern buildings, new instruments had to be developed and utilized. Some of these are described in this paper.

First of all, a multipurpose thermoconductometer was designed to measure samples of real structures in steady, quasi-steady, and unsteady states under laboratory conditions. Its main purposes are as follows:

1. Determination of the steady-state thermal resistance of the wall.
2. Determination of the temperature wave attenuation ($\nu = A_e/A_{ip}$) and of the phase lag (ϵ) of harmonic temperature waves, passing through the measured wall from the outer air to its inner surface.
3. Determination of the cooling rate of the element from the steady state, when an adiabatic condition is applied at the warmer surface of the element. The characteristical parameter achieved by this type of test is called "the relative descent of the inner surface temperature ϑ."

In contrast with the resistance R, the parameters ν, ϵ, and ϑ have not been frequently experimentally determined, though analytical methods of evaluating all these three parameters (ν, ϵ, ϑ) have been previously described. Practice indicates, especially for multilayer walls and thin and geometrically

---

*The abstract of this paper appears on page 461.

complicated elements, that direct measurements are an effective complement to analysis. All of the above mentioned parameters (their maximum or minimum allowed magnitudes) are prescribed by the Czechoslovak Standard CSN 730540 [1].

THE MULTIPURPOSE THERMOCONDUCTOMETER

The thermoconductometer, AVH-2, was designed as a double-box apparatus with separate controls. In comparison with the usual conductometers it differs in three main aspects. They are:

1. The air temperature in one or both boxes can be regulated. The harmonic variation of the outer-input air temperature t_e, when measuring the ν and ϵ parameters, or the descending course of the inner-air temperature t_i, when measuring the ϑ parameter, are controlled automatically. During the ν and ϵ tests, the input side temperature is imposed on the system, while the output (inner-indoor) air temperature t_i is held constant. The mean input air temperature is commonly higher than the output one. When measuring the relative descent ϑ, higher temperature can be applied either at the inner or the outer wall surfaces, so that the sample can respond in opposite directions during two different tests. This is important with non-homogeneous or asymmetrical elements.

2. When measuring the R value, the effect of the heat-transfer coefficients α can be neglected since they are acting as separable series connected resistances. In the case of steady periodic and transient heat flow conditions, the interdependence of R and α must be considered. Therefore both boxes of the AVH-2 equipment are provided with adjustable air driving systems. In the prototype set, a vertical convective air stream was applied to the surfaces. In later models, a highly modernized redeveloped system using Roots ventilators for a more sensitive regulation of the α factors was utilized. The air velocities in the prototype set, as well as in the new models, called MTV-50, are variable over a wide range and can be adjusted automatically by means of an α-meter device, which is described below.

    The real α values can be measured directly or they can be evaluated "a posteriori" from the record of the mean $(t_e - t_{ep})$ and $(t_{ip} - t_i)$ temperature differences.

3. For the purpose of measuring the inner surface temperature descent the adiabatic condition at the input surface must be realized. This means that the input surface heat flux during the descent period must be zero. This adiabatic condition is maintained by a special electronic control device, consisting of a surface heat flux meter (thermotransitometer), linked by feedback to the heating controls.

In order to fulfill all the above described functions, the input (hot) box of the thermoconductometer is provided with automatically programmed heating and cooling systems, capable of maintaining a harmonic variation (the period being normally 24 hrs). The temperature changes may also be other than sinusoidal, e.g., irregular, but always smooth (without sudden variations). The mean input air temperature during the ν and ϵ tests is normally maintained at 42.5°C, while the A_0 amplitude is fixed at ±17.5°C. The device also permits a steeper, or a gentler mean gradient. During this type of test the output air temperature, t_i, is maintained near 273°K to reduce the external heat losses. Since the attenuation ability of various samples differs, within broad limits, it is necessary to adjust the cooling system for each separate test.

When measuring the R value, the orientation of the sample is not important and the steady gradient $(t_e - t_i)$ is adjusted to 298°K-258°K. The resistance may be checked directly by means of the surface heat-flux meter, attached to the sample, or it also may be determined indirectly from the electrical energy consumption needed for heating the hot measuring box. For this purpose the inner hot box is framed by a compensating space. The condition of zero heat flux between these two spaces (the measuring and the protective chambers) is checked by heat flux meters.

When measuring the R value, the conditioning system in the measuring hot box is closed. During the ν, ϵ, and ϑ test it is "opened" and the heat flux through the measured sample can be determined only by means of the surface heat flux meter.

During a period of nine years, a long series of tests under very different conditions has been performed. No difficulty was found in generating the harmonic variation of the input air temperature. The deviations from the harmonic functions did not exceed ±0.8 K with the heavy apparatus, and ±0.25 K with the light apparatus. The reproducibility of the maximum and minimum points, however, was much better, being less than ±0.2 K. In the present setup, a new continuous electronic control of the input air temperature is under development which should reduce the deviations to less than ±0.1 K.

It is rather troublesome to adjust the prescribed α values in both boxes, especially during the initial tests. (The CSN Standard prescribes $\alpha_i = 8.14$ W/m²K and $\alpha_e = 23.26$ W/m² K for the winter and $\alpha_e = 15.1$ W/m²K for the summer season). With the prototype equipment, the deviations of the experimental values were a maximum of ±40 percent from the prescribed ones. In recent tests, a checking of the propellers' revolutions by means of an electronic stroboscope reduced the differences to not more then ±30 percent. It appears possible that with help of the direct α-meter, the correspondence between the specified and adjusted α coefficients could be better. However, especially at the inner surface, ($\alpha_i = 8.14$ W/m²K) a smaller difference than ±20 percent, requires a highly sensitive device. A mathematical analysis of the influence of the magnitudes of α_i and α_e on ν, nevertheless, shows that the deviations should not exceed $\alpha_i = \pm 9$ percent and $\alpha_e = \pm 12$ percent. However, for direct comparison of the experimentally achieved ν value with the analytically calculated ν factor, it is better that α_i should be kept within the cited limit. The test can then be repeated with two or more different α_e values to bracket the correct value.

For quick information on whether the sample does or does not comply with the Standard ($\nu \geqslant 8.34$ for winter), the α factors should be adjusted to

$$\alpha_i < 8.14 \text{ W/m}^2\text{K}$$
$$\alpha_e > 23.26 \text{ W/m}^2\text{K}$$

First, the experimental results were compared with computed values of ν when the actual test values of α were used in the calculation. The measured ϑ values were also compared with values obtained when the Standard prescribed α values were used. In spite of the deviations in the α factors, the difference between the calculated ϑ factors and the measured ones did not exceed ± 20 percent. Similar correspondence was achieved in examining the phase lag.

Experience with the prototype of the thermoconductometer, although its development was accompanied by a series of complications, showed that the experimental investigation of the steady periodic and transient state characteristics was effective and that laboratory methods proved to be applicable. The apparatus and both methods are protected by a CS patent.

THE α-INDICATOR

For the purpose of measuring α values, a new temperature-compensated α-indicator has been developed. In contrast to other models of such sensors, this indicator has been designed to be responsive to both the convective and the radiative components of the surface heat transfer. The resulting α-meter consists of two sensors. The temperature of one is maintained at the environmental temperature. The second sensor is heated to a constant higher temperature and the temperature difference is maintained to an accuracy of ± 0.001 K. The heat losses from the second sensor are measured and indicate the deficiency caused by both types of the surface heat transfer.

The α-meter has an operating range of 0–50 W/m²K with a resolution of 0.5 W/m²K. This high resolution, and the accuracy of maintaining the temperature difference, has been achieved by means of rather sophisticated electronic differential amplifiers and multipliers. The device is equipped with a linear recorder and can be employed in several ways. It has been installed in the thermoconductometer for automatic feedback control of the α factors, and for checking the distribution and homogeneity of the α values inside the boxes.

The α-meter can also be used for direct measurements as a portable apparatus. It can also be modified to measure only the radiative component, which is not too large under the room conditions. It is more complicated to shield the thermal radiation, so that the device measures only the convective component of heat transfer, but this is also possible. The sensor may either be installed in an open space or it can be attached to a wall surface.

During the past year, a series of measurements inside rooms, in corners, near windows, etc., have been carried out. The results achieved are rather interesting in that they have resulted in new information about the distribution of α values near walls, openings, etc. The α-wattmeter (designed by Dr. Tydlitát, J. Jadavan, and the author) is protected by a patent.

THE HEAD-MODEL HEAT-LOSS SENSOR

The third apparatus developed recently is an artificial head designed on principles similar to the α-meter. This equipment was designed pursuant to a request of the Technical University of Bratislava, Civil Engineering Faculty, Chair of Building Physics. The principal component is a spherical sensor with a diameter of 100 mm, freely hanging in space. This metal, cloth-covered, sphere is divided into four independent vertical segments and consequently it is oriented directionally. The sensor is heated to a constant temperature of 36°C. The heat input (equal to the heat losses) can be determined in the entire head, or in all the four segments independently, in two ranges: either 0–300 W/m^2K or 50–150 W/m^2K. The resolution is less than 1 W/m^2K. The sensor reacts to temperature differences of 0.005 K and to very slight changes of radiation, so that the effect of a lighted cigarette or of blowing smoke in a normal room is distinguishable. The sensor can be fixed in a stable position, or can be moved automatically vertically between 300 and 1,800 mm above the floor by means of a special support. The main purpose of this equipment is to study the effect of the thermal environment, e.g., large windows, the influence of furnaces, draughts, etc., with respect to the orientation.

NOMENCLATURE

R Thermal resistance $m^2 K/W$

ν Attenuation of the harmonic temperature wave. The ratio of the outer air temperature amplitude A_e to the inner surface temperature amplitude A_{ip}. Dimensionless

A_e Amplitude of the outer air temperature changes [K]

A_{ip} Amplitude of the inner surface temperature changes [K]

ϵ phase lag of the harmonic temperature wave, passing from the outer air to the inner surface of the outer wall [sec]

ϑ Relative descent of the inner surface temperature. The ratio of the inner surface-outer air temperature difference after 8 hr of adiabatic cooling to the same temperature difference, existing at the beginning of the descent, dimensionless

α Surface heat transfer coefficient/film/factor [$W/m^2 K$]

t_i Inner air temperature [K]

t_{ip} Inner surface temperature [K]

t_{ep} Outer surface temperature [K]

t_e Outer air temperature [K]
α_i Inner surface heat transfer coefficient [W/m² K]
α_e Outer surface heat transfer coefficient [W/m² K]

REFERENCES

1. CSN 730540: Standard Specification, Thermotechnical properties of constructions.
2. Augusta, I.: Konstrukce zarizeni pro tepelne technicka mereni sten, systemem AVH-2. The Final Report of the Programme CC 3008/1965-C2052/1966, SZP-VVU, Praha, 1966.
3. Augusta, I.: Studium sireni a utlumu harmonickych teplotnich vln stavebnimi konstrukcemi. Internal Research Report, VVU–SZP, Praha, 1974.
4. Tydlitat, V.: Mereni stacionarnich a dynamickych tepelnych vlastnosti dvou druhu zdiva z tvarovek CDK-Fortuna. Internal Research Report VVU–SZP, Praha, 1972.
5. Augusta, I.: A new conductometer for studying thermotechnical properties of building constructions. Paper presented at the XIV International Congress of Refrigeration, Moscow, 1975.

ABSTRACTS

1. CONVECTIVE HEAT AND MASS TRANSFER

1.1 Investigation of Natural Convection Heat Transfer in Gas-Filled Porous Media at High Pressures

V. S. Avduyevsky, V. N. Kalashnik, and R. M. Kopyatkevich
Moscow

At large Rayleigh numbers, using the boundary layer approach, a solution is given of the problem of natural convection near a vertical nonisothermal flat plate and isothermal surfaces of a cylinder and sphere bounded by a porous medium. The solution is obtained by numerical and approximate integral methods, and criterial relations for heat transfer are found. Experimental data on the thermal conductivity of a porous-insulation flat layer are given.

English translation will appear in Heat Transfer—*Soviet Research*

1.2 A Study of the Nonisothermal Near-Wake behind an Axisymmetric Self-Propelled Body

V. E. Aerov, B. A. Kolovandin, G. G. Starobinets, N. N. Luchko, and Yu. M. Dmitrenko
Luikov Heat and Mass Transfer Institute, Minsk

A model is developed for predicting, with sufficient accuracy, nonisothermal turbulent jet flows with zero excess momentum by a prescribed initial distribution of turbulence characteristics. An experimental study is made in the near-wake behind an axisymmetric self-propelled body to formulate the initial conditions and to check the validity of the predictions performed by the above model.

English translation will appear in Heat Transfer—*Soviet Research*

1.3 On Convective Heat Transfer with Turbulent Flow in Channels of a Superconducting Cable (SCC)

I. T. Aladiev, K. D. Voskresenskiy, E. S. Turilina, and A. A. Ivlev
Krzhizhanovskiy Power Engineering Institute, Moscow

This paper deals with the problems of heat transfer in the channels of a superconducting cable on the basis of semiempirical concepts. Rather simple expressions were obtained for the heat transfer to liquid helium. The results predicted on state parameters in the supercritical region are compared with experiment.

English translation will appear in Heat Transfer—*Soviet Research*

1.4 The Influence of the Temperature Factor on Heat Transfer Involving Turbulent High-Temperature Gas Flows in the Engrance Region of a Circular Tube

A. Ambrazevicius, P. Valatkevicius, and P. M. Kezelis
Institute of Physical and Technical Problems of Energetics
Lithuanian Academy of Sciences, Kaunas

Experimental results on local turbulent heat transfer in a tube 21.5 mm dia with cold walls and $1 < 1/d < 16$ at temperatures of air or nitrogen from 500 to 6000 K have shown that for all peripheral angles at the tube entrance (0, 45, 90, 180°) the temperature factor (the ratio of gas temperatures on the tube axis to wall temperatures which is less than or equal to 15) does not appreciably influence the heat transfer. The measured gas velocities and temperatures along the axis of the entrance region are shown to be constant for $1/d < 10$. Therefore, the experimental data were treated in a similar manner as the external problem. Relations to calculate local heat transfer in the entrance region of the tube with various entrance conditions are given.

English translation will appear in Heat Transfer—*Soviet Research*

1.5 Electrical Convection in Dielectric Liquids

M. S. Apfelbaum, T. N. Baranova, A. M. Severov, N. O. Skuratovskiy, and F. I. Yantovskiy
Moscow

The jet flow of dielectric liquids from a high-voltage electrode with heat exchange (electrical convection) is considered. Using electrical field-dependent thermal conductivity, the dependence of the velocities, temperatures, and Nusselt numbers on the voltage supplied and the electrical properties of a liquid is obtained. Jets above a free surface as well as those immersed in a transparent cell fixed in the projector were observed.

English translation will appear in Heat Transfer—*Soviet Research*

1.6 The Combined Effect of Mach and Reynolds Numbers on the Halfcone Leeward Heat Exchange in a Hypersonic Flow

M. M. Ardasheva, V. Ya. Borovoy, P. I. Gorenbukh, and M. V. Ryzhkova

The Zhukovsky Central Aerohydrodynamic Institute, Moscow

The data for heat exchange on the flat surface of a blunt halfcone in helium flow at $\alpha = 0$ and large Mach numbers (up to 23.5) are presented. It is shown that at large Mach numbers on the plane (upper) surface of the halfcone, narrow regions of intensive heat exchange appear, provided the Reynolds number is sufficiently large. The measured maximum heat transfer coefficients for wide ranges of the Mach and Reynolds numbers ($M_\infty = 3 \div 23.5$ and $Re_{\infty_a} = 10^4 \div 4 \times 10^5$ is a bluntness radius) are correlated.

1.7 An Experimental Study of Recirculation Currents in Sodium

J. C. Astegiano, D. Grand, R. Martin, and Ph. Vernier

Centre de TRI, Grenoble, France

The paper presents the results of an experimental study of mixed convection (recirculation current) of liquid sodium in a two-dimensional rectangular cavity, whose three walls were heated with a constant heat flux, in the range of the Reynolds number $10^4 <$ Re $< 10^5$ and the mixed convection parameter $0.1 < G/Re^2 < 12$ (G is the Grashof number). If $G/Re^2 > 1$, then the molecular diffusion was found to dominate the exchange mechanism.

1.8 Evaluation of Measurements of the Local and Total Heat Transfer from Smooth and Rough Surface Cylinders in Cross Flow

E. Achenbach

Institut für Reaktorbauelmente, Jülich, West Germany

The flow associated with local and total heat transfer from smooth and rough circular cylinders in an air cross flow has been examined in the Reynolds number range $2 \cdot 10^4 <$ Re $< 4 \cdot 10^6$. The roughness parameter k_s/d was varied over a range of $0 < k_s/d < 900 \cdot 10^{-5}$. Experimental results are presented on local static pressure, local heat transfer coefficients, total heat transfer, and flow resistance. By increasing the roughness parameter, heat transfer is improved by a factor of 2.7. Transition from a laminar to a turbulent boundary layer, as well as the percentage contribution of heat, transferred from the front portion, to the total heat, is considered.

This paper appears on page 1, this volume

1.9 Heat and Mass Transfer with Free Convection in Closed Vessels Partially Filled with Water

N. U. Bakirov, G. M. Solov'yev, I. Kh. Khayrullin, and I. A. Tsvetkov
Kazan Chemical Engineering Institute

Optical and thermocouple measurements are used to analyze processes in the liquid and vapor-gas phases of water and *i*-octane heated from the walls. Correlations are given for calculation both of the pressure rise in the container and of the temperature increase on the free liquid surface.

English translation will appear in Heat Transfer—*Soviet Research*

1.10 The Peculiarities of Heat Transfer in a Turbulent Boundary Layer on a Porous Plate and in a Film Cooling Region

Yu. V. Baryshev, A. I. Leontyev, and N. K. Peyker
Institute of High Temperatures, USSR Academy of Sciences, Moscow

Results are presented of an experimental study on the effectiveness of the film cooling region and heat transfer behind different-length porous plates, in a subsonic turbulent boundary layer.

English translation will appear in Heat Transfer—*Soviet Research*

1.11 Thermal Convection of a Ferromagnetic Fluid in a Nonuniform Magnetic Field

V. G. Bashtovoy, S. V. Isayev, M. P. Pavlinov, V. K. Polevikov, and V. E. Fertman
Luikov Heat and Mass Transfer Institute, Minsk

The criteria for heat transfer in a ferromagnetic fluid are introduced. The conditions, with regard to magnetic field disturbances, are obtained. The numerical and experimental results show that ferromagnetic fluid-controlled cooling of current conductors is possible.

English translation will appear in Heat Transfer—*Soviet Research*

1.12 A Study of Heat Transfer, with a Two-Dimensional Gas Jet Falling Downwards, onto an Obstacle

V. Yu. Bezuglyi, and N. M. Belyaev
Dnepropetrovsk State University

A numerical study is made of heat transfer between a two-dimensional impact gas jet, flowing out of a hole at low subsonic velocities, and a flat obstacle. A difference scheme, based on disturbed difference operators, approximates a system of elliptical partial differential equations of a second-order of accuracy. The difference boundary-value problem is solved by the Gauss-Seidel method. An analysis is made on the effect of the temperature factor, the relative distance from the outlet section of the hole to the obstacle, and other similarity criteria on the intensity of heat exchange between the jet and the obstacle.

1.13 Hydrodynamics and Heat Exchange between a Low-Temperature Plasma Jet and a Flat Plate

I. A. Belov, I. P. Ginzburg, G. F. Gorshkov, and V. S. Terpigor'yev
Leningrad Institute of Mechanics

Some results of an experimental investigation of gas dynamic and heat transfer parameters of free and impinging low-temperature plasma jets are given. It is shown that for free jets, the profiles of momentum and heat content are similar. An expression for the stagnation-point velocity gradient is also presented. The latter value is then used to calculate stagnation-point heat flux, taking into account the effect of jet turbulence on heat transfer.

English translation will appear in Heat Transfer–*Soviet Research*

1.14 An Investigation of Heat Transfer in a Reverse Flow Region near a Flat Plate Subjected to Normal Impingement by a Composite Jet

I. A. Belov, I. P. Ginzburg, G. F. Gorshkov, and V. S. Terpigor'yev
Leningrad Institute of Mechanics

The results of an experimental investigation of the dynamic and heat transfer characteristics of a composite jet with an outer velocity maximum which impinges normally on an infinite flat plate, are presented. It is shown, that compared with exit uniform jet flow impingement, such a jet interaction with the plate is characterized by elevated heat fluxes.

1.15 Some Problems of Convective Heat Transfer in a Ferromagnetic Fluid

B. M. Berkovskiy, and A. N. Vislovich
Luikov Heat and Mass Transfer Institute, Minsk

Within the framework of a ferromagnetic-fluid model with an asymmetric stress tensor, an expression is obtained for the mass force due to a dissipative part of the magnetization vector in the case of slowly alternating fields. A study is made of the effect of uniform steady and rotating fields on convective heat transfer in a vertical channel.

English translation will appear in Heat Transfer—*Soviet Research*

1.16 An Experimental Study of the Turbulent Heat and Mass Transfer Rate in the Starting Length of a Tube at High Prandtl Number

R. D. Borisova, A. A. Gukhman, V. V. Dilman, and B. A. Kader
State Institute for Nitrogen Industry, Moscow Chemical Engineering Institute

Electrochemical techniques have been used to study turbulent mass transfer in the starting length of a circular duct. Experiments were made at Prandtl numbers from 600 to 36,000. Reynolds number from $5 \cdot 10^3$ to $200 \cdot 10^3$ and $1/d$ from $3 \cdot 10^{-3}$ to 8. Close agreement was obtained between the predicted and the experimental results for rates of turbulent transfer at the starting length for $Pr \gg 1$. The experimental results showed that the Nusselt number varies with 1/3 power of the Prandtl number.

English translation will appear in Heat Transfer—*Soviet Research*

1.17 A Study of Combined Free and Forced Convection from a Vertical Porous Plate

P. M. Brdlik and V. G. Sugak
Moscow Wood Technology Institute

The method of integral relations is used to solve the differential equations which describe combined convection near flat surfaces, with variation of the free (stream) velocity, injection (suction) and surface temperature: $U_0 \sim x^m$, $V_w \sim x^m$, $T_w \sim T_0 \sim x^n$ (x is the distance along the plate from a leading edge). The analytical relationships of the output characteristics of a boundary layer are obtained when $m = (n + 1)/2 = 2m + 1$ for a vertical and $m = (2n + 1)/5 = 2m + 1$ for a horizontal surface. The results are compared with those of other authors.

1.18 The Effect of Heat Transfer on the Stability and Transition of Freon and Water Boundary Layers

A. P. Wazzan, H. Taghavi, and W. C. Hsu
University of California, Los Angeles

The paper presents the results of an experimental study of the effect of heat transfer on transition from a laminar boundary layer flow to a turbulent one for water and Freon-14. The influence of the wall temperature on the critical Reynolds number R_c is analyzed for flows with Hartree β = 1.0, 0.0, and −0.1988. Predictions are made based on the linear instability theory of Tollmien-Schlichting. The variation of the critical Reynolds number with the heating of water and Freon is explained by disturbance energy calculations.

1.19 Heat Transfer in Separated Flows at Low Reynolds Numbers

A. A. Vasil'yev and P. G. Itin
Moscow

An experimental study is made of the flow pattern and heat flux distribution of a supersonic rarefied gas flow past a flat plate with projections in the form of cylinders and protrusions at h/d = 0.08 ÷ 20. Relationships for separation zone lengths and for the shape of lines of maximum heat fluxes are obtained.

English translation will appear in Heat Transfer—*Soviet Research*

1.20 The Calculation of Pulsation Heat Fluxes and the Rate of Temperature Fluctuations in Turbulent Nonisothermal Incompressible Liquid Flows in a Rotating Cylindrical Tube

I. A. Vatutin, O. G. Martynenko, and I. V. Skutova
Luikov Heat and Mass Transfer Institute, Minsk

A theoretical study is made of the effect of a cylindrical tube rotation relative to its axis on the distribution of pulsation heat fluxes and temperature fluctuation intensities over a tube section in a fully developed nonisothermal turbulent incompressible liquid flow.

English translation will appear in Heat Transfer—*Soviet Research*

1.21 The Local Heat Transfer from a Plate with Repeated-Rib Roughness Elements

A. Veski and R. Kruus
Tallin Polytechnical Institute

Experimental data concerning local heat transfer rates for ribbed plates are obtained. The influence of pitches between the roughness elements on the heat transfer rates is shown. It is determined that the average heat transfer coefficient of the ribs is 2–3 times greater than that of the plane parts between the ribs.

English translation will appear in Heat Transfer—*Soviet Research*

1.22 The Effects of Varying Physical Properties on the Heat Transfer of a Turbulent Air Flow in an Annular Channel

J. Vilemas and M. Nemira
Institute of Physical and Technical Problems of Energetics
Lithuanian Academy of Sciences, Kaunas

Local heat transfer was studied experimentally for either outer or inner wall heating in the range of Re from 10^4 to 5×10^5. The wall to bulk temperature ratio was varied from 1.05 to 3. The inner to outer tube diameter ratio was 0.373. The influence of varying physical properties on heat transfer was found to be significantly lower than in circular tubes, and different for inner and outer wall heating.

English translation will appear in Heat Transfer—*Soviet Research*

1.23 Transient Natural Convection of a Heat Generating Fluid within an Enclosure

R. Viskanta and R. O. Johnson
Purdue University, West Lafayette, Indiana

The effects of internal heat generation on natural convection in a fluid enclosed in a horizontal rectangular region heated from below and cooled from above are investigated. The fluid dynamics equations of motion based on the Boussinesq approximation are solved numerically using the A.D.I. technique. Results have been obtained for a range of Grashof numbers, Prandtl numbers, and the heat generation parameter, and are presented graphically in dimensionless form. Good qualitative agreement is obtained, and both analyses and experiments on similar problems which have available results, are presented.

This paper appears on page 5, this volume

1.24 Heat and Mass Transfer in Turbulent Jets Submerged in a Sound Vibration Field

E. V. Vlasov, A. S. Ginevskiy, and L. N. Ukhanova
Moscow

The influence of sound disturbances of varying intensity and frequency on the propagation of an isothermal air jet, a nonisothermal air jet, and a helium jet into an air medium is investigated experimentally. The results show that acoustic phenomena either intensify or attenuate turbulent mixing. Hence the possibility arises of controlling heat and mass transfer processes in jets.

1.25 An Experimental Investigation of Film Cooling Effectiveness in Supersonic Axisymmetric Nozzles

E. P. Volchkov, V. K. Koz'menko, and V. P. Lebedev

Thermophysics Institute, Siberian Department of the USSR Academy of Sciences, Novosibirsk

An investigation was made of film cooling effectiveness in nozzles under adiabatic conditions both with and without the separation of the boundary layer in the supersonic zone of the nozzle. The maximum value of the acceleration parameter K ranged from 2.3 · 10^{-6} to 32 · 10^{-6}. The wall stagnation temperature was used to calculate film cooling effectiveness. The results are compared with the prediction and other experimental data on convergent ducts.

English translation will appear in Heat Transfer– *Soviet Research*

1.26 The Stability of Convective Motions in a Cylinder due to Electric Current

Vu Duy Quang

Hanoi Polytechnical Institute, Vietnam

The stability of steady convective motion due to Joule heat generation with the passage of electric current through an electrically conducting liquid filling a vertical cylinder is considered.

Some simplifying assumptions concerning the convection equations are made. The distributions of velocity, temperature and pressure at convective steady state are obtained. The stability of such a flow is investigated by the Galerkin–Bubnov method. The stability boundary is determined.

1.27 Heat Transfer in Turbulent Gas Flows with High–Frequency Pressure Oscillations

B. M. Galitseyskiy and Yu. A. Ryzhov

Moscow Aircraft Institute

Based on a semiempirical turbulence model, heat transfer with high-frequency oscillations in a gas flow in channels is considered. The results of experimental investigations of heat transfer with high-frequency pressure oscillations in the air flow in different size channels are presented. Under the action of oscillations, heat transfer is shown to vary from 5–6 times, as compared with steady-state flows. The calculations obtained using the proposed model are in good agreement with experimental results.

English translation will appear in Heat Transfer–*Soviet Research*

1.28 Local Heat Transfer from a Cylinder in Cross Flow

G. J. Gimbutis, V. J. Shapola, and V. J. Shimkevichus
Kaunas Polytechnical Institute

Experimental data on local heat transfer of water and transformer oil from a cylinder in cross flow are presented. The thermal boundary condition of a constant heat flux is specified.

English translation will appear in Heat Transfer–*Soviet Research*

1.29 Transient Heating of a Plate-System in a Gas Flow

V. Hlavačka
National Research Institute for Machine Design, Prague, Czechoslovakia

Assuming a harmonic change in the inlet gas temperature, the transient heat transfer in a system of plates has been solved analytically as a function of amplitude ratio and phase lag of the gas temperature. A comparison is carried out using previously reported data for circular rods and packed beds of spheres. To illustrate the practical application of the results, the paper illustrates its application to the measurement of the heat transfer coefficients in a rotary regenerator matrix.

This paper appears on page 15, this volume

1.30 Combined Cooling of Perforated Plates in a Turbulent Gas Flow

V. V. Glazkov, M. D. Guseva, and B. A. Zhestkov
Central Institute of Aircraft Materials, Moscow

The experimental results on heat and mass transfer with the cooling of perforated plates are presented. The formulae for calculating a heat transfer coefficient are derived, and a method of determining the plate temperature is proposed. Cases of the applicability of the calculation method are considered.

1.31 Heat Transfer and Pressure Drop of In-Line Banks of Tubes with Artificial Roughness

H. G. Groehn and F. Scholz
Institut für Reaktorbauelemente, Jülich, West Germany

Using banks of straight corrugated tubes, artificially rough tube walls are shown to improve the heat transfer coefficient without increasing the drag factor. The ratio St/ζ, for heat exchangers, was found to be 15–16 times greater for the bank of corrugated tubes than for the same bank of smooth tubes. Small, longitudinal ribs, attached to the tubes of the heat exchanger effected a rise in heat transfer of 30% with about the same increase in the drag coefficient.

This paper appears on page 21, this volume

1.32 An Investigation of the Formation of Temperature Fields in a Partially Filled Spherical Vessel with Natural Convection in a Field of Variable Mass Forces

A. A. Gukhman, A. A. Zaytsev, G. M. Solov'yev, N. G. Styushin, and S. A. Chizhikov

Moscow Institute of Chemical Engineering

Experimental results on the formation of temperature fields in natural convection in a partially filled vessel in a nonuniform field of mass forces, are presented. The nonuniform field of mass forces is created by rotating the vessel around a vertical axis other than its axis of symmetry. Temperature distributions in the walls, gas and the liquid were obtained at various angular speeds of rotation, height of the liquid and heat flow rates. The influence of the parameters under investigation on the increase in pressure within the vessel was investigated. The experiments were made with n-butanol and glycerin.

1.33 A Study of a Vertical Nonisothermal Turbulent Jet of an Incompressible Liquid

K. E. Dzhaugashtin

Leningrad

Some results of an investigation of mean (by the local similarity method) and fluctuation (by the method of pulsation energy balance) characteristics of the vertical nonisothermal axisymmetric turbulent jet of an incompressible liquid are presented.

1.34 Statistical Turbulence Characteristics for Use in Convective Heat Transfer Calculation

E. P. Dyban and E. Ya. Epik

Engineering Thermophysics Institute, Ukrainian Academy of Sciences, Kiev

A method of calculating the outer flow turbulence influence on the process of convective heat transfer intensification, assuming an additional turbulent viscosity formation in the boundary layer, is proposed. The turbulent viscosity can be determined by measuring at least six statistical turbulence characteristics (three velocity pulsation components, $\overline{u'v'}$ correlations, the transverse pulsation component spectrum and the intermittent coefficient).

English translation will appear in Heat Transfer–*Soviet Research*

1.35 Local (Longitudinal and Peripheral) Heat Transfer in a Plane Channel with Different Turbulence Levels at the Entrance

E. P. Dyban, E. Ya. Epik, and V. E. Filipchuk

Engineering Thermophysics Institute, Ukrainian Academy of Sciences, Kiev

The results of an experimental investigation of heat transfer in a plane channel with a bell-mouthed entrance of different turbulence levels (0.6 + 6%) and a "sharp edge" are presented.

The significant nonuniformity of the distribution of the heat transfer coefficients along the length and perimeter is attributed both to the development of transition and to the flow laminarization in the corner regions.

English translation will appear in Heat Transfer—*Soviet Research*

1.36 The Effect of Surface Orientation on Separated Flow Characteristics

V. E. Yevenko and A. K. Anisin

Institute of Transport Engineering, Bryansk

The results of heat transfer and aerodynamic drag experimental research for three types of surfaces at varying angles of surface orientation to the air flow direction are presented.

It is shown that the efficiency of a heat transfer surface in separated flows greatly depends on the surface orientation relative to the direction flow.

English translation will appear in Heat Transfer—*Soviet Research*

1.37 A Study of a Thermogasdynamic Interaction between a Supersonic Jet and an Inclined Permeable Plate

I. K. Yermolayev, V. G. Puzach, and V. A. Fadeyev

Institute of High Temperatures, USSR Academy of Sciences, Moscow

This paper deals with an experimental investigation of heat and mass transfer on permeable and impermeable surfaces, effected by a high-temperature ($T \approx 3000°K$) axisymmetric supersonic superexpanded jet ($Ma = 2.32$; $n = 0.8$) under essentially nonisothermal conditions ($\varphi = 0.06 \div 0.2$) and at a nearly constant wall temperature. Measurements were made of maximum thermal stress at different distances ($l = 3 + 6d_a$) from the nozzle and at different inclination angles ($\beta^a = 30° + 90°$) to the incident jet axis. Plate regions in the jet-spreading zone are identified. Engineering calculation methods are developed.

English translation will appear in Heat Transfer—*Soviet Research*

1.38 Heat and Mass Transfer in Transition to Turbulent Conditions on a Permeable Surface

V. M. Yeroshenko, A. A. Klimov, and Yu. N. Terent'yev
Krzhizhanovsky Power Engineering Institute, Moscow

Experimental data and some semiempirical generalizations from a study of transient mass transfer above a permeable surface from laminar to turbulent conditions are analyzed. A transient phenomenological model is considered and an interpolation formula is presented for the calculation of the paramaters which determine the transient processes. The relationships for the analysis and calculation of heat and mass transfer under transient conditions with the injection of different gases are obtained.

The experiments were carried out on a gasdynamic installation using interferometric and thermoanemometric techniques.

English translation will appear in Heat Transfer—*Soviet Research*

1.39 A Study of the Effect of the Frequency of an Electric Field on Heat Transfer

V. A. Zheltukhin, Yu. K. Solomyatnikov, D. M. Mikhaylov, and
A. G. Usmanov
Chemical Engineering Institute, Kazan

This paper deals with an experimental study of the effect of electric field frequency on heat transfer to organic liquids in natural convection and boiling point. It has been found that the field frequency depends on the dipole moment and dielectric liquid permeability. The maximum values of the heat transfer coefficient were identified for acetone boiling at 27–65 Hz.

English translation will appear in Heat Transfer—*Soviet Research*

1.40 Heat Transfer Efficiency of Tube Bundles in Crossflow at Critical Reynolds Numbers

A. Zukauskas and R. Ulinskas
*Institute of Physical and Technical Problems of Energetics,
Lithuanian Academy of Sciences, Kaunas*

Experimental results of average heat transfer and hydraulic drag are analyzed for three in-line and three staggered tube bundles 1.25×1.25, 1.50×1.50, and 2.0×2.0, in a crossflow of water at Re from 5×10^4 to 2×10^6 and at Pr from 3 to 7. A relation is suggested for calculating the heat transfer from the tube bundles in the critical flow regime of the crossflow. The efficiency of the tube bundles is analyzed relating to unit heat transfer volume.

English translation will appear in Heat Transfer—*Soviet Research*

1.41 Heat Transfer from Tube Bundles in Crossflow at Low Reynolds Numbers

A. Zukauskas, R. Ulinskas, E. Bubelis, and C. Sipavičius

*Institute of Physical and Technical Problems of Energetics,
Lithuanian Academy of Sciences, Kaunas*

The flow structure has been studied in a crossflow of aviation fuel on in-line tube bundles of 2.0 X 2.0 at Re = 500. Distributions of velocity, turbulence level and temperature around one of the tubes at various distances from the surface along with the frequency of vortex separation and local values of the heat transfer.

The effects of hydrodynamic parameters on the local heat transfer along the tube perimeter are considered.

The results may be used for the study of flow dynamics with heat transfer, for designing new heat exchangers of high efficiency, and for creating some new calculation techniques for separated flows.

English translation will appear in Heat Transfer—*Soviet Research*

1.42 The Effects of Turbulence on Heat Transfer of a Cylinder in Crossflow at Critical Re

A. Zukauskas, J. Ziugzda, and V. Survila

*Institute of Physical and Technical Problems of Energetics
Lithuanian Academy of Sciences, Kaunas*

The experimental results of a study of local and average heat transfer are presented for flows of water and air at Re from 4×10^4 to 1.2×10^6, Pr from 0.7 to 6 and Tu from 1.0% to 15%.

It has been found from distribution changes in local heat transfer coefficients that the location of the transition point is a function of both Re and Tu, such that it is moved upstream as they increase. An increase in turbulence causes an increase in heat transfer, at the front point of the cylinder. The power of Re in the equation of mean heat transfer is ~ 0.8 in the critical flow. At the forward stagnation-point, the heat transfer is proportional to Re to the power of 0.6.

English translation will appear in Heat Transfer—*Soviet Research*

1.43 A Study of the Physics of Turbulent Transport Phenomena

Z. Zarić

Boris Kidric Institute, Belgrade, Yugoslavia

The results of a conventional statistical analysis of experimentally determined velocity and temperature fluctuations in zones of high intensity turbulence are presented. The analysis indicates the presence of intermittent phases superimposed on the background turbulence but it is unable to describe them quantitatively. The results of an analysis demonstrate that the evolution of the probability distributions of velocity and temperature fluctuations could well be explained by the statistical behavior of the intermittent phases present in the flow. The role of these intermittent phases, which are in fact, coherent, quasi-ordered structures in turbulent transport processes is emphasized. The inadequacy of Reynolds type statistics and the need for more adequate analysis is discussed.

1.44 An Investigation of Thermal Boundary Layer Development in Turbine Blades

L. M. Zysina-Molozhen, A. A. Dergach, M. A. Medvedeva, and E. G. Roost
The Polzunov Central Boiler and Turbine Institute, Leningrad

This paper describes the results of an experimental investigation of the influence of turbulence, rotation and pressure gradient on the behavior of the thermal boundary layer in both stationary and rotating turbine blades. Some empirical relations for Nu number determination are proposed.

English translation will appear in Heat Transfer—*Soviet Research*

1.45 Investigations of Heat Transfer in Turbine Vanes at Operating Conditions

L. M. Zysina-Molozhen, M. M. Ivashohenko, A. A. Dergach, and
Ya. M. Feldshtein
The Polzunov Central Boiler and Turbine Institute, Leningrad

A system of third-kind boundary conditions in an industrial turbine is determined by solving some steady-state heat transfer problems based on experimental data. It is shown that design relations for the transient boundary conditions can be obtained by assuming quasi-steady–state conditions and by taking into consideration the initial turbulence and the relaminarization of the boundary layer.

English translation will appear in Heat Transfer—*Soviet Research*

1.46 A Numerical Solution for Heat Transfer from the Rough Surfaces of Electromotor Cooling Channels

M. Jícha and Z. Ramík
Brno Institute of Technology, Czechoslovakia

This paper deals with heat transfer from a rough surface to an air flow in the cooling channels of an electromotor. A finite-difference DuFort–Frankel method is suggested and developed in order to solve a time-averaged partial differential energy equation for the turbulent boundary layer. The equation is solved for both smooth and rough surfaces of circular channels. The results of this theoretical solution are substantiated by experiments. The roughness under investigation is quite irregular and random. The order of its height is 0.01 mm. The Reynolds number was varied between 10,000 and 100,000.

1.47 Heat Transfer and Fluid Friction for Water Flow in Tubes with Supercritical Pressures

S. Ishigai, M. Kaji, and M. Nakamoto
Faculty of Engineering, Osaka University, Osaka, Japan

The pressure drop and heat transfer in a supercritical water flow in vertical and horizontal tubes up to 400 ata and 500°C were simultaneously measured, with an emphasis on the near-critical region. In spite of a similarity in the behavior of the friction factor and of the heat transfer coefficient, the nonexistence of a similarity law is proven by a departure of Colburn's j factor from the friction factor. A modified factor of $j = St \cdot Pr^{0.2}$ is proposed for a pseudo-critical domain in correlating the friction factor and the heat transfer coefficient. A dimensionless equation is derived for the vertical upward flow heat transfer by using not the Nusselt number, but the Stanton number, which gives a simpler form.

This paper appears on page 25, this volume

1.48 Heat Transfer in Combined Free and Forced Convection in Closed Volumes

E. K. Kalinin, G. A. Dreytser, and A. S. Neverov
Moscow Aviation Institute

The results of an experimental investigation of free convection heat transfer in a vertical cylindrical tank (partly filled with a liquid) with gas injection and liquid displacement are presented.

It is established that under these conditions the heat transfer coefficient differs considerably from the values calculated using the generally accepted relations. It is shown that this difference is due to the instability of boundary conditions.

English translation will appear in Heat Transfer—*Soviet Research*

1.49 A Generalization of the Results of an Experimental and Theoretical Investigation of Unsteady Convective Turbulent Heat Transfer in Channels

E. K. Kalinin and G. A. Dreytser
Moscow Aviation Institute

A generalization of the results of an experimental and theoretical investigation of unsteady heat transfer in pipes, with gas and liquid flows, and with time-variation of the wall temperature and the flow rate is presented.

It is shown that the difference between the unsteady heat transfer coefficient and quasi-steady one is due to a superposition of unsteady heat conduction on steady convective heat transfer (this superposition is essential for liquids and insignificant for

English translation will appear in Heat Transfer—*Soviet Research*

gases), this being determined by the flow turbulent structure and its acceleration or deceleration. The influence of these three factors on heat transfer can be analyzed using three parameters of thermal and hydrodynamic instabilities.

1.50 An Experimental Study of Heat Transfer with a Fluid Flow in a Rotating Radial Confuser

V. M. Kapinos, V. N. Pustovalov, A. P. Rud'ko, and L. A. Gura

Kharkov Polytechnical Institute

An experimental study of heat transfer in an air flow between two parallel disks rotating at the same speed was made. The motivation for this study came from a need to describe the heat transfer boundary conditions on the surface of turbine rotors. Similarity equations for fully developed turbulent and transient flows were formulated. The similarity equation governing transition to a fully developed turbulent flow was also determined.

English translation will appear in Heat Transfer—*Soviet Research*

1.51 Free-Convective Heat Transfer Characteristics of Cryogenic Liquids Stored in Closed Volumes

Yu. A. Kirichenko, V. N. Shchelkunov, V. N. Timonkin, and L. Yu. Radchenko

Cryogenics Institute, Kharkov

In this paper, problems concerning temperature stratification in a closed spherical volume partially filled with a cryogenic liquid are considered. A relation describing the time-dependent temperature of the liquid-vapor interface has been obtained. The existence of a heat-flux threshold density below which there is no liquid stratification, has been found experimentally.

English translation will appear in Heat Transfer—*Soviet Research*

1.52 Some Experimental Studies of Convective Heat Transfer in Toroidal Vessels

Yu. A. Kirichenko and Zh. A. Suprunova

Cryogenics Institute, Kharkov

Temperature fields in toroidal vessels filled with cryogenic and high boiling-point liquids under Neumann's boundary conditions are studied. A generalized relation permits the calculation of the temperature at the liquid surface at Rayleigh numbers $Ra = 5 \cdot 10^8 \div 7 \cdot 10^{12}$.

English translation will appear in Heat Transfer—*Soviet Research*

1.53 An Experimental Study of Turbulent Transport in a Viscous Sublayer Using the Electrochemical Method

M. Kh. Kishinevskiy, T. S. Korniyenko, and A. V. Loginov
Voronezh Institute of Technology

Electrochemical measurements were made of fully developed mass transfer in a hydro-dynamically developed turbulent flow. The data were correlated and an average deviation of 3.4% is found.

English translation will appear in Heat Transfer—*Soviet Research*

1.54 A Numerical Solution to the Natural Convection Problem with Suction and Blowing

A. Kumar Kolar and V. M. K. Sastri
Indian Institute of Technology, Madras, India

The influence of suction and blowing on the heat transfer and skin friction at the surface of an isothermal flat vertical plate is studied numerically. The relevant laminar boundary layer equations in a modified von Mises form are solved by an implicit, stable finite difference scheme and a "Marching Solution" is obtained. Results are obtained for several Prandtl Numbers and a wide range of blowing parameters, including strong blowing and suction.

The results confirm the finding that blowing decreases heat transfer and skin friction, whereas suction has the opposite effect.

A comparison of the present results with those available in the literature is found to be very satisfactory.

1.55 A Description of Statistical Transfer Characteristics in an Incompressible Liquid

B. A. Kolovandin
Luikov Heat and Mass Transfer Institute, Minsk

Based on a finite number of equations for the moments of inhomogeneous fields of temperature and velocity fluctuations, a system is proposed for an asymptotic model governing large-scale momentum and heat transfer with inhomogeneous turbulence at large turbulent Reynolds and Peclét numbers.

1.56 Characteristics of Natural Convection Development in Cylindrical Layers

E. P. Kostogorov, E. A. Shtessel, and A. G. Merzhanov
Institute of Chemical Physics, Moscow

Some unsteady characteristics of the onset and development of natural convection in horizontal cylindrical sublayers are experimentally studied.

1.57 Ferrohydrodynamic Convection in a Flow Past a Cylinder

G. E. Kroŋkalis, E. J. Blums, and M. M. Mayorov
Institute of Physics, Riga

Thermomagnetic convection near a horizontal cylinder in a magnetic fluid has been studied experimentally. A considerable increase in heat transfer is found in the case of a nonuniform magnetic field.

English translation will appear in Heat Transfer—*Soviet Research*

1.58 The Effect of Turbulence Characteristics on Heat Transfer from a Cylinder in Crossflow

P. Kubes
Research Institute of Food Technology, Prague, Czechoslovakia

Results are presented of a study carried out to determine the effect of turbulence characteristics on heat transfer from a cylinder in a wind tunnel for a range of Reynolds number of 20,000 to 75,000. A relationship is established between the heat transfer properties and the turbulent pulsations of the stream.

1.59 An Investigation of Transient Heat Transfer and Hydrodynamics in Tube Bundles with Large Disturbances

Yu. N. Kuznetsov, V. N. Oyvin, and V. I. Pevzner
The Dzerzhinskiy All-Union Thermal Engineering Institute, Moscow

The results of an experimental investigation of transient heat transfer and hydrodynamics in tube bundles are presented. The experimental apparatus is described. The experimental results are compared with predictions made by the usual practical engineering methods of the calculation of transient processes with large disturbances.

English translation will appear in Heat Transfer—*Soviet Research*

1.60 Investigations of the Turbulent Transport Properties of Nonisothermal Jets

V. I. Kukes and L. P. Yarin

Naval Engineering Institute, Leningrad

The results of experimental investigations of the turbulent transport properties of nonisothermal jets are presented.

1.61 A Numerical Study of Natural Convection Heat Transfer near a Porous Horizontal Cylinder

V. S. Kuptsova and V. G. Malinin

Wood Technology Institute, Moscow

The results of a numerical investigation of natural convection near a porous horizontal cylinder are presented. The region in which the boundary layer equations apply is found. The results show a change in the heat transfer coefficient with air suction and injection. A formula is given for calculation of the heat transfer coefficient.

English translation will appear in Heat Transfer—*Soviet Research*

1.62 The Effect of Thermocapillary Forces on Heat and Mass Transfer near a Free Surface in a Horizontal Turbulent Thermal Gravitational Convective Liquid Layer

S. S. Kutateladze, A. G. Kirdyashkin, and V. S. Berdnikov

Thermophysics Institute, Siberian Department of the USSR Academy of Sciences, Novosibirsk

Heat transfer by free convection in an evaporating alcohol layer is examined in the experimental range of Ra numbers $3 \cdot 10^6 \div 0.92 \cdot 10^7$. Measurements of the temperature distribution on the liquid-gas interface show that the eddy cell flows in the neighborhood of a free liquid surface are of both a buoyant and a surface-tension-driven nature.

English translation will appear in Heat Transfer—*Soviet Research*

1.63 Laser Beam Propagation in Turbulent Free Convection

S. S. Kutateladze, N. A. Rubtsov, V. A. Bazanov, and M. N. Dulin

Thermophysics Institute, Siberian Department of the USSR Academy of Sciences, Novosibirsk

The propagation of a narrow laser beam in a horizontal layer of liquid with free convection is considered for the case of a random inhomogeneous medium and from the point of view of geometrical optics.

The experimental data of the mean and mean-squared deflections of the beam passed through the medium are used to determine the properties of the medium, the isothermality, and the static homogeneity of the central core of the layer.

With an additional measurement of the dispersion of the temperature fluctuations in the region of the beam propagation, this method permits an estimation of the characteristic dimensions of thermal inhomogeneities.

English translation will appear in Heat Transfer—*Soviet Research*

1.64 Thermal Free Convection in a Vertical Slot under Turbulent Flow Conditions

S. S. Kutateladze, V. P. Ivakin, A. G. Kirdyashkin, and A. N. Kekalov
Thermophysics Institute, Siberian Department of the USSR Academy of Sciences, Novosibirsk

A study is made of the turbulent free-convective liquid flow in a slot with isothermal walls. Experimental results on heat transfer and hydrodynamics are reported. In this case the behavior of heat transfer is found to be similar to that on an isothermal vertical plate.

English translation will appear in Heat Transfer—*Soviet Research*

1.65 A Study of the Stability of Mass Transfer with Free and Forced Convection in Horizontal Rectangular Channels

M. E. Lago
Universidad de Los Andes, Merida, Venezuela

A steady laminar flow with forced convection between infinite parallel horizontal flat plates with a fully developed velocity profile where an electrolysis is produced is studied. A secondary convectional flow is established in the system and this gives rise to a "roll cell" with a velocity profile similar to that of the "Goertler vortices."

A system of equations of linear stability is solved using the method of Galerkin. Re (eigenvalue of the system), is obtained for a range of variables and the Ra numbers are calculated to obtain a unique value for each group of parameters (Ra = 6700) characterizing the phenomenon under study. In addition, graphs of the relationships between the distinct variables and the given parameters are presented.

1.66 Heat Transfer to a Rod Bundle in the Entrance Region of an Axial Flow

V. L. Lelchuk, K. F. Shuyskaya, A. G. Sorokin, and O. N. Bragina
The Dzerzhinskiy All-Union Thermal Engineering Institute, Moscow

Wall temperature distributions and heat transfer coefficients on the surface of a cluster of seven rods arranged in a hexagonal pattern witnin a channel of a hexagonal cross section and cooled by a longitudinal air flow were determined.

English translation will appear in Heat Transfer—*Soviet Research*

1.67 An Experimental Investigation of Heat Transfer from a Flat Porous Surface with Alternating Injection and Suction

A. I. Leont'yev, V. I. Rozhdestvenskiy, Yu. A. Vinogradov, and
V. I. Sysoyev
Institute of Mechanics, Moscow University

The results of an experimental investigation of supersonic (M ~ 2) flow heat transfer from a flat porous surface with alternating injection and suction in the flow direction are given. An empirical relation is obtained for calculating the heat transfer of the injection sections at various intensities of suction.

English translation will appear in Heat Transfer—*Soviet Research*

1.68 Mean and Fluctuating Characteristics of a Turbulent Boundary Layer and Heat Transfer in a Retarded Flow

A. I. Leont'yev, E. V. Shishov, V. M. Belov, and V. N. Afanas'yev
Bauman Institute of Technology, Moscow

The results of an experimental study of heat transfer and the distribution of the fluctuating characteristics of a thermal boundary layer developed in a retarded equilibrium turbulent one are discussed. It is found that an adverse pressure gradient has a considerable effect on the distribution of the mean temperature and the fluctuational characteristics in the retarded turbulent boundary layer. At the same time it is shown that if the scales of velocity and temperature in St and Re_T numbers are chosen properly, the heat transfer law applies and the mean temperature profiles in the wall region remain the same as in the case of a flat plate. It is also found that the hydrodynamic and thermal boundary conditions affect the distribution of the turbulent Prandtl number in the boundary layer.

English translation will appear in Heat Transfer—*Soviet Research*

1.69 Laminar Heat Transfer with Varying Physical Properties

V. Makarevičius
*Institute of Physical and Technical Problems of Energetics,
Lithuanian Academy of Sciences, Kaunas*

Based on analytical transformations of the differential boundary-layer equations and on computer analysis of numerical data for boundary-layer heat transfer, a relation has been obtained which predicts heat transfer over a wide range of physical properties.

The solution of the "suspended" differential energy equation yields a general heat transfer law of prime importance.

The Prandtl number ratio for the boundary layer limits allows for the effect of variable physical properties on heat transfer.

English translation will appear in Heat Transfer—*Soviet Research*

1.70 A Prediction of Turbulent Heat Transfer at Stabilized Flow in Tubes

P. L. Maksin, B. S. Petukhov, and A. F. Polyakov
Institute of High Temperatures, USSR Academy of Sciences, Moscow

A computation scheme is presented for turbulent heat transfer in a stabilized liquid flow in a tube over a range of $Pr = 10^{-3} \div 10^{2}$ and $Re = 10^{4} \div 10^{6}$. A correlation was made between temperature and velocity fluctuations. A method of determining the constants of approximation is suggested. The experimental results are compared with numerical ones.

English translation will appear in Heat Transfer—*Soviet Research*

1.71 Heat Transfer from a Gas Jet to a Porous Barrier

S. K. Matveyev and G. V. Kocheryzhenkov
Institute of Mathematics and Mechanics, Leningrad State University

The interaction of a gas jet with a porous barrier in the presence of filtration is considered. Using a special difference method, the temperature and pressure distributions are obtained. An analysis of the data shows that the application of the difference method for such problems is appropriate.

English translation will appear in Heat Transfer—*Soviet Research*

1.72 Reynolds Analogy in a Turbulent Boundary Layer with a Pressure Gradient

V. K. Migai
The Polzunov Central Boiler and Turbine Institute, Leningrad

Some values of the Reynolds analogy coefficient, $\eta = 2St/c_f$, for confuser ($\eta < 1$) and diffuser ($\eta > 1$) flows are obtained by solving a turbulent heat conduction equation.

Calculations of the sub-separated state of a boundary layer are given.

English translation will appear in Heat Transfer—*Soviet Research*

1.73 The Effect of Large Temperature Difference on the Turbulent Heat and Momentum Transfer in an Air Flow inside a Circular Tube

T. Mizushina, T. Matsumoto, and S. Yoneda
Kyoto University, Japan

From the results of analytical and experimental investigations of the local behavior of turbulent heat and momentum transfer in a hot air flow through a cooled circular tube with a steep radial temperature gradient, it was concluded that, in the turbulent core region, the profiles of mixing length are the same as those of the isothermal ones. It was assumed that the behavior of the mixing length near the wall could be expressed by extending the van Driest model to include a nonisothermal flow of gas. The correlation between the analytical results and the experimental ones was satisfactory. In cooling a turbulent low Reynolds number gas flow, the deviations of the Nusselt number and friction factor from the isothermal case were negligibly small when the bulk temperature was used to evaluate the physical properties.

This paper will appear on page 34, this volume

1.74 Convective Heat Transfer from a Tube Bundle in an Asymmetrical Flow

I. Mikk, A. Veski, and R. Kruus
Tallin Polytechnical Institute

An experimental study of the heat transfer of an in-line tube bundle was made for two cases: a) when the angle between the flow direction and the axes of the tubes differs from 90°; b) when the angle between the flow direction and the plane of some of the tubes differs from zero.

English translation will appear in Heat Transfer—*Soviet Research*

1.75 The Influence of an Unheated Starting Length on Heat Transfer in Subsonic and Supersonic Turbulent Boundary Layers with Different Flow Histories

B. P. Mironov, V. N. Vasechkin, and N. I. Yarygina
Thermophysics Institute, Siberian Department of the USSR Academy of Sciences, Novosibirsk

Local impermeable cylindrical surface heat transfer coefficients are found to be in good agreement with the usual heat transfer law for a flat plate, if the distance from an initial point on the heated surface to an experimental area is used as an effective linear dimension in the Reynolds number.

A change in the velocity profile in the region upstream from the heated surface did not influence the results.

English translation will appear in Heat Transfer—*Soviet Research*

1.76 Heat Transfer through an Annulus with a Rotating Inner Cylinder

A. A. Mosyak, B. G. Rykova, P. D. Kostov, and G. I. Gruzintsev
Moldavgidromash Corporation, Kishenev, Moldavia, USSR

The experimental results of a study of the influence of peripheral rotation speed and axial flow velocity on heat transfer from an external cylinder surface with water and transformer oil flow inside the circular gap are presented.

English translation will appear in Heat Transfer—*Soviet Research*

1.77 The Effect of Periodically Variable Wall Temperature on Heat Transfer in Liquid Metals

J. Mošnerová
*National Research Institute for Machine Design, Prague - Bechovice
Czechoslovakia*

This paper presents a description of heat transfer in liquid metals flowing through a tube with time and space dependent wall temperatures. The energy equation for the time-dependent heat convection is solved numerically by the alternating direction method. The boundary condition is assumed at the wall as the temperature undergoing a slow step change with periodic movement along a certain length of the tube. Computations of the dependence of the Nusselt number on the frequency of the wall temperature pulsations, the length of the pulsation region, and the Reynolds and Prandtl numbers are presented.

This paper appears on page 47, this volume

1.78 Heat and Mass Transfer in a Radial Liquid Jet Impinging on the Surface

V. E. Nakoryakov, B. G. Pokusayev, and E. N. Troyan
*Thermophysics Institute, Siberian Department of the USSR Academy of
Sciences, Novosibirsk*

The hydrodynamics and heat and mass transfer of a liquid jet impinging on a surface are investigated both theoretically and experimentally. Simple analytical solutions are obtained which are compared with the experimental data.

English translation will appear in Heat Transfer—*Soviet Research*

1.79 A Nonclassical Model of Turbulent Heat Transfer

G. Naue, W. Schmidt, and W. W. Schmidt
Merseburg Institute of Technology, German Democratic Republic

A steady, two-dimensional turbulent flow in a flat channel is investigated. Dimensionless temperature distributions are calculated by means of finite difference and variational methods.

The turbulent processes are characterized by a set of kinematic variables including velocity, first-order spin, and the turbulence intensity of higher-order spins. This theory is based on the possibility of asymmetric stresses.

This paper appears on page 54, this volume

1.80 Heat Transfer in a Separated Flow Region Downstream of a Roughness Element

N. Nishiwaki, T. Sakuma, H. Tanaka, and A. Tsuchida
University of Tokyo, Japan

Heat transfer in a separated flow region downstream from single roughness element placed near the leading edge of a two-dimensional wedge, onto which an air jet impinges was experimentally investigated. The local heat-transfer coefficient downstream from the roughness element shows a sharp maximum at the reattachment point. The entire physical phenomenon, as typified by this maximum of heat transfer, seems to be determined by a characteristic Reynolds number which is defined by $Re^* = U^*Z/V$, where U^* is the free stream velocity at the roughness element measured outside the boundary layer in the case of an undisturbed flow, and Z is the distance from the roughness element to the reattachment point.

This paper appears on page 61, this volume

1.81 On the Effect of Film Cooling in the Separation Zone

N. Nishiwaki, N. Numata, H. Kato, and R. Fujii
University of Tokyo, Japan

This paper deals with the effect of film cooling at a zone of separation. Hot air was vertically injected into the main flow at a separated zone behind a fence placed on a heated flat plate.

The injection of air affects not only the region downstream from the air injection slit, but also the upstream region.

The reduction of heat transfer in the case of air injection was one third of that in the case of no injection.

This paper appears on page 67, this volume

1.82 Statistical Characteristics of Temperature Fluctuations in Turbulent Water Flow

B. V. Perepelitsa
Thermophysics Institute, Siberian Department of the USSR Academy of Sciences, Novosibirsk

Measurements of the statistical characteristics of the temperature fluctuations in a nonisothermal turbulent water flow in a rectangular channel for $Re = 10^4 \div 6 \cdot 10^4$ are presented. Temperature fluctuations were measured using a thermocouple of a lateral dimension equal to 10μ, which records the temperature fluctuations near the wall, including those in the viscous sublayer.

A computer was used to record and analyze the experimental results.

English translation will appear in Heat Transfer—*Soviet Research*

1.83 Turbulent Heat and Mass Transfer near a Rough Surface

V. N. Pilipenko
Institute of Mechanics, Moscow University

A study was made of a violation of Reynolds analogy in flows near rough surfaces. An analysis of experiments on flow visualization near rough surfaces leads to a conclusion that this violation may be explained by a difference in dimensionless shear stress and heat flux distributions near a rough surface. A method for calculating the heat and mass transfer near a rough surface using arbitrary values of a dimensionless roughness height and the Prandtl and Reynolds numbers is proposed.

English translation will appear in Heat Transfer—*Soviet Research*

1.84 Natural Convection Heat Transfer from a Horizontal Cylinder to CO_2 at Supercritical Pressures with Cooling

V. P. Petrov, S. P. Beschastnov, and V. I. Belozerov
Obninsk Branch of Moscow Engineering Physics Institute

A study is made of convective heat transfer with cooling. Measurements of heat transfer coefficients and heat fluxes at pressures of 70–100 bar and CO_2 temperatures of 40–80°C are presented. The data for the pressure effect on heat transfer are given. A critical analysis of the experimental data reveals a considerable discrepancy between them and existing calculation recommendations.

English translation will appear in Heat Transfer—*Soviet Research*

1.85 An Experimental Study of Heat Transfer in Pipes with a Variable Specific Wall Heat Flux

B. S. Petukhov, V. S. Grigor'yev, A. F. Polyakov, and
S. V. Rosnovskiy
Institute of High Temperatures, USSR Academy of Sciences, Moscow

Local heat transfer has been studied experimentally in water and gaseous-nitrogen turbulent flow in a circular duct with a linear increase or decrease in the specific wall heat flux and small changes in physical properties. A correlation of local heat transfer in the thermal entrance region of a pipe has been obtained for $Pr = 0.7 \div 100$ and $Re = 10^4 \div 5 \cdot 10^5$. It is shown that under such conditions the superposition principle for the calculation of heat transfer is applicable.

English translation will appear in Heat Transfer—*Soviet Research*

1.86 Laminar Entry Length Heat Transfer in Ducts of Rectangular Cross-Section with Boundary Conditions of the Second Kind

V. Preiningerová
Research Institute of Electrical Engineering, Praha-Bechovice, Czechoslavakia

Heat transfer in the thermal entrance region of rectangular ducts with laminar flow of a viscous fluid is numerically determined with boundary conditions of the second kind. The influence of cross-section aspect ratio and of the number and arrangement of heated and unheated walls on the mean and local heat transfer along the duct axis and periphery is studied.

This paper appears on page 74, this volume

1.87 Use of Structural Methods and Laplace Transforms for Solving Problems of Heat Transfer in Pipes and Channels of Complex Sections

V. L. Rvachev, A. P. Slesarenko, and V. I. Popivshchiy
Institute of Mechanical Engineering Problems, Kharkov

An application of the structural method and Laplace transforms is considered for solving the problem of laminar heat transfer in the entrance region of pipes of complex cross sections. Also considered are the conjugated problems arising in the cells of heat generating elements.

English translation will appear in Heat Transfer—*Soviet Research*

1.88 The Influence of Gas Dynamic and Geometrical Parameters on the Heat Transfer of Supersonic Jets on an Obstacle

Yu. M. Rudov
Leningrad Institute of Mechanics

The results of an investigation of heat transfer on flat and spherical obstacles are presented. It is found that heat transfer or a flat obstacle exposed to a supersonic jet of 1) steady and 2) unsteady flow depends on the position of the obstacle relative to the nozzle. An analysis of a temperature-dependent inner wave structure at different points on the obstacle is given. The effect of an increase in temperature with decreasing distance between the nozzle and obstacle is studied. The relationships between the heat flux at the central point of the obstacle and the distance, degree of expansion, and Mach number at the nozzle exits are reported.

The unsteady flow regime was also investigated.

English translation will appear in Heat Transfer—*Soviet Research*

1.89 Convective Heat and Mass Transfer in a Hypersonic Near-Wake

L. I. Skurin and A. V. Urkov
Leningrad State University

The laminar near-wake behind a blunt axisymmetric cone with a flat base in a hypersonic flow is considered. A computational scheme based on a model of viscous-inviscid flow interaction is used.

The resulting calculations are compared with the available experimental data on base pressure.

The numerical data for the base injection and cone wall temperature effect on the thermal and dynamic characteristics of a near-wake are presented.

English translation will appear in Heat Transfer—*Soviet Research*

1.90 Heat and Mass Transfer of a Condensible Component from Binary Gas Mixtures in Porous-Wall Ducts with Suction

B. M. Smolskiy and V. S. Bogachev
Luikov Heat and Mass Transfer Institute, Minsk

A numerical study is made of the transfer processes in a laminar binary vapor-gas mixture flow in porous-wall ducts with the suction of a condensible component through these walls. Results obtained using a flat channel, round tube, and a duct formed by flat discs and mixtures of air–H_2O, H_2–H_2O, He–H_2O, and CO_2–H_2O are presented. The transfer processes are shown to depend on varying physical properties, on the vapor-gas mixture, and on both the longitudinal and the transverse flows.

English translation will appear in Heat Transfer—*Soviet Research*

1.91 The Influence of Environment Compressibility on Thermoconvective Processes in a Closed Region

N. M. Stankevich and G. B. Petrazhitskiy
Institute of Applied Mathematics and Cybernetics, Gorkiy

On the basis of a numerical solution of an equation system written in the physical variables u, p, and T, and taking into account compressibility and dissipative processes, an investigation is carried out on the influence of the compressibility criterion $1/F$ on the nature of the thermoconvective processes in spherical layers. The behavior of the flow and the temperature fields with $1/F$ varying in the range of $0.01 \leqslant 1/F \leqslant 0.5$ is considered. The intensity of heat convection is shown to increase with an increase in the compressibility criterion.

English translation will appear in Heat Transfer—*Soviet Research*

1.92 Heat Transfer in the Initial Section of a Plane Channel with Different Turbulence Scales in the Inlet Flow

A. S. Sukomel, D. F. Gutsev, and V. I. Velichko
Moscow Power Institute

An experimental study is made of local heat transfer in the initial section of a plane channel with an air flow under various flow conditions in the boundary layer.

Equations have been obtained for determining heat transfer taking into account the turbulence scale in the inlet flow, as well as the turbulence attenuation in the flow core along the channel length.

1.93 A Study of Large Amplitude Cellular Convection

A. J. Suo-Anttila and I. Catton
University of California, Los Angeles

A study of the nonlinear advective processes occurring in a layer of fluid being heated uniformly from below and cooled from above (cellular convection) with the Rayleigh number exceeding a critical value was undertaken. The amplitude of the resulting cellular convection and of the heat flux (Nusselt number) is determined as a function of the thermal potential (Rayleigh number). An explanation for the discrete heat flux transitions, and a description of the structure of the thermal boundary layers are given. The existence of multiwavenumber solutions at all Rayleigh numbers above the critical wavenumber is confirmed. The predicted value of the Nusselt number is in close agreement with the experimentally determined value.

1.94 Confined-Jet Convective Heat Transfer in a Centrifugal Force Field

E. P. Sukhovich
Institute of Physics, Riga

Convective heat exchange between a hot air jet and a coaxial rotational flow in a tube was studied. The centrifugal forces in a rotational flow with a negative temperature gradient cause a stratification of the flow into temperature layers and a reduction in heat transfer.

It is proposed that a cold rotational flow may be used to protect the internal surface of the tube from a high-temperature flow.

1.95 The Effect of Variable Fluid Properties on Turbulent Transport

J. R. Tyldesley
University of Glasgow, Great Britain

Based on inter entity response equations and using an entity model for a turbulent fluid, a study is made of the effect of variable fluid properties on the turbulent transport of momentum and energy. The turbulent relaxation phenomena associated with a variation in the temperature field and the effect of this variation on heat transfer within turbulent boundary layers are discussed.

1.96 A Study of the Turbulent Heat Transfer in a Nonisothermal Axisymmetric Jet

B. P. Ustimenko, V. N. Zmeykov, and A. A. Shishkin
Kazakh Power Engineering Research Institute, Alma-Ata

A study of the effect of velocity and temperature fluctuation, the correlation coefficient and frequency spectra in a nonisothermal axisymmetric free jet at initial temperatures of between 3 and 500°C at the nozzle outlet is presented.

1.97 The Unsteady Turbulent Boundary Layer in the Initial Length of a Permeable Wall Pipe

A. V. Fafurin and K. R. Shangareyev
Kazan Aviation Institute

The results of an analytical and experimental investigation of a problem concerning the development of the velocity and thermal boundary layers in the starting length of an axisymmetric duct with permeable walls are presented.

1.98 Unsteady Thermal Convection in Horizontal Cavities Heated from below with Periodic Heat Supply to the Lateral Walls

V. E. Fertman, Yu. I. Barkov, and A. R. Bayev
Luikov Heat and Mass Transfer Institute, Minsk

An experimental study was made of unsteady convective motion in a horizontal rectangular cavity with periodic temperature modulation at the lateral boundaries, over a range of the Rayleigh number Ra = 0-10^4. Some characteristics of the interaction of thermal convective waves propagating in a cavity are given.

1.100 A Study of Convective Heat Transfer in Complex Channels of Triangular, Rectangular, Semicircular, and Trapezoidal Cross Sections

C. Chiranjivi
Andhra University, Waltair, India

A comparative study of the laminar and turbulent flow characteristics, with regards to pressure drop and convective heat transfer rates, under conditions of constant wall temperature, in complex channels of triangular, rectangular, circular-segmental, and trapezoidal cross sections is presented. Laminar and turbulent flow friction factors, computed from the experimental results obtained in the aforementioned study have been compared with theoretically predicted friction data. The experimental laminar flow forced convective heat transfer data are presented and compared with values computed from the analytical expressions available in the literature. The development of a general empirical equation describing the heat transfer by both forced and free convection in noncircular channels is presented.

1.101 Film Cooling Effectiveness of a Rotating Surface

I. T. Shvets, V. M. Repukhov, K. A. Bogachuk-Kozachuk
Engineering Thermophysics Institute, Ukrainian Academy of Sciences, Kiev

Using the first integral of the energy equation based on relative motion, expressions for wall film cooling effectiveness are obtained. It is shown that with some correction, the methods for calculating the effectiveness in an absolute coordinate systems can be applied in the case of rotating coordinates. Comparison of the experimental and theoretical data obtained for the cooling of rotors of axial and centripetal gas turbines is presented.

1.102 Turbulent Flow in Coiled Tubes and the Influence of Prandtl Number on the Associated Heat Transfer

R. Schiestel and J. Gosse
Conservatoire National des Arts et Métiers, Paris

The turbulent flow and associated heat transfer in a coiled tube is studied using a tridimensional turbulence model. The mathematical description of turbulence employed involves partial differential transport equations for double velocity correlations and for two effective length scales. The results obtained permit an estimation of the influence of the Prandtl number on heat transfer for a fixed Reynolds number and for three values of the radius of curvature. A satisfactory agreement between numerical predictions and experimental results is found.

This paper appears on page 91, this volume

1.103 Heat Transfer Calculations for Rough Wall-Surfaces

M. -R. Drižius and A. Slančiauskas
Institute of Physical and Technical Problems of Energetics, Lithuanian Academy of Sciences, Kaunas

In a study of turbulence characteristics, heat transfer was found to affect large-scale motions and turbulent core decay while wall surface properties did not. This finding allows the application of a universal calculation method of boundary-layer equations for both smooth and rough walls. Based on a hypothesis whereby mixing length describes turbulence, the method is applied for the prediction of heat transfer from a wall over a range of roughness and for different Prandtl numbers. An increase in friction resulting from the shape factor of the roughness elements is accounted for by an appropriate redistribution of tangential stresses near the wall.

1.104 An Investigation of the Flow and Heat Transfer in the Entrance Region of a Tube with Initial Swirling

V. K. Shchukin, A. A. Khalatov, A. V. Koznevnikov, and V. G. Letyagin
Kazan Aviation Institute

A study is made of the hydrodynamics and heat transfer in a porous tube, 12.5 diameters long, with a central body at the entrance and the injection of a coolant into a heated air flow which is swirled by a blade turbulizer.

A determination of the characteristics of porous-fibrous materials (the coefficients of viscous and inertial resistances, internal heat transfer, and thermal conductivity) is made.

2. HEAT AND MASS TRANSFER INVOLVING CHEMICAL CONVERSIONS

2.1 The Statement and Solution of Some New Conjugated Problems of Unsteady-State Heat and Mass Transfer between Reacting Bodies and a Reacting Gas Stream

V. M. Agranat, V. N. Bertsun, A. M. Grishin, and V. I. Zinchenko
Research Institute of Applied Mathematics and Mechanics, Tomsk

The problem of heat and mass transfer between a reacting gas stream and carbon graphite solids in the neighborhood of a (forward stagnation) point is considered, and the effect of chemical reactions on the characteristics of unsteady-state heat transfer is analyzed. By means of a dimensional analysis, computer calculations and the methods of dynamic systems theory, the development of unsteady-state conditions of heat and mass transfer is found to be prolonged. In a number of cases an oscillating regime is effected as the fields of temperature and the species concentrations change periodically with time.

English translation will appear in Heat Transfer—*Soviet Research*

2.2 Heat and Mass Transfer in Chemically Reacting Systems with Gas Formation

G. A. Akselrud, A. I. Dubynin, and Ya. M. Gumnitskiy
Lvov Polytechnical Institute

Heat and mass transfer in a solid-liquid system with chemical reactions followed by gas formation is studied. The limit of the analogy with boiling heat transfer is determined. Simple critical equations are obtained for the estimation of mass transfer coefficients and the temperature drop at the phase interface.

English translation will appear in Heat Transfer—*Soviet Research*

2.3 Multiple Steady States of Heat and Mass Transfer in a Reactive Stream

L. Yu. Artyukh, E. A. Zakarin, V. P. Kashkarov, and S. M. Orlova
Kazakh State University, Alma-Ata

Theoretical steady states and the transient characteristics of a premixed reactive system in a one-dimensional laminar gas stream were studied using approximate graphic-analytical and exact numerical methods. A comparison of the results obtained using both methods shows that the approximate graphic-analytical method can give a satisfactory prediction of the critical transient conditions.

2.4 Heat and Mass Transfer of an Fe-C System of Particles Interacting with Oxygen

V. I. Baptizmanskiy, V. A. Fedoseyev, V. B. Okhotskiy, K. S. Prosvirin, Yu. A. Ardelyan, V. V. Kuryatnikov, and A. N. Kovsik
Institute of Metallurgy, Dnepropetrovsk

The burning of metal particles in an Fe-C system in oxygen was studied experimentally. A heat and mass transfer model of the process has been developed. Numerical calculations based on this model are in good agreement with the experimental data.

2.5 The Burning of Hydrocarbons in a Boundary Layer

V. L. Batiyevskiy and G. T. Sergeyev
Luikov Heat and Mass Transfer Institute, Minsk

The results are presented of a study of heat and mass transfer with propane injection through a semiinfinite penetrable plate in an air flow. The problem is solved for diffusion burning with the reducing and oxidizing zones of the process being described by a two-front reaction model. Calculations based on the theory are found to be in good agreement with the experimental results.

2.6 A Theoretical and Experimental Study of the Destruction of Graphite in a Dissociating Air Flow

T. A. Bovina, Yu. V. Zvyagin, N. V. Markelov, and Yu. V. Chudetskiy
Moscow

A method of describing the heating of carbon materials and their erosion in a high-temperature air flow is suggested. On the assumption of chemical equilibrium at the carbon surface, and taking into account the heat effects of carbon combustion and sublimation, a calculation method is developed. The equilibrium evaporation of carbon in the form of C_1, C_2, and C_3 molecules is considered. The results of the calculations are in good argeement with the experimental data obtained for a rather wide range of temperature and stagnation pressure.

2.7 Research into Local Heat Transfer to Carbon Acid Vapors Accounting for Their Dimer-Monomer Interactions under Different Flow Conditions

V. P. Bolshov, E. S. Sergeyenko, V. I. Skomorokhov, and A. G. Usmanov
Kirov Institute of Chemical Engineering, Kazan

This paper deals with an experimental study of the local heat transfer to dissociating vapors of formic and acetic acids in a vertical tube flow. The influence of the major variables on the rate of heat transfer is studied. Some correlations are determined which may be used to calculate the heat transfer in reactors (for Lewis numbers from 0.8 to 1.2).

English translation will appear in Heat Transfer—*Soviet Research*

2.8 Some Studies of Active Heat Protection in Hypersonic Hot-Shot Tunnels

B. V. Boshenyatov, I. G. Drucker, L. Ya. Treyer, and M. I. Yaroslavtsev
Institute of Theoretical & Applied Mechanics, Novosibirsk

This paper presents an experimental study of heat and mass transfer in hypersonic high-enthalpy facilities of short duration. The experimental data on the active heat protection obtained using a uniform blowing-in of gaseous and liquid coolants in a hot-shot tunnel are also discussed. Good agreement with numerical computations in a range of small blowing confirms the usefulness of hot-shot tunnels for studying heat and mass transfer in a hypersonic range of flight.

2.9 Heat and Mass Transfer in a Turbulent Boundary Layer with Suction under Nonisothermal Conditions

E. P. Volchkov, E. I. Sinayko, and V. I. Terekhov
Thermophysics Institute, Siberian Department of USSR Academy of Sciences, Novosibirsk

According to the asymptotic theory of S. S. Kutateladze and A. I. Leont'yev, relative coefficietns of heat and mass transfer are obtained with uniform suction of a non-isothermal turbulent boundary layer. Experimental results are given of the effect of suction and nonisothermal conditions on the heat and mass transfer in a boundary layer at a chemically active surface. The temperature factor was measured to be 7.5, a relative mass velocity on the wall being up to $3 \cdot 10^{-3}$.

English translation will appear in Heat Transfer—*Soviet Research*

2.10 An Experimental Investigation of Heat and Mass Transfer Conditions as Applied to Controlled Heat-protecting Coatings of Gas-turbine Nozzle Blades

G. N. Delyagin and E. M. Shavartsshteyn
Institute of Mineral Fuels and the Dzerzhinskiy All-Union Thermal Engineering Institute, Moscow

The heat and mass transfer between a cooled gas-turbine blade and an SiO_2-aerosol is experimentally investigated in a temperature range from 1300 to $1960°K$ and a particle concentrations from 10^{-4} to $3 \cdot 10^{-3}$ (Re = $4 \cdot 10^4$ and 10^5, S = 2.66). The conditions for the formation of sintered cracked-off deposits as well as for rough, smooth sintered and melted deposits, are determined.

English translation will appear in Heat Transfer—*Soviet Research*

2.11 Thermal Explosions in a Reacting Variable Viscosity Fluid Flow between Coaxial Cylinders

P. V. Zhirkov, S. A. Bostandzhiyan, V. I. Boyarchenko, and Zh. A. Zinenko
Kuibyshev Polytechnical Institute

A Couette flow of a reactive fluid between two cylinders is considered. The fluid viscosity depends on the temperature and the conversion of the material. Heat generation due to both a chemical reaction and a dissipation of energy, is taken into account. The plots of the critical conditions vs. mechanical parameters, and of temperature and hydrodynamic characteristics vs. time, are presented.

2.12 Heat and Mass Exchange in High-Temperature Oxidation of Metals

M. V. Zake, V. N. Kovalev, V. E. Liepinya, and V. K. Melnikov
Institute of Physics, Riga

Heat, mass and charge transfer involving chemical reactions on metal surfaces with transient heating in a low-temperature argon-oxygen plasma flow has been studied. The kinetic constants in the Arrenius equation are obtained for Ti and Zr. The effect of chemical reactions on the heating of the metal, the convective heat flux, and the radiation energy transfer is demonstrated. For the ignition process, emittance ϵ is found to be nearly equal to unity. A chemoluminescence band in the radiation spectrum is observed. A nonequilibrium emission of electrons from the plasma-metal interface is found.

English translation will appear in Heat Transfer—*Soviet Research*

2.13 An Investigation of Heat Transfer in a Regenerator with Chemically Reacting Coolant

G. E. Zenkovich, B. I. Lomashev, A. A. Mikhalevich, V. A. Nemtsev, and V. B. Nesterenko
Institute of Nuclear Power Engineering, Minsk

An experimental and theoretical investigation was carried out on the heat transfer in a heat exchanger with a chemically reacting gas coolant, N_2O_4. The kinetics of the chemical reaction have a considerable effect on both the hot and the cold sides.

English translation will appear in Heat Transfer—*Soviet Research*

2.14 A Numerical Investigation of the Heat and Mass Transfer in a Turbulent Flow of N_2O_4 in a Circular Tube and a Rod Bundle

T. I. Mikryukova, V. B. Nesterenko, V. Yu. Petrovich, B. E. Tverkovkin, and N. N. Tuchin, A. P. Yakushev
Institute of Nuclear Power Engineering, Minsk

The major results of a numberical investigation of the heat and mass transfer in a flow of nitric tetroxide in smooth and rough tubes, and in a rod bundle are presented. The suggested mathematical model for a circular tube is experimentally verified. A relation is given for the calculation of Nu for a N_2O_4 flow in rough tubes. A circumferential distribution of temperature and concentration is calculated for different pressures.

English translation will appear in Heat Transfer—*Soviet Research*

2.15 A Prediction of the Properties of Chemically Reacting N_2O_4 Turbulent Flow in a Tube

M. D. Mikhaylov, V. B. Nesterenko, K. I. Spasov, and B. E. Tverkovkin
*Centre of Applied Mathematics, Sofia, Bulgaria and
Institute of Nuclear Power Engineering, Minsk*

An approach is suggested and the results of a numerical solution to the momentum, energy, and mass conservation equations for turbulent N_2O_4 flow in a circular tube with variable thermophysical properties are presented. The results are compared with those calculated using the Patankar–Spalding method. The effect of the Reynolds number, the pressure of the mixture, and the entrance gas properties on the distribution of gas parameters in the tube is demonstrated.

2.16 An Approximate Theory of Heat and Mass Transfer with Condensation of Chemically Reacting Gas Inside a Vertical Tube

A. A. Mikhalevich and A. E. Sinkevich
Institute of Nuclear Power Engineering, Minsk

The condensation of a chemically reacting gas $N_2O_4 \rightleftarrows 2\,NO_2 \rightleftarrows 2\,NO + O_2$, of a nonequilibrium composition in a vertical tube is considered. An approximate distribution of temperature, gas composition, intensity of condensation and other parameters of heat transfer and hydrodynamics along the condensation region was obtained.

2.17 An Analytical Study of the Heat and Mass Transfer in a Chemically Nonequilibrium Mixture (2NO$_2$ $\rightleftarrows$ 2 NO + O$_2$) in a Laminar Plane Duct on an Inert Impermeable Surface at Constant Temperature

L. V. Mishina and G. Z. Serebryanyy
Institute of Nuclear Power Engineering, Minsk

Errors due to the linearization of the mass source in a specific system are analyzed. The Fourier method has been used to solve a system of linearized partial differential equations. Analytical expressions are obtained for the main heat and mass transfer properties of a chemically nonequilibrium mixture. The application of a zonal calculation procedure gives a quite satisfactory agreement with the computer solution of a system of nonlinear differential equations governing heat and mass transfer in the system under consideration.

2.18 Energy and Mass Transfer in a Dissociating Laminar Air Boundary Layer with Finite-Rate Chemistry

A. Oyegbezan and J. Algermissen
Institut für Thermodynamik der Luftund Raumfahrt der Universität Stuttgart, West Germany

A numerical analysis of the energy and mass transfer processes in a laminar air boundary layer is presented. Employing a finite-difference model to describe the chemical process, the governing nonlinear partial differential equations of a two-dimensional viscous flow are solved by a finite-difference procedure based on the Du Fort–Frankel scheme. Predictions of the energy flux to the wall, the wall skin-friction coefficient, velocity, temperature, and the mass fractions of chemical species in the boundary layer are obtained. The influence of the degree of wall cooling on the predicted boundary-layer variables and of the wall catalycity on the obtained energy fluxes to the wall is determined.

2.19 Heat and Mass Transfer in a Dissociating and Vibrationally Relaxing Turbulent Gas Flow in a Two-Dimensional Channel

M. S. Povarnitsyn
Novosibirsk

The heating and turbulent flow of a diatomic gas in a two-dimensional channel with heat-emitting catalytic walls, are discussed, along with the nonequilibrium processes of dissociation and vibrational relaxation. A system of equations is closed by the transfer equation for total viscosity. The temperature, concentration, total viscosity, and other functions are calculated for the case of hydrogen.

2.20 Heat and Mass Transfer in a Flat Channel with the Deposition of Products of Homogeneous Chemical Reactions on the Wall

V. P. Popov and Yu. S. Skoropanov
Luikov Heat and Mass Transfer Institute, Minsk

A numerical study is made of heat and mass transfer with chemical deposition from a gas phase in the presence of homogeneous chemical reactions. The formation of a mono-crystal layer of Ga As by $Ca(CH_3)_3$ pyrolis in a mixture of H_2 and AsH_3 and the deposition of a dielectric SiO_2 film by SiH_4 oxidation in a mixture of O_2 and N_2 are used as examples.

2.21 An Investigation of Thermal Self-Ignition under Conditions of Natural Convection

K. V. Pribytkova, S. I. Khudyayev, and E. A. Shtessel
Institute of Chemical Physics, Moscow

Using numerical methods and an electronic digital computer, the behavior of the convective motions in liquids with volumetric chemical heat sources, as well as the effect of the developed convection on the reaction, is studied.

2.22 A Theoretical Investigation of the Heat and Mass Transfer between Reacting Solid Components Separated by a Gaseous Film

S. S. Rybanin and V. A. Strunin
Institute of Chemical Physics, Moscow

The heat and mass transfer between two reacting solids separated by a gaseous film of their products is investigated. An analytical relationship is obtained for the effect of the Reynolds number and the ratio between the solid's gasification rates on the Nusselt number. Various chemical reactions between the components are considered.

2.23 Heat Exchange with a Barrier at Varying High-Temperature Gas Jet Parameters

V. L. Sergeyev and V. P. Veselov
Luikov Heat and Mass Transfer Institute, Minsk

The results are presented of a study on the heat transfer between a high-temperature gas jet and a barrier in the region of a stagnation point with the main gas stream parameters varying with time. The process is shown to be quasi-stationary in the range of parameters studied.

English translation will appear in Heat Transfer—*Soviet Research*

2.24 An Experimental Study of Radiant Flux Attenuation by a Gaseous Layer Formed by the Products of Material Destruction Resulting from Radiation

Yu. B. Sokolova, G. Ya. Umarov, R. A. Zakhidov, and B. B. Petrikevich
The Starodubtsev Engineering Physics Institute, Tashkent

An experimental investigation is carried out on the effect of a 30–105 W/cm^2 radiant flux on glassplastics, using as a source of radiation a radiating stove "Uran-1," having an arc-discharge xenon lamp of 1000 W. The dynamics of the formation of a screening aerosol layer and its influence on the heat and mass transfer of the material tested has been analyzed. Some characteristics of the radiant destruction were established.

2.25 A New Model of Turbulent Combustion

D. B. Spalding
Imperial College of Science and Technology, London

The paper presents a new model of turbulent chemical reaction which is applicable to the turbulent diffusion flame, to premixed flames and to combustion phenomena having features in common with both these extreme cases. The new model has been incorporated into a computer program for predicting turbulent flame phenomena. The rate reaction is found to be proportional to the mean rate of strain of the fluid, density of the fluid and concentration of fuel. The suggested model appears to be more general than earlier methods and to have a simplicity which renders it especially convenient for practical use.

This paper appears on page 81, this volume

2.26 On the Effect of Increased Diffusion and Conduction in Reacting Turbulent Flows

R. S. Tyulpanov
Leningrad Technological Institute

Using an analysis of the effects of increased diffusion and conduction, an expression is obtained for the coefficients of heat and mass transfer in reacting turbulent flows. These coefficients depend on the turbulence of the flows. For this reason a change in turbulence can sometimes lead to a considerable change in chemical kinetics.

2.27 Heat and Mass Transfer Involving a Reaction between a Solid Fuel and an Oxidant

Yu. A. Finayev and G. T. Sergeyev
Luikov Heat and Mass Transfer Institute, Minsk

Heat and mass transfer involving a reaction between an oxidant and a solid fuel of high volatility is studied. In contrast with current opinion, distinct and successive stages in the burning of volatiles and coke residues are not observed. Pseudostages only are found. The mechanism and conditions required for the ignition of fuel particles as well as the rate of burning of the separate components of some elementary fuels are considered.

2.28 The Effect of Natural Convection on Mass Transfer in Chemical Transport

E. A. Shtessel and L. V. Kalashnikova
Institute of Chemical Physics, Moscow

The effect of natural convection on the rate of metal transfer in chemical transport reactions is studied experimentally. A formula for predicting the rate of formation in the presence of natural convection is obtained.

3. HEAT AND MASS TRANSFER WITH PHASE CHANGES

3.1 A Study of Heat and Mass Transfer in a Heat Pipe using a Mathematical Simulation Method

I. N. Avakyan, I. I. Kulagin, and O. V. Shelud'ko
Northwestern Polytechnical Institute, Leningrad

A mathematical model is developed for a coaxial heat pipe operation. A system of interrelated differential heat and mass transfer equations is solved using the method of characteristics in time together with the method of elimination. The results are employed to identify an optimum regime of start-up in a sodium coaxial heat pipe.

3.2 Some Aspects of Heat and Mass Transfer in an Injector-Condenser

S. I. Aladiev
Krzhizhanovsky Power Engineering Institute, Moscow

The effect of heat and mass transfer in an injector on the temperatures, void fraction, and pressure fields is studied. The transfer of mass, energy, and momentum in drops of different sizes and velocities is considered.

English translation will appear in Fluid Mechanics–*Soviet Research.*

3.3 Interaction of Molten Tin with Water: Effect of Temperature Stratification in the Coolant

V. H. Arakeri, I. Catton, and W. E. Kastenberg
University of California, Los Angeles

It has been found experimentally that the thermal constraints required to produce vapor explosions when molten tin is dropped into a water bath can be relaxed by introducing a stable thermal stratification in the coolant. In the present work, the constraints could be relaxed sufficiently to show that achievement of limiting coolant superheats associated with spontaneous nucleation is not the only mechanism by which vapor explosions in liquid–liquid systems are possible. In addition, multiflash photography with extremely short duration exposure times per flash has shown that transition or nucleate boiling is a possible triggering mechanism for vapor explosions.

This paper appears on page 102, this volume.

3.4 Some Characteristics of Heat and Mass Transfer with Condensation of Metal Additives from Mixtures of Gases

K. M. Aref'yev, V. M. Borishanskiy, N. I. Ivashchenko, N. M. Fishman,
B. M. Khomchenkov, and B. F. Remarchuk
*Kalinin Polytechnical Institute and the Polzunov Central Boiler and
Turbine Institute, Leningrad*

Some specific features of the heat and mass transfer with condensation of metal vapors from moving mixtures of gases involving diffusion and mist formation are considered.

Experimental data on diffusion coefficients are used to obtain a relationship for a diffusion of molecules of mixtures of monoatomic metal vapors interacting with He, Ar, and N_2.

English translation will appear in Heat Transfer—*Soviet Research*

3.5 Transport Characteristics and Some Operating Features of a Heat Pipe with Water, Methanol, and Freon-113

J. Asakavičius, V. Gaigalis, and V. Eva
*Institute of Physical and Technical Problems of Energetics
Lithuanian Academy of Sciences, Kaunas*

Experimental characteristics of a heat pipe with a tightly pressed screened wick are presented. A circumferential temperature variation of the wick and temperature hysteresis effects are observed with a decrease in the heat load.

3.6 Experimental Investigation of Convective Boiling of Ammonia at High Pressures

G. Barthau
*Institut für Thermodynamik und Wärmetechnik
Stuttgart, West Germany*

Heat transfer coefficients for a fully developed nucleate boiling flow of ammonia (upward flow) at pressures of $34 \leqslant p \leqslant 108$ bar ($0.30 \leqslant p/p_{cr} \leqslant 0.95$) were determined. The mass velocity was $\dot{m} \approx 1000$ kg/m² sec, the heat fluxes, $5 \leqslant q \leqslant 130$ W/cm², and the heat transfer coefficients obtained, $9 \leqslant \alpha \leqslant 38$ W/cm²K. The dnb-points were determined for each pressure investigated. The test section consisted of a gold-plated nickel-tube with an inner diameter, $d_i = 30.34$ mm, an outer diameter, $d_o = 40.00$ mm and a length, $1 = 474$ mm. The tube was heated from the outside by a condensing stream.

This paper appears on page 106, this volume

3.7 An Investigation of the Heat and Mass Transfer Crisis in Closed Two-Phase Thermosyphons with the Cooling of Metallurgic Furnace Elements

M. K. Bezrodniy, S. N. Faynzilberg, A. I. Beloivan, E. A. Kondrusik, and N. Y. Koloskova
Kiev Polytechnical Institute

Experimental results are presented of the heat transfer capacity of annular and circular thermosyphons having an insert for a condensate return into a heat supply zone. Use of the insert is found to markedly increase the heat transfer capacity of the thermosyphon.

English translation will appear in Heat Transfer—*Soviet Research*

3.8 Static and Dynamic Characteristics of a Gas Heat Pipe under Complex Heat Effects

A. P. Belyakov and E. S. Platunov
Leningrad Institute of Precision Mechanics and Optics

Equations are obtained for the evaluation of some static and dynamic errors of thermostating under the simultaneous and disturbing effects of a gas reservoir, an outer heat agent and an object of stabilization. The transfer functions obtained allow the estimation of the output time of the "gas heat pipe–object" system for the prescribed level of thermostating as well as the determination of the allowable amplitude and frequency values of the fluctuations of the destabilizing effects.

3.9 Pool Boiling Maximum Heat Flux Correlations

J. Berghmans
Afdeling Toegepaste Mechanika en Energiekonversie
Katolieke Universiteit Leuven, Heverlee, Belgium

A consistent hydrodynamic theory for the prediction of the maximum heat flux (critical heat flux) during pool boiling is presented. An analysis of the interfacial stability of the vapor columns formed as the maximum heat flux is approached is performed. Two different cases are distinguished. In the first, the gravitational forces are small compared with the surface tension forces. The classical hydrodynamic approach (1), based upon the concept of vapor column breakup, cannot explain the occurrence of the maximum heat flux in this case. For cylindrical and spherical heaters a new model is proposed which satisfactorily explains the observed dependence of the critical heat flux upon the parameters involved. In the second case the gravitational forces are large compared with the surface tension forces. For this case it is shown that the same correlation for the critical heat flux as previously reported can be obtained, based upon the results of a column stability analysis.

This paper appears on page 111, this volume

3.10 Heat Transfer and Pressure Drop in Once-Through Steam Generators

G. Bianchi and M. Cumo
Centro di Studi Nucleari Della Casaccia, Rome

An experimental and theoretical research program on once-through steam generators has been carried out at the CNEN Laboratorio Tecnologie Reattori, the main purpose being to provide designers with "ad hoc" and reliable correlations for thermal design. Both a straight, vertical tube, and a coiled one were imployed in a 4 MW water loop. Different regions of heat transfer regimes were investigated and a set of correlations has been determined, each correlation representing a single characteristic region or regime. Large differences in the burnout characteristics in the pressure ranges of $50 < p < 90$ kg/cm^2 and $110 < p < 180$ kg/cm^2 were found. The pressure drop design correlations were also investigated. A homogeneous model was found to have an intrinsic validity and very simple correction functions, depending on the pressure, are suggested.

This paper appears on page 120, this volume

3.11 Heat Transfer Enhancement with Film Condensation Affected by an Electric Field

M. K. Bologa and A. B. Didkovskiy
Institute of Applied Physics, Kishinev

An experimental study of heat transfer and fluid dynamics with condensation of pure vapors of dielectric liquids as affected by electric fields has revealed a pronounced enhancement of heat transfer due to a decrease in the condensation film thickness. Heat transfer is found to depend on the strength of the electric field. It is shown that heat transfer is enhanced by the use of polar liquids as heat-transfer agents. A general correlation is proposed between heat transfer and the condensation affected by an electric field.

English translation will appear in Heat Transfer—*Soviet Research*

3.12 A Study of a Gas Admixture with Condensation in a Vertical Tube

V. M. Borishanskiy, D. I. Volkov, N. I. Ivashchenko, O. P. Krektunov, and L. A. Vorontsova, G. A. Makarova, and N. M. Fishman
The Polzunov Central Boiler and Turbine Institute and the Leningrad Shipbuilding Institute

An experimental study is carried out on heat transfer with partial condensation of pure steam and vapor from a vapor-gas mixture in a vertical tube at mixture pressures of $P = 0.8$ and 3.0 MPa, heat loads of $(15–450) 10^3$ W/m^2, inlet velocities of 2–50 m/s, outlet vapor contents of $x_2 \leqslant 0.5$ and volume gas concentrations of up to 2.5%.

The effect of the gas on heat transfer with condensation of the vapor-gas mixture in the tube is considered. Some calculations of the heat transfer with steam condensation from the vapor-gas mixture are suggested.

English translation will appear in Heat Transfer—*Soviet Research*

3.13 An Investigation of Vibrational Effects on Heat Transfer and Fluid Dynamics with Steam Condensation on a Single Tube

Yu. M. Brodov, R. Z. Savel'yèv, V. A. Permyakov, V. K. Kuptsov, and A. G. Galperin
Ural Polytechnical Institute, Sverdlovsk and the Polzunov Central Boiler and Turbine Institute, Leningrad

The results of an analytical and experimental investigation of convective heat transfer with film condensation of motionless steam on horizontal and vertical vibrating tubes are presented.

English translation will appear in Heat Transfer—*Soviet Research*

3.14 Heat Transfer and Gas Dynamics with Hot Vapor Discharge into a Condenser through a Supersonic Nozzle

A. V. Buyevich, E. I. Lavrov, A. Z. Rosinskiy, O. I. Sokolov, and G. G. Shklover
Kaluga Turbine Plant

A vapor-discharge arrangement of the ejector type is considered with the Laval nozzle positioned in a tube bundle of a condenser.

It is found that the cooling of a discharged vapor is possible without the use of cooling water. The effect of some geometric and operational-parametric characteristics on the cooling efficiency of a high-speed vapor flow is considered.

English translation will appear in Heat Transfer—*Soviet Research*

3.15 An Investigation of Cryogenic Gas-Controlled Heat Pipes

L. L. Vasil'yev and S. V. Konev
Luikov Heat and Mass Transfer Institute, Minsk

The operational characteristics of cryogenic gas-controlled heat pipes are analyzed. The advantage of using a noncondensing gas in a cryogenic heat pipe in the presence of an additional reservoir is considered.

Experimental results are obtained for a liquid nitrogen heat pipe with various amounts of a heat-transfer agent.

English translation will appear in Heat Transfer—*Soviet Research*

3.16 Molecular Aspects of a Liquid Boiling Process
I. P. Vishnev
Moscow

The vaporization of liquids is considered from the point of view of molecular interactions. A generalized equation is presented along with some macro- and microcharacteristics of the boiling process. An energy system describing the molecules often used in vaporization is presented.

It is noted that the physical (energetic) properties of substances during boiling have a quasi-periodical dependence on the nature of the intermolecular bonds as well as on the relative molecular mass of the substance.

Relationships are proposed for the prediction of heat transfer with nucleate boiling of various liquids from helium to the liquid metals in a pressure range from low to supercritical.

3.17 An Investigation of Heat Transfer with Helium Boiling on Surfaces with Varying Orientations

I. P. Vishnev, I. A. Filatov, Ya. G. Vinokur, V. V. Gorokhov, and G. G. Svalov
All-Union Research Institute of Helium Engineering and All-Union Research Institute of the Cable Industry, Moscow

It is experimentally shown that the heat transfer coefficient and burnout resulting from nucleate boiling of helium are affected by the orientation and nonmetallic coating of a heated surface.

English translation will appear in Heat Transfer—*Soviet Research*

3.18 Heat Transfer Intensification during Dropwise Condensation of Organic Liquids

V. T. Vylkov
Research Division, Energoproyekt, Sofia

Transition from film to drop condensation was experimentally studied by means of various lyophobilizers for 21 organic liquids with surface tension at the saturation temperature between 0.0176 and 0.036 N/m. The film to drop condensation is quantitatively investigated for a number of liquids under heat flux increase conditions. An increase in the heat transfer coefficient is established experimentally for drop condensation of ethylenediamine and phenol. An experimental study of the vapor-to-wall heat transfer coefficient is made with both drop and film condensation over a horizontal tube. The experimental data are used to determine a critical dependence which is compared with that for steam.

This paper appears on page 134, this volume

3.19 Heat Transfer with Boiling of Aqueous Solutions of Electrolytes

N. I. Gelperin, B. M. Gurovich, L. N. Taktayeva, G. Sh. Polishchuk, and
A. Ya. Gorelov
Lomonosov Institute of Light Chemical Technology, Moscow
Beruni Polytechnical Institute, Tashkent

The experimental results of a study on heat transfer with boiling of aqueous solutions of
$NaCl$, $NaNO_3$, Na_2SO_4, $MgCl_2$, $CaCl_2$, $CuSO_4$, KNO_3, $K_2Cr_2O_7$, NH_4NO_3, NH_4Cl,
$(NH_4)SO_4$, $Ca(NO_3)_2$ at atmospheric pressure and natural circulation in a vertical tube
and with pool boiling of $NaCl$, $NaNO_3$, Na_2SO_4, NH_4NO_3, KNO_3 solutions are
presented.

The dependence of the heat transfer coefficient in a developed boiling region on the
physical-chemical properties of liquids and on the heat load is considered along with the
dependence of the mean heat transfer coefficient on the physical-chemical properties of
solutions, heat load, height of the boiler tube, and the apparent liquid level. The pool
boiling heat transfer coefficient is presented as a function of the thermophysical
properties of solutions and the heat load.

English translation will appear in Heat Transfer–*Soviet Research*

3.20 Heat Transfer with Liquid Boiling in a Dispersed Bed of Solid Particles

Z. R. Gorbis and M. I. Berman
Odessa Institute of Refrigeration Technology

A model and some relationships derived from it are presented for heat transfer with
liquid boiling in a dispersed bed of particles. The predicted results are compared with the
experimental data on the boiling of water, ethanol, and an $NaCl$ solution.

English translation will appear in Heat Transfer–*Soviet Research*

3.21 An Experimental and Theoretical Investigation of a Heat Pipe Operation

I. G. Goryachko and G. V. Zhizhin
Leningrad

The results of a study of the temperature along a sodium heat pipe and its heat fluxes
under supersonic flow conditions are presented. The experimental data are compared with
the results predicted by a unidimensional steady-state theory for a delivery nozzle with a
dry vapor flow.

English translation will appear in Heat Transfer–*Soviet Research*

3.22 Generalization of Boiling Characteristics using the Thermodynamic Similarity Method

M. A. Gotovskiy, A. V. Borishanskaya, and G. P. Danilova
*The Polzunov Central Turbine and Boiler Institute and the
Institute of Refrigeration Technology, Leningrad*

Experimental data on departure diameters are correlated with pool boiling heat transfer. Formulae describing the dependence of the departure diameters and boiling heat transfer coefficients on dimensionless thermodynamic variables are obtained.

3.23 The Enhancement of Heat Transfer with Low-Temperature Liquid Boiling at Small Heat Flux Rates

G. N. Danilova, E. I. Guigo, A. V. Borishanskaya, V. G. Bukin,
V. A. Dyundin, A. A. Kozyrev, L. S. Malkov, and G. I. Malyugin
Institute of the Refrigeration Technology, Leningrad

The experimental results of a study on various ways of advancing heat transfer with Freon and ammonia boiling on horizontal tubes and in vertical channels are presented. A high rate of boiling heat transfer is established on porous and screened coatings, in a falling-down film, and in slots.

English translation will appear in Heat Transfer—*Soviet Research*

3.24 Heat Transfer and Hydraulic Resistance in High-Quality Steam-Water Mixtures Flowing in an Annular Channel

Yu. I. Dzarasov
Daghestan Department of Energetics Research, Makhachkala

The results of a complex investigation into the heat transfer coefficient, pressure drop, water flow rate in the wall layer, and flow patterns in an annular evaporating channel with internal heating at pressures of 35 and 70 bar are presented. The investigation is primarily concerned with the region of high vapor fractions and mass velocities between 1000 and 3000 $kg/sec \cdot m^2$, where two-phase convection-controlled heat transfer predominates. The experimental data are found to be in good agreement with those of other studies in which annular tubes were used. The convective two-phase heat transfer coefficient can be sufficiently correlated using a proposed equation. A dispersed-annular flow model is postulated. The model assumes that heat is transferred through a very thin laminar liquid film on the wall as a homogeneous steam-droplet mixture flows through the core.

English translation will appear in Heat Transfer—*Soviet Research*

3.25 A Determination of the Boundaries between the Annular and Dispersed Patterns of a Two-Phase Flow

V. E. Doroshchuk, L. L. Levitan, and F. P. Lantzman
The Dzerzhinskiy All-Union Thermal Engineering Institute, Moscow

Based on systematic investigations of burnout due to the drying out of a wall liquid film (burnout of the second type), the boundary between the annular and dispersed patterns of a two-phase flow is experimentally determined in the pressure range from 1.0 to 16.7 MN/m^2 and with mass flow rates between 750 and 3000 kg/m^2 s. Simple semiempirical and empirical equations are obtained generalizing the experimental results with sufficient accuracy.

English translation will appear in Heat Transfer—*Soviet Research*

3.26 Local Void Fraction Variations for Two-Phase Flow Constraints

G. M. Zaki and M. M. Marwan
Atomic Energy Establishment, Cairo, United Arab Republic

The pressure loss and recovery due to a sudden contraction and expansion of a flow channel are considered for a two-component two-phase flow. The radial void fraction distribution is measured for water and air flow rates of up to 4.2m^3/g and 8.7m^3/g, respectively, with tube diameters ratios of 0.28 for expansion and 3.57 for contraction.

The local void fraction data have a symmetrical radial profile inside the tubes and show a decrease in the void fraction value after expansion and an increase in it after contraction of the cross section.

The Petrick correlation shows a reasonable agreement when compared with the void measurements within a deviation of $+14\%$ to -6% for expansion and $+18\%$ to -9%, for contraction. A relationship for calculating the local variation of the void fraction is of the form

$$\eta(\theta, z, t) = C\, e^{i(k\,z\,+\,m\theta)}\, e^{i\omega t}$$

An investigation was carried out on various parameters affecting the pressure gradient of a two-phase flow.

English translation will appear in Fluid Mechanics—*Soviet Research*

3.27 A Mathematical Model for Boiling Water Reactors

G. M. Zaki, M. M. Farahat, and M. El-Genki
Atomic Energy Establishment, Cairo, United Arab Republic

A theoretical model is presented for the prediction of some local parameters of the boiling channels of nuclear reactors. The model is based on steady-state infinite-difference equations which describe mass, energy, and momentum conservation under thermal nonequilibrium conditions.

Employing an iteration technique, two-phase flow parameters (the true void fraction of vapor, steam quality, evaporation rate, and true phase velocities ratio) for a 59 MW ship boiling reactor operating at 49.27 atm are considered.

The mathematical model shows good agreement with the experimental results.

3.28 On the Prediction of the Heat Transfer with Surface Boiling of Water in Tubes

A. F. Zaletnev, A. F. Akselrod, and A. V. Tikhonov
Leningrad Institute of Precision Mechanics and Optics

A model of a heat transfer process is proposed. An analysis of some of the thermohydraulic characteristics of the surface boiling of liquids in tubes is considered.

English translation will appear in Heat Transfer—*Soviet Research*

3.29 An Experimental Investigation of Sodium Boiling Heat Transfer and Pressure Losses in a Vertical Tube

Yu. Zeygarnik and V. D. Litvinov
Institute of High Temperatures, USSR Academy of Sciences, Moscow

The heat transfer and pressure losses in a sodium boiling upward flow are investigated. The experimental data on the friction resistance are described by the Martinelli–Lockhart correlation with a correction factor for the flow rate.

The measured heat transfer coefficients are found to be very high confirming the hypothesis that sodium boiling heat transfer occurs by vaporization from the liquid film surface.

English translation will appear in Heat Transfer—*Soviet Research*

3.30 An Analytical and Experimental Study of Heat Transfer with Vapor Condensation on Finned Surfaces

N. V. Zozulya, V. A. Karkhu, and V. P. Borovkov
Engineering Thermophysics Institute, Ukrainian Academy of Sciences, Kiev

On the basis of a proposed model, a differential equation for a condensate film moving between the fins of a surface if presented.

A relationship between the heat transfer rate and the vapor condensation on horizontal finned tubes is proposed, and can be used in the design of heat exchange facilities.

English translation will appear in Heat Transfer—*Soviet Research*

3.31 The Problem of Internal Thermal Transfer in Steam-Liquid Flows

V. A. Zysin, E. L. Kitanin, F. R. Latypov, and B. E. Ivanov
Leningrad Polytechnical Institute

A comparison is made between the calculated diabatic and adiabatic coefficients of heat transfer and the outflow of three different liquids close to the saturation state.

Based on an analysis of the internal heat transfer in similar flows, the boiling-up regime with outflow through a diaphragm is studied.

English translation will appear in Fluid Mechanics—*Soviet Research*

3.32 An Investigation of Heat and Mass Transfer in the Suspended Moisture in a High-Temperature Vapor-Gas Flow

V. A. Zysin, V. V. Nevinskiy, E. L. Pludovskaya, and V. I. Rosenblum
Leningrad Polytechnical Institute

The problem of nonequilibrium heterogeneous suspended moisture moving in a vapor-gas flow is formulated. The calculation of some thermodynamic properties of the mixture components and the flow characteristics in a channel is described. The contribution of interphase heat transfer to the processes of vapor-liquid flow is evaluated using simplified models. The numerical results are compared with the experimental data on a heterogeneous vapor-liquid flow. A numerical analysis of two-phase systems movement with large thermal nonequilibrium is shown to be possible.

English translation will appear in Fluid Mechanics—*Soviet Research*

3.33 Heat Transfer with Refrigerant Condensation inside Channels

O. P. Ivanov, C. N. Danilova, V. O. Mamchenko, and Yu. N. Shiryayev
Technological Institute of Refrigeration Technology, Leningrad

Theoretical and experimental results of a study on hydrodynamics and heat transfer with condensation of refrigerants inside channels are presented.

On the basis of a local characteristic analysis, a thermohydrodynamic method is proposed for calculations involving through condensers.

3.34 A Conventional Signal Quantification-Based Analysis of the Fluctuations of Liquid Superheat during Boiling

L. Iović and N. Afgan
Boris Kidrič Institute, Belgrade, Yugoslavia

Liquid superheat near a wall during boiling represents an unsteady value which, in addition to an incidental character, possesses a governing unsteady character. The incidental character of the change is attributed to stochastic fluctuations of the parameters affecting the liquid superheat, while the nonincidental character is related to the dynamics of the onset of a vapor phase in a two-phase boundary layer.

An analysis of the fluctuation of liquid superheat with boiling, based on the conventional quantification of a recording of temperature fluctuation in a two-phase boundary layer is presented. A statistical analysis of the signal corresponding to various systems is performed giving a temperature distribution for each system under consideration. The method employed allows the mean vapor content to be determined at each point along the two-phase boundary layer. A variation in the superheat in the surrounding liquid and the superheated layer, which depends on the distance from the heated surface, causes nonaccidental temperature variations in the two-phase boundary layer which are explained by a mechanism of heat supply at a phase boundary.

3.35 Unsteady Steam Condensation on a Thick-Walled Surface

T. V. Iordanova and V. P. Pavlov
Science Division, Energoproyekt, Sofia, Bulgaria

Pure steam condensation on a thick-walled surface is considered.

The problem is formulated for both conjugated-unsteady condensation and unsteady heat conduction. A mathematical model of the process is proposed and the system of equations obtained is computed with the aid of a numerical method.

3.36 An Investigation of Cryogenic Liquid Boiling

Yu. A. Kirichenko, M. L. Dolgoy, N. M. Levchenko, V. V. Tsybulskiy, L. A. Slobozhanin, and N. S. Shcherbakova
Cryogenics Institute, Kharkov

Some internal and integral characteristics of cryogenic liquid boiling are investigated in a wide range of saturation pressure. It is shown that the change in the boiling characteristics with pressure is different for "low" and "high" pressures and is controlled by quasi-static and dynamic regimes of vapor bubble departure.

Formulas are determined for the calculation of nucleate boiling heat transfer coefficients, the first critical heat fluxes and the first critical temperature.

English translation will appear in Heat Transfer—*Soviet Research*

3.37 Effect of the Geometrical Factor on Critical Heat Flux in a Rod Bundle

T. Kobori, A. Kikuchi, T. Obata, and M. Matsuo
Power Reactor & Nuclear Fuel Development Corporation, Japan

Steady-state critical heat flux measurements were made in order to study the effect of rod bundle eccentricity on a boiling water flow using a full scale rod bundle in a pressure tube reactor.

The experiment was performed at 70 bars in a full scale rod bundle whose heated length was 3.7 m, and whose bundle eccentricity was 0–1.0 mm. Other conditions were as follows: mass velocity, 1000–2300 kg/m^{2}s, and inlet subcooling, 40–600 J/g.

Experimental data show that bundle eccentricity of between 0.6 and 1.0 mm lowers the critical heat flux by 20–30%. These phenomena can be explained by a subchannel distribution of the local steam quality deviation in the rod bundle.

This paper appears on page 142, this volume

3.38 A Study of the Heat Transfer Crisis in Subcooled Forced-Convection Water Boiling on a Finned Surface

D. Ya. Derevyanko, I. I. Dolgintsev, M. V. Makhalova, V. S. Sinelnikov, and S. A. Kovalev
Leningrad

The experimental method and results of a study of the critical heat fluxes in water boiling on finned surfaces of various geometries over a wide range of subcooling and at various rates are presented.

Generalized dependences are proposed to describe the data on critical heat fluxes in pool and forced-motion boiling.

3.39 An Investigation of the Effect of Surface Roughness on Boiling Heat Transfer

I. Z. Kopp
Leningrad

Based on a nucleation model, the effect of surface roughness on boiling heat transfer is shown. Equations giving a statistical description of a real surface are used to determine the number of nucleation sites.

The experimental results on mercury, boiling heat transfer on surfaces of different roughness, and specific heat fluxes of up to 3 • 10^6 W/m^2 are presented. The experimental data show satisfactory agreement with the analytical model.

English translation will appear in Heat Tranfer—*Soviet Research*

3.40 A Study of Light Hydrocarbons and Ethylene-Ethane Mixture Boiling

V. A. Kravchenko, Yu. N. Ostrovskiy, and L. F. Tolubinskaya
The Gas Institute and the Engineering Thermophysics Institute
Ukrainian Academy of Sciences, Kiev

Boiling heat transfer with light hydrocarbons of ethane, ethylene and their mixture is investigated.

Calculation equations are proposed for boiling heat transfer coefficients for pure hydrocarbons and ethylene–ethane mixtures.

English translation will appear in Heat Transfer—*Soviet Research*

3.41 An Analytical and Experimental Study of Air-Transpiration Cooling of Cylindrical Channels

V. V. Kudryavtsev, I. I. Rats, and A. I. Surovov
Moscow

The heat and mass transfer of gas-liquid flows in horizontal and vertical cylindrical channels under vacuum conditions is considered.

On the basis of the results obtained, a method for designing heat-exchangers with air-transpiration cooling is elaborated.

3.42 Acoustic Spectra in Critical Boiling Flows of Liquids

E. V. Lykov
Rostov-on-Don

Acoustic spectra in critical flows of surface boiling liquids are investigated.

Based on the experimental data, a concept of "thermoacoustic turbulence" in the region of the critical heat fluxes is formulated.

3.43 A Study of Drop Evaporation in Two-Phase Flow

J. Madejski
Polish Academy of Sciences, Warsaw

A study was made of local drop evaporation on channel walls involving multiple drop impingement. The condition necessary for such a situation is $(Re)_0 \geq (Re)$ where $(Re)_0$ is the Reynolds number of the impinging drop and (Re), is that of the rejected one. Putting $(Re)_0 = (Re)$, the actual value of the wall superheat can be estimated.

3.44 An Investigation of Heat and Mass Transfer Involving Two-Phase Transpiration Cooling inside a Porous Heat-Producign Element

V. A. Mayorov and L. L. Vasil'yev
Luikov Heat and Mass Transfer Institute, Minsk

An analytical model of two-phase transpiration cooling of a porous heat-producing element is developed. Based on the model, relationships between fluid dynamics and heat and mass transfer in a coolant flow evaporating inside a porous heat-producing element are studied.

The dimensions of the various regions of the coolant flow are established and the relative pressure drops in each region are determined.

3.45 The Disturbance Effect on Vapor-Water Flow Characteristics Resulting from the Sudden Expansion of a Channel Cross Section

S. Marković and N. Afgan
Boris Kidrič Institute, Belgrade, Yugoslavia

The pressure growth, irreversible pressure loss, and recovery length in a vapor-water flow are experimentally determined for various expansions of the cross. section. A generalized relationship between the flow recovery length, the degree of channel cross-section expansion and two-phase flow parameters is obtained. Based on laws of mass, energy, and momentum conservation, and employing a correlation for the flow recovery length, an expression is obtained for a coefficient of irreversible pressure loss due to the abrupt expansion of the vapor-water flow.

Experiments were carried out on a pressure loop with channel section diameters of σ = 0.250, σ = 0.347, and σ = 0.440, at an operating pressure from 5 to 15 bar, with mass flow rates between 500 and 2000 $kg/m^2 s$ and vapor content of up to 20%.

3.46 The Effect of Sodium Chloride Additions on the Heat Transfer Crisis in Upward and Downward Water Flows

L. S. Midler and M. E. Shitsman
Krzhizhanovskiy Power Engineering Institute, Moscow

Earlier studies have found that small additions of NaCl to supercritical pressure water flows completely change the heat transfer nature in upward and downward flows for certain combinations of the working parameters. Experimental data are presented on the heat transfer crisis in upward and downward water and NaCl solution flows for W_p = 1000 $kg/m^2 s$ and P = 100 $\cdot$ 10^5 Pa. In the range of parameters explored, the flow direction, i.e., buoyancy, does not affect the q and x values at the crisis point. However, the addition of salt to the water (0.13–3.8 g/l) results in a decrease in the boundary qualities as compared with those for pure water.

English translation will appear in Heat Transfer—*Soviet Research*

3.47 The Relationship between the Local and Average Values of Heat Transfer Coefficients and Hydraulic Resistances with Steam Condensation inside Channels

Z. L. Miropolskiy and L. R. Khasanov
Krzhizhanovskiy Power Engineering Institute, Moscow

The effect of a change in the heat flux along a channel on the local and average values of the heat transfer coefficients and on the hydraulic resistances, with steam condensation inside the channels, is considered.

English translation will appear in Heat Transfer—*Soviet Research*

3.48 Heat and Mass Transfer Involving Phase Conversions of Solutions Dispersed in a High-Temperature Gas Flow

I. L. Mostinskiy and D. I. Lamden
Institute of High Temperatures, USSR Academy of Sciences, Moscow

The evaporation of drops moving in a hot gas flow is considered. Analytical solutions are obtained for all stages of the process (including melting and heating) taking into account the mutual effect of a change in the size and velocity of the drops, the concentration of the solution therein, the vapor outflow, and various other factors. Expressions for the change in the evaporated admixture and solvent fractions along the flow in the case of evaporation of the dispersed system of drops are derived. The observation of a mutual sliding of drops and a time difference between the onset and the completion of evaporation of drops with different initial sizes is accounted for.

English translation will appear in Heat Transfer—*Soviet Research*

3.49 The Interrelation of Thermal and Hydrodynamic Characteristics in a Two-Phase Nonequilibrium Flow

E. I. Nevstruyeva and V. V. Tyutyayev
Institute of High Temperatures, USSR Academy of Sciences, Moscow

The results of an experimental study of true void fractions and temperature distributions in two-phase nonequilibrium flows in both heat-generating and unheated channels are presented.

It is found that measurements of the true void fractions and the liquid phase temperatures in a nonequilibrium two-phase flow allow the estimation of the true mass and void fractions as well as the slip ratio. These are quantities which cannot be determined by other known methods.

English translation will appear in Heat Transfer—*Soviet Research*

3.50 Some Aspects of Boiling Physics and Methods of Enhancing Heat Transfer with Phase Changes

E. I. Nesis and V. V. Chekanov
Stavropol Pedagogical Institute

Some problems on boiling physics are formulated, the solution of which allows methods of heat transfer enhancement to be outlined. On the basis of a proposed mechanism of active sites, an analytical expression is obtained which describes the number of active sites as a function of recess parameters on the surface. A relationship showing the contribution of various factors to boiling heat transfer on the surface is presented. The possibility of controlling ferrofluid boiling with an external magnetic field is analyzed. Some experimental results on the magnetic field effect on liquid superheating, growth rate, and frequency of bubble departure, are presented.

English translation will appear in Heat Transfer—*Soviet Research*

3.51 An Investigation of the Liquid Distribution and Mass Transfer Rate between the Core and a Liquid Film in a Disperse Annular Flow

B. I. Nigmatulin, V. I. Milashenko, V. I. Alekseyev, V. E. Nikolayev, and S. I. Ivandayev
All-Union Thermal Engineering Institute and the All-Union Correspondence Polytechnical Institute, Moscow

The results of a systematic experimental investigation of the liquid flow rate in a film for an upward steam-water mixture flowing in a vertical unheated tube of 13 mm ID, at pressures between 10 and 100 bar and specific mixture flow rates between 500 and 4000 kg/m² sec are presented.

Based on the available experimental data on the liquid flow rate variation in the film along a channel, relationships between the droplet entrainment from the film and the moisture deposition on it are determined.

English translation will appear in Heat Transfer—*Soviet Research*

3.52 The Effect of Interphase Heat and Mass Transfer on the Unsteady Outflow of a Metastable Fluid

R. I. Nigmatulin, B. I. Nigmatulin, A. I. Ivandayev, A. A. Gubaydullin, I. Kh. Rakhmatulina, and N. S. Khabeyev
Institute of Mechanics, Moscow State University

Nonequilibrium metasable fluid outflow from high-pressure vessels is investigated. Unsteady outflow (macroprocess) is found to be controlled by interphase heat and mass transfer. A procedure of numerical calculation is developed and some results are presented.

English translation will appear in Fluid Mechanics—*Soviet Research*

3.53 Vapor Bubble Growth on a Heated Surface

V. S. Novikov
Engineering Thermophysics Institute, Ukrainian Academy of Sciences, Kiev

A new mathematical model of the vapor bubble growth on a heated surface is formulated.

An approximate analytical solution of the proposed model is obtained permitting, for the first time, a determination of the laws governing the variation of the radius, with time, and the variation in temperature, pressure and vapor mass in the bubble with its growth. These depend on the physical properties of the liquid, the external pressure, and the intensity of the heat supply to the bubble.

3.54 An Analysis of Variation in the Critical Heat Load in Annular Channels with Large Coefficients of Nonuniform Heat Generation along a Channel

A. P. Ornatskiy, V. A. Chernobay, A. F. Vasil'yev, and G. V. Struts
Kiev Polytechnical Institute

The results of a study of a heat transfer crisis with uniform increases and decreases in heat generation are presented. A procedure for determining the increment of critical heat flux is given. The critical heat flux increment is shown to be independent of both the inlet heat flux and the nature of the nonuniform load. The application of the two critical heat flux increment value for the analysis and generalization of experimental data on boiling heat transfer crisis is considered.

English translation will appear in Heat Transfer—*Soviet Research*

3.55 Heat and Mass Transfer with Interactions between a Vapor-Gas Mixture and a Water Drop

V. P. Pavlov, V. T. Vylkov, and G. A. Apostolova
Research Division, Energoproyekt, Sofia, Bulgaria

Some criterial relations between interrelated processes are presented based on an analysis of the differential equations describing the transfer processes in a two-component flow at boundary conditions for heat and mass transfer between the flow and a water drop surface. Some results on steam condensation of falling drops of different sizes in a vapor-gas mixture in a monodisperse flow are presented.

3.56 An Experimental Investigation of Transient Heat and Mass Transfer and Separation in Bubbling

V. I. Pevzner, V. P. Tolchanov, Yu. N. Kuznetsov, and G. L. Urusov
All-Union Thermal Engineering Institute, Moscow

The experimental results of a study of the transient processes in a bubbler-type vessel with a built-in thin chevron separator, for large disturbances in the steam flow rates at the inlet and outlet of the test section are presented.

English translation will appear in Fluid Mechanics—*Soviet Research*

3.57 Heat and Mass Transfer in a Rotory Heat Pipe with the Rotation Axis Parallel to the Pipe Axis

F. Polašék and P. Schulz
Statní Vyzkumný Ustav pro Stavbu Strojù, Preha-Béchovice, Czechoslovakia

A method of calculating the thermal characteristics of a rotary heat pipe with the axis parallel to the rotation axis is presented. The calculation is based on heating efficiency as a function of the amount of working substance, the geometry of the system, and the centrifugal acceleration. Both the balance of pressure losses as the working substance flows from the condenser to the evaporator in the pipe and the effect of centrifugal acceleration on a liquid film in these areas are considered. The area of wetting of the heat-transfer surface of the pipe by the working substance is determined for different values of centrifugal acceleration.

The theoretical results are verified for aluminum water-filled pipes in a range of centrifugal acceleration between 10 and 1600 m sec^{-2}.

3.58 A Theory of the Formation and Evaporation of a Microlayer

E. M. Puzyrev, A. D. Gorbunov, A. V. Kuzmin, and V. V. Salomatov
Tomsk Polytechnical Institute

Some hydrodynamic and thermophysical problems of a pool boiling bubble microlayer model are considered. Relationships describing the microlayer existence situation give important information concerning the formation and subsequent evaporation of the microlayer, the surface temperature fluctuations, and the radius of the dry zone.

Analytical expressions for bubble growth rates with regard for the thermophysical properties of the heater material satisfactorily describe the experimental data in a range of Jacob numbers from 30 to 800.

English translation will appear in Heat Transfer—*Soviet Research*

3.59 The Effect of Surface Forces on the Film Condensation of a Vapor

V. G. Rifert, P. A. Barabash, A. B. Golubev, G. G. Leont'yev, and
S. I. Chaplinskiy
Kiev Polytechnical Institute

To substantiate the effects of the geometry of finning condenser tubes on the intensification of vapor condensation, the dimensions of the flooding area in the bottom part of the horizontal tubes as well as the parameters of the film flowing down (into) the grooves of vertical tubes are determined.

A theoretical solution of the problem of vapor condensation on the profiled surfaces is presented. The experimental data on the condensation of the vapors of various liquids substantiate this solution.

English translation will appear in Heat Transfer—*Soviet Research*

3.60 The Rate of Heat Transfer and Its Mechanism with Vapor Formation in a Liquid Film Flowing Down and Along a Profiled Surface

V. G. Rifert, P. A. Barabash, V. Yu. Zadiraka, and B. V. Andreyev
Kiev Polytechnical Institute

The wall-to-vaporizing film heat transfer coefficients are obtained for electrically heated profiled surfaces. The effect of the profiling parameters heat flux densities and watering on the heat transfer rate is determined. The optimum profiling characteristics, providing a maximum intensity of the process as compared with that for a smooth surface are identified. Relationships for calculating the average heat transfer coefficients with vapor formation in a film on profiled surfaces are presented.

English translation will appear in Heat Transfer—*Soviet Research*

3.61 Film Boiling of Freon–114 Drops at Pressures up to Critical

T. R. Rhodes and K. J. Bell
Oklahoma State University, Stillwater, Oklahoma

Film boiling of Freon-114 drops up to the critical pressure (32.3 atm) was studied experimentally and analytically. The experimental portion of this study consisted of photographic determination of diameter histories and lifetimes at reduced pressures of 1/8, 1/4, 1/3, 1/2, 3/4, and 1. Surface temperatures up to 270°C were studied and the Leidenfrost temperature was determined as a function of pressure. A distinct Leidenfrost point was obtained at the critical pressure.

The experimental results were compared to the predictions of the Gottfried-Lee-Bell model which attempts to predict the fluid mechanics and heat and mass transfer rates for the Leidenfrost Phenomenon. Agreement was generally satisfactory up to and including a reduced pressure of 3/4.

This paper appears on page 148, this volume

3.62 Heat Transfer Crisis and Pressure Drop in a Rod Bundle with Heat Transfer Intensificators

A. N. Ryabov, F. T. Kamen'shchikov, V. N. Philipov, A. F. Chalykh, T. Yugay, E. V. Stolyarov, T. I. Blagovestova, V. M. Mandrazhitskiy and A. I. Emel'yanov
Kurchatov Atomic Energy Institute and Krzhizhanovskiy Engineering Institute, Moscow

A short survey concerning the methods of heat enhancement in rod bundles is presented.

Experimental data on temperature regimes, heat transfer crisis and pressure drop in a rod bundle with different intensificators of heat transfer are presented.

It is found that the heat transfer intensificators employed permit an increase in the critical heat flux of up to 60%.

English translation will appear in Heat Transfer—*Soviet Research*

3.63 An Investigation of Boiling Heat Transfer in Low-Temperature Two-Phase Closed Thermosiphons

G. A. Savchenkov and Z. R. Gorbis
Odessa Institute of Refrigeration Technology

An experimental investigation of boiling heat transfer in a closed thermosiphon is presented.

Criterial equations generalizing the experimental data for a wide range of variation in the geometric parameters, the filling rates and properties of the heat agent, the orientation in space, and the presence of noncondensing admixtures are determined.

English translation will appear in Heat Transfer—*Soviet Research*

3.64 A Study of Temperature Fields in Thin Shells in Contact with Liquid Droplets

A. L. Satanovskiy
Engineering Institute, Ukrainian Academy of Sciences, Kiev

The temperature fields in heated thin shells cooled by the evaporation of liquid drops contacting them are investigated. Relationships for the approximate calculation of the temperature fields in such objects are obtained.

English translation will appear in Heat Transfer—*Soviet Research*

3.65 Forced Convection Laminar Film Condensation on Arbitrary Heat-Flux Surfaces

K. N. Seetharamu and M. V. Krishnamurthy
Department of Mechanical Engineering, Indian Institute of Technology, Madras, India

The forced convection laminar film condensation on an arbitrary-heat-flux surface is analyzed using boundary layer theory. The wall heat flux is assumed to be an arbitrary function of the distance along the surface. Solutions are obtained in the form of perturbation equations for an isothermal case. The differentials of the wall heat flux are employed as perturbation elements, allowing the deviation of universal functions for steam and other refrigerant vapors. These universal functions can be used to estimate the wall temperature variation for any type of wall heat-flux variation. Using these universal functions, solutions for some specific wall heat-flux variations are derived.

This paper appears on page 154, this volume

3.66 Heat Transfer with Thermal Distillation of Sea Water in Thin-Film Distillation Plants

V. N. Slesarenko, G. A. Gudakov, V. M. Saverchenko, V. S. Nagorny, A. E. Rudakova, and S. D. Ugryumova
Far Eastern Polytechnical Institute, Vladivostok

The process of heat transfer in distillation plants with upward air-sea water, vapor-sea water, and horizontal-film flows is analyzed and some empirical and critical equations related to it are proposed.

3.67 Maximum Heat Fluxes in Boiling under Various Special Conditions

G. F. Smirnov
Odessa Institute of Refrigeration Technology

Based on an approximate analysis of two-phase boundary layer stability, formulas generalizing the experimental data on maximum heat fluxes in boiling under special conditions are proposed. A correlation of the experimental data is presented.

English translation will appear in Heat Transfer—*Soviet Research*

3.68 Boiling Heat Transfer in Narrow Channels, Capillaries, and under Other Constraining Conditions

G. F. Smirnov, A. L. Koba, B. A. Afanas'yev, and V. V. Zrodnikov
Odessa Institute of Refrigeration Technology

An experimental and analytical study is carried out on boiling heat transfer, inner characteristics and limiting heat fluxes in narrow channels. Formulas for the calculation of heat transfer coefficients and limiting heat fluxes are proposed.

English translation will appear in Heat Transfer—*Soviet Research*

3.69 Post-Dryout Heat Transfer trom Smooth and Artificially Rough Surfaces of Steam-Generation Tubes

M. A. Styrikovich, A. I. Leont'yev, V. S. Polonskiy, and I. I. Malashkin
Institute of High Temperatures, USSR Academy of Sciences, Moscow

Based on an asymptotic theory of the turbulent boundary layer, an analytical model for the calculation of post-dryout heat transfer coefficients is proposed. Artificially rough surfaces are found to have a large effect on the enhancement of heat transfer in a disperse two-phase-flow. An analysis of previous experiments in accordance with the proposed model shows a satisfactory agreement between the experimental and the predicted data.

English translation will appear in Heat Transfer—*Soviet Research*

3.70 Heat Transfer with Condensation of Low-Pressure Vapor in Grid-Type Slit Channels of Plate Condensers

L. L. Tovazhnyanskiy, V. I. Atroshchenko, and M. S. Kedrov
Lenin Polytechnical Institute, Kharkov

Heat transfer with condensation of low-pressure vapor in grid-type slit channels of plate condensers is investigated. The effect of a condensing vapor of specific heat flux, vapor pressure, and velocity on heat transfer is shown. The design parameters of plate vacuum condensers are determined based on the generalized experimental data.

English translation will appear in Heat Transfer—*Soviet Research*

3.71 A Study of the Principles of the Kinetics of Boiling Phenomena

V. M. Tokarev
Chelyabinsk Polytechnical Institute

Problems of nuclear-to-film boiling transition are considered in terms of collective phenomena kinetics. A method of estimating the relative values of the critical heat fluxes under all possible conditions is proposed.

A model of the mechanism of boiling heat transfer is presented in order to explain the behavior of the experimental boiling curve, i.e., the Nukijama curve.

English translation will appear in Heat Transfer—*Soviet Research*

3.72 On an Investigation of the Dynamics of Heat-Generating and Heat-Transfer Elements

V. M. Tokarev
Chelyabinsk Polytechnical Institute

The dynamic properties of heat-generating and heat-transfer elements are examined using stability theory methods. The effect of wall thickness and the thermal physical properties of the heater material on the first critical heat flux value is analyzed using the proposed methods.

Recommendations concerning the enhancement of heat removal in heat-exchangers with a boiling heat transfer agent are proposed.

English translation will appear in Heat Transfer—*Soviet Research*

3.73 Boiling Heat Transfer Intensity and Crisis under Free-Motion Conditions

V. I. Tolubinskiy
Engineering Thermophysics Institute, Ukrainian Academy of Sciences, Kiev

Generalized relations for boiling heat transfer intensity and crisis under free-motion conditions are substantiated.

It is shown that by applying the internal characteristics of the process, the two major problems of the approximate theory of boiling heat transfer—the determination of the heat transfer coefficient and the critical heat flux—can be solved on the basis of the general similarity equation.

3.74 Heat Transfer Crisis with Boiling in Concentric and Eccentric Annular Slots

V. I. Tolubinskiy, E. D. Domashev, A. K. Litoshenko, and A. S. Matorin
Engineering Thermophysics Institute, Ukrainian Academy of Sciences, Kiev

Experimental results on the regularities of heat transfer crisis occurrence in annular slots are presented. Experimental data are also obtained for eccentric slots with turbulators on a heat nontransferring surface. Empirical formulas generalizing the experimental data with sufficient accuracy are proposed.

English translation will appear in Heat Transfer—*Soviet Research*

3.75 An Investigation of Water Boiling inside a Vertical Channel under Various Heating Conditions

L. A. Feldberg, A. L. Dobkes, L. V. Zysin, and A. G. Sazhenin
The Polzunov Central Boiler and Turbine Institute, Leningrad

Optical methods are employed in order to study a flow structure with repeated boiling. A quantitative shadow method is applied to obtain the temperature fields of the liquid. A holographic method is used to determine the concentration and size of the steam bubbles in the flow.

The real form of the vapor bubbles is determined from photos taken of a near wall region. The investigations were carried out on a vertical plane-parallel channel.

English translation will appear in Heat Transfer—*Soviet Research*

3.76 Maximum and Minimum Heat Flux in near Critical Pool Boiling

E. Hahne and G. Feurstein

Lehrstuhl und Institut für Thermodynamik und Wärmetechnik
Universität Stuttgart, West Germany

An experimental investigation of the maximum and minimum heat fluxes in near critical pool boiling was performed on horizontal platinum wires of different diameters (d = 0.05; 0.1; 0.2; and 0.3 mm) in a pressure vessel filled with either carbon dioxide (CO_2), trifluormonochlormethane (R 13; CF_3Cl) or trifluormethane (R 23; CHF_3) and in a pressure range of $p/p_{critical}$ = 0.5 to 0.99. The experimental results are compared with correlations available in the literature. This substantiates the evidence for the tendency of the critical heat fluxes to decrease with increasing pressure to be correctly represented in all correlations. However, the quantitative agreement between our high pressure experimental results and the correlations is unsatisfactory in most cases.

This paper appears on page 159, this volume

3.77 Steam Condensation in a Dynamic Two-Phase Layer

A. N. Khose, Yu. V. D'yachenko, and A. S. Zakharov

Laboratory for Cooling Systems, Institute of Electrical Engineering
Novosibirsk

Some experimental results of a study on steam condensation on rarefied staggered bundles of tubes in a foam layer are presented. Some calculation dependences are proposed for the mean volumetric gas contents and the hydraulic resistance of the layer. The total heat and mass transfer process in the foam layer is considered.

English translation will appear in Heat Transfer—*Soviet Research*

3.78 A Model for the Calculation of Heat Transfer Coefficients in an Annular Two-Phase Flow

J. Huhn

University of Technology, Dresden

A model representing both the flow boiling and the condensation of a gas-liquid annular flow in vertical tubes at high gas velocities is presented. The equations for momentum and heat transfer through a tubulent liquid film on a wall are used to calculate the heat transfer coefficient. The eddy viscosity is determined using the Szablewski method which takes into account the effect of the liquid/vapor interphase on the eddy viscosity dependence. Approximate equations are used to describe the model and are recommended for the practical application of it. A comparison is made between the experimental data and the predicted results.

3.79 A Mathematical Model of Two-Phase Annular Flow

G. F. Hewitt and P. Hutchinson
Thermodynamics Division, Atomic Energy Research Establishment Harwell, England

A mathematical model is described which allows prediction of the local behavior of flow properties in annular two-phase flow. By calculating the rate of droplet entrainment from the liquid film, and the rate of deposition of these droplets back onto the film, the variation of entrained liquid flow along the tube can be obtained. The method also leads to prediction of the local variation of other parameters of the flow such as pressure drop, void fractions, and wall film thickness. The fundamental relationships used in the model are presented and examples given of their experimental verification.

This paper appears on page 171, this volume

3.80 Heat Transfer with Steam Condensation on a Vertical Cylindrical Surface at Temperatures below the Solid-State Phase Transition

P. A. Tselishchev, T. M. Bogacheva, and G. G. Abayev
Krzhizhanovskiy Power Engineering Institute, Moscow

The problem of steady heat transfer with vapor condensation at a temperature above the triple point on a vertical cylindrical surface with a constant temperature being below that of the phase transition of a liquid to the solid state is considered.

Experiments are carried out to study the steam condensation in a vertical tube for temperatures of the cooling liquids to $-30°C$.

English translation will appear in Heat Transfer—*Soviet Research*

3.81 Heat Transfer in a Two-Phase Ammonia Flow in Horizontal-Tubular Apparatus of Refrigerating Installations

I. G. Chumak, A. V. Gordiyenko, and A. I. Kokhanskiy
Odessa Institute of Refrigeration Technology

As a result of the experimental study performed, relationships are proposed for the calculation of heat transfer in horizontal-tubular apparatus of ammonia refrigerating installations.

It is shown that the heat transfer intensity depends on the behavior of the vapor-liquid mixture and the geometric characteristics of the object under investigation.

3.82 A Study of the Heat Transfer and Pressure Drop for Boiling Nitrogen in a Horizontal Tube

D. Steiner and E. U. Schlünder

Lehrstuhl und Institut für Thermische Verfahrenstechnik der Universität Karlsruhe, West Germany

The local heat transfer and pressure drop for boiling liquid nitrogen has been studied experimentally in a 14 mm I.D. smooth horizontal copper tube.

The measured local heat transfer coefficients in nucleate boiling depend on the heat flux, the quality of the vapor and the mass flow rate. An empirical equation for correlation of the heat transfer data is presented and the application of this equation to the data of earlier studies yields a good agreement.

The measured total pressure drop illustrates the well-known interdependence of the vapor quality, mass flow rate and heat flux. Using a recently published correlation by Bandel, the predicted frictional pressure drop is found to agree reasonably well with the experimental data.

3.83 Bubble Growth Rates in Aqueous Binary Systems at Subatmospheric Pressures

S. J. D. van Stralen, W. Zijl, and D. A. de Vries

Laboratory for Fluid Dynamics and Heat Transfer, Eindhoven University of Technology, Eindhoven, The Netherlands

A growing vapor bubble is blown up in accordance with the Rayleigh solution. Advanced bubble growth is governed by heat and mass diffusion. The combination of both mechanisms during the transition period results in oscillations in the equivalent bubble radius. Experimental bubble growth data on water and aqueous binary systems (water-ethanol, water-1-butanol, and water-2-butanone) are in quantitative agreement with a theoretical model of Van Stralen et al., which combines the Rayleigh solution with a diffusion-type solution, accounting for both a relaxation and an evaporation microlayer. Experimental data on the oscillations of growing and imploding vapor bubbles in water are in quantitative agreement with a numerical solution of the basic conservation equations.

This paper appears on page 188, this volume

3.84 Interferometric Investigation of Liquid Surface Evaporation and Boiling due to Depressurization

S. S. Grewal, C. Shin, and M. M. El-Wakil

University of Wisconsin, Madison, Wisconsin

An initially subcooled liquid which becomes superheated under sudden depressurization, has been observed to undergo free surface evaporation and boiling, and the expulsion of a two-phase mixture. This phenomenon is investigated using interferometry and high-speed photography. The investigation shows the formation of a heterogeneous mixture of subcooled and superheated liquid at and just below the surface. In the system studied, Freon–11, initially at high atmospheric pressure and 21°C, was depressurized to 20.3 cm Hg vacuum, corresponding to a saturation temperature of 15.3°C. The superheat of 5.7°C and subcoolings of more than 5.9°C were measured. It is this mixture of superheat and subcooling that is now believed to account for the violent two-phase expulsion phenomenon at large superheats.

This paper appears on page 207, this volume

3.85 Heat Transfer in Boiling Mixtures

V. D. Yusufova and A. I. Chernyakhovskiy
Azerbaydzhan Research Institute for Power Energetics, Baku

Experimental results for heat transfer in binary mixtures at high pressures are presented. Some peculiar features of boiling mixtures are considered from a modern perspective.

English translation will appear in Heat Transfer—*Soviet Research*

3.86 Heat Exchange, Hydrodynamics, and Thermal Nonequilibrium in Dispersed Film Boiling Flows of Hydrogen, Nitrogen, and Argon

S. A. Yarkho, N. V. Philin, A. P. Inkov, A. A. Statiyev, and V. N. Utkin
Scientific and Industrial Association for Cryogenic Machine Building and All-Union Correspondence Institute of Machine Building, Moscow

Experimental results of an investigation of the heat exchange, hydrodynamics, and thermal nonequilibrium of two-phase hydrogen, nitrogen, and argon flows are presented. The investigation is based on combined experimental and theoretical methods.

An analysis of the mechanism and a generalization of the experimental data are given. A one-dimensional model with a separate description of phases is found to be the most advantageous way of investigating heat fluxes with phase changes.

English translation will appear in Heat Transfer—*Soviet Research*

4. HEAT AND MASS TRANSFER IN TWO-PHASE FLOWS

4.1 Heat Transfer and Hydrodynamics of Sieve Trays with Tube Bundles

L. S. Akselrod, N. I. Vorotnikova, and A. A. Kozlov
Moscow Chemical Engineering Institute

A local model is used to study the rate of heat transfer from the wall of a single heat exchange element with various arrangements of tube bundles, to a bubbling bed on the trays of a mass transfer column with both cross- and counterflow.

Criterial relations are obtained which correlate the results. The effects of the tube bundle arrangement on tray operation, hydraulic resistance, and heat transfer rate are determined.

English translation will appear in Heat Transfer—*Soviet Research*

4.2 Temperature and Concentration Fields as Applied to the Study of Heat and Mass Transfer in Industrial Rectification Columns

Y. A. Alekseyev, Yu. G. Myasishchev, and S. G. Mazina
Krasnodar Branch of the All-Union Petrochemicals Institute

New methods of determining the kinetic coefficients of rectification columns using their temperature and concentration profiles are suggested and developed. This allows for a more accurate determination of the kinetic coefficients. Using the proposed methods, relationships correlating the scale-up coefficients for kinetic equations are obtained under the conditions of a pilot plant. Methods are derived for the static optimization of industrial equipment.

English translation will appear in Heat Transfer—*Soviet Research*

4.3 Heat and Mass Transfer from a Rotating Tube to a One-and-Two Phase Internal Flow

R. Z. Alimov and E. L. Kaspin
Tupolev Aviation Institute, Kazan

Heat transfer from the internal wall of an electrically heated rotating tube to an inner airflow was experimentally investigated. Mass transfer with a thin solid evaporating water film is studied. The results are given in the form of ordinary criterial relationships.

The effectiveness of the heat and mass exchanger apparatus tested is analyzed using the experimental data obtained on the hydraulic resistance of the system.

English translation will appear in Heat Transfer—*Soviet Research*

4.4 A Theoretical and Experimental Study of a Falling Drop Temperature and the Concentration Variation in Apparatus with Solution Concentration by Pulverization

A. Artikov, I. I. Yunusov, and N. R. Yusupbekov
Tashkent Polytechnical Institute

From an analysis of the mechanism of distillation with water vapor, a mathematical model with parameters distributed over time and along the apparatus height is constructed for the pulverization of a liquid with a nonvolatile component. The model allows for the evaporation of a volatile component through the self-evaporation of a heat-transfer gas mixture. An evaporation coefficient is introduced and is estimated experimentally. The model allows a determination of the curves of the pulverized liquid falling drop concentration and the temperature distribution along the height of the apparatus.

4.5 Heat and Mass Transfer between a Liquid Film and A Gas in A Downward Flow

V. N. Babak, T. B. Babak, L. P. Kholpanov, V. A. Malyusov, and N. M. Zhavoronkov
Institute of New Chemical Problems and the Kurnakov Institute of General and Inorganic Chemistry, Moscow

The heat and mass transfer between a liquid film and a gas under downward flow conditions was investigated. The results are presented in the form of graphs which allow the calculation of a dimensionless temperature (concentration) for various heat and mass transfer parameters.

4.6 Equations of Crystallization with Heat Transfer

Tibor Blickle and Zsuzsanna Halász
Research Institute for Industrial Chemistry, Veszprém, Hungary

Based on data reported in the available literature, a mathematical model describing crystallization kinetics is proposed. An algorithm uniquely correlating the mathematical model for different cases is presented. This algorithm allows the calculation of the characteristics of the resulting flow and the sizes of the particles if the initial data are known.

4.7 Balance Equations Describing Two-Phase Transfer Elements

Tibor Blickle and B. Chukas
Research Institute for Industrial Chemistry, Vezprém, Hungary

A theoretical investigation of a system of operative units and the balance equations that describe them results in a general algorithm prescribing a mathematical model for a given element and vice versa. An algorithm relating two-phase heat and mass transfer systems and their mathematical models is proposed.

4.8 Simultaneous Heat and Mass Transfer in Rectification

V. E. Bogoslovskiy, A. N. Krasilnikov, and V. S. Krylov
Research Institute of Synthetic Resins, Vladimir, and Institute of Electrochemistry, Moscow

An experimental study of the interphase heat exchange in one-and-two-component mixtures under bubbling is carried out. Heat transfer rates in binary acetone-water, methanol-water, and acetone-methanol systems are found to be much higher than in one-component systems. The data are theoretically accounted for by the mutual influence of heat and mass transfer processes.

4.9 Heat Transfer and Crisis Conditions in Vertical Channels with a Falling Liquid Film and Downward Two-Phase Annular Flow

B. G. Ganchev, A. E. Bokov, and A. B. Musvik
Bauman Institute of Technology, Moscow

Local heat transfer to a falling liquid film and a two-phase downward annular flow was studied experimentally taking into account the wave motion characteristics of the film surface and the hydraulic resistance. Relationships for the calculation of the main film characteristics along the channel and the local heat transfer coefficients are proposed.

It was found that the critical vapor content can be as high as 90 percent for the liquid film cooling of a channel at atmospheric pressure.

English translation will appear in Heat Transfer—*Soviet Research*

4.10 Heat Transfer Laws in Melting and Crystallization

N. I. Gel'perin, G. A. Nosov, V. D. Parakonnyy, and Ya. V. Tikhonenko
Moscow Institute of Light Chemical Technology

Crystal growth on a cooled plate is examined theoretically under third-kind boundary conditions for various models of heat transfer in a melt. Relations for the calculation of the required temperature for solid phase cooling are presented. Melting of a semiinfinite body is studied to elucidate the nature of convective and conductive heat transfer in a melt.

4.11 Freezing Processes and the Kinetics of Freeze Drying

A. S. Ginzburg, V. I. Syroyedov, and V. I. Kovnatskiy
Moscow Institute of Food Technology

The influence of crystallization on the structure and properties of frozen coffee extracts is studied along with the effect of supercooling on crystal form and arrangement. The dependence of the coordinate of the freeze drying zone, ξ, on the time, τ, was determined analytically. The total transfer coefficient can be determined from the analytic expression, using a known value of ξ for the time, τ.

English translation will appear in Heat Transfer—*Soviet Research*

4.12 Vapor Desublimation from a Vapor-Gas Mixtue in a Vacuum

A. A. Gukhman, A. Z. Volynets, A. V. Zhuchkov, V. E. Matkhanova, and V. K. Safonov
Moscow Chemical Engineering Institute

Vapor desublimation from laminar flow of a vapor-gas mixture in channels is considered. The thermal resistance of a layer of ice and the dissipative effects are assumed to be insignificant. Solutions are obtained for a flow in flat and cylindrical tubes, for both constant and parabolic profiles of the longitudinal velocity. A satisfactory agreement is observed between the analytical and experimental results even in the case of a significantly thick ice layer. The analytical solutions are also compared with the experimental results from an investigation of alkaline metal vapor condensation.

4.13 Heat Transfer to Turbulent Upward Flowing Gas-Liquid Mixtures in a Vertical Twisted-Tape Tube

I. V. Domanskiy and V. N. Sokolov
Lensovet Institute of Technology, Leningrad

Heat transfer to a gas-liquid swirled flow was analyzed based on a semiempirical theory of turbulent transfer. A simple expression for the calculation of the heat transfer coefficient is obtained.

English translation will appear in Heat Transfer—*Soviet Research*

4.14 The Hydrodynamics and Heat and Mass Transfer of Unsteady Liquid Film Flows

V. A. Elyukhin and L. A. Kalimulina
Chelyabinsk Polytechnical Institute

The effect of interface mass and heat transfer and surfactants on the linear and nonlinear stability of liquid films has been studied.

Analytical relationships for the phase velocity, amplitude of the nonlinear waves, velocity fields, and the temperatures of unsteady liquid film flows have been obtained.

English translation will appear in Fluid Mechanics—*Soviet Research.*

4.15 An Estimation of the Efficiency of High-Rate Heat-and-Mass Exchangers with Swirled Flows

A. I. Yershov and M. F. Shnayderman
Minsk Technological Institute

The efficiency of a high-rate heat and mass exchanger with phase interaction in a swirled co-current flow has been calculated and compared with that of other known mass exchangers on the basis of engineering and economic advantages.

4.16 Possibilities of the Intensification of Heat and Mass Transfer in Two-Phase Film Boiling

J. Zemanek
*National Research Institute for Mechanical Engineering, Praha
Czechoslovakia*

An analysis of a two-phase gas-liquid system shows that the intensification of heat and mass transfer in liquid film evaporation at forced convection of the gas phase is limited by flooding (hydraulic overloading) and by the nonuniform wetting of the system with a liquid film at subcritical hydraulic loading.

An arrangement of wavy capillary-porous sheets provides not only partial suction of the liquid film from its surface (in the case of overloading), but also uniform wetting of this surface (in the case of subcritical loading), thus eliminating the above difficulties and at the same time intensifying the heat and mass transfer.

4.17 A Study of the Temperature Distribution in Continuous Flow Mixing Vessels

V. V. Kafarov, V. M. Barabash, L. N. Braginskiy, O. N. Mankovskiy and V. I. Begachev
Leningrad Chemical Engineering Institute

A mathematical model, based on the hydrodynamics and turbulent transfer in unbaffled mixing vessels, is suggested for the calculation of the temperature field in a mixing volume. Simple relationships are obtained which estimate temperature drops in the apparatus.

4.18 The Transient Flow of Multiphase Media Taking into Account the Kinetics of the Phase Changes in Channels

P. M. Kolesnikov, A. A. Karpov, and V. M. Petrenko
Luikov Heat and Mass Transfer Institute, Minsk

The heat and mass transfer in a transient channel flow of a multiphase media with vapor bubbles is studied. A description of the experimental apparatus and some experimentally obtained characteristics of the flow is presented. Some peculiarities of the flow, taking into account the dynamics of the vapor bubbles, are considered. Some analytical solutions of the model are presented.

English translation will appear in Fluid Mechanics—*Soviet Research*

4.19 A One-Dimensional Condensation Model and Some Calculations of Condensation of Gas-Vapor Mixtures

R. Krupiczka, H. Herman, and J. Aerts
Institute of Chemical Engineering, Gliwice, Poland

On the basis of a previously presented one-dimensional model, numerical calculations of vapor condensation in the presence of an inert gas in parallel-flow exchangers are presented. There is a sufficient agreement between the calculation results and the experimental data.

4.20 Heat Transfer to Thin Liquid Films Flowing Down a Polymer Surface

Yu. E. Lukach, L. B. Radchenko, Yu. M. Tananayko, and A. D. Petukhov
Kiev Polytechnical Institute

The problem of a gravitational liquid film flowing down a moving polymer surface has been solved analytically. This solution along with an experimental investigation of hydrodynamics and heat transfer allows simple design equations to be determined. A system of liquid cooling for the production of blow-extruded film has been worked out.

English translation will appear in Heat Transfer—*Soviet Research*

4.21 The Contribution of Thermal Effects to Mass Transfer in Adiabatic Rectification

Ya. D. Zelvenskiy, S. A. Malinov, and V. A. Shalygin
The Mendeleyev Chemical Engineering Institute, Moscow

A tracer atoms technique is used to determine the contribution of the thermal component to the overall mass-transfer coefficient for the rectification of three model binary mixtures with approximately equal coefficients of the component surface tension, but with different boiling points. It is found that the diffusion component of the mass-transfer coefficient for rectification can be computed from data on the physical absorption of gases. The experimental results are satisfactorily described by a correlation allowing for the superposition of thermal and diffusional flows.

4.22 The Effect of Heat Transfer on Binary Mixture Rectification Efficiency

V. A. Malyusov, V. A. Lotkhov, E. V. Bychkov, and N. M. Zhavoronkov
Institute of General and Inorganic Chemistry, Moscow

Temperature and concentration profiles in the cross section of a film-rectification column are presented. A method of calculating the rectification kinetics, taking into account both evaporation and condensation, is suggested for the case of a saturated vapor.

4.23 The Intensification of Processes in Liquid-Solid Systems by Superimposing Modulated Impulsive Input Vibrations with Vapor Condensation in the Apparatus

V. I. Mamontoy, G. I. Shkurupiy, and V. M. Zadorskiy
The Dzerzhinskiy Chemical Engineering Institute, Dniepropetrovsk

The problems of vapor phase impulsive input into a liquid and of bubble condensation in it are considered. It is shown that the impulsive input of a vapor phase involves modulated vibrations which intensify the mass exchange in a solid-liquid system in the apparatus.

4.24 Simultaneous Film Evaporation and Condensation of a Two-Directionally Inclined Surface

D. Moalem, S. Sideman, and R. Semiat
Tel-Aviv University and Israel Institute of Technology, Haifa, Israel

Horizontal evaporator-condensers in water desalination units are studied. A theoretical analysis of the effects of inclination on the heat transfer rate during the simultaneous film evaporation on one side and film condensation on the other side of a solid flat plate is presented. The decrease in the evaporating film thickness due to the two-dimensional nature of the flow is relatively small. Hence, the corresponding increase in the overall transfer coefficient is practically insignificant. For a system in which the evaporating film is sustained by a vertically downward flowing liquid, the main effect of inclining the plane from the horizontal results from the formation of an unwetted area. Some practical conclusions are suggested.

4.25 Simultaneous Heat and Mass Transfer in the Design of Countercurrent or Cocurrent Direct Contact Equipment

J. C. Mora, R. Bugarel, and M. Prevost
Institute of General Chemistry, Toulouse, France

A general method for the design of direct contact heat and mass transfer equipment is proposed. Three particular applications are presented, along with a comparison between the experimental and theoretical results. The method appears to be a convenient way of obtaining a specific model of the process and can be applied to gas-liquid, liquid-liquid, solid-liquid, or solid-gas contacts.

4.26 Heat and Mass Transfer in a Spray Evaporator with Multicomponent Solutions

V. N. Mudrikov and L. M. Damskiy
Engineering Thermophysics Institute, Ukrainian Academy of Sciences, Kiev

A system of integral-differential transfer equations was solved by a computer. The results show a deterioration in the heat and mass transfer at the initial apparatus length due to a levelling of the temperature, and the drop and flow velocities. These findings are further substantiated experimentally.

English translation will appear in Heat Transfer—*Soviet Research*

4.27 Heat and Mass Transfer in a Cylindrical Tube with Combined Twist of One- and Two-Phase Liquid-Gas Flows

Yu. I. Osipenko
Tupolev Aviation Institute, Kazan

Heat transfer from the internal wall of an electrically heated vortex tube with an additional local screw insert to an air flow in it is experimentally investigated. Mass transfer is studied in the case of a fine solid film of water evaporating from the internal wall of the tube. The results are described by ordinary criterial relations. The effectiveness of this kind of a mass and heat transfer apparatus is analyzed using experimental data on hydraulic resistance.

English translation will appear in Heat Transfer—*Soviet Research*

4.28 Heat and Mass Exchange between a Liquid and a Gas Involving Interface Formation

L. M. Pikkov, E. K. Siirde, and P. A. Tint
Tallin Polytechnical Institute

Heat transfer and diffusion in a fluid deformed by some mechanical force are studied. Mathematical relations are derived for the transfer intensity, the deformation rate, and the minimum necessary energy. The theoretical results are compared with experimental data on oxygen absorption during water dispersion in an air flow.

English translation will appear in Heat Transfer—*Soviet Research*

4.29 Heat Transfer in a Disk Crystallizer

V. G. Ponomarenko, A. I. Kalmychkov, N. A. Gavrya, Yu. A. Kurlyand, K. P. Tkachenko, and T. D. Bogomaz
Kiev

Heat transfer in a disk crystallizer is studied experimentally. A criterial equation is derived for the calculation of the coefficient of heat transfer from a heat transfer surface to a suspension moving down the channels of the apparatus and being mixed by scraper-stirrers.

4.30 Heat and Mass Transfer in a Two-Phase Counterflow with Free Fall of the Disperse Material

S. M. Reprintseva and N. V. Fedorovich
Luikov Heat and Mass Transfer Institute, Minsk

A mathematical model of the heat and mass transfer in a two-phase counterflow with free fall of the disperse material is presented. The algorithms for estimating the height of a vertical column in the apparatus and the distribution of the moisture content of the disperse materials along the column height are determined.

English translation will appear in Heat Transfer—*Soviet Research*

4.31 The Laws of Heat and Mass Transfer in Rotating Bubble Beds

A. I. Safonov and V. S. Krylov
Research Institute of Chemical Engineering and Institute of Electrochemistry, Moscow

The construction of an apparatus with a rotating bubble bed is described and some advantages of this apparatus over other types of heat and mass transfer equipment are discussed. Experiments on gas-phase controlled mass transfer are carried out for a very small (of an order of magnitude equal to the separated bubble diameter) thickness of the liquid bed. Mass transfer is investigated for the case of ammonia absorption by water from an air-ammonia mixture. Using the same apparatus, the heat transfer rates in an air-liquid nitrogen system are measured over a wide range of hydrodynamic parameters. The results indicate the dominating role played by the hydrodynamic entrance region and corresponding to the stage of bubble formation found in interphase heat and mass transfer.

English translation will appear in Heat Transfer—*Soviet Research*

4.32 Heat Transfer and the Kinetics of Drying in a Magnetic Field

A. M. Somov, S. V. Chunayev, G. I. Nikolayev, P. G. Romankov,
D. T. Mitev, N. B. Rashkovskaya
Lensovet Institute of Technology, Leningrad

A consideration is made of the heat and mass transfer at high-rate hydrodynamic modes in two-phase gas-solid particle systems with the addition of heat transfer elements and the superposition of a variable magnetic field. An increase in the heat transfer coefficient with the addition of finer particles is studied.

English translation will appear in Heat Transfer—*Soviet Research*

4.33 An Experimental and Theoretical Study of Heat Transfer with Liquid Evaporation in a Falling Film

E. I. Taubman and Yu. I. Kalishevich
Odessa Institute of Refrigeration Technology and Odessa Polytechnical Institute

A method of calculating the heat transfer coefficient of an evaporated liquid film is presented. This method accounts for liquid superheating due to the friction of the vapor against the film surface. A theoretical relation is obtained. Heat transfer in a long-tube film evaporator is experimentally studied. A comparison of the calculated and experimental data shows a satisfactory qualitative agreement.

English translation will appear in Heat Transfer—*Soviet Research*

4.34 Direct-Contact Heat Exchange between Two Nonmixing Liquids

I. A. Shchuplyak, A. N. Verigin, M. V. Mikhalev, and M. V. Aleksandrov
Lensovet Institute of Technology, Leningrad

The primary laws governing the direct-contact heat exchange between two nonmixing liquids are elucidated through experimental and theoretical studies. A method of evaluating the heat transfer efficiency of column-type equipment is proposed.

English translation will appear in Heat Transfer—*Soviet Research*

4.35 Heat and Mass Transfer in an Impinging-Jet Apparatus as Applied for the Protection of the Environment

I. T. Elperin, I. S. Turovskiy, V. L. Meltser, L. L. Pavlovskiy, and
L. L. Goldfarb
Luikov Heat and Mass Transfer Institute, Minsk

The impinging-jet method was considered as a means of purifying the surrounding environment. The transfer processes in impinging-jets at sites of dust-collection, evaporation of salt waste water, and drying of mechanically dehydrated waste water deposits are discussed.

5. HEAT AND MASS TRANSFER IN CAPILLARY-POROUS BODIES

5.1 Water Evaporation from a Capillary-Porous Plate at High Temperatures

N. V. Antonishin and V. S. Nikitin
Luikov Heat and Mass Transfer Institute, Minsk

The experimental results of a study of water evaporation from the surface of a capillary-porous plate having its lower portion in water and its upper portion in contact with a high-temperature fluidized finely dispersed layer, are presented.

5.2 Statistical Methods of the Study of Porous Structures

R. I. Ayukayev, V. K. Kivran, and M. E. Aerov
Civil Engineering Institute, Kuibyshev and Research Institute for Synthetic Alcohols and Intermediates, Moscow

The methods of statistical physics are employed in this study of the energetic (thermal, physical, and chemical) interaction between the components of dispersed systems. A mathematical model of the structure of a dispersed system is proposed and is used to calculate the radial distribution of elements.

5.3 A Thermodynamic Analysis of the Effects of an External Pulse on the Heat and Mass Transfer in Moist Capillary-Porous Bodies

B. N. Bodkova
Ural Research and Design Institute of Building Materials, Chelyabinsk

The effect of an external pulse on the heat and mass transfer in capillary-porous bodies is analyzed based on principles of nonequilibrium thermodynamics and thermomechanics.

5.4 Heat and Moisture Transfer in Violently Heated Building Constructions

V. N. Bogoslovskiy and V. M. Roytman
Kuibyshev Civil Engineering Institute, Moscow

A method of studying heat and moisture transfer in violently heated capillary-porous bodies is described. Curves of the temperature, moisture-content, and excessive pressure variation over a cross section of a body subjected to one-sided heating in accordance with a standard fire regime are plotted. The concept of a moisture content coefficient at high-rate drying is used to analyze the mechanism of heat transfer under the conditions considered. A mathematical model of the transfer processes is suggested in order to solve problems.

5.5 On the Physical Nature of the Mass-Transfer Potential and the Potentiometric Curve of a Capillary-Porous Media

Yu. M. Volfkovich, V. E. Sosenkin, E. I. Shkolnikov, and V. S. Bagotskiy
Institute of Electrochemistry, Moscow

The physical nature of the moisture-transfer potential and a specific isothermal moisture-capacity found in capillary-porous media, as put forth by drying theory, are established. It is shown that the potentiometric relation between the moisture-content of a porous body and the potential can be completely determined by radius- (or capillary pressure)-distribution curves for the body examined and some porous standard. The capillary-pressure gradient is found to be the driving force behind the moisture transfer in liquid-filled pores.

English translation will appear in Heat Transfer—*Soviet Research*

5.6 Heat and Mass Transfer at Structure Formation of Colloidal Capillary Porous Bodies

N. I. Gamayunov, A. E. Afanas'yev, N. D. Kozinnikov, and L. G. Amusin
Kalinin Polytechnical Institute

Some heat and mass transfer problems of model and capillary-porous bodies are studied and the influence of a variation in thermophysical and mass-exchange characteristics on structure formation is shown.

5.7 Heat and Mass Exchange Involving Phase Separation with Inner and Outer Field Interaction

A. S. Ginzburg
Moscow Institute of Food Technology

Molecular, kinetic, and thermodynamic methods have been used to analyze the complex phenomena taking place in a multiphase system. Detailed studies of the system's characteristics are presented and a law governing an observed change in molecular structure with the interaction of inner and outer fields is proposed.

English translation will appear in Heat Transfer—*Soviet Research*

5.8 Heat and Mass Exchange Optimization in a Colloidal Capillary-Porous Body with Radiation-Convection Heating

M. K. Goroshenko, I. A. Makhlina, and E. A. Kuleshov
All-Union Correspondence Institute of Food Technology, Moscow

This work deals with an application of the Pontryagin principle to the optimization of processes of heat and mass exchange. An algorithm describing optimum conditions is obtained.

5.9 A Mathematical Model of Internal Heat and Mass Transfer Processes in Droplets

A. A. Dolinskiy, G. K. Ivanitskiy, L. M. Damskiy, and K. D. Maletskaya
Engineering Thermophysics Institute, Ukrainian Academy of Sciences, Kiev

A physical model of the heat and mass transfer through solid crusts on the surfaces of intensely evaporating drops is suggested. A system of differential equations is used to describe dehydration while taking into account the continual transition from a moist to a dry state.

5.10 Acoustic Effects on Heat and Mass Transfer in Rocks and Living Tissues

S. A. Yefimova, E. V. Karus, O. L. Kuznetsov, E. M. Simkin,
M. L. Surguchev, M. Ya. Khodas
All-Union Research Institute of Geophysics and Geochemistry, Moscow

Some experimental and theoretical studies of heat transfer intensification in rocks and living tissues are reviewed. A tenfold increase is achieved in the thermal dissasivity of porous materials in a field of elastic waves. The elastic field is found to intensify gas hydrate decomposition. A possible mechanism is suggested. Possible applications of the results (e.g., gas hydrate production, selective destruction of cancer cells) are discussed.

5.11 A Study of the Temperature Field inside an Evaporating Drop

I. S. Yefremova and M. S. Smirov
All-Union Correspondence Institute of Food Technology, Moscow

A study of the temperature field in a liquid drop (where the linear law of evaporation surface movement holds) is presented.

The results may be used to analyze the behavior of an airodispersed system with spray drying.

English translation will appear in Heat Transfer—*Soviet Research*

5.12 Heat and Mass Transfer in Moist, Electrically Heated, Capillary-Porous Bodies

A. S. Zelepuga and V. S. Karpenko
Luikov Heat and Mass Transfer Institute, Minsk

The results of an experimental study of temperature and moisture fields in flat, electrically heated, capillary-porous samples are presented.

Preliminary electric heating of the samples is shown to intensify the external mass transfer without affecting the quality of the samples. A transport mechanism for the electric heating of capillary-porous materials is proposed.

English translation will appear in Heat Transfer—*Soviet Research*

5.13 Heat and Mass Transfer in Capillary-Porous Materials Exposed to Infrared Radiation

S. G. Iliasov and V. V. Krasnikov
Institute of Food Technology, Moscow

A simple method allowing a rather accurate analytical calculation of $E_0(x)$ and $q(x)$ in selectively absorbing and dissipating materials is proposed. A variation in the spectral composition and spatial distribution of radiation with selective coordinate absorption and nonisotropic dissipation is explained using the characteristics of a coordinate-variable optical material.

5.14 Heat and Mass Transfer in Model Capillary Systems

O. A. Kiseleva, Ya. I. Rabinovich, V. D. Sobolev, V. M. Starov, and N. V. Churayev
Institute of Physical Chemistry, Moscow

A theory of nonisothermal mass transfer at the free portion of a conical capillary is developed. Thermocapillary and viscous film flow are considered along with vapor diffusion. Equations describing the liquid and vapor flows are correlated under conditions of local adsorption equilibrium. Experimental data on thermocapillary dodecane film flow rates over the surfaces of quartz capillaries are presented.

English translation will appear in Heat Transfer—*Soviet Research*

5.15 Unsteady Heat Exchange with Liquid Motion Through Underground Permeable Strata

O. A. Kremnev, A. V. Shurchkov, N. N. Aronova, and E. M. Kozlov

Engineering Thermophysics Institute, Ukrainian Academy of Sciences, Kiev

Differential physical and mathematical models of unsteady heat transfer with liquid motion through underground permeable strata are discussed. A mathematical model of the heat exchange in granular and cracked collectors is constructed. Both exact and approximate analytical solutions for temperature fields are given. Heat transfer laws are analyzed and the influence of various factors on the heat transfer intensity is estimated.

English translation will appear in Heat Transfer—*Soviet Research*

5.16 Moisture Transfer and Segregational Ice-Formation in Freezing and Thawing Soils

V. A. Kudryavtsev, E. D. Yershov, V. G. Cheverev, Yu. P. Lebedenko, and L. V. Shevchenko

Moscow State University

Moisture transfer in freezing and thawing soils has been studied and the results are presented. A new mechanism for the formation and growth of cryogenic textures, taking into account the thermophysical and physiomechanical conditions of heat and mass exchange, is suggested.

English translation will appear in Heat Transfer—*Soviet Research*

5.17 An Approximate Solution of the Generalized Stefan Problem in a Porous Medium with Variable Thermal Properties

I. J. Kumar and L. N. Gupta

Scientific Analysis Group, R and D Organization, New Delhi, India

The generalized Stefan problem for coupled heat and mass transfer in a porous medium has been formulated in an earlier paper. Its solution gives the position of a variable evaporation front as well as a distribution of the temperature and moisture in a porous body, assuming a negligible thermal diffusion coefficient and constant thermal properties. In the present paper the above problem has been solved in the case of variable thermal properties taking into account the thermal diffusion coefficient. It is found that an increase in the P_n (Posnov number), reduces the nondimensional mass transfer potential and the velocity on an evaporation surface.

This paper appears on page 214, this volume.

5.18 A Numerical Study of Variable Mass Particle Motion in an Eddy Flow

P. S. Kuts and N. N. Grinchik
Luikov Heat and Mass Transfer Institute, Minsk

The motion of constant- and variable-mass particles in a three-dimensional eddy flow has been studied numerically over a range of Reynolds numbers, and taking into account the force of gravity. The results are compared with experimental data.

English translation will appear in Fluid Mechanics—*Soviet Research.*

5.19 A Theoretical Study of Heat and Mass Transfer in Capillary-Porous Bodies in a Vacuum

P. S. Kuts, I. F. Pikus, and V. D. Kononenko
Luikov Heat and Mass Transfer, Minsk

A mathematical model of internal heat and mass transfer with high-temperature treatment of moist capillary-porous materials in a vacuum is proposed.

A solution describing the unsteady temperature and pressure fields in the case of an infinite cylinder is presented.

5.20 The Effect of Energy Supply on the Sublimation Dehydration Mechanism and the Formation of a Capillary-Porous Structure under Vacuum

P. D. Lebedev, D. P. Lebedev, and V. V. Uvarov
All-Union Bioengineering Institute, Moscow and Institute of Agricultural Engineering, Moscow

The mechanism and physics of sublimation dehydration in a vacuum with either a contact or a thermoradiational energy supply are considered.

English translation will appear in Heat Transfer—*Soviet Reseasrch*

5.21 A Study of a Binary Gas Mixture Flow with Evaporation in a Capillary

V. G. Leytsina, N. V. Pavlyukevich, and G. I. Rudin
Luikov Heat and Mass Transfer Institute, Minsk

A slow binary gas mixture flow in a cylindrical capillary with evaporation at the bottom is considered. The flow is studied using a kinetic equation for the binary gas mixture of the Hamel form. A system of two linear integral equations is obtained for the calculation of the component mass velocities and an expression for the diffusion coefficient is found. For a continuum regime, the above equations are found to reduce to differential equations of the second order and solutions of these equations are obtained. The extreme cases of small and large concentrations of the evaporating component were studied.

5.22 Heat and Mass Transfer in Capillary-Porous Peat Systems at Low Temperatures

I. I. Lishtvan, G. P. Brovka, and P. N. Davidovskiy
Peat Institute, Minsk

An installation for the study of heat and mass transfer in peat systems under conditions of freezing is described. Methods of calculating thermophysical properties are discussed.

The thermal conductivity and heat capacity of valley peat of varying moisture content were studied.

5.23 An Analysis of the Fundamental Equations of Intensive Heat and Mass Transfer Kinetics

P. P. Lutsik
Kiev Technological Institute of the Light Industry

The fundamental equations of intensive heat and mass transfer kinetics are given. A relationship between capacitance coefficients and mass content on the one hand, and the average surface and volume temperatures of a body on the other, is established. The effect of the Rebinder and Mikhailov numbers on heat and mass flows is analyzed.

5.24 Heat and Mass Transfer in Colloidal Capillary-Porous Bodies with Heating

A. V. Lykova, A. M. Medvedev, and V. D. Kosoy
Moscow Institute of Meat and Dairy Technology

A method of solving the problem of unsteady-state thermal conductivity with a heat source due to evaporation is discussed. The three-dimensional problem is reduced to a two-dimensional one. The main laws of heat and mass transfer are defined.

English translation will appear in Heat Transfer—*Soviet Reseasrch*

5.25 The Invariant-Group Method of Solving a Nonlinear Heat and Mass Transfer Problem for Capillary-Porous Bodies

S. G. Romanovskiy, K. D. Lukin, and L. I. Margolin
Luikov Heat and Mass Transfer Institute, Minsk

A nonlinear system of differential heat and mass transfer equations is classified into groups.

Based on this classification, some invariant-group solutions are obtained and the boundary-value problem is solved for the system under consideration.

5.26 Heat and Mass Transfer in Multicomponent Coatings on Heat Transfer Surfaces

S. G. Romanovskiy, Yu. M. Martinchik, and T. V. Rabchuk
Luikov Heat and Mass Transfer Institute, Minsk

The methods of heat and mass transfer theory are used to obtain a solution to the problem of momentum transfer fields in multicomponent coatings on heat transfer surfaces heated by electromagnetic induction. Equations for the engineering calculation of process kinetics are derived.

5.27 Heat and Mass Transfer in Polymer Materials having a Rigid Capillary-Porous Structure

B. S. Sazhin and N. E. Shadrina
All-Union Research and Design Institute of Chemical Engineering, Moscow

On the basis of a complex analysis of the sorption, structural and thermophysical properties of the most important polymer materials, a classification of materials depending on their thermal and diffusion resistances to heat and mass transfer is developed. A concept of standard states is proposed for the calculation of heat and mass transfer in capillary-porous polymer materials, taking into account the integral effects accompanying technological process with model materials. A nomograph is composed and can be used to estimate intensity and duration of the heat and mass transfer given certain characteristics of the material.

5.28 Heat and Mass Transfer in Capillary-Porous Multicomponent Film Materials with Combined Heat Transfer

L. S. Slobodkin, G. N. Pshenichnaya, M. N. Barskaya, and
L. E. Kiselshteyn
Luikov Heat and Mass Transfer Institute, Minsk

The effect of temperature and concentration fields on the properties of multicomponent capillary-porous systems with combined heat transfer is considered in terms of a general heat and mass transfer theory.

A mathematical model of the heat and mass transfer in a system with internal heat sources resulting from chemical conversions and infrared radiation is proposed. The model allows a prediction of the behavior of the process and the properties of the material under the given conditions.

5.29 Heat and Mass Transfer in a Condensible Vapor-Gas Mixture Flowing through Porous Materials

B. M. Smolskiy, P. A. Novikov, V. A. Yeremenko, E. K. Snezhko, V. I. Balakhonova, and L. Ya. Lyubin
Luikov Heat and Mass Transfer Institute, Minsk

The results of a study of the heat and mass transfer in a condensible vapor-gas mixture flow through porous materials of different thermophysical properties and at pressures below the triple point are presented.

5.30 Drop Evaporation in a Two-Component Flow

E. G. Tutova, G. A. Kuvshinov, and T. V. Kuchko
Luikov Heat and Mass Transfer Institute, Minsk

The interaction between a two-component flow (gas-solid particles) and a single drop is considered. The analytical and experimental results on the influence of solid particles on the drop evaporation rate are compared.

English translation will appear in Heat Transfer—*Soviet Research*

5.31 A Study of the Mechanism of Sorption-Contact Mass Transfer in a Vacuum

E. G. Tutova and R. I. Feldman
Luikov Heat and Mass Transfer Institute, Minsk

Experimental data on the influence of contact mass transfer on the rate of material dehydration in a vacuum are presented. Some specific properties of the mechanism of moisture removal by sorbents are revealed. The multicomponent material-sorbent structure is considered as a means of effecting an increase in efficiency.

English translation will appear in Heat Transfer—*Soviet Research*

5.32 A Solution of an Internal Heat and Mass Transfer Problem for a Capillary-Porous Body with Forced Convection

Jiřina Houšova
Institute for Food Industry, Prague, Czechoslovakia

Some problems of internal forced-convection heat and mass transfer in a capillary-porous dispersed material under thermal treatment are considered. The experimental data are applied to a theory of radiation energy supply, allowing a mathematical model of the process to be proposed for simple bodies with known thermophysical properties. The theoretical solution is compared with the experimental data.

5.33 Heat and Mass Exchange between a Rheologically Compound Thin-Wall Capillary-Porous Body and Its Surroundings

L. B. Tsimermanis and F. Kh. Tsimermanis
Ural Research and Design Institute of Building Materials, Chelyabinsk

A system of phenomenological equations describing the heat and mass exchange between a thin wall capillary-porous body and its surroundings, taking into account the structure formation in the body, is given.

Relationships for the estimation of the temperature drop and hydration potential between the body and its surroundings are proposed and can account for the dependence of various parameters of the body on its structural and rheological characteristics.

5.34 An Analysis of the Heat and Mass Transfer Laws for Capillary-Porous Bodies Depending on Material-Moisture Bonds

L. G. Chernaya and E. A. Raskina
Luikov Heat and Mass Transfer Institute, Minsk

The thermodynamic mass transfer parameters for adsorptional, capillary and osmotic moisture in colloidal capillary-porous bodies are analyzed.

Expressions are obtained for the coefficients C_m and δ_ρ in a form illustrating their functional dependence on moisture content and temperature.

5.35 The Calculation of Heat and Mass Transfer Involving the Movement of a Phase-Conversion Boundary

G. S. Shubin
Moscow Wood Technology Institute

A study of the processes involved in the movement of a phase-conversion front of a cylinder is presented.

Correlation equations describing, in a general form, the nature of the temperature field are determined.

Computer results are presented and compared with experimental data.

English translation will appear in Heat Transfer—*Soviet Research*

5.36 A Calculation of Transport Coefficients for a Decreasing Rate of Drying

S. Endrenyi and D. Szentivanyi
Budapest, Hungary

A model based on the time variation of temperature and concentration on material surface is presented. The heat and mass transfer coefficients thus calculated also vary with time. A numerical example of the model is presented and the results are compared with experimental data.

5.37 Some Equilibrium Problems of Steady Evaporation on Porous Surfaces

S. Endrenyi and B. Palancz
Institute for Industrial Economy, Organization and Computation, Budapest, Hungary

A study of liquid evaporation from the surfaces of porous bodies with a steady state of sorption equilibrium providing a continuous supply of liquid to the surface is presented.

The psychrometric evaporation laws may be extended to include the study of porous surfaces. Some specific features of the evaporation and equilibrium states above porous surfaces which are different from those in the case of evaporation from solid surfaces, are considered.

6. HEAT AND MASS TRANSFER IN DISPERSED SYSTEMS

6.1 Unsteady-State Heat Transfer between a Plate having a Heat Source and a Dispersed Layer

N. V. Antonishin, M. A. Geller, L. R. Gurvich, and A. L. Parnas
Luikov Heat and Mass Transfer Institute, Minsk

The problem of unsteady heat transfer between a plate having a heat source and a dispersed layer has been solved analytically. The results are used to analyze the operation of heat sensors designed to measure heat transfer coefficient pulsations in a fluidized bed.

English translation will appear in Heat Transfer—*Soviet Research*

6.2 Heat and Mass Transfer in a Vibro-Rotational

N. V. Antonishin, V. S. Nikitin, O. G. Martynov, and G. F. Puchkoy
Luikov Heat and Mass Transfer Institute, Minsk

A study of the heat exchange between cylindrical and flat bodies, and a vibro-rotational bed is presented. The parameters of rotational speed, frequency, and position of the bodies are considered.

6.3 A Study of the Hydrodynamic and Heat and Mass Transfer Characteristics of a Filtration Process using a Separated Flow Model

R. I. Ayukayev, E. V. Badatov, A. A. Volkov, and M. E. Aerov
Civil Engineering Institute, Kuibyshev, Institute of Catalysis, Novosibirsk, and Research Institute of Synthetic Alcohols and Intermediates, Moscow

A study of the hydrodynamic and heat and mass transfer situation in a free volume in a granular bed is presented. A pattern of "outside flow-around" is considered for the elements of granular beds and a model of the separation flows in a Betchelor–M. Lavrentiyev bottom trench is proposed. The influence of the geometry of the bed is shown.

English translation will appear in Heat Transfer—*Soviet Research*

6.4 On the Possibility of Directly Measuring the Thermophysical Properties of Dispersed Systems with a Solid Phase

V. E. Babenko, M. B. Grinbaum, and G. B. Panter
Moscow

Methods of studying solid particle heating are developed based on a drastic change in the magnetic properties of materials near the Curie point, rather than by measuring the temperature in different zones of the apparatus. It is shown that the methods can be used to trace the motion of a single particle in a fluidized bed apparatus.

English translation will appear in Heat Transfer—*Soviet Research*

6.5 Some Fluidized Bed Heat Transfer Problems

A. P. Baskakov, B. V. Berg, Yu. M. Goldobin, A. M. Dubinin, A. A. Zharkov, G. Ya. Zakharchenko, S. V. Zvyagin, O. M. Panov, A. V. Sokolov, N. F. Filippovskiy, and V. V. Khoroshavtsev
Ural Polytechnical Institute, Sverdlovsk

Heat transfer in a high-temperature fluidized bed is studied with an emphasis on its conductive-convective and radiative components. The effect of various factors on the heat transfer from horizontal 125 × 220 mm cylinders is considered, along with some specific features of heat transfer and hydrodynamics in a fluidized bed sectioned with large-piece packings. Heat transfer from the rarefied to the dense phase of the fluidized bed is studied by supplying the rarefied phase over the bed space with horizontal jets of hot air. The possibility of fluidized-bed cooling of the heat transfer surfaces of the apparatus is discussed.

English translation will appear in Heat Transfer—*Soviet Research*.

6.6 The Effective Radial Thermal Conductivity of a Packed Bed of Particles with Various Shapes

R. Bauer and E. Schlünder
Institut für Thermische Verfahrenstechnik, Karlsruhe, West Germany

The effective radial thermal conductivity of a packed bed of spheres, cylinders, Rashig rings, and crushed particles at temperatures of up to 1000°K has been experimentally studied. Based on the experimental data, the methods of calculating the effective thermal conductivity of a packed bed of particles of any shape are developed.

6.7 A Study of Heat Transfer in Disperse Systems Affected by an Electric Field

M. K. Bologa, M. P. Zhelyaskov, and V. V. Pushkov
Institute of Applied Physics, Kishinev

Heat transfer in gas-solid and fluid-solid suspensions with free and forced convection under the influence of an electric field has been studied experimentally. The basic laws of local heat transfer for a gas suspension flow in an annular channel with the application of electric fields of varying intensities are presented. A mechanism of the effect of the electric field on the heat transfer to a suspension flow is proposed and relationships yielding satisfactory calculations are presented.

6.8 Heat and Mass Transfer in Nonisothermal Fluidized-Bed Reactors

V. A. Borodulya
Luikov Heat and Mass Transfer Institute, Minsk

Some models of the noncatalytic processes in fluidized beds are considered taking into account longitudinal mixing in the dense phase and mass exchange between phases. Methods of calculating the coefficients of longitudinal gas mixing with interphase heat and mass transfer are analyzed, and a study of these coefficients is undertaken for a model reaction (ozone decomposition on an alumosilicate catalyzer) and a tracer (with the application of an input signal of arbitrary shape). The models developed can be used for the study of industrial fluidized-bed apparatus.

6.9 A Study of Heat and Mass Transfer to Bodies of Various Shapes Immersed in a Fixed or Weakly Fluidized Granular Bed

Yu. A. Buyevich and D. A. Kazenin
Institute of Problems of Mechanics, Moscow and Institute of Chemical Engineering, Moscow

Exterior heat and mass transfer with a Darcy flow past bodies in a granular bed is considered in terms of a one-phase dispersion model. An accepted form of the transfer equation for axisymmetric and plane flows with an anisotropic dispersion tensor is presented. For large Peclét numbers, an effective approach to the problem, based on nonclassic perturbation method, is proposed. Heat transfer distributions on isothermal sphere and cylinder surfaces at various ratios of the molecular and convective dispersions are considered.

English translation will appear in Heat Transfer—*Soviet Research*

6.10 Some Problems of Disperse Fluids Application in Heat Exchangers

L. K. Vasanova, A. I. Karpenko, I. D. Larionov, A. I. Safronov, V. N. Linetskiy, E. M. Chizhevskaya, and A. G. Chulanova
Ural Polytechnical Institute, Sverdlovsk

The results of an experimental study of the stabilized heat transfer in a liquid fluidized bed with a turbulent jet supply are given. The possibility of employing a fluidized bed and gas suspension in heat exchangers is considered.

English translation will appear in Heat Transfer—*Soviet Research*

6.11 Experimental Study of Heat Transfer to Embedded Tube and Coil Surfaces in a Fluidized Bed Combustor

V. N. Vedamurthy and V. M. K. Sastri
Indian Institute of Technology, Madras, India

The heat transfer to a tube and a coiled surface submerged in a fluidized shallow-bed combustor in which crushed particles of briquetted lignite are burnt in an inert bed of coal ash, is considered. The particles vary in size from 3.15 to 6.3 mm and the fluidizing velocity ranges from 0.2 to 0.45 m/s. The experimental results are compared with calculations made assuming the emulsion packet at the wall to be an absorbing and emitting medium. Wall heat transfer is approximated by the model for the case of the emulsion being a black body. The embedded surface heat transfer is more accurately predicted by a model in which the packet is assumed to be an absorbing and emitting medium.

This paper appears on page 222, this volume

6.12 Heat and Mass Transfer between a Disperse System and a Flow Swirled around a Vertical Axis

Z. Viktorin
Research Institute of Mechanical Engineering, Prague, Czechoslovakia

The possibility of heat and mass transfer intensification in a gas-solid particle system is considered. The energy efficiencies of the major types of dispersed systems are compared.

6.13 Heat Exchange between a Fluidizing Agent and Fluidized Solid Particles in a Multisectional Apparatus with Boundary Conditions of the First and Third Kinds

N. I. Gelperin, Z. V. Yermakova, and V. G. Aynshteyn
Institute of Light Chemical Technology, Moscow

The problems of interphase heat and mass transfer as applied to a monodisperse fluidized bed in a multisectional apparatus are considered. The ideal mixing of solids in each section, and the movement of fluid in a piston flow regime are analyzed. Relationships for calculating the temperature of solid particles as they leave each section (specifically, the mean bulk and surface temperature of the particles) and for calculating the temperature of the fluidizing agent, are determined.

6.14 The Structure and Heat Transfer Ability of a Turbulent Gas-Suspension Flow

N. I. Gelperin, V. G. Aynshteyn, L. I. Krupnik, Z. N. Memedlyayev, and S. V. Yefimova
Institute of Light Chemical Technology, Moscow and the Severodonetsk Branch of State Institute of Nitrogen Industry

An experimental study has been made of gas suspension motion and the laws of heat transfer. It is shown that the velocity profile of the carrying fluid in horizontal and vertical flows can be described by a universal logarithmic "wall law." On this basis, expressions for the calculation of the thickness of the viscous sublayer and the shear stress of the carrying fluid are proposed.

Turbulent momentum and heat transfer characteristics are analyzed and the relative contribution of the major factors in the heat exchange between a gas suspension and the bounding surface is determined.

English translation will appear in Heat Transfer—*Soviet Research*

6.15 Influence of Velocity and Particle Mass on the Heat Exchange in a "Fluidized" Bed Air Cooler

S. P. Dichev
Institute of Food Technology, Plovdiv, Bulgaria

Heat exchange in a new type of "fluidized" bed air cooler is investigated under laboratory conditions. The studies are carried out in a closed wind tunnel with air cooling by an intermediate coolant of granulated polyethylene. The intermediate coolant is cooled directly by the evaporation of Freon-12 in the tunnel. The influence of the fluidizing air flow velocity and the total bed mass on the heat exchange is established.

This paper appears on page 229, this volume

6.16 Hydrodynamics and Heat Transfer in Modified Spouting Bed Systems

S. S. Zabrodskiy, I. T. Elperin, A. F. Dolidovich, and V. S. Yefremtsev
Luikov Heat and Mass Transfer Institute, Minsk

The results of experimental studies of modified spouting-bed systems for the treatment of dispersed materials over a wide range of particle size are presented. The following cases are considered: 1) a bed of finely dispersed particles in apparatus having a relative increase in size of the entrance orifice, 2) a spouting bed having a twisted jet core, and 3) a system with the two-phase dispersed medium. Problems of bed structure and interphase heat exchange are briefly discussed.

English translation will appear in Heat Transfer—*Soviet Research*

6.17 Heat Transfer Characteristics of Dispersed Systems

R. Z. Ivanova, N. F. Gurin, and U. I. Ivanov
*All-Union Research Institute of Nonmetallic Building Materials and
Hydromechanization, Toliyatti*

The role of particle pulsation in a fluidized bed with increasing heat transfer (heat conductivity) is examined quantitatively. The influence of elementary processes and various parameters on the heat transfer increase is investigated and expressions for calculating the effective thermal conductivity of the fluidized bed are proposed.

6.18 The Temperature Distribution and Heat Transfer of a Moving Packed Blow-Down Bed with Heat Generation

V. A. Kalenderian and Yu. F. Nerushev
Odessa Institute of Refrigeration Technology

A model of the heat transfer in a packed blow-down bed with a cocurrent rodlike flow of components in parallel-plate channels with nonuniform heat generation in the gas phase, is proposed. Relationships for the temperature distribution and heat transfer coefficient are determined. The rate of heat transfer of such a two-component system is found to increase with increasing heat generation.

6.19 Heat Transfer Characteristics of Two-Stage Rectangular Fluidized Bed Apparatus

G. V. Kryukov, O. V. Gabeskiriya, M. V. Lykov, and I. S. Stutyagin
Research Institute for Fertilizers, Insecticides, and Fungicides, Moscow

The heat transfer characteristics of rectangular two-stage fluidized-bed apparatus under conditions of external control of heat transfer are considered. The heat transfer rate is found to increase two-fold when the ratio of the sides is increased to 7:1. The experimental data showed a satisfactory agreement with the calculated values and were verified under industrial conditions.

English translation will appear in Heat Transfer—*Soviet Research*

6.20 The Heat Transfer in an Apparatus with a Spherical Layer Vibrationally Moving along a Closed Contour

V. M. Malchenko and V. M. Bograd
Nikolayev

The rate of heat transfer between a spherical vibrationally moving layer and a gaseous medium blown through it is studied using a transient method. Relationships for calculating the degree of heat regeneration in a heat exchanger with a spherical layer vibrationally moving along a closed contour over a wide range of operating conditions are determined.

6.21 Heat Transfer from Vibrating Surfaces to Packed Beds of Spheres under Conditions of Atmospheric Pressure and Vacuum

E. Muchowski and E. U. Schlünder
Institut für Thermische Verfahrenstechnik der Universität Karlsruhe, West Germany

A coefficient of heat transfer between an electrically heated vibrating plate and packed beds of glass spheres at atmospheric pressure and in a vacuum has been experimentally determined. The influence of the frequency of vibration and the pressure in the apparatus on the heat transfer coefficient is studied.

6.22 Some Heat Exchange Problems in Two-Phase Finely Dispersed Flows

A. T. Nikitin, V. V. Sapozhnikov, and V. V. Bogachev
Krasnodar Polytechnical Institute

Some heat transfer problems in gas streams with microsolid particles at high temperatures are discussed. The experimental data on the local heat transfer in an air-graphite suspension in annular channels are correlated with the results of a study of the behavior of the average heat transfer as the solid particle concentration varies from 0.1 to 4 kg/kg.

English translation will appear in Heat Transfer—*Soviet Research*

6.23 Heat Transfer from a Flat Plate to a Gas Suspension Flow

V. S. Nosov, I. D. Larionov, and V. V. Mamayev
Ural Polytechnical Institute, Sverdlovsk

The results of an experimental investigation of the local and overall heat transfer from a flat plate to a low-concentration gas suspension flow are presented. The influence of gas suspension flow rate pulsations on heat transfer intensity in a vertical tube is discussed.

6.24 A Study of the Similarities between the Transfer Processes in a Fluidized Bed and the Heat Transfer and Drag of Moving Bodies

R. B. Rozenbaum and O. M. Todes
Leningrad Mining Institute

A similarity has been found between the mechanism of heat transfer from a heating surface immersed in a fluidized bed and the mechanism of momentum transfer from a body moving through the bed. The transfer is achieved by "packets" of solid particles being periodically replaced by gas "bubbles." It is shown that the effective viscosity of the bed depends on the density of the particle material.

English translation will appear in Heat Transfer—*Soviet Research*

6.25 Heat Transfer between a Surface and a Vibrating Fluidized Bed

A. E. Ryzhkov, A. P. Baskakov, and V. A. Munts
Ural Polytechnical Institute, Sverdlovsk

Heat transfer from a calorimeter to a vibrofluidized bed up to 500 mm high has been investigated. The experimental results are compared with results from a study of bed hydrodynamics.

English translation will apper in Heat Transfer—*Soviet Research*

6.26 The Local Heat Transfer between a Vibrating Fluidized Bed and Horizontal Tube Bundles

B. G. Sapozhnikov, E. G. Reshetnikov, G. D. Kosenko, and
N. M. Kharisova
Ural Polytechnical Institute, Sverdlovsk

The results of an experimental investigation of the local heat transfer between a vibrofluidized bed and horizontal tube bundles influenced by vibrational parameters, particle size, bed height, number of tube rows and tube diameter and arrangement (in-line or staggered) are presented. The effect of various factors on the value and position of minimum and maximum local heat transfer coefficients is examined. On the basis of an analysis of coefficient fields, patterns of the gas flows which develop in a vibrofluidized bed formed in vessels having permeable walls, are examined.

English translation will appear in Heat Transfer—*Soviet Research*

6.27 Heat Exchange with Granulation of a Solution in an Apparatus with Local Spouting in a Fluidized Bed

A. A. Sokolovskiy, G. L. Groshev, and S. Sh. Gadzhiyev
All-Union Research Institute of Vitamins, Moscow

The aerodynamic characteristics of a local flame inside a fluidized bed have been studied and analytical expressions for critical speeds have been obtained. The heat and mass exchange during granulation of an ammonium nicotinate solution with local spouting in a fluidized bed is examined. Thermograms and temperature fields of the flame and bed are determined and heat and mass transfer coefficients are calculated.

English translation will appear in Heat Transfer—*Soviet Research*

6.28 Radiation-Convection Heat Transfer in a Turbulent Gas Suspension Flow

F. E. Spokoynyy, Z. R. Gorbis, N. V. Svyatetskiy, and R. V. Zagaynova
Odessa Institute of Refrigeration Technology, and
Odessa State University

A mathematical description of the combined heat transfer to an axisymmetrical gas flow having a small concentration of solid particles is given. The integrodifferential equations are solved for conditions in which either the longitudinal or the transverse flow temperature distribution dominates. Some experimental results are presented and discussed.

English translation will appear in Heat Transfer—*Soviet Research*

6.29 An Experimental Investigation of the Heat Transfer between Tube Walls and a Flowing Aerosol

A. S. Sukomel, V. A. Rozhko, R. V. Kerimov, and F. F. Tsvetkov
Moscow Power Engineering Institute

Measurements of local coefficients of heat transfer between stainless steel tube walls and an aerosol are presented. The aerosol was formed by adding solid particles to a flowing gas (helium, nitrogen, or argon).

English translation will appear in Heat Transfer—*Soviet Research*

6.30 The Local Instantaneous Intensity of External Heat Transfer in a Nonuniform Fluidized Bed

N. I. Syromyatnikov, V. M. Kulikov, V. S. Nosov, and V. N. Korolev
Ural Polytechnical Institute, Sverdlovsk

The fluctuations in surface temperature porosity, and gas velocity in a nonuniform fluidized bed have been experimentally investigated. The fluctuations in heat flux and the local coefficients of heat transfer from a vertical plate are found to be small with respect to their average values, while the intensity of the heat transfer to gas bubbles is found to be very large. It is proposed that an external heat transfer model be based on a study of convective heat and mass transfer.

English translation will appear in Heat Transfer—*Soviet Research*

6.31 Heat Transfer between a Horizontal Staggered Tube Bundle and a Fluidized Bed

A. I. Tamarin, S. S. Zabrodskiy, and Yu. G. Yepanov
Luikov Heat and Mass Transfer Institute, Minsk

A 35 × 45 cm bed of crushed shamotte (particles of 3 mm) and millet (particles of 3 mm) was fluidized with air (in a rectangular 45 × 25 cm column) at room temperature.

A steady-state method was used to measure the coefficient of heat transfer between the bed and a horizontal tube 30 mm in diameter located in a staggered bundle of tubes having interstices of various sizes.

English translation will appear in Heat Transfer—*Soviet Research*

6.32 The Influence of Boundary Conditions on the Hydrodynamics and External Heat Exchange in a Fluidized Bed

L. I. Frenkel, N. B. Kondukov, and V. P. Tarov
Tambov Chemical Engineering Institute
Moscow Chemical Engineering Institute

The qualitative behavior of a fluidized bed with varying parameters is examined using differential thermodynamic equations. Changes in the fluidized bed hydrodynamics due to the introduction of a heat transfer surface and a variation in the heat transfer coefficient along its height are found. The results of an investigation of the gas flow distribution effect on the rate of external heat transfer are discussed.

6.33 A Vibro-Fluidized Bed Method for the Thermal Treatment of Loose Materials

M. Hoč
Statni Vyzkumny Ustav pro Stavbu Stroju, Praha-Běchovice, Czechoslovakia

Basic relationships describing the effect of a vibratory motion of a substrate on a granular bed are presented. Experimental results are obtained for the case of a vibro-bed on an impermeable substrate in a longitudinal air flow and for the case of a vibro-bed on a grid in a transverse air flow. The pressure variation in the air gap between the substrate and the bed was measured using pressure strain gauges. Laws of heat exchange between the air and the bed were determined by the method of sublimation. Granulated naphthalene was used as a test material.

6.34 Fluidized-Bed Drying of Inert Bodies

B. Csukás, K. Pataki, and Z. Ormós
MTA Mükki, Vesprem, Hungary

The drying of microcrystal materials of high moisture content (pasty materials) of inert bodies in a fluidized bed is considered. The advantages of the process, possible applications and an approximate mathematical model are discussed.

6.35 A Study of the Heat and Mass Transfer in a Fluidized Bed Based on Jet Theory

N. A. Shakhova, G. N. Lastovtseva, and V. K. Lukashev
Moscow Chemical Engineering Institute

Based on jet theory, heat and mass transfer in a fluidized bed was theoretically and experimentally studied. Relevant correlations were determined and engineering methods of calculating the heat and mass transfer in a fluidized bed are developed.

English translation will appear in Heat Transfer—*Soviet Research*

6.36 A Study of the Temperature Field in a Blow-Down Regenerator Taking into Account Heat Loss through the Side Surfaces

V. S. Shvydkiy, F. R. Shklyar, and E. D. Lekomtseva
All-Union Research Institute of Metallurgical Thermal Engineering, and Ural Polytechnical Institute, Sverdlovsk

A solution of the problems of heat transfer in a fixed bed unidirectionally blown with gas and in a countercurrent gas medium is presented taking into account heat loss in a direction perpendicular to the gas flow. A satisfactory agreement was found between the numerical calculation and the experimental results. The results may be applied in order to estimate the influence of heat loss on the temperature fields in beds of various sizes and with varying thermal parameters.

6.37 Turbulent and Pseudoturbulent Transfer in Two-component Flows

A. A. Shrayber
Engineering Thermophysics Institute, Ukrainian Academy of Sciences, Kiev

The results of a study of the transverse migration of particles in gas-suspension flows caused by turbulent diffusion and influenced by interparticle collisions are presented. A model of heat transfer in gas-suspension flows is constructed on the basis of an analysis of the influence of the particles on the laws of continuum flow near the wall. This model can account for most of the known experimental data over a wide range of characteristic parameters.

English translation will appear in Heat Transfer—*Soviet Research*

6.38 Heat Transfer in Two-Phase Gas-Suspension Flow with Particle Precipitation on the Channel Walls

V. K. Shchukin, V. A. Filin, A. I. Mironov, N. S. Idiatullin,
N. N. Kovalnogov, A. A. Yakshin, N. A. Nadyrov, and I. Kh. Fakhrutdinov
Tupolev Aviation Institute, Kazan

Some experimental results of a study of the local convective heat transfer between a turbulent gas-solid particle (1–32 μm aluminum oxide particles) suspension flow and the walls of the subcritical section of a nozzle and a rectangular curvilinear channel under conditions of cooling the flow with relatively low dust content are presented. Heat transfer is found to be considerably intensified in a region of inertial particle precipitation. The correlation between the relative increase in heat transfer and the density of particle precipitation on the walls is discussed.

English translation will appear in Heat Transfer—*Soviet Research*

6.39 An Experimental Study of the Solid Particle Distribution and Local Intercomponent Gas-Solid Particle Heat Exchange in a Monodisperse Two-Phase Flow

G. I. El'kin and Yu. B. Timofeyev
Odessa Institute of Food Technology

A method for the experimental study of the local characteristics in a counterflow of a monodisperse gas-suspension is suggested. Empirical relations between the solid particle distribution and the local heat transfer coefficient in the flow are determined over a wide range of various factors.

English translation will appear in Heat Transfer—*Soviet Research*

7. HEAT AND MASS TRANSFER IN RHEOLOGICALLY COMPLEX SYSTEMS

7.1 Unsteady Natural Convection in Nonlinear Viscous Liquids

V. I. Baykov and E. A. Zaltsgendler
Luikov Heat and Mass Transfer Institute, Minsk

A theoretical study of unsteady natural convection in nonlinear viscous ("power-law") liquids is presented. Temperature-dependent rheological coefficients are taken into account. A statement of the problem for both external and internal asymptotic expansions is presented. The influence of the non-Newtonian behavior index and heat-dependent parameters on the heat transfer and hydrodynamics of the system is determined.

English translation will appear in Heat Transfer—*Soviet Research*

7.2 Similarity Problems Concerning the Rotation of Axisymmetric Bodies in Non-Newtonian Power Fluids

V. N. Beshkov and P. Mihchka
Bulgarian Academy of Sciences, Sofia and Institute of Chemical Engineering Theory, Prague, Czechoslovakia

Laminar boundary layer equations for the rotation of axisymmetric bodies in non-Newtonian power-law fluids are analyzed. The relationship $\bar{r} = A\eta^m$, where $\bar{r}$ is the dimensionless radius of the transverse body curvature, and η is the dimensionless abscissa measured along a meridian from the body apex is applied. It is found that similarity solutions to the problems of heat and mass transfer may be obtained for $m = 1/3$. Exact numerical solutions to such problems may be used to estimate the accuracy of various approximate methods of calculating boundary layers with the rotation of axisymmetric bodies in non-Newtonian fluids.

7.3 A Theoretical Investigation of the Processes of Radical Polymerization and Dissolution of Polymer Systems

V. I. Boyarchenko, P. V. Zhirkov, V. I. Yankov, and V. I. Kernitskiy
Moscow

Based on mathematical models, heat and mass transfer regularities with radical polymerization in a tube reactor and in a polymer system dissolving in an autogeneous continuous solvent are studied. The dependence of viscosity on temperature and polymer concentration in the first case and on temperature and shear stress in the second is taken into account. Temperature curves and some pressure-flow rate characteristics of the reactor are presented. For the second case, a comparison is made between the mathematical and experimental results.

English translation will appear in Heat Transfer—*Soviet Research*

7.4 Heat Transfer with Viscoelastic Flow along a Semiinfinite Plate

O. V. Wein and P. Mihchka
Institute of Chemical Engineering Theory, Prague, Czechoslovakia

The Blasius similarity problem is formulated in terms of a very general model for a viscoelastic body. Relationships for determining the coefficients of heat and mass transfer are obtained from an analysis of the equations for a boundary layer.

7.5 The Effect of Dissipation and Wall Heat Flux on the Temperature Distribution in a Circular Tube

K. Wichterle and O. K. Dakhin
Institute of Chemical Engineering Theory, Prague, Czechoslovakia

The problem of heat transfer in a highly viscous non-Newtonian fluid flow in a circular tube with a constant wall heat flux is solved including a dissipative factor. A solution is obtained by the method of separation of variables in the form of a series of eigenfunctions. Its convergence is examined for ten terms of the series and for various non—Newtonian fluids. It is found that within those ranges, where the convergence is not satisfactory, the problem may be solved by the thermal boundary layer method.

7.6 A Study of Temperature-Dependent Electrorheological Effects and Dielectric Parameters in Hydrocarbon Suspensions having a Hydrated Dispersed Phase

Yu. F. Deynega, K. K. Popko, and N. Ya. Kovganich
Institute of Colloidal and Water Chemistry, Kiev

An investigation of the temperature-dependent electrorheological effect (ERE) and various dielectric parameters with high intensities of an electric field on hydrocarbon suspensions of starch, polyvinyl alcohol, and silica gel is presented. It was found that as the temperature in suspensions having a hydrated dispersed phase increased, the ERE and dielectric constant passed through a maximum and the electric conductivity of the system increased. It is theorized that the polarization transition to electric conductivity increase as the temperature and moisture of the dispersed phase and the electric field intensity increase is due to the formation of hydration layers on the particle surface and their particular electric properties.

English translation will appear in Heat Transfer—*Soviet Research*

7.7 A Mathematical Model for the Two-Dimensional Nonisothermal Process of Calendering Viscoelastic Polymers

Yu. G. Dmitriyev and E. A. Sporyagin
Lvov Polytechnical Institute

The present work deals with some characteristics of a two-dimensional nonisothermal flow of viscoelastic polymer materials in the gap between two rolls. The rheological White–Metzner equation describes the behavior of a polymer melt with deformation. A mathematical model of the process with regard to the initial and boundary conditions is solved by substitution, resulting in functions relating velocity components and geometrical and rheological parameters, and a correction coefficient which accounts for the effect of the two-dimensionality of the problem and the elastio viscosity of the material on the velocity field in the gap between the rolls. An Algol program is developed to calculate the power and energy parameters of the process. The computer "Minsk-22" is thus programmed to determine the nature of the dependence of the temperature field of a polymer with deformation on its rheological and thermophysical characteristics, the amount of heat released due to internal friction, and various conditions of heat transfer between the roll surface and the polymer melt.

English translation will appear in Heat Transfer—*Soviet Research*

7.8 A Study of the Nonisothermal Processes of Calibration and Cooling of Thermoplast Sheets on Sheeting Calenders

S. I. Dobronogova, Yu. E. Lukach, and L. I. Ruzhinskaya
Kiev Polytechnical Institute

The present paper deals with some characteristics of the simultaneous calibration and cooling of thermoplast sheets on sheeting calenders. A mathematical model for simultaneous calibration and cooling of thermplast sheets is proposed. A finite-difference method is used to solve the mathematical model and a computer algorithm of the problem is developed.

English translation will appear in Heat Transfer—*Soviet Research*

7.9 The Nonisothermal Flow of Viscous and Non–Newtonian Fluids on a Centrifugal Apparatus Rotor

N. H. Zinnatullin, F. M. Gimranov, I. V. Flegontov, and N. A. Kashcheyeva
Chemical Engineering Institute, Kazan

Some problems of nonisothermal film flow of viscous and non-Newtonian fluids on the rotor surface of a centrifugal apparatus are solved. Analytical relations are obtained allowing the determination of the meridional velocity, and the thicknesses of the thermal boundary layer and the liquid film. These analytical relations are verified experimentally.

English translation will appear in Heat Transfer—*Soviet Research*

7.10 The Structural Effect of Boundary Phases on Transfer Processes in Disperse Rheological Systems

U. I. Ivanov, V. A. Minayeva, and R. Z. Ivanova
Chemical Engineering Institute, Kazan

The influence of structure on the transfer processes in a system is investigated. A quantitative description of various properties of the system over a wide range of parameters, based on an analysis of boundary phase properties and the system structure, is found to be accurate.

English translation will appear in Heat Transfer—*Soviet Research*

7.11 Convective Heat Transfer in Turbulent Flows with Polymer Additives

V. A. Ioselevich and V. N. Pilipenko
Institute of Mechanics, Moscow University

Some laws of turbulent heat transfer in dilute polymer solutions characterized by a reduction in drag are considered. The influence of the degree of drag reduction, and Prandtl and Reynolds numbers on a decrease in heat transfer is determined. A formula for the Stanton number containing a skin friction coefficient and various parameters of the logarithmic relations for the velocity and temperature profiles is obtained. The influence of the polymer additive on the thermal starting length is studied.

English translation will appear in Heat Transfer—*Soviet Research*

7.12 A Model of the Enrichment Process with Discrete Particles in Accelerated Flows of Viscoelastic Media and Its Application in Chemical Engineering

R. Kärmer and E. -O. Reher
Merseburg Institute of Technology, West Germany

The Uebler effect is considered by examining the nature of the flow prior to a diaphragm. The White–Metzner model is taken to be the rheological state equation. A formula to determine the critical precipitated-particle size, depending on the flow Weissenberg number, is proposed. The applicability of this model to engineering calculations has been confirmed by a study of the precipitation of gas bubbles and solid particles.

7.13 A Theory of Convective Heat Transfer with the Mixing of Non-Newtonian Fluids

V. V. Konsetov and V. G. Ushakov
Plastpolymer Research and Production Association, Okhtinsk, Leningrad

Theoretical solutions of convective heat transfer problems are suggested for non-Newtonian fluids flowing in both channels and agitated vessels at $Pr \geqslant 1$. The generalized dimensionless equations, developed for calculating various heat transfer factors, can be used in engineering practice.

English translation will appear in Heat Transfer—*Soviet Research.*

7.14 Experimental Friction Factors for Fully Developed Flow of Dilute Aqueous Polyethylene-Oxide Solutions in Smooth Wall Triangular Ducts

W. J. Leonhardt and T. F. Irvine, Jr.
State University of New York, Stony Brook

Friction factors for fully developed flow of a Newtonian fluid (ordinary water) and viscoelastic polymer solutions (polyethylene-oxide) in smooth wall ducts are experimentally determined over a range of $100 < \mathrm{Re_{dh}} < 30,000$. Four ducts are studied having isosceles triangular cross sections and apex angles of 10, 15, 50, and 60 degrees. The concentrations studied are 10, 15, 20, and 100 wppm. Comparisons are made with existing Newtonian triangular tube data and the polymer data of several circular tube investigators.

This paper appears on page 236, this volume

7.15 The Heating of Viscous and Non-Newtonian Fluids on a Rotating Disk

G. I. Lepekhin, G. V. Ryabchuk, and N. V. Tyabin
Volgograd Polytechnical Institute

This paper deals with the nonisothermal flow of viscous and non-Newtonian fluids on the surface of a flat rotating disk. The first stage of an iteration method of solving the problem, providing that rheological constants are independent of temperature, has been considered. The energy equation has been solved for an arbitrary radial temperature gradient, but the dissipative effect has not been taken into account. The distribution of radial and tangential velocities and the temperature field over the film thickness formed by the fluid flowing along the surface of the rotating flat disk is determined.

English translation will appear in Heat Transfer—*Soviet Research*

7.16 A Study of the Strength of Peat Structures as Related to Heat and Mass Transfer Processes

I. I. Lishtvan, N. N. Bityukov, and A. A. Terent'yev
Peat Institute, Minsk

The deformation state of peat systems is considered. The relationship between structural properties and the amount of bound water is determined along with various qualitative indices of lump peat. Vibrorheological studies have been made. The influence of vibration on packing and peat system viscosity is shown.

7.17 The Nonisothermal Flow of Polymer Melts in Operating Units of Extruders

S. A. Matkovskiy, V. N. Kocherov, N. V. Zhuk, and E. A. Sporyagin
Lvov Polytechnical Institute

The joint effect of velocity and temperature fields, taking into account secondary (circulation) flows, on the stability of polymer melt flows in rotational devices and in disk and combined extruders is considered. Theoretical and experimental results show the absence of a simple shear in the rotational devices. A curve of neutral stability is determined, allowing a distinction between stable and unstable values of α and Re. In order to optimize the melting process, computer "Minsk–22" programs are worked out using mathematical models of the processes.

7.18 A Nonisothermal Method for Studying the Rheological Properties of Flowing Systems in Rotational Viscosimetry

A. G. Merzhanov, A. M. Stolin, and B. N. Shatalov
Moscow

A nonisothermal method of viscosimetric measurement is discussed. A nonisothermal rotational viscosimeter is developed permitting the application of this method. The experiments were performed with castor oil (a Newtonian fluid) and polyester filled with aeroforce (a non-Newtonian system).

English translation will appear in Heat Transfer—*Soviet Research*

7.19 A Numerical Study of the Hydrodynamics and Heat Transfer Characteristics in Polymer Extrusion

V. P. Pervadchuk and V. I. Yankov
Perm Polytechnical Institute

The present paper deals with various hydrodynamic and heat transfer problems arising from the treatment of polymers by the extrusion technique. A finite-difference method is used to solve a system of differential equations describing the flow of anomal-viscous fluids in a screw channel. The motion of high- and low-viscous polymers is studied under the first and third kinds of temperature boundary conditions. A comparison between different extrusion models is presented.

English translation will appear in Heat Transfer—*Soviet Research*

7.20 Turbulent Heat Transfer in Dilute Surfactant Polymer Solutions

I. L. Povkh, A. B. Stupin, S. N. Maksyutenko, P. V. Aslanov,
A. P. Simonenko, and A. I. Musiyenko
Donetsk State University

A new model of turbulent heat transfer in dilute solutions of polymers is proposed, and a formula for determining the universal temperature profile is obtained. The resulting calculations are compared with the experimental data from a study of a turbulent flow having polyacrilamide additives. The velocity and temperature profiles of the turbulent flow of aqueous surfactant solutions are determined using a laser anemometer and thermocouples. It is found that the velocity and temperature profiles in a transition region of an aqueous surfactant solution change more sharply than the universal profile for water.

English translation will appear in Heat Transfer—*Soviet Research*

7.21 On the Mechanism of the Influence of Polymer Additives on a Shear Turbulent Flow Lacking Fixed Boundaries

N. A. Pokryvaylo, A. S. Sobolevskiy, and V. V. Kulebyakin
Luikov Heat and Mass Transfer Institute, Minsk

A possible mechanism for the effect of elastic polymer solutions on free shear turbulence is discussed. It is shown that a decrease in turbulence may be controlled by a decrease in the frequency of destruction of large vortices which are a permanent source of turbulence in the flow periphery.

English translation will appear in Heat Transfer—*Soviet Research*

7.22 The Influence of Variable Viscosity on the Laminar Heat Transfer and Friction in a Structural-Viscous Flow

V. I. Popov
Thermophysics Institute, Siberian Department of the USSR Academy of Sciences, Novosibirsk

An approximate analytical solution of the problem of forced convection heat transfer and friction in a steady laminar nonisothermal wall-spilling flow of structural-viscous fluids obeying the linear law of fluidity is given. A solution is obtained for the initial thermal region of flat and circular channels with boundary conditions of the first and second kind.

English translation will appear in Heat Transfer—*Soviet Research*

7.23 The Control of Transfer Processes in a Turbulent Liquid Flow

B. M. Smolskiy, I. T. Elperin, and L. I. Levental
Luikov Heat and Mass Transfer Institute, Minsk

A turbulent activated liquid flow is described on the basis of the three-component Karman–Prandtl model. Based on Reynolds analogy and using a description of the turbulent activated liquid flow, a relationship for calculating the heat transfer in tubes is obtained. The calculated values are compared with existing experimental data.

7.24 The Rheological Characteristic of a Ferromagnetic Suspension in a Magnetic Field

D. Spasoević and N. Afgan

Boris Kidrič Institute, Belgrade, Yugoslavia

The forced circulation of a ferromagnetic suspension in a horizontal channel 750 mm long, of rectangular cross section 20 X 2 mm was studied. A uniform magnetic field oriented normally to the fluid shear surfaces varied from 0 to 9000 Gauss. The ferromagnetic suspension was made up of silicon oil (of viscosity 80–240 c_p) and iron particles (of size 2.5–3.5μm). The particle concentration ranged from 0 to 5 percent by volume. The variables studied were: shear rate, external magnetic field strength, the volume of the ferromagnetic particles, the viscosity of the carrying fluid, and temperature. Theoretical and experimental data are compared with the results of an analysis of secondary effects on the measurement accuracy.

7.25 Critical Conditions for a Hydrodynamic Thermal Explosion in a Power-law Fluid Flow

A. M. Stolin, S. A. Bostandzhiyan, and N. V. Plotnikova

Institute of Chemical Physics, Moscow

Analytical expressions for the critical conditions needed for a hydrodynamic thermal explosion in a power-law fluid in both plane and cylindrical models of a rotational viscometer under various thermal conditions of the cylinder walls are derived. The steady-state temperature and velocity fields in the subcritical region are determined.

English translation will appear in Heat Transfer—*Soviet Research*

7.26 Rheological Investigations of Heat Processes in the Blood of Healthy People and That of People Suffering from Miocardial Infarction

A. N. Sundukov, M. A. Postnikova, K. V. Zvereva, and P. A. Ivanov

Gorkiy

Some rheological characteristics of the blood of healthy people and that of people suffering from miocardial infarction have been determined under conditions of microcirculation in capillaries, arterioles, and small veins. The relationship between the heat processes in blood and its rheological properties is determined. The activation energy of a viscous flow at velocities characteristic of actual blood circulation is estimated. A difference in the rheological characteristics of the blood of healthy males and females has been found and sheds light on the observation of a higher incidence and earlier onset of miocardial infarction in males than in females.

English translation will appear in Heat Transfer—*Soviet Research*

7.27 Heat and Mass Transfer Involving a Chemical Reaction with a Gas Bubble Moving in Power-Law Fluids

G. A. Fishbeyn and B. M. Abramzon
All-Union Petrochemical Research Institute, Leningrad

A numerical solution of equations describing the heat and mass transfer around a single gas bubble in a slow non-Newtonian fluid flow has been obtained. Local and mean Sherwood numbers have been calculated for $1 \leqslant Pe \leqslant 1000$. The influence of a fast homogeneous second-order reaction on the rate of mass transfer has been analyzed by a boundary-layer consideration.

English translation will appear in Heat Transfer—*Soviet Research*

7.28 A Generalized Theory of Nonisothermal Laminar Flow and the Heat Transfer of Non–Newtonian Fluids with Internal Heat Sources in Pipes

G. B. Froysheter
All-Union Research and Design Institute of Petroleum Refineries and Petrochemical Industry, Kiev

A mechanism of laminar heat transfer of non-Newtonian fluids with internal heat sources is discussed. A new parameter is introduced which permits the generalization of heat transfer data from various rheological models with heat sources arbitrarily distributed over a cross section and along the flow. This parameter is found to play an important role in nonisothermal flows. A formula is derived to calculate heat transfer and hydraulic resistances. The results of the theoretical calculations are correlated with the experimental data.

English translation will appear in Heat Transfer—*Soviet Research*

7.29 Rheodynamics and Heat Transfer in Rheologically Complex Nonlinearly Viscoelastic and Viscoplastic Systems with Dynamic Contact Effects

V. V. Kharitonov, B. P. Bateyev, V. I. Moskalev, V. F. Novikov, and A. A. Voyno
Byelorussian Institute of Railway Transport Engineers, Gomel

Problems of heat transfer and the rheodynamics of frictional processes in metal–polymer pairs with a lubricant on the contacting surfaces are discussed. The influence of disperse metal fillers on the viscous and thermophysical characteristics of greases and the thermostability of the friction unit is studied.

7.30 Influence of Diameter on Heat Transfer and Friction Factor for Non-Newtonian Fluids in Turbulent Pipe Flow

J. P. Hartnett, K. S. Ng, and J. E. Rasson
University of Illinois at Chicago Circle, Chicago

Experimental heat transfer and pressure results are presented for aqueous solutions of carbopol and attagel in turbulent pipe flow through tubes of 2.18 cm and 1.30 cm inside diameter. The solutions studied are purely viscous non-Newtonian fluids with values of the flow behavior index varying from 0.4 to 1.0. No influence of pipe diameter on the dimensionless heat transfer and friction factor was found. The pressure drop results are in good agreement with the predictions of Dodge and Metzner while the measured heat transfer values are in good agreement with the proposed correlation of Hartnett and Yoo. The hydrodynamic and thermal entrance lengths for the purely viscous non-Newtonian fluids were of the same order of magnitude as those for Newtonian fluids.

This paper appears on page 251, this volume

7.31 Energy and Mass Transfer in a Structure-Forming System with Chemical Reactions and Dissipative Processes Interaction

L. B. Tsimermanis and D. I. Shtakelberg
Ural Research and Design Institute of Building Materials, Chelyabinsk, and Riga Polytechnical Institute

An entropy expression for the fundamental Gibbs equation is proposed taking into account the elementary work in dissipative processes, i.e., structure formation and chemical reaction. A system of mass, energy, and velocity equations for structure formation and chemical reaction, is derived and describes the state of the processes near thermodynamic equilibrium. Criteria for the interaction of dissipative processes are obtained.

7.32 Laminar Entrance Flow of a Pseudoplastic Fluid in a Square Duct

R. Chandrupatla and V. M. K. Sastri
Department of Mechanical Engineering, Indian Institute of Technology, Madras, India

The laminar entrance flow of an incompressible, power-law, pseudoplastic fluid in a square duct is numerically analyzed using finite difference representations of the nonlinear equations of motion. The mathematical model incorporates axial momentum, continuity equations, and an equation relating the transverse velocities. A dimensionless variable viscosity is defined in terms of the second invariant of the irrotational strain-rate tensor. The power-law model is used to describe a stress strain-rate relationship for pseudoplastic fluids. Distributions of the velocity in the entrance region are determined by the extrapolated Liebmann method. The correlation between the friction factor and the Reynolds number is obtained. The friction factor-Reynolds number product for the square duct in a fully developed flow region is compared with that found in the literature and excellent agreement is noted.

7.33 The Effect of a Magnetic Field on the Convective Heat Transfer and Thermophysical Characteristics of Ferromagnetic Suspension

Z. P. Shulman, V. I. Kordonskiy, L. N. Novichenok, T. V. Kunevich, S. A. Demchuk, and I. V. Bukovich
Luikov Heat and Mass Transfer Institute, Minsk

Using a mass-transfer analogue, the effect of a nonuniform magnetic field on convective heat transfer in a turbulent wall boundary layer of a ferromagnetic suspension is studied experimentally. The heat and mass transfer intensification is correlated with an increase in wall friction. The effective heat transfer coefficients and thermal diffusivity are measured for various orientations of a uniform magnetic field and heat flux in the ferromagnetic suspension. It has been found that heat transfer increases with parallel orientation of the magnetic fields and heat flux, while it decreases with an inter-perpendicular orientation.

English translation will appear in Heat Transfer—*Soviet Research*

7.34 The Contribution of an Electrorheological Effect to the Heat Transfer in a Coaxial Cylindrical Channel

Z. P. Shulman, E. V. Korobko, A. D. Matsepuro, and B. M. Khusid
Luikov Heat and Mass Transfer Institute, Minsk

Based on Shulman's three-parametric rheological model, the problem of convective heat transfer in rheological suspensions with flow in a horizontal coaxial channel with boundary conditions of the first and second kinds is solved. A satisfactory agreement between the predicted and experimental data substantiates the structural concepts incorporated in the model.

English translation will appear in Heat Transfer—*Soviet Research*

7.35 Heat Transfer Involving Forced Convection Flow of Bingham-Plastics

A. N. Shcherban, Yu. P. Zolotarenko, and V. P. Chernyak
Engineering Thermophysics Institute, Ukrainian Academy of Sciences, Kiev

Experimental results on turbulent and laminar heat transfer of clay solutions in straight circular, concentric, and eccentric annuli are given. The empirical formulae obtained allow the calculation of heat transfer in deep boreholes.

English translation will appear in Heat Transfer—*Soviet Research*

8. RADIANT HEAT TRANSFER AND COMPLEX HEAT TRANSFER

8.1 On the Problem of Developing a Generalized Radiation Heat Transfer Method with Regard for the Anisotropy of Dissipation

V. N. Adrianov
Moscow Institute of the Graphic Industry

The processes of radiation heat transfer in anisotropic radiating and dissipating systems are analyzed in light of generalized concepts according to which an effective surface of particles is substituted for the medium.

A zonal method of calculating the radiation heat transfer is proposed. This can account for the anisotropy of bulk and surface dissipation as well as for the relationship between radiation characteristics of the surface and direction.

8.2 The Influence of Pressure on the Radiation Characteristics of Gases and Their Mixtures

N. Kh. Akhunov, E. A. Khakimov, L. M. Vasil'yeva, K. B. Panfilovich, and
A. G. Usmanov
Kazan Chemical Engineering Institute

Measurements of the total emissivity of steam at pressures of up to 21 bar for beds 0.16 m, 0.265 m, and 0.4 m thick, in a temperature range of 530 to 1015 K, and of the total emissivity of propane, butane, and propane-nitrogen mixtures at atmospheric pressure for a bed 0.3 m thick in a temperature range of 480 to 725 K are presented.

English translation will appear in Heat Transfer—*Soviet Research*

8.3 An Experimental Investigation of Thermal Radiation in Sulphur Pyrite Kiln Aggregates

A. G. Blikh, F. F. Zigmund, V. I. Sagadeyev, M. A. Taimarov,
M. M. Kocherov, V. A. Kuzmin, and V. G. Chaikovsky
The Polzunov Central Turbine and Boiler Institute, Leningrad and Civil Engineering Institute, Kazan

A technique and apparatus for the determination of the radiant characteristics of a gas burnt in a working kiln aggregate are considered. The results of this determination of the radiant characteristics of burnt gas are presented.

English translation will appear in Heat Transfer—*Soviet Research*

8.4 Complex Heat Transfer in Short Channels

A. Kh. Bokovikova, Y. I. Shcherbinin, and F. R. Shklyar
All-Union Research Institute of Metallurgical Thermal Engineering, Sverdlovsk

The laws of heat transfer in laminar and turbulent flows in short cylindrical and flat channels are investigated by a numerical solution of the energy equation.

Radiation fluxes have been determined taking into account the temperature distribution along the channel cross section and length. Zonal and one-dimensional methods were used in the case of a cylindrical channel while gray and selectively radiating gases were studied in the case of a flat channel. The effect of axial radiation fluxes on the channel heat transfer is studied.

8.5 Radiation-Convection Heat Exchange with Gas Combustion in Cellular-Perforated Systems

O. N. Bryukhanov and V. G. Kharyukov
Kaliningrad State University

Radiation-convection heat exchange with gas burning in cellular-perforated systems is considered under steady-state conditions.

Simple equations and formulae are obtained for engineering calculations of the radiating elements of cellular-perforated systems.

8.6 An Investigation of Transient Radiant-Conductive Heat Transfer in Selectively Absorbing Media

A. L. Burka, N. A. Rubtsov, P. I. Stepanenko, and A. D. Khripunov
Thermophysics Institute, Siberian Department of the USSR Academy of Sciences, Novosibirsk

A theoretical and experimental investigation of the temperature fields in flat layers of selectively absorbing media (glass, plexiglass) with nongray boundaries affected by external radiation is presented.

There is satisfactory agreement between the theoretical and experimental results.

English translation will appear in Heat Transfer—*Soviet Research*

8.7 Intensities in the ν_3-Fundamentals of CO_2 and N_2O

Prasad Varanasi
Department of Mechanical Engineering, State University of New York, Stony Brook, New York

Discrepancies between the results of various studies on the intensities of the ν_3-fundamentals of CO_2 and N_2O have prompted a remeasurement. The remeasurement was carried out on a double-beam instrument having a higher resolution than a commercial, Ebert-mounted, grating spectrometer, with varying pressure at a temperature of $296°K$. The absolute intensities were estimated to be 2452 ± 72 cm^{-2} $(atm^{-1})_{STP}$ for the CO_2 band and 1411 ± 54 cm^{-2} $(atm^{-1})_{STP}$ for the N_2O band.

This paper appears on page 265, this volume

8.8 Complex Heat Transfer in Various Gas Flow Patterns in the Working Space of Furnaces

M. A. Denisov, F. R. Shklyar, and A. Kh. Bokovikova
All-Union Research Institute of Metallurgical Thermal Engineering, Sverdlovak

Complex heat transfer in emitting and absorbing fluids flowing in the working space of heating furnaces is investigated. The influence of combustion recirculation selectivity and turbulence on heat transfer to a metal is determined. On the basis of calculations, a comparison is made between the efficiency of roof firing with flat-flame burners and that of end-wall firing of the furnace. The calculated and experimental results are compared. A method of separating the components of complex heat transfer is developed and applied.

English translation will appear in Heat Transfer–*Soviet Research*

8.9 The Irradiation of Bodies Accounting for the Characteristics of a Medium

S. P. Detkov, N. N. Ponomarev, and V. P. Kondrat'yev
Sverdlovsk

A generalized method of calculating surface irradiation taking into account the surface geometry and spectrum of the medium is suggested.

8.10 Integral Equations for Additive Function of Markovian Processes and Their Application to the Theory of Molecular and Radiative Transport

I. S. Zhitomirskiy and A. M. Kislov
Cryogenics Institute, Kharkov

The transport of particles (either molecules or photons) in nonisothermal systems with sources and sinks is considered to be a Markovian process in a phase space of particle states. The local density is determined by averaging additive functions (the particle lifetime as a function of the number of collisions in the region under study) which are determined on a variety of random trajectories. Integral equations for averaging the additive functions are derived and exact solutions are found for some specific case, these equations can be solved by the Monte Carlo numerical method. Exact solutions of the considered transport problem are presented.

8.11 Unsteady-State Heat-Transfer Problems on Graphs and Methods of Solution

I. S. Zhitomirskiy, V. I. Borisenko, L. A. Ishchenko, and V. A. Pestryakov
Cryogenics Institute, Kharkov

A mathematical formula for the general problem of complex heat-transfer is plotted. All available concrete models are divided into three classes according to their range of Strouhal number. For quasi-stationary models (Sh $\ll$ I) the problem is reduced to successive solving of a linear algebraic equation system for the unknown thermodynamic parameters in the apexes of the graph. A specific example of this class of model is considered. In the class corresponding to the values Sh $\sim$ I, a subclass of equibaric models for which the problem is also substantially simplified is considered.

8.12 The Thermal State of a Radiation-Heated Permeable Wall

B. A. Zhestkov and V. P. Lukash
Central Institute of Aviation Engines, Moscow

The results of an experimental study of the thermal state of a perforated plate under conditions of combined heat transfer with injection of air, argon, and helium are presented. The surface of the plate was heated by a self-contained source of radiative energy. The existing methods of calculating the temperatures of permeable walls are analyzed. The distribution of the calculated data, depending on the injection rate, heat capacity, and degree of coolant heating is considered.

English translation will appear in Heat Transfer—*Soviet Research*

8.13 The Calculation of Air Radiation Cooling behind Strong Shock Waves Employing Averaged Optical Characteristics

V. P. Zamurayev, I. I. Maslennikova, and R. I. Soloukhin
Institute for Pure and Applied Mechanics, Novosibirsk

Some averaged optical characteristics of air at temperatures from 10,000 to 20,000°K and pressures from 0.1 to 100 atm are obtained. Using these characteristics, the radiation air cooling behind strong shock waves is calculated. The influence of various parameters on the temperature and gas density distribution behind a shock wave is determined. A consideration is also made of the effect of radiation absorption, in an incident flow, on gas parameters in the shock layer.

English translation will appear in Heat Transfer—*Soviet Research*

8.14 Inverse Problems of Radiation-Conduction Heat Transfer as Related to Temperature Field Reconstruction by Heat Radiation

G. Ya. Zeliger and A. A. Men
The Mendeleyev Research Institute of Metrology, Leningrad

Temperature field reconstruction by means of emerging heat radiation in flat and cylindrical semitransparent samples is considered with optic characteristics independent of temperature. Equations describing various schemes of measurement are given, methods of solving the equations are shown, and the numerical results of these methods are discussed.

8.15 Unsteady-State Radiation Heat Transfer in a Flat Layer of a Scattering and Absorbing Medium

A. P. Ivanov, S. M. Reprintseva, and V. L. Dragun
Institute of Physics, Minsk and Luikov Heat and Mass Transfer Institute, Minsk

The problem of unsteady-state radiation heat transfer in flat absorbing and scattering media considered to be gray and possessing selective characteristis is solved. The role of optimal characteristics in describing the radiation transfer in such systems is considered. Based on experimental investigations, the radiation transfer characteristics are determined.

8.16 Heat and Mass Transfer and the Stability of a Laminar Flame in a Magnetic Field

St. Kalitzin
Technische Hochschule, Madgnburg, German Democratic Republic

An expression for the entropy production in a laminar flame is proposed taking into consideration heat conduction, convection, diffusion, and radiant heat transfer in the presence of a perpendicular magnetic field. On the basis of the expression for entropy production, the conditions for stability of the laminar flame are defined.

8.17 An Investigation of the Radiative Heat Transfer of Heterogeneous Combustion Products in Finite Cylinders

V. Ya. Klabukov, E. I. Maratkanova, V. A. Kuzmin, and
L. T. Grebenshchikov
Kirov Polytechnic Institute

A numerical solution for an integro-differential equation of radiative transfer is obtained for two-dimensional cylindrical absorbing, emitting and scattering media. The effects of fundamental parameters on the radiative heat fluxes are discussed.

8.18 The Effect of a Polydisperse Microstructure on Heat Transfer in Channels

L. A. Konyukh, F. B. Yurevich, N. T. Keda, and E. D. Sergiyevsky
Luikov Heat and Mass Transfer Institute, Minsk

The problem of heat transfer by convection, conduction, and radiation in a dusty liquid flow in a flat channel is solved. The effect of particle size, the nature of the radiation scattering, and the particle concentration in the flow on heat transfer is studied.

English translation will appear in Heat Transfer—*Soviet Research*

8.19 Unsteady-State Heat Transfer in a Low-Temperature Plasma Jet

L. Krejci, V. Dolinek, and J. Fogel
Ustav Termomechaniky CSAV, Prague, Czechoslovakia

A method of measuring the unsteady-state heat flux from a plasma jet to a normally inclined plate with a temperature field in the plate is proposed. It follows from the experimental data that, in a number of cases, the Newton formula and Fourier equation cannot be used to describe heat transfer intensity. This is attributed to the unsteady state of the process, specifically, when the jet converts from a laminar to a turbulent structure.

8.20 Thermal Behavior of Confined Arcs with Local Fluid Constriction

C. J. Cremers and H. S. Hsia
University of Kentucky, Lexington

The behavior of blown and confined electric arcs with local fluid constriction is reported. The local fluid constriction, which is caused by blowing radially inward on the arc, causes the arc column to constrict with a concomitant increase in local electric field strength and plasma temperatures. Data are presented which show how the wall heat flux, static pressure distribution, and wall electric potential are affected by the radial blowing. Isotherms for the blowing region are also presented.

This paper appears on page 269, this volume

8.21 Simultaneous Radiative, Convective, and Conductive Heat Transfer in an Extended Surface

T. Kunitomo and S. Tanaka
Kyoto University, Kyoto, Japan and Fukui Technical College, Fukui

This paper presents some heat transfer characteristics of a longitudinally finned plane surface and cylinder and a circumferentially finned cylinder, where radiation, convection and conduction interact. The radiant heat exchange is calculated by the Monte Carlo method and a numerical solution of the heat transfer is obtained by iteration of the energy equation. The heat transfer is considered in terms of system effectiveness, the total heat transfer rate and the heat transfer increase per unit fin volume. Design parameters for optimal heat transfer are examined in various combinations. These parameters include: the emissivity, Biot number, convection–radiation, parameter, the fin spacing to fin height ratio, the fin height to fin thickness ratio, and the specular component of reflection.

This paper appears on page 276, this volume.

8.22 Radiation Heat Transfer at Cryogenic Temperatures

S. S. Kutateladze, N. A. Rubtsov, Ya. A. Baltsevich, and G. P. Yeremenko
Thermophysics Institute, Siberian Department of the USSR Academy of Sciences, Novosibirsk

The problem of radiation and complex heat transfer at cryogenic temperatures is examined. The results of an investigation of temperature fields resulting from radiation-conduction heat transfer in a continuous, semitransparent, uniform media, and a system of screens (metal, semitransparent, and mixed) are presented. A proposed method of approximate calculation of the thermal stabilization of the system of metal screens is found to be in good agreement with the experimental results.

English translation will appear in Heat Transfer—*Soviet Research*

8.23 Hydrodynamic Structure and Heat Transfer in a Plasma Jet with Varying Electric Arc Stabilization

J. Kučera and J. Dundr
Ustav Termomechaniky CSAV, Prague, Czechoslovakia

Relationships characterizing jet structure with the vortex stabilization of a discharge of varying intensity with an axial argon flow in the discharge chamber are proposed. The dependence of the radial and longitudinal pressure and temperature profiles of the jet on various discharge parameters is determined experimentally. Heat flux to the wall is found to depend on the flow pattern.

8.24 An Analysis of the Influence of Various Thermal Factors on External Pollution

P. L. Magidey
Leningrad Polytechnical Institute

Data on the thermal resistance of furnace screen pollution are analyzed. The experimental data cover a wide range of furnace processes, under various operating conditions. The dependence of thermal resistance on various characteristics of the furnace heat, work, and fuel properties is established. The results are plotted and calculation expressions for the furnace heat transfer are recommended. The calculations are found to have good accuracy.

English translation will appear in Heat Transfer—*Soviet Research*

8.25 Radiation Heat Transfer in a Gas Flowing over a Heated Surface

M. M. Melman and A. S. Nevskiy
All-Union Research Institute of Metallurgical Thermal Engineering, Sverdlovsk

The radiation heat transfer in a layer of gray gas flowing on a heated surface and bounded by this surface and an adiabatic lining has been considered. A theoretical solution is obtained. The effect of the Boltzmann and Bouguer criteria on the heat transfer and the dome temperature has been investigated.

The above situation is compared with the case of a gas flow along a heating surface. The nature of the dependence of the heat transfer on the radiation parameter is the same in both cases.

English translation will appear in Heat Transfer—*Soviet Research*

8.26 The Absorptivity of Contaminated Radiation-Receiving Surfaces of Boilers

I. R. Mikk and T. B. Tiikma
Tallin Polytechnical Institute

Possible approaches for defining the absorptivity of ash-covered heating surfaces of boilers are analyzed. Devices for the measurement of total absorptivity and special emissivity are described. The spectral emissivities of oil shale furnace deposits are determined.

English translation will appear in Heat Transfer—*Soviet Research*

8.27 Radiative Heat Transfer and Evaporation from Bodies in Hypersonic Flow

V. N. Mirskiy and V. P. Stulov
Institute of Mechanics, Moscow University

The problem of a radiating gas flow over the front portion of a body with intensive evaporation is solved. The flow region consists of two layers; a shock layer and a vapor

8.31 An Investigation of Heat Transfer by Radiation and Conduction in a Plane Layer of a Condensed Medium with Selective Optical Properties

V. A. Petrov and S. V. Stepanov
Institute of High Temperatures, USSR Academy of Sciences, Moscow

A new method of calculating radiative and conductive heat transfer equations using the Green function to describe the radiative heat flux is proposed.

The method is used to investigate the influence of the reflective coefficient and its angular component, the optical thickness, as well as the effect of the dimensionless thermal conductivity and the character of the spectral absorption coefficient approximation on the heat transfer in a plane layer of a condensed nongray medium with opaque selective boundaries.

English translation will appear in Heat Transfer—*Soviet Research*

8.32 The Application of Optical-Geometrical Functions to the Analysis of Nondiffuse Radiation Heat Transfer

S. P. Rusin
Moscow Institute of the Graphic Industries

Several mathematical models for the numerical calculation of nondiffuse radiation heat transfer are presented.

8.33 Optimization of the Spectral Composition of Radiation in Cameras with IR-Heating

L. N. Ryzhkov and V. I. Rychkov
Moscow Institute of the Graphic Industries

Problems of correlation between the spectral radiation composition and optical properties of an object are considered. The influence of reflection, the organization of the field of radiation, and spectral properties of the reflectors are investigated in order to optimize infrared heating.

It is shown that the velocity of the process may be increased several times over by proper functioning of the various factors influencing the spectral composition of the radiation. Equations for the evaluation of efficiency are given.

English translation will appear in Heat Transfer—*Soviet Research*

8.34 Radiation Transfer in a Real Spectrum

V. G. Sevastyanenko
Institute of Pure and Applied Mechanics, Novosibirsk

A new method for calculating the integral spectral characteristics of a radiation field is proposed. The method is verified over a wide range of temperatures, pressures, and

layer, divided by a contact surface. Radiative heat flux is absorbed partly by the vapor layer. The remainder reaches the body surface and causes evaporation. A similarity law for the radiative heat transfer coefficient at the stagnation point is formulated.

English translation will appear in Heat Transfer—*Soviet Research*

8.28 A Study of the Radiation Structure of Steam-Boiler Furnaces

V. V. Mitor and I. N. Konopelko
The Polzunov Central Boiler and Turbine Institute, Leningrad

The results of a study on the integral and spectral emissivity of ash deposits on the walls of large-capacity steam boilers and the spectral composition of furnace radiation with the combustion of various fuels are presented. The importance of spectral radiation characteristics in making precise thermal calculations of large-capacity boiler furnaces is shown.

English translation will appear in Heat Transfer—*Soviet Research*

8.29 The Role of Phenomenology, Statistics, and Analogy in the Theory of Molecular and Radiant Transfer

G. L. Polyak
Krzhizhanovskiy Power Engineering Institute, Moscow

Based on radiation theory, equations of molecular and radiation transfer along with boundary conditions considered on kinetic and phenomenological levels, are obtained.

Analogies between the processes of molecular and radiation transfer, as well as between their coefficients, including accomodation, sorption, condensation, and absorptivity, are established allowing the results obtained in one particular field to be generalized to another.

8.30 Thermal Radiation of Diatomic Gases

Yu. A. Popov
All-Union Research Institute of Metallurgical Thermal Engineering, Sverdlovsk

A calculation is made of the monochromatic emissivity and total emissivity of two atom gases taking into account anharmonic molecules near the rigid rotor and using Elsasser's model for absorption bands.

The calculation results for CO, NO, OH are presented. There is a satisfactory agreement between the predicted and experimental results.

English translation will appear in Heat Transfer—*Soviet Research*

dimensions of the medium. The calculation of heat transfer by the proposed method insures high accuracy and requires negligible computer time.

English translation will appear in Heat Transfer—*Soviet Research*

8.35 Thermal Radiation Models for the Atmospheres of Jupiter and Saturn

R. D. Cess
State University of New York, Stony Brook, New York

Model atmosphere calculations for Jupiter and Saturn are reviewed and then employed to suggest possible models for ammonia ice clouds within the atmospheres of both planets. It is shown that the cloud models are qualitatively consistent with available observational information.

This paper appears on page 294, this volume

8.36 Heat Transfer with the Burning of a Gas in a Boundary Layer

B. S. Soroka
Gas Institute, Ukrainian Academy of Sciences, Kiev

The available information on combustion heat transfer at a phase boundary is analyzed. A radiative-convective-diffusional heat transfer in a system of two infinite plates with one exposed to a flow of burning gases is considered. Based on theoretical and experimental evidence, the possibility of enhancing the heat transfer to a wall when the temperature of the latter exceeds the ignition temperature of a combustive mixture with the reaction in a flow not yet completed is considered

English translation will appear in Heat Transfer—*Soviet Research*

8.37 Radiative Heat Transfer in a High Power Diesel Engine Cylinder

M. V. Stradomskiy, E. A. Maksimov, V. A. Asmalovskiy, and
V. S. Malyarov
Engineering Thermophysics Institute, Ukrainian Academy of Sciences, Kiev

Some methods of measuring the radiative heat flow in a high power marine diesel cylinder are presented. Taking into consideration the geometric size of the hollow sensor and the combustion chamber, the radiative heat flux being supplied to the cylinder cap is determined. A comparison of experimental data with the calculated results is presented.

English translation will appear in Heat Transfer—*Soviet Research*

8.38 Some Basic Problems of Radiant Heat Transfer Theory

Yu. A. Surinov
Institute of Economics and Statistics, Moscow

Some theoretical fundamentals of a new interaction-zone method for solving integral radiation equations allowing a determination of radiant heat transfer characteristics both local and averaged within appropriate boundary and volume zones in radiating systems filled with absorbing and scattering media are presented. New radiant heat transfer characteristics are introduced and the theory on which the above method is based, is formulated.

8.39 A Study of Heat and Mass Transfer in a Granule of Chemico-Pharmaceutical with a Radiative Heat Supply

N. V. Fedorovich and S. M. Reprintseva
Luikov Heat and Mass Transfer Institute, Minsk

The present paper presents the results of an experimental study of the infrared radiation on the internal heat and mass transfer processes in separate granules of chemico-pharmaceuticals.

9. THEORY OF HEAT CONDUCTION

9.1 A Qualitative Study of Some Nonlinear Boundary-Value Heat Conduction Problem Solutions

A. A. Aleksashenko
Moscow Institute of the Graphic Industries

A method is proposed to qualitatively investigate some nonlinear boundary-value heat conduction problems. A qualitative picture of the graph of the solution can be obtained without actually solving the problem. The method is elaborated for a one-dimensional problem.

9.2 Gradient Methods for Recovering a Boundary Heat Regime

O. M. Alifanov
Moscow Aviation Institute

Various types of inverse unsteady heat conduction problems are systematized. A method for interpreting the results of heat tests is based on solutions of boundary-value inverse problems of the generalized heat conduction equation. The problem formulated is an external one with a limitation on the discrepancy level. Iteration schemes, suggested for its solution, are based on methods of the steepest descent and conjugate gradients.

9.3 The Solution of a Nonlinear Inverse Heat Conduction Problem

O. M. Alifanov, E. A. Artyukhin, and B. M. Pankratov
Moscow Aviation Institute

A numerical solution to the nondimensional nonlinear generalized inverse unsteady heat conduction problem with a heat source (sink) and a convection term is considered. A solution is obtained for a domain with moving boundaries based on explicit and implicit approximation schemes. The resulting system of nonlinear difference equations is used to construct the Tikhonov-regularization algorithm. The results of solutions of examples are presented.

English translation will appear in Heat Transfer—*Soviet Research*

9.4 The Contribution of the Thermal Separation Effect to Heat Transfer in Multicomponent Gas Mixtures

A. F. Bogatyrev, Yu. I. Zhavrin, N. D. Kosov, and V. F. Kryuchkov
Kazakh State University

In view of the necessity of accounting for the thermodiffusion effect on heat transfer in gas mixtures, a formula is proposed for calculating thermodiffusional separation in a multicomponent mixture using the separation values in corresponding binary mixtures.

The validity of the formula is confirmed by experimental results on tertiary and quaternary gas mixtures.

English translation will appear in Heat Transfer—*Soviet Research*

9.5 Derivation of the Wave Heat Conduction Equations

B. A. Bubnov
Moscow Machine-Building Institute

Two methods are presented for the derivation of the wave heat conduction equations: 1) the introduction of a functional dependence on time into the formula for heat flux and 2) an analysis of isotherm behavior.

9.6 Heat-Propagation Velocity in the Heat Conduction Theory

A. V. Bulyga
Institute of Physics of Solids and Semiconductors, Minsk

Applying the conservation law in the form proposed by A. A. Vlasov to the distribution function of identical particles, a differential hyperbolic-type heat conduction equation containing two relaxation times is obtained. One relaxation time governs the quasi-steady state heat conduction process while the other corrects the first and accounts for the unsteady state character of the process.

9.7 An Analysis of Heat Transfer Dynamics in State Space

E. N. Boot and D. F. Simbirskiy
Kharkov Aviation Institute

The dynamics of the heat transfer of linear, steady state, and transient objects is considered from the point of view of state space allowing the application of matrix methods of the general theory of dynamic systems. To this end, initial heat transfer equations are approximated by means of differential-difference models.

Relevant concepts and terms are introduced, the most important being concepts concerning state and output vectors and the transfer matrix.

Formulas for defining a state vector of the body are presented.

9.8 An Application of the Method of Integral Characteristics to Inverse Problems of Thermophysics

V. V. Vlasov, Yu. S. Shatalov, E. N. Zotov, A. S. Labovskaya,
N. P. Puchkov and A. A. Churikov
Tambov Chemical Engineering Institute

Based on some integral characteristics of experimental field data a method for solving inverse problems of engineering thermophysics is proposed.

9.9 A Method of Determining Thermophysical Transfer Potential-Dependent Characteristics Taking into Account Expansion Effects

V. V. Vlasov, S. V. Mishchenko, N. P. Fedorov, and Yu. S. Shatalov
Tambov Chemical Engineering Institute

Exact analytical expressions for the determination of the thermophysical characteristics of materials taking into account thermal expansion effects are obtained.

Thermophysical characteristics of solids are defined in terms of temperature, moisture content, and heat and moisture flow as a function of time in a cross section of a given sample.

9.10 On the Possibility of Employing the Unsteady Doppler Effect for the Determination of the Heat Capacity of Coatings

V. I. Derban, I. P. Zhuk, V. I. Krylovich, and V. K. Serikov
*Luikov Heat and Mass Transfer Institute, and the Byelorussian
Institute of Agricultural Mechanization and Electrification, Minsk*

The possibility of determining the heat capacity of coatings by means of the unsteady acoustic Doppler effect is considered. Calculations are made for the case of a thin wire.

English translation will appear in Heat Transfer—*Soviet Research*

9.11 Some Singular Perturbation Methods Applied to Problems of Heat Conduction with Moving Sources

N. V. Diligenskiy, Yu. V. Mikheyev, and B. Z. Chertkov
Kuibyshev Polytechnical Institute

This paper deals with the construction of compound asymptotic solutions to problems of heat conduction with moving sources.

The solutions are of a form suitable for the analysis of heat exchange.

9.12 An Approximate Analysis of Temperature Fields in Systems of Solids with Energy Sources

G. N. Dulnev and A. Yu. Potyagaylo
Leningrad Institute of Precision Mechanics and Optics

A novel analytical method of approximation is proposed for calculating the temperature fields in systems of solids having internal energy sources.

An approximate solution is determined on the basis of the local effect principle. A statistical analysis of the error factor in the method is conducted.

9.13 An Experimental Study of Mass Transfer in Frozen Disperse Materials

S. S. Yefimov and L. M. Nikitina
Institute of Physical and Engineering Problems of the North, Yakutsk

The mass transfer characteristics of coals to be transported in bulk are discussed in connection with the building of the Baikal-Amur railway and the development of the South-Yakutian engineering district.

The dependence of the diffusion coefficients on the temperature and moisture content of the material is determined.

Safe moisture content values of coal from the "Moshchny" seam (Neryungrinsky deposit) are obtained.

English translation will appear in Heat Transfer—*Soviet Research*

9.14 Heat and Mass Transfer in a Region having a Moving Boundary

I. S. Yefremova and M. S. Smirnov
All-Union Correspondence Institute of Food Technology, Moscow

Based on a system of differential heat and mass transfer equations for a region with a moving boundary, an equivalent system for a region with a fixed boundary is proposed.

After simplification, the equations in this system may be solved in various ways as the boundary moves according to parabolic or linear laws.

9.15 The Asymptotic Heat Conduction Theory for Thin Bodies

I. E. Zino
Leningrad Polytechnical Institute

Singular perturbation methods are applied for some boundary-value heat conduction problems of thin bodies with convective heat transfer on their surfaces and an asymptotic integration procedure is developed. The zero approximation accounts for the main direction of heat flux in the body (the quasi-one-dimensional solution), while higher approximations describe the boundary effects resulting from a difference between the actual heat conduction problem and an ideal one-dimensional model.

9.16 Heat Transfer at the Solid-Gas Interface: Thermal Accommodation Coefficients for Helium on Gas Covered Tungsten

B. J. Jody and S. C. Saxena
University of Illinois at Chicago Circle, Chicago, Illinois

An improved hot wire column instrument is developed to determine several thermal properties of gases, solid metal wires, and gas-solid interface over a wide temperature range. Here, the thermal accommodation coefficients for helium on gas covered tungsten are reported in the temperature range 700-2300 K as determined from steady state heat transfer rate data taken in the temperature-jump regime. The available information on the adsorption of gases on tungsten is employed in conjunction with the thermal accommodation coefficients for helium on clean and gas covered tungsten to estimate the adsorbate coverage and to formulate a semiquantitative model that relates the thermal accommodation coefficient to the adsorbate coverage.

This paper appears on page 302, this volume

9.17 The Correlation between a Rigorous Kinetic Theory for the Thermal Conductivity of Gas Mixtures and Experiment Performed at Lower Temperatures

A. G. Karpushin, N. D. Kosov, and Kh. S. Seitov
Kazakh State University

The correlation between the rigorous kinetic theory for the thermal conductivity of gases and the experimental data for He-N$_2$, H$_2$-N$_2$, H$_2$-Ar systems in a temperature range of 90-300 K at a pressure of 1 bar is investigated

English translation will appear in Heat Transfer—*Soviet Research*

9.18 An Integral Method for the Solution of Generalized-Type Boundary-Value Unsteady Heat Conduction Problems

E. M. Kartashov and V. M. Nechayev

Moscow Textile Institute, and Voskresensk Computing Centre

The heat potential method is modified in order to solve boundary-value heat conduction problems in a region with moving boundaries. The resulting method provides a concise and simplified solution of the problem for the case of uniform boundary motion.

Certain boundary conditions have been chosen in order to determine the appropriate Green functions needed to solve the boundary-value problems in a region with arbitrarily moving boundaries.

9.19 Application of a Trial-and-Error Method for the Solution of Reverse and Inverse Nonlinear Heat-Conduction Problems

L. A. Kozdoba

Engineering Thermophysics Institute, Ukrainian Academy of Sciences, Kiev

A procedure of electrical simulation is described in order to solve reverse, inverse, inductive, and inverted nonlinear unsteady heat-conduction problems. The results of solutions of control and actual reverse and inverse problems by means of a trial-and-error method are presented. Heat fluxes for steel tempered in oils are determined, and some thermophysical characteristics of high-temperature heat insulation are given.

9.20 Heat Field Propagation in Nonlinear Media

P. M. Kolesnikov, V. N. Abrashin, G. F. Gromyko, L. V. Grishanova, N. S. Kolesnikova, and N. G. Zhadayeva

Luikov Heat and Mass Transfer Institute, Minsk

A modulus method of constructing a program package for the solution of nonlinear partial differential equations describing the propagation of heat fields in nonlinear media under various boundary conditions is proposed. Automatization of the construction of the program package is considered along with the possibility of effective computer calculation of concrete, nonlinear equations. The solutions of some concrete nonlinear equations are analyzed, and the dependence of the solutions on parameters in the equations is investigated.

9.21 An Investigation of Some Heat and Mass Transfer Problems Involving Phase Changes with Moving Boundaries

P. M. Kolesnikov, L. V. Grishanova, V. N. Abrashin, L. N. Degtereva,
N. S. Kolesnikova, and V. A. Tsurko
Luikov Heat and Mass Transfer Institute, Minsk

Some problems involving phase changes are considered. Stefan-type problems and those on bubble pulsation with moving boundaries are investigated. Using numerical methods, nonlinear equations for heat and mass transfer with moving boundaries are solved. The influence of characteristic parameters on the solutions is investigated. Calculation results are presented in the form of tables and graphs.

English translation will appear in Heat Transfer—*Soviet Research*

9.22 The Simulation of Thermophysical Processes by the Contraction of Human Muscular Tissue

F. A. Krivoshey
Engineering Thermophysics Institute, Ukrainian Academy of Sciences, Kiev

A mathematical model of the thermal behavior of muscular tissue during short contractions is proposed. A joint study of temperature fields, metabolic heat generation, and heat losses due to moisture diffusion is presented.

9.23 The Temperature Field in Plates with Piecewise-Constant Heat Transfer Coefficients

A. N. Kulik and V. S. Didyk
Institute of Mathematics, Lvov State University

A method of determining temperature fields in plates with piecewise-constant heat transfer coefficients is presented. Heat transfer coefficients are presented by means of single functions and in terms of the corresponding coordinate functions. This yields differential equations incorporating step function coefficients. Two heat conduction problems illustrate the proposed method: heating of thin plates with either periodical nonthrough cylindrical or band heat sources.

9.24 Solution Methods and Some Practical Applications of Temperature Wave Problems

I. N. Manusov, N. M. Belyayev, and A. N. Andriyko
Dneprodzerzhinsk Industrial Institute

Approximate methods are presented for calculating the temperature fields and temperature stress fields in solids during cyclic heating, taking into account heat transfer coefficient fluctuations.

An analytical method of investigating the temperature waves in heat exchangers with porous separating walls is described.

9.25 Concerning Identification in Nonlinear Heat Conduction Problems

Yu. M. Matsevity, A. V. Malyarenko, and A. V. Multanovskiy
Institute of Machine-Building Problems, Kharkov

The application of the optimal dynamic filtration method is considered for solving direct and inverse multidimensional heat conduction problems in both linear and nonlinear forms. The finite-difference equations are taken to be initial ones, thus eliminating the need to integrate the initial system in order for transient matrices to be determined.

9.26 Application of the Algometric Language "Formal" for the Analytical-Numerical Computation of Some Heat and Mass Transfer Problems

M. D. Mikhaylov and D. D. Aleksandrov
Center for Applied Mathematics, Sofia, Bulgaria

Some heat and mass transfer problems (steady-state heat transfer in a tube, quasi-steady solutions of heat conduction, the freezing of liquid flowing onto a plane wall, the heating of viscous liquid flowing in a tube) are solved on an electronic computer by analytical-numerical iterative algorithms employing a new man-to-computer language "FORMAL." The possibility of using an electronic computer for problem solving with the aid of analytical methods is discussed.

9.27 Temperature Fields in Elements of Heat Exchangers with Small-Finned Surfaces

A. A. Mikhalevich, V. I. Peslyak, A. V. Novoselskiy, and L. I. Demina
Institute of Nuclear Power Engineering, Minsk

Using a variational method, heat conduction in elements of heat exchangers with small-finned surfaces is investigated.

It is found that, for this type of surface, the temperature of the fin base, the interfin surface, and the nonfinned side wall is not constant

9.28 Thermoeffects due to Mutual Diffusion in Liquids

R. Ya. Ozols
Institute of Physics, Riga

Analytical solutions for the temperature profiles in a diffusion cell when the temperature gradient is governed by the Dufour effect are presented.

At small Fourier numbers (< 0.1) these results can be used to estimate the Dufour coefficient provided the Lewis number is known.

9.29 An Application of Asymptotic Programming Methods for Heat Conduction Problems

A. N. Panchenkov and M. N. Borisyuk
Siberian Power Engineering Institute, Irkutsk

Asymptotic programming methods are applied for the investigation of various heat conduction problems.

An initial problem is reduced to an extremal problem which is solved using various algorithms of asymptotic programming.

9.30 A Solution of the Equation $\overline{v}^2 u + \lambda^2 u = 0$ in an Arbitrary, Finite, and Simply Connected Plane Domain with Third-Kind Boundary Conditions and Piecewise-Continuous Coefficients

F. P. Plachco
Universidad de los Andes, Mérida, Venezuela

Based on appropriate independent functions, the solution to an expansion of the Helmholtz equation is found for a complex, real or imaginary λ. The boundary condition considered is of the type

$$a(p) \cdot u(p) + b(p) \frac{\partial u(p)}{\partial n} = f(p), \quad p \in L$$

where the coefficients $a(p)$ and $b(p)$ and the function $f(p)$ are complex, real or imaginary, continuous-piecewise, and independent of each other. The domain is assumed to be plane, finite, arbitrary, and simply connected, with contour L smooth by parts.

9.31 Solution of Some Classical Transfer Problems in Wedge Regions

F. P. Plachco
Universidad de los Andes, Mérida, Venezuela

With the aid of the Sommerfeld–Malyuzhints integral, a Green function is obtained for transfer problems in infinite wedge regions under combined first-, second-, and third-kind conditions. The boundary-value problem is reduced to a solution of the functional Malyuzhints–Tuzhilin equations by means of the Malyuzhints function.

9.32 The Generalized Similarity of the Temperature Fields of Bodies having Different Profiles with Variable Boundary Conditions

V. G. Prokopov and Yu. V. Sherenkovskiy
Kiev Polytechnical Institute

The possibility of comparing physical processes in dissimilar regions is considered in both a strict and an approximate formulation. The approximate formulation is based on the conservative nature of the desired function under some of the governing parameters.

To illustrate, a solution of a heat conduction problem is considered.

9.33 Nonlinear Statistical Methods in Heat and Mass Transfer Theory

B. S. Sadykov
Moscow Institute of The Graphic Industries

The nonlinearity mechanism and dependence of the transfer coefficient L_{ik} on the intensity of electrical and other fields x_i cannot be adequately explained by the phenomenological theory. A statistical method is therefore proposed. It is shown that L_{ik} values are even functions of x_i. General formulas are obtained to relate L_{ik} and x_i and the symmetry relations are given.

9.34 A Solution of the Stefan Problem in Contacting Bodies of Complex Geometry for a Multidimensional Case

P. S. Samoylenko
Saratov

An approximate method of solving the Stefan problem for a multidimensional case in contacting bodies of complex configuration is proposed, taking into account actual contact between the surfaces.

9.35 Steady Heat Conduction in Contacting Bodies of Complex Geometry

P. S. Samoylenko
Saratov

An approximate method of solving the steady heat conduction equation under third-kind boundary conditions for contacting bodies of complex configuration is proposed taking into account contact phenomena between the surfaces.

Using the proposed method, a solution for the temperature field in contacting bodies is found.

9.36 The Small Parameter Method, Laplace Transforms, R-Functions Method and Variational Methods in Nonlinear Unsteady Heat Conduction Problems

A. P. Slesarenko
Institute of Machine-Building Problems, Kharkov

The application of the small parameter method, Laplace transforms, the R-functions method and variational methods is considered for the solution of a nonlinear unsteady heat conduction problem.

Numerical results from a sample problem are compared with experimental data.

English translation will appear in Heat Transfer—*Soviet Research*

9.37 A Man-to-Computer Regime of Computer Calculation of the Temperature Fields in Construction Elements of Complex Geometry Using a Cathode-Ray Tube

A. P. Slesarenko and V. K. Bogorodskiy
Institute of Machine-Building Problems, Kharkov

A man-to-computer regime for the computer calculation of temperature fields in elements of complex configuration by means of the structural method, the Ritz method, and a cathode-ray tube is proposed. The exact geometry of the element is prescribed by R-functions in analytical form.

The results of the calculation of temperature fields are presented for two construction elements.

9.38 A Multiparametric Method of Solving a One-Dimensional Heat Conduction Problem for a Layer with a Moving Boundary

E. M. Smirnov
Leningrad Polytechnical Institute

A universal equation for the influence of arbitrary time-dependent temperature on immovable boundaries and heat flux on movable boundaries of a plane layer formed by freezing is determined and integrated.

To obtain the solution for a specfic case the first-order differential equation, which in particular cases is integrated in quadratures, should be used.

9.39 One Approach to the Solution of the Stefan Problem

I. A. Solov'yev
All-Union Correspondence Institute of Food Technology, Moscow

A statement of the Stefan problem on the basis of the hyperbolic heat conduction equation with discontinuity coefficients is given. The Stefan condition is found to follow from the basic equation. A formula for the interphase velocity is obtained.

9.40 On the Solution of Nonlinear Heat Conduction Equations

G. A. Surkov
Luikov Heat and Mass Transfer Institute, Minsk

By means of a Fourier expansion of the appropriate functions, a nonlinear system of equations is reduced to a linear one.

A simple solution having a high degree of accuracy, is thereby obtained.

9.41 Some Integral Methods for the Solution of Inverse Heat Conduction Problems

A. G. Temkin
Riga Polytechnical Institute

General considerations necessary for the solution of inverse heat conduction problems are presented.

English translation will appear in Heat Transfer—*Soviet Research*

9.42 Concerning the Calculation of Unsteady Heat Conduction Problems with Heat Sources

W. Fratzscher
Merseburg Institute of Technology, German Democratic Republic

Considering the effective density of released flukes a solution of a differential unsteady heat conduction equation for cylindrical fuel elements of nuclear reactors is obtained. The solutions of simple time functions of the density are presented in graphical form.

9.43 A Calculation of the Multidimensional Temperature Fields in Constructions

V. S. Khokhulin and B. M. Pankratov
Moscow Aviation Institute

A method of investigating the temperature distribution in constructions with various heat loads is considered. The model represents a combination of separate elements united in a graph of the "skeleton" construction structure. A determination of the "skeleton" structure temperature allows the calculation of the temperature distribution in separate elements of the construction.

9.44 Concerning the Application of the Collocation Method for Solving Diffusion Equations

W. Zwick
Zentralinstitut für Mathematik und Mechanik, Berlin, German Democratic Republic

An unsteady-state boundary-value problem is formulated with regard for convection terms having space and time-dependent diffusion coefficients. This problem can be solved by the method of collocation. It may be shown that with a step decrease, the approximation error of the equations tends to zero. An *a priori* prediction reveals that the approximate equations are stable, possess a unique solution, and that their solution can be reduced to the solution of an analytical problem.

9.45 Temperature Fields and Stresses in Multidimensional Bodies and Envelopes

P. V. Tsoy, U. Kh. Kamarov, M. U. Mairansayev, A. D. Gunashev, O. B. Kianovskiy, and A. I. Ismatullayev
Tadjik Polytechnical Institute, Dushanbe

An approximate analytical method of determining unsteady temperature fields inside a convex multidimensional body for a prescribed symmetric and nonsymmetric temperature distribution on the surface of a cone, an ellipsoid, and a cylindrical body with a triangular section is presented.

Simple and sufficiently precise functions of the temperature fields and stresses in a plate, a hollow cylinder, and a spherical envelope under thermal impact are obtained.

9.46 On the Solution of Nonlinear Contact Heat Conduction Problems

P. V. Cherpakov, L. S. Milovskaya, and A. A. Kosarev
Voronezh State University

The solution of the first nonlinear contact heat conduction problem by a finite difference method is presented. The stability and convergence of the difference scheme is shown. A solution is obtained for the case of a net integrator.

9.47 On the Construction of Approximate Solutions for Unsteady Heat Conduction Problems using the Cauchy Method

A. A. Shmukin
Institute of Mechanics, Dnepropetrovsk

The Cauchy method is applied in order to construct approximate solutions of direct and inverse unsteady heat conduction problems.

Sufficiently simple algorithms are obtained which can adequately solve a wide class of heat conduction problems including those in regions with moving boundaries.

9.48 Numerical Integration of the Heat Conduction Equation with High-Accuracy Difference Analogs Substituted for the Second-Kind Boundary Conditions

P. P. Yushkov
Leningrad Institute of Refrigeration Technology

Explicit difference analogs of the second-kind boundary condition are obtained making it possible, by performing a numerical integration of the heat conduction equation, to calculate surface temperature and errors of the third and fourth order of smallness relative to the particular step on the space axis.

10. EXPERIMENTAL METHODS AND MEASUREMENTS

10.1 Three New Devices for Measuring Heat Transfer in Buildings

I. Augusta
*Research and Development Institute of the Prague Building Trust,
Prague, Czechoslovakia*

The paper describes a multipurpose thermoconductometer for the experimental measurement of the thermal resistance, attenuation and phase lag of a harmonic temperature wave the rate of cooling under adiabatic conditions in samples in real structures. A temperature-compensated sensor with a resolution of 0.5 $W/m^2 K$, used for measuring the surface heat transfer coefficient and sensitive to both the radiative and convective components of heat flow is also described along with a spherical, human head modelling device (with a resolution of 1 $W/m^2 K$) used to measure biological comfort.

This paper appears on page 315, this volume

10.2 Measurement of the Temperature of a Highly Heated Two-Phase Flow

V. E. Alemasov, V. K. Maksimov, V. I. Sagadeyev, E. S. Sergeyenko, and M. A. Taymarov
Tupolev Aviation Institute, Kazan

The effect of dispersion on the accuracy of temperature measurement by the method of absolute intensity of spectral lines depending on the concentration and size of aluminum oxide particles is considered. The concentration ranged from 0 to 16%. Dispersion effects were found at concentrations as high as 6 percent with the influence of particle size being insignificant.

English translation will appear in Heat Transfer—*Soviet Research*

10.3 An Automatically Controlled System for Processing the Data of Heat Experiments

O. M. Alifanov, V. S. Kuznetsov, B. M. Pankratov, and I. M. Ukolov
Moscow Aviation Institute

The problem of constructing an automated system for processing the data of unsteady heat experiments is considered. A hybrid of the digital and analog systems is investigated with a set of special software.

10.4 New Equipment and Methods of Measuring the Temperature and Velocity Fluctuations and Temperature Profiles in a Thin Layer of the Sea Surface

E. G. Andreyev, V. V. Gurov, and G. G. Khundzhua
Moscow State University

A quick-setting electronic complex used for investigating the structure of a thin boundary sea layer is described. The instrument is used for measuring the thermal conductivity near the sea surface, the total heat flow from the sea surface, and its variation over a 24 hour cycle.

10.5 The Measurement of Surface Temperatures by Film Thermocouples

A. Ya. Anikin, L. S. Grigor'yev, and D. F. Simbirskiy
Kharkov Aviation Institute

A generalization is made of the efforts to create a high temperature film thermocouple (HTFT) designed for the experimental study of heat transfer in heat-and-power devices, their principal positive feature being a negligibly low level of systematic errors. The suggested HTFTs are fit for measurement of surface temperatures of parts made of electroconductive and nonelectroconductive materials in the range of temperatures from subzero to +1200–1400°C.

English translation will appear in Heat Transfer—*Soviet Research*

10.6 Methods of Investigation and Measuring Equipment for Defining Finite Transfer Rates and Local Heat and Mass Fluxes in Capillary-Porous Bodies with Pulse and Static External Forces

B. N. Bobkova and E. A. Sbornikov
Ural Research and Design Institute for Building Materials, Chelyabinsk

Methods are presented for experimental study of finite heat and mass transfer rates in capillary-porous bodies under the action of different external pulse forces. A capacity probe is described allowing distant continuous recording of weight decrease of samples under temperature and dynamic stresses (heat and vacuum pulses).

10.7 Optimal Planning of Experiments on the Determination of Heat Transfer Conditions on Boundaries

V. G. Bogdanov and S. V. Yepifanov
Kharkov Aviation Institute

A method is suggested of optimal planning of experiments making it possible, with minimum information available, to choose the number of thermal receivers, their position, frequency of recording, and the errors of instruments for obtaining the necessary accuracy in the measurements of heat transfer on a boundary. As an illustration, an a priori analysis was performed of determining the local heat transfer coefficients on the boundaries of the cross section of a turbine blade.

English translation will appear in Heat Transfer—*Soviet Research*

10.8 A Control-Measuring Scheme for a Hyperthermal Procedure

A. I. Brazgovka, S. B. Minkin, S. A. Nekrasov, N. P. Potershuk, V. E. Ulashchik, and B. I. Fyodorov
Luikov Heat and Mass Transfer Institute, Minsk

A control-measuring scheme is described which was used for multipoint measurements of patient temperature with an accuracy not less than $0.1°C$. Measurements were also made of arterial blood pressure, systole frequency, the frequency of respiration, and other vital parameters during controlled artificial hyperthermia.

English translation will appear in Heat Transfer—*Soviet Research*

10.9 The Use of an Electrodiffusional Method of Simulation and Diagnostics for Investigations of Hydrodynamic and Transport Processes

A. P. Burdukov, V. E. Nakoryakov, G. G. Kuvshinov, N. V. Valukins, and B. G. Kozmenko
Thermophysics Institute, Siberian Department of The USSR Academy of Sciences, Novosibirsk

Application of an electrodiffusional method for the investigation of the mechanism of a gas-liquid flow with a small gas content is considered. Data are obtained on fluid velocity profiles and spectral characteristics of friction for a "microturbulent" flow regime of the mixture in the transition region. The first attempts on the use of the above method for the investigation of free-convective and boiling heat transfer are described.

English translation will appear in Heat Transfer—*Soviet Research*

10.10 Experience in the Application of a System of Self-Recording Experimental Data in Heat Transfer Investigations

O. S. Vasilevich, V. T. Derov, N. S. Zaporezhets, L. I. Kolykhan V. N. Solov'yev, and E. A. Tyrkich
Institute of Nuclear Engineering, Minsk

Considered are the circuits, experience in the creation and maintenance of a system of self-recording of primary data and the parameters of experimental devices of different technological schemes and designs. The system has been used for automatic recording of experimental data on heat transfer in a one-pulse flow, with boiling and condensation of dissociating nitrogen tetroxide.

10.11 Application of the Methods of Mathematical Planning of an Experiment when Studying Heat Transfer in Turbulent Swirled Flows

I. A. Vatutin, V. A. Voznesenskiy, V. Ya. Kersh, O. G. Martynenko, and V. K. Popov
Luikov Heat and Mass Transfer Institute, Minsk

Statistical models are presented which describe the temperature distribution in a turbulent swirled flow. Optimization of models permits one to obtain the conditions providing a parabolic temperature distribution in a flow over the whole length of a channel.

10.12 New Means and Methods of Thermal Control

O. A. Gerashchenko, T. G. Grishchenko, L. V. Dekusha, I. G. Neverov, V. N. Pakhomov, S. A. Sazhina, A. A. Stepkin, and V. G. Fyodorov
Engineering Thermophysics Institute, Ukrainian Academy of Sciences, Kiev

Information is presented on new methods and means of medico-biological diagnostics of thermal radiation, heat and mass transfer, and of thermophysical properties. All apparatus are based on the application of highly sensitive heat flow probes.

English translation will appear in Heat Transfer—*Soviet Research*

10.13 Investigation of Temperature and Concentration Fields by the Holographic Interferometry

T. I. Gural, O. A. Yershov, A. G. Sorits, and B. A. Fomenko
Leningrad

Hydrodynamic and thermal disturbances in a stable stratified fluid are examined experimentally by interferometric hologram techniques. The structure, amplitudes and lengths of the internal waves generated by a body moving horizontally in a fluid with a constant salinity gradient are investigated at different values of the Froude number, the behavior of convection at a heated vertical cylinder in a stratified fluid is described.

English translation will appear in Heat Transfer—*Soviet Research*

10.14 Application of Optical Holography for Studies of Heat Transfer in Liquid Media

P. L. Gusik, O. A. Yershov, L. A. Oborin, A. G. Sorits, and M. E. Strzhalkovskiy
Leningrad Institute of Civil Engineering

As an example of the investigation of the temperature field of a constant power cylindrical probe (instrument measuring the thermal conductivity of liquids), an application of optical holography for the studies of heat transfer in transparent liquid media is considered. Experimental data on temperature fields in water and glycerine-water are obtained with a holographic interferometer.

English translation will appear in Heat Transfer—*Soviet Research*

10.15 On the Theory of an Electric-Thermal Analogy

A. A. Gukhman and D. A. Kazenin
Moscow Chemical Engineering Institute

A problem is considered of finding an electrical potential field from a given temperature field in two contacting bodies possessing electric conductivity and comprising a thermocouple. account is made of transient conditions of the temperature field as well as the Thomson and Seebeck effects. Averaging for two- and one-dimensional bodies is made. An effective method of solution of the problem by the Green's function is presented. It is proposed to use an integral form of solution of the problem to obtain the solution of the reverse problem on recovering the temperature field from the measured electrical data.

10.16 An Interferometric Technique for the Analysis of Local Statistical Characteristics of Turbulent Flows

V. M. Yeroshenko, A. A. Klimov, A. V. Sevostianov, Yu. N. Terent'yev,
E. I. Yakubovich, and L. S. Yanovskiy
Krzhizhanovskiy Power Engineering Institute, Moscow

An interferometric technique has been developed for the experimental study of turbulent coefficients of heat and mass transfer as well as for determining the most important statistical characteristics of turbulent flows. The technique is based on the analogy between the integral interferometric signal from a photodetector and the signal of a thermoanemometer with a long filament of the probe. Correlations between integral optical signals with light passing through a turbulent flow in different directions makes it possible to relate measured quantities to local parameters of the flow.

10.17 Experimental Determination of Heat Transfer from Surfaces

L. Imre, Gy. Danko, and P. Niedemayer
Budapest Polytechnical University, Hungary

Examination of operational conditions of complicated thermal systems requires the knowledge of interaction between the system and the environment. To describe it mathematically, the overall heat transfer coefficient is used, particularly when the method of a heat flux network is applied. In cases of complicated geometrical flow conditions, one has to determine the heat transfer coefficient by measurement. This paper discusses the procedure and equipment of the measurement.

10.18 Effect of an Incompressible Flow Temperature on the Cooling of a Hot Wire

P. Ionash

Institute of Thermomechanics, Prague, Czechoslovakia

The cooling of a hot wire of a thermoanemometer is studied in a wide temperature range of the main flow (from about 10 to 60°C) and at different temperatures of the wire. Appreciable systematic errors are found to occur when characteristics are measured of a turbulent flow with critical temperatures varying in a wide range. A method is proposed providing the necessary accuracy of measurement in an isothermal flow.

10.19 Measurement of One-Point Second-Order Correlations in Nonisothermal Turbulent Flows using a Hot-Wire Anemometer

H. Kitzing

Central Institute of Mathematics and Mechanics, Berlin, German Democratic Republic

Determination of heat and momentum transfer in three-dimensional flows requires the measurement of all one-point second-order correlations. Using a single-wire probe these correlations are obtained as a solution of a system of linear equations. The construction of this system is described and the dependence of the solvability upon the probe-type and the angle of attack is investigated.

10.20 Mathematical Planning of Heat and Mass Transfer Experiments

A. I. Lyubarskiy

Minsk Branch of the Mekhenergokhimprom Production Association

Possible fields of successful application of the mathematical planning of heat and mass transfer experiments on the basis of two systematic examples are considered. Logical application and combination of these methods with similarity theory ensures an optimum study both of the model and the mechanism of heat and mass transfer processes.

10.21 Thermal Conductivities of Helium and Xenon Mixtures at High Temperatures

I. Mashtovskiy

Institute of Thermomechanics, Prague, Czechoslovakia

Experimental and theoretical thermal conductivities are given for helium and xenon mixtures at temperatures up to 6000°K and atmospheric pressure. New experimental data on thermal conductivities of 0.1 He–0.9 Xe and 0.5 He–0.5 Xe are obtained in a shock tube by an unsteady-state method. Predictions of thermal conductivities of 0.1 He–0.9 Xe; 0.25 He–0.75 Xe; 0.5 He–0.5 Xe; 0.75 He–0.25 Xe, and 0.9 He–0.1 Xe are performed according to the molecular-kinetic theory of gases applying the Lennard–Jones potential 12–6. Theoretical and experimental results are compared with other reported data.

10.22 On Application of an Electrodiffusional Method to Measure Convective Heat Transfer Characteristics in Various Liquids

N. A. Pokryvaylo, A. K. Nesterov, A. S. Sobolevskiy, T. V. Yushkina, and Yu. E. Zverkhovskiy
Luikov Heat and Mass Transfer Institute, Minsk

Discussion is presented of the use of an electrodiffusional method for measurement of mean values Nu, St, U, as well as of pulsation components and their spectra. An approximate solution is given of the equation of a diffusional boundary layer of a measuring electrode accounting for the effect of mass turbulent diffusion into a diffusional current of a friction probe and recommendations are given regarding the choice of its parameters. The theory is considered of velocity measurements by an electrodiffusional probe with conical and wedgelike working surfaces. On this basis, calculation formulae are presented to determine amplitude-frequency characteristics. Calibrating characteristics of probes in different liquids are given. A comparison is made between theoretical and experimental data.

English translation will appear in Heat Transfer—*Soviet Research*

10.23 Methods of Cybernetic Diagnostics in Experimental Heat Transfer Research

D. F. Simbirskiy
Kharkov Aviation Institute

Basic considerations of cybernetic diagnostics of linear and nonlinear dynamic heat objects are presented. The diagnostics involve receiving optimal estimates of a vector of unknown heat flow by comparing a real physical situation of directly measured temperatures at some points of the object with their values predicted by their mathematical models. An optical Kalman filter and optimal estimation by the method of least squares are proposed for the use as algorithms.

10.24 Method of Measurement of Velocity and Temperature Fluctuations in High Temperature Gas Flows and Flames

B. P. Ustimenko, V. N. Zmeykov, A. A. Shishkin, and B. O. Rivin
Research Institute for Power Engineering, Alma-Ata

A method of measurement of turbulent characteristics such as fluctuations of velocity, temperature as well as correlation coefficients in heated gas flows and flames with a cooled film probe and a microcouple is presented.

10.25 Measurement of the Thermal Conductivity of Thin Solid Layers by a Radiometric Method

Yu. A. Chistyakov
All-Union Research Institute for Metrology, Leningrad

A new noncontact method is presented for the measurement of thermal conductivity of thin layers of solids. A thin layer is used as a vane of a radiometer. The method is based on the relationship between a radiometric force acting on the vane and the temperature difference of its opposite surfaces. The method has been verified experimentally on glass and stainless steel specimens. The diameter of the specimens was 10 mm and the thickness was 0.2 and 0.4 mm. The discrepancy between the measured thermal conductivities and those given in the literature does not exceed 10–15 percent.

English translation will appear in Heat Transfer—*Soviet Research*

AUTHOR INDEX